He suffers so you don't have to

Always read your operator's manual for each machine, and closely follow the instructions that they contain. Keep machine shields and guards in place, and don't work on a machine that is being operated. Take time to read and follow all safety warning messages that are on your machine and in your operator's manual.

Agricultural Safety

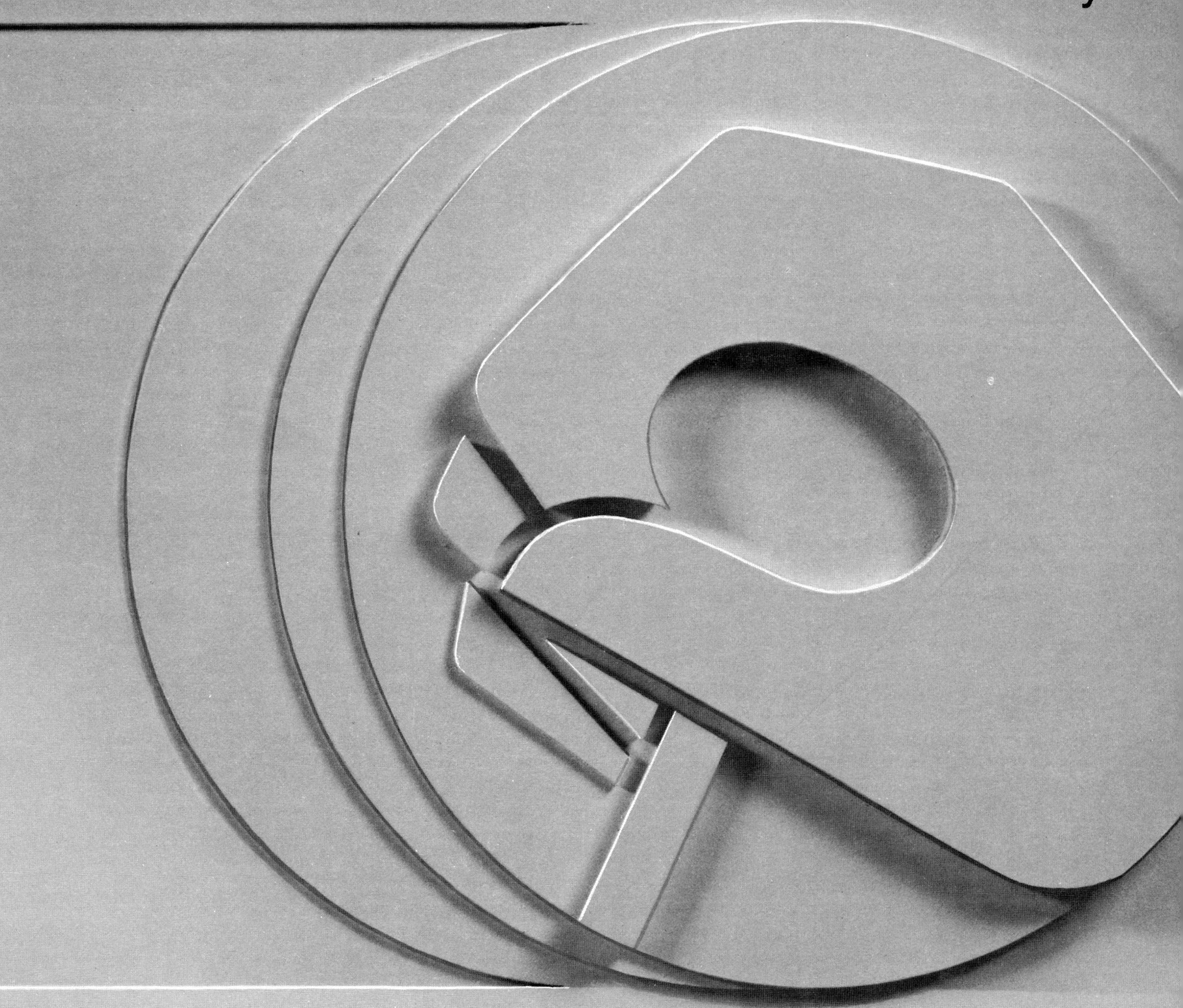

PUBLISHER

Fundamentals of Machinery Operation (FMO) is a series of manuals created by Deere & Company. Each book in the series is conceived, researched, outlined, edited, and published by Deere & Company. Authors are selected to provide a basic technical manuscript which is edited and rewritten by staff editors.

PUBLISHER: DEERE & COMPANY SERVICE TRAINING, Dept. 333, John Deere Road, Moline, Illinois 61265; DIRECTOR OF SERVICE: Robert A. Sohl.

FUNDAMENTAL WRITING SERVICES EDITORIAL STAFF

Managing Editor: Louis R. Hathaway
Editor: Larry A. Riney
Promotions: Annette M. LaCour

TO THE READER

SPECIAL ACKNOWLEDGEMENTS

The editors wish to thank *K. L. Pfundstein, K. C. Anderson,* and *P. L. Bellinger,* of the Deere & Company Product Safety Department, *W. M. Van Syoc,* of the Deere & Company Product Engineering Department, for their assistance in preparing this book. Also we wish to thank a host of other John Deere people who gave extra assistance and advice on this project.

Agricultural Safety is a group project by Deere & Company and the faculty and staff of the Agricultural Engineering Department of Michigan State University.

MICHIGAN STATE UNIVERSITY STAFF

PROJECT COORDINATOR: *Howard J. Doss,* Coordinator, Agricultural Training.

PROJECT CONSULTANTS: *Richard G. Pfister,* Professor and *C. J. Mackson,* Professor.

AUTHORS: *Richard H. Bittner,* Assistant Professor; *Clint Bolton,* Research Associate; *Roy E. Childers, Jr.,* Area Agricultural Engineer—Cotton Mechanization (Agricultural Extension Service, Texas A & M University); *Howard J. Doss,* Coordinator, Agricultural Training; *Harold A. Hughes* (currently Assistant Professor, Agricultural Engineering Department, Virginia Polytechnic Institute and State University); *Leroy K. Pickett,* Assistant Professor; *Robert H. Wilkinson,* Assistant Professor. CONTRIBUTING WRITERS: *Charles Hausmann, Thomas Paige, Russell Parker.*

CONSULTING EDITORS:

Thomas A. Hoerner, Associate Professor, Agricultural Mechanics, Iowa State University. *Keith R. Carlson,* General Manager of Agri-Education, Inc.

CONTRIBUTORS

Robert Aherin, University of Minnesota; David E. Baker, University of Missouri; Dennis J. Murph, Penn State University; W. M. Van Syoc, American Society of Agricultural Engineers; L. D. Baker; Goodyear Tire & Rubber Co.; Gulf Oil Chemicals Company, Merriam, Kansas; National Safety Council—Farm Department; Velsicol Chemical Corporation, Chicago, Ill.

THE SCOPE OF THIS BOOK

This text and its supporting materials are intended only as an educational media and should not be considered a substitute for operator's manuals for specific machines. Always refer to the pertinent operator's manual for specific operating procedures and safety precautions.

FOR MORE INFORMATION

This text is part of a series of texts and visuals on agricultural machinery entitled Fundamentals of Machine Operation (FMO). An instructor's kit is also available for each subject. For information, request a free Catalog of Agricultural Teaching Materials. Send your request to: John Deere Service Publications, Dept. F., John Deere Road, Moline, Illinois 61265.

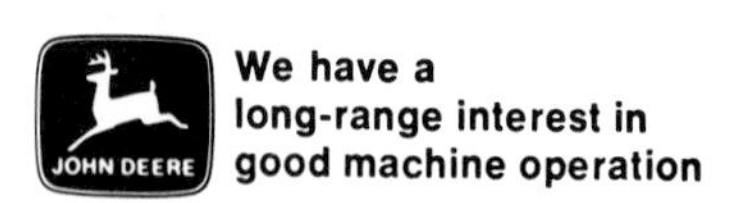

ISBN 0-86691-095-6

Contents

1

SAFE FARM MACHINERY OPERATION

2

HUMAN FACTORS

3

RECOGNIZING COMMON MACHINE HAZARDS

7

CHEMICAL EQUIPMENT

8

HAY AND FORAGE EQUIPMENT

9

GRAIN HARVESTING EQUIPMENT

10

OTHER HARVESTING EQUIPMENT

11

MATERIALS HANDLING EQUIPMENT

12

FARM MAINTENANCE EQUIPMENT

GLOSSARY AND LAWS

SUGGESTED READINGS

INDEX

1
Target: Safe Farm Machinery Operation

ZEROING IN ON SAFETY

- **Accidents hurt and kill!**
- **Accidents cost!**
- **Accidents can be avoided!**

ACCIDENTS HURT AND KILL!

In 1985 there were 1,600 deaths and over 170,000 disabling injuries due to agricultural related accidents according to the National Safety Council. In 1985, agriculture was one of the most dangerous occupations in the country.

The farm worker not only faces immediate machine and animal caused injury, they face long-term health problems from improperly handled chemicals such as fertilizers and pesticides. Government studies are pointing to chemicals as a major cause of cancer and genetic damage.

Don't become a safety fatality statistic. Beat the odds and help lower the incidence of injury and long-term health problems by always being conscious of safety fundamentals.

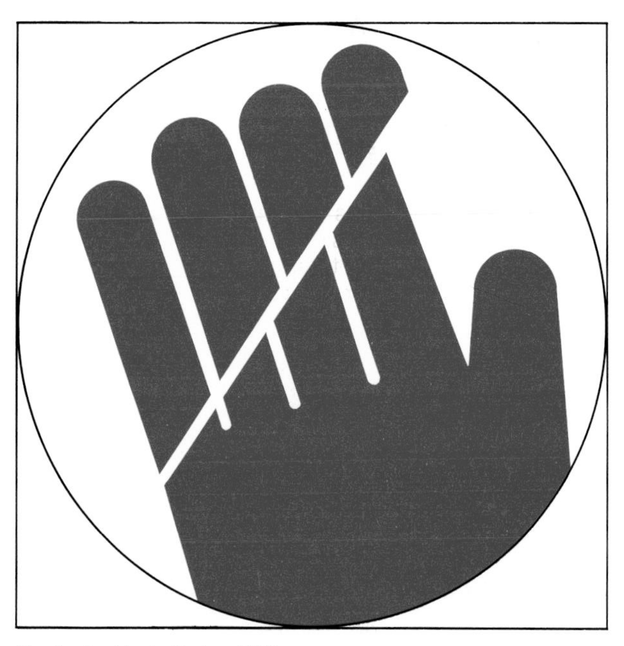

Fig. 1—Accidents Hurt and Kill!

Fig. 2—Can You Afford An Injury?

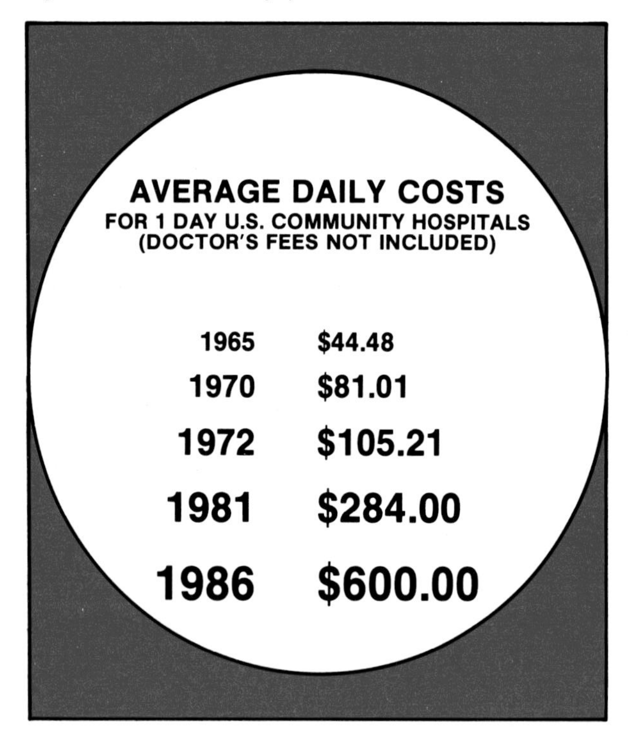

ACCIDENTS COST

Safety is too expensive to learn by accident. Accident costs reduce the profit margin of your operation.

Time = Money

Each farm accident costs farmers days of lost time. How much time can you lose? What happens to your crops or livestock during that time? Will you ever be able to do the same work you did before?

Medical Attention = Money

Hospital bills, doctor bills, medical supplies, and rehabilitation costs can be a big financial burden (Fig. 2).

Hired Help = Money

Will you have to hire somebody to do your work?

Property Damage = Money

Nobody knows the cost of farm machinery and equipment better than you.

Intangible Losses = More Than Money

A permanent handicap or the loss of your health can't be measured in dollars and cents. What effects would a disability have on you? On your family? Your earning power?

ACCIDENTS CAN BE AVOIDED!

Accidents are reduced by incorporating safe work practices into farm management programs. Safety must be thought of as a normal part of the farm management process just as land costs, fertilizer, and equipment are.

Colleges and universities, farmers organizations, farm equipment manufacturers, and safety regulations (Fig. 3) provide practical safety standards. It is up to individual farmers to incorporate these standards into their operations.

WHY WORRY ABOUT SAFETY?

Agriculture is one of the most hazardous occupations in the United States.

Farm and ranch residents in the United States have hundreds of thousands of accidents every year. And thousands of these accidents are fatal! About half the injuries to rural residents occur on the farm or ranch. And of all farm accidents, many involve machinery (Fig. 4).

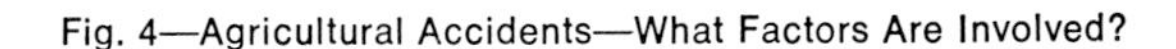

Fig. 4—Agricultural Accidents—What Factors Are Involved?

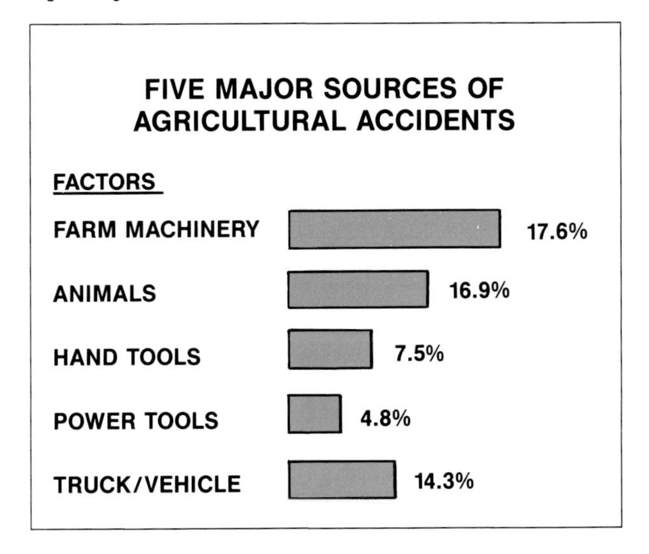

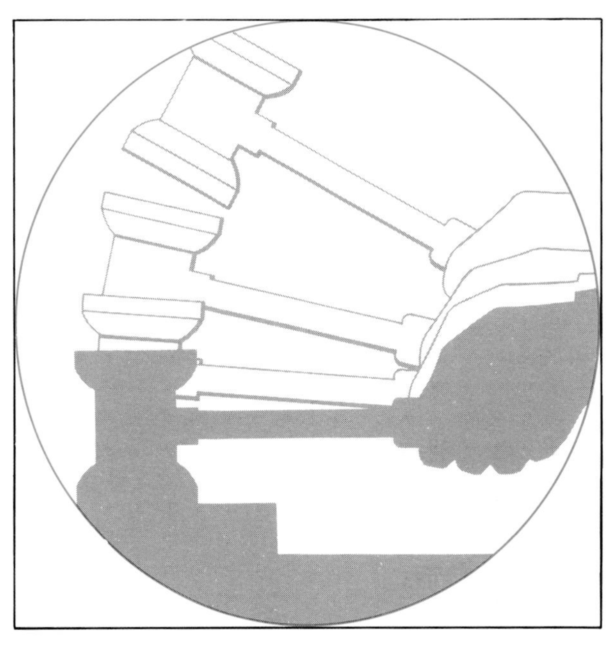

Fig. 3—Practical Safety Standards and Regulations are Provided — Farmers Must Adhere to Them

ZEROING IN ON A CHALLENGE TO OPERATORS

1. Know your limitations. Machinery is designed to work longer and harder than humans. Consider your limitations and allow time for rest when you operate farm machinery. Many accidents happen because the operator is tired.
2. Learn about common machine hazards. Avoid practices like dismounting from a tractor with the engine running to adjust attached equipment. Such shortcuts lead to injuries.
3. Always read and follow your operator's manual when you operate your machine. These manuals are provided to help you operate your machinery safely and efficiently.
4. Keep all safety signs in good condition on your machines. Replace those that cannot be read.

Fig. 5—How Many Safety Symbols Do You Recognize?

BE YOUR OWN SAFETY DIRECTOR

Many industrial organizations have safety directors or engineers who work continuously at keeping their plants or workshops safe.

There is no safety director on the "back forty." *You* must do that job when you work on a farm or ranch. Start by learning safe work practices for farm machinery. Finish the job by practicing safe work practices.

Fig. 6—Safety-Alert Symbol. It Means: Attention! Become Alert! Your Safety Is Involved!

Also, read the machine operator's manuals. These manuals help you learn safe operating and servicing procedures.

ZEROING IN ON COMMUNICATION

Good, clear communication between machinery operators and helpers reduces accidents. Safety symbols, warning decals, operator's manuals, and instruction labels, help communicate safety information.

Let's look at the kind of communication used in agriculture.

SYMBOLS THAT COMMUNICATE

Everywhere you look there are symbols that communicate. Take a look at Fig. 5. Do you recognize all of the symbols? Do you know what they mean? Try to identify each one to see if you understand it. Compare your answers to those below.

Answers: (1) Compulsory way for pedestrians. (2) Dangerous curve. (3) Beware of animals. (4) No U-turns. (5) No stopping. (6) Speed limits for light and heavy motor vehicles. (7) No passing. (8) Road narrows. (9) Slippery road. (10) Compulsory minimum speed.

Safety-Alert Symbol

The safety-alert symbol (Fig. 6) was designed to say: *Attention! Become alert! Your safety is involved!*

You'll find this symbol on safety signs on most agricultural machines and in operator's manuals. It's used to alert you to hazards.

Signal Words

The safety-alert symbol is often used with the signal words *Danger, Warning,* and *Caution to draw attention to potentially unsafe areas (Fig. 7 & 8). Learn these signal words and let them become your "think triggers."*

DANGER means that one of the most serious potential hazards is present. Exposure to these hazards would result in a high probability of death or a severe injury if proper precautions are not taken.

WARNING means the hazard presents a lesser degree of risk of injury or death than that associated with Danger.

CAUTION is used to remind the operator of safety instructions that must be followed and to identify some less serious hazards.

⚠DANGER

Disengage harvesting unit drive and shut off engine before unclogging machine.

⚠CAUTION

1. **Keep all shields in place.**
2. **Stop engine and shift transmission to PARK before dismounting.**
3. **Disengage motor power and wait for moving parts to stop before servicing machine.**
4. **Keep hands, feet, and clothing away from power-driven parts.**

Fig. 7—Safety Signs Use Signal Words To Alert

Engage gate lock valve before working on or around raised gate. Stand clear before unlocking gate lock valve.

Fig. 8—Safety Signs Use Color And Pictures To Warn

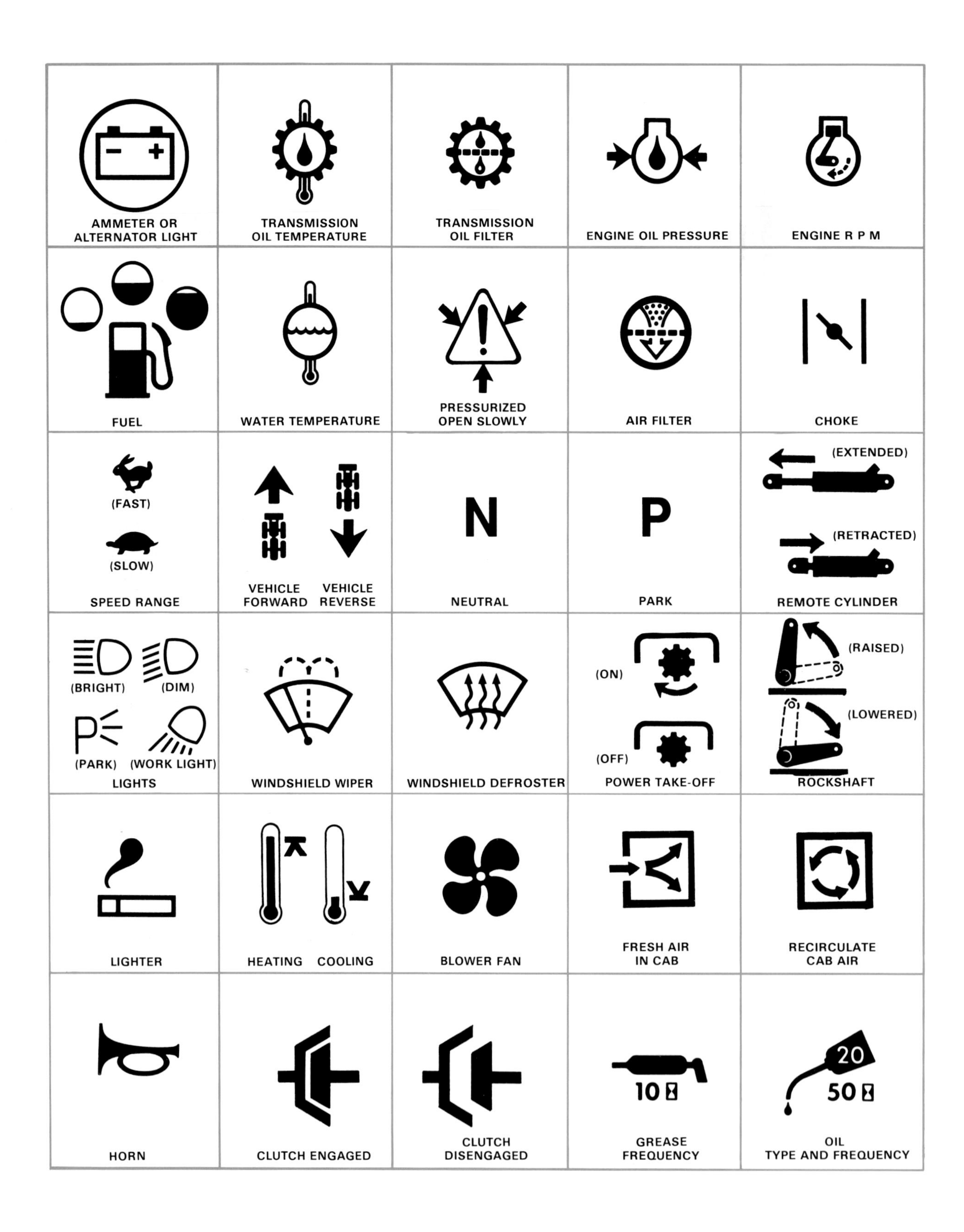
AMMETER OR ALTERNATOR LIGHT
TRANSMISSION OIL TEMPERATURE
TRANSMISSION OIL FILTER
ENGINE OIL PRESSURE
ENGINE R P M
FUEL
WATER TEMPERATURE
PRESSURIZED OPEN SLOWLY
AIR FILTER
CHOKE
(FAST)
(SLOW)
SPEED RANGE
VEHICLE FORWARD
VEHICLE REVERSE
N
NEUTRAL
P
PARK
(EXTENDED)
(RETRACTED)
REMOTE CYLINDER
(BRIGHT)
(DIM)
(PARK)
(WORK LIGHT)
LIGHTS
WINDSHIELD WIPER
WINDSHIELD DEFROSTER
(ON)
(OFF)
POWER TAKE-OFF
(RAISED)
(LOWERED)
ROCKSHAFT
LIGHTER
HEATING
COOLING
BLOWER FAN
FRESH AIR IN CAB
RECIRCULATE CAB AIR
HORN
CLUTCH ENGAGED
CLUTCH DISENGAGED
10
GREASE FREQUENCY
20
50
OIL TYPE AND FREQUENCY

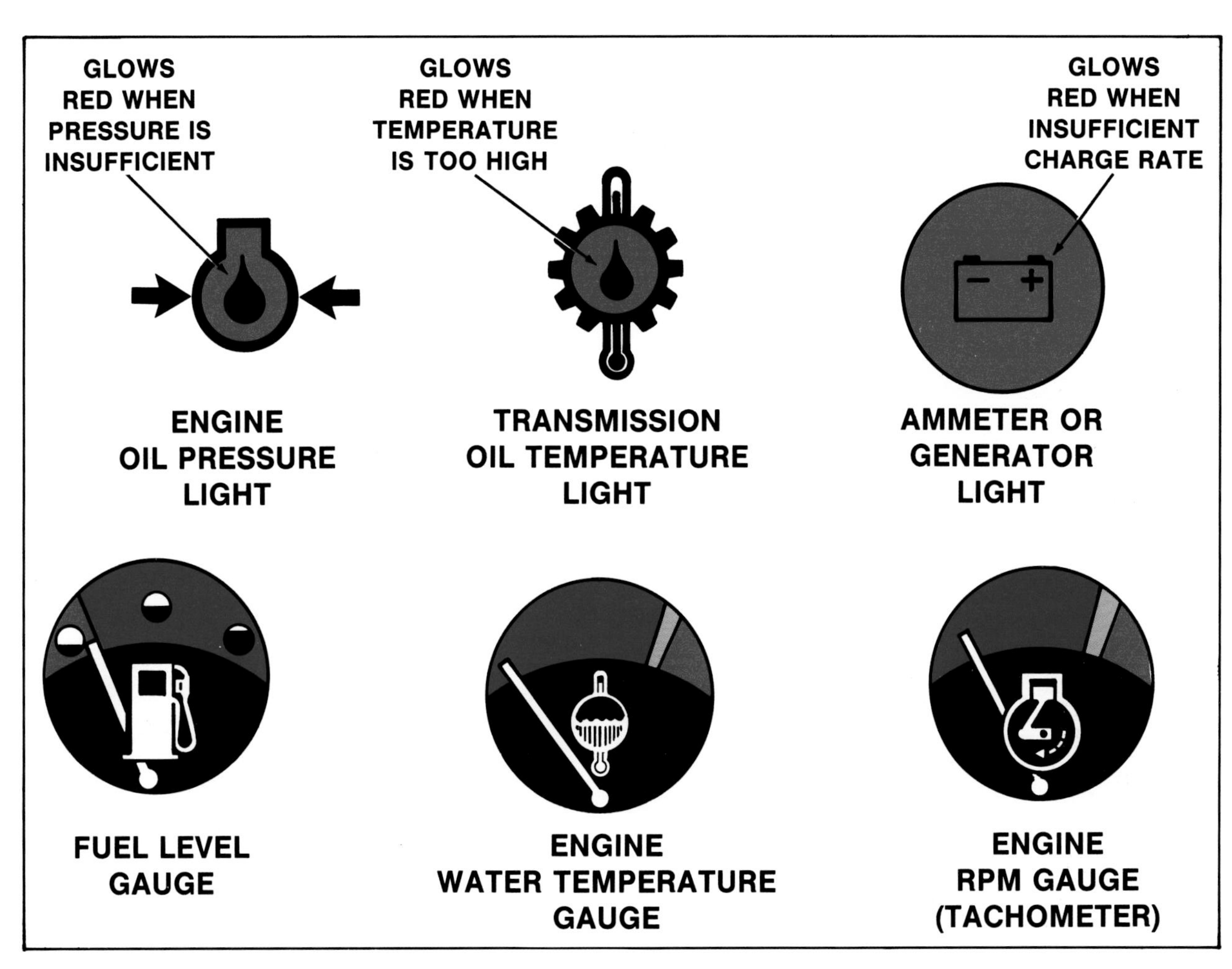

Fig. 10—Typical Use Of Color With Universal Symbols

Pictorials on safety signs improve the operator's recognition and understanding of hazards. A good pictorial should identify the hazard and portray the potential consequences of failure to follow instructions.

Safety Signs And Their Colors

Red and white are the colors used with the word Danger. Black and yellow are found on signs carrying the words Caution or Warning (Figs. 7 and 8).

Universal Symbols

Most agricultural machinery uses universal symbols to help identify controls. These symbols should be recognized by operators all over the world (Fig. 9).

If you learn these, you will understand machinery controls better. Color is often used with these universal symbols to indicate operating conditions (Fig. 10).

Fig. 9—Typical Universal Symbols For Controls

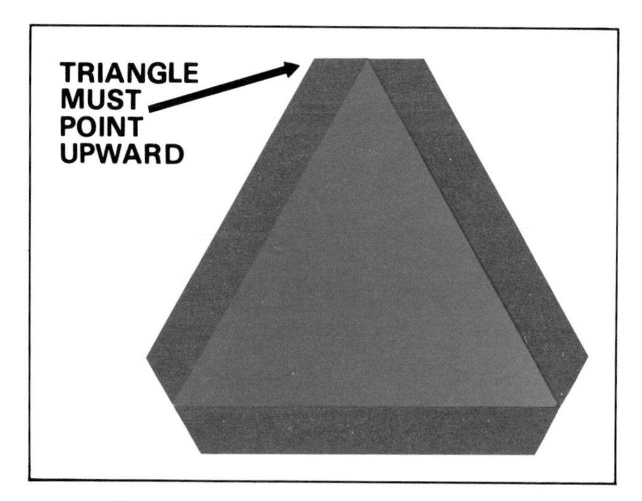

Fig. 11—Slow-Moving-Vehicle Emblem Identifies Machinery That Normally Travels Slower Than 25 MPH

SMV Emblem

Another universal symbol known to both automobile drivers and agricultural people is the slow-moving-vehicle (SMV) emblem (Fig. 11). This emblem is attached to the rear of agricultural machinery to identify slow-moving vehicles.

Use SMV emblems on machinery that normally travels at speeds slower than 25 mph (40 kph). Check your state's laws concerning slow-moving farm machinery on public roads.

Keep your SMV emblems clean, bright, and properly positioned. Replace them when they fade. They don't reflect well after they fade.

Instruction Labels

Labels communicate too. Reading a container label like the one in Fig. 12 can give you information that could prevent serious injury.

The chemical user who does not read the label is headed for trouble. He can permanently damage his health and the health of others. He can also harm crops and livestock, and he may not get the results he expected when he bought the chemical. Take time to read the label to prevent serious consequences.

HAND SIGNALS

Most farmers use hand signals when trying to communicate with machinery operators or helpers. These work fine *if* everyone understands the signals.

To help keep your signals straight, the American Society of Agricultural Engineers (ASAE) has published a set of hand signals for farm machinery operators (Figs. 13 and 14). Use them and teach others to use them. They can save time and prevent mistakes when noise or distance prevent talking.

Fig. 12—A Typical Label For A Chemical Container

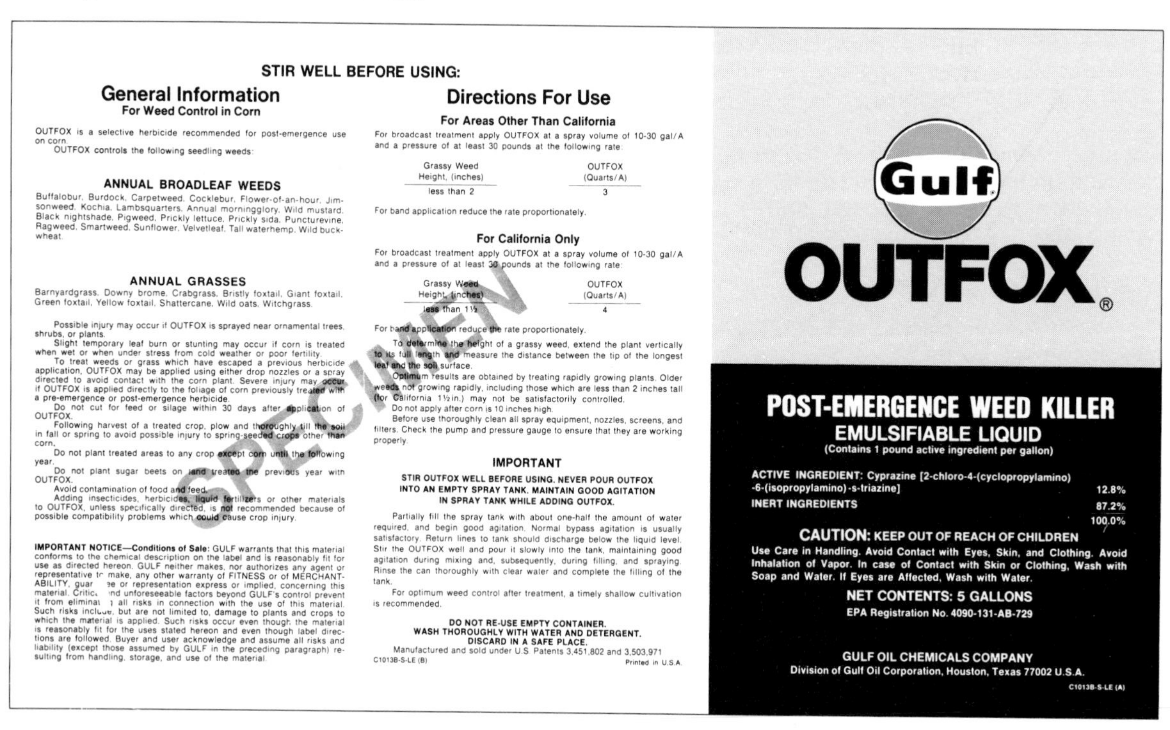

STIR WELL BEFORE USING:

General Information
For Weed Control in Corn

OUTFOX is a selective herbicide recommended for post-emergence use on corn.

OUTFOX controls the following seedling weeds:

ANNUAL BROADLEAF WEEDS

Buffalobur, Burdock, Carpetweed, Cocklebur, Flower-of-an-hour, Jimsonweed, Kochia, Lambsquarters, Annual morningglory, Wild mustard, Black nightshade, Pigweed, Prickly lettuce, Prickly sida, Puncturevine, Ragweed, Smartweed, Sunflower, Velvetleaf, Tall waterhemp, Wild buckwheat.

ANNUAL GRASSES

Barnyardgrass, Downy brome, Crabgrass, Bristly foxtail, Giant foxtail, Green foxtail, Yellow foxtail, Shattercane, Wild oats, Witchgrass.

Possible injury may occur if OUTFOX is sprayed near ornamental trees, shrubs, or plants.

Slight temporary leaf burn or stunting may occur if corn is treated when wet or when under stress from cold weather or poor fertility.

To treat weeds or grass which have escaped a previous herbicide application, OUTFOX may be applied using either drop nozzles or a spray directed to avoid contact with the corn plant. Severe injury may occur if OUTFOX is applied directly to the foliage of corn previously treated with a pre-emergence or post-emergence herbicide.

Do not cut for feed or silage within 30 days after application of OUTFOX.

Following harvest of a treated crop, plow and thoroughly till the soil in fall or spring to avoid possible injury to spring-seeded crops other than corn.

Do not plant treated areas to any crop except corn until the following year.

Do not plant sugar beets on land treated the previous year with OUTFOX.

Avoid contamination of food and feed.

Adding insecticides, herbicides, liquid fertilizers or other materials to OUTFOX, unless specifically directed, is not recommended because of possible compatibility problems which could cause crop injury.

IMPORTANT NOTICE—Conditions of Sale: GULF warrants that this material conforms to the chemical description on the label and is reasonably fit for use as directed hereon. GULF neither makes, nor authorizes any agent or representative to make, any other warranty of FITNESS or of MERCHANTABILITY, guar[illegible]ee or representation express or implied, concerning this material. Critic[illegible]nd unforeseeable factors beyond GULF's control prevent it from eliminat[illegible]g all risks in connection with the use of this material. Such risks include, but are not limited to, damage to plants and crops to which the material is applied. Such risks occur even though the material is reasonably fit for the uses stated hereon and even though label directions are followed. Buyer and user acknowledge and assume all risks and liability (except those assumed by GULF in the preceding paragraph) resulting from handling, storage, and use of the material.

Directions For Use

For Areas Other Than California

For broadcast treatment apply OUTFOX at a spray volume of 10-30 gal/A and a pressure of at least 30 pounds at the following rate:

Grassy Weed Height, (inches)	OUTFOX (Quarts/A)
less than 2	3

For band application reduce the rate proportionately.

For California Only

For broadcast treatment apply OUTFOX at a spray volume of 10-30 gal/A and a pressure of at least 30 pounds at the following rate:

Grassy Weed Height, (inches)	OUTFOX (Quarts/A)
less than 1½	4

For band application reduce the rate proportionately.

To determine the height of a grassy weed, extend the plant vertically to its full length and measure the distance between the tip of the longest leaf and the soil surface.

Optimum results are obtained by treating rapidly growing plants. Older weeds not growing rapidly, including those which are less than 2 inches tall (for California 1½ in.) may not be satisfactorily controlled.

Do not apply after corn is 10 inches high.

Before use thoroughly clean all spray equipment, nozzles, screens, and filters. Check the pump and pressure gauge to ensure that they are working properly.

IMPORTANT

STIR OUTFOX WELL BEFORE USING. NEVER POUR OUTFOX INTO AN EMPTY SPRAY TANK. MAINTAIN GOOD AGITATION IN SPRAY TANK WHILE ADDING OUTFOX.

Partially fill the spray tank with about one-half the amount of water required, and begin good agitation. Normal bypass agitation is usually satisfactory. Return lines to tank should discharge below the liquid level. Stir the OUTFOX well and pour it slowly into the tank, maintaining good agitation during mixing and, subsequently, during filling, and spraying. Rinse the can thoroughly with clear water and complete the filling of the tank.

For optimum weed control after treatment, a timely shallow cultivation is recommended.

DO NOT RE-USE EMPTY CONTAINER.
WASH THOROUGHLY WITH WATER AND DETERGENT.
DISCARD IN A SAFE PLACE.

Manufactured and sold under U.S. Patents 3,451,802 and 3,503,971

C1013B-S-LE (B) Printed in U.S.A.

SPECIMEN

Gulf

OUTFOX®

POST-EMERGENCE WEED KILLER
EMULSIFIABLE LIQUID
(Contains 1 pound active ingredient per gallon)

ACTIVE INGREDIENT: Cyprazine [2-chloro-4-(cyclopropylamino)-6-(isopropylamino)-s-triazine]	12.8%
INERT INGREDIENTS	87.2%
	100.0%

CAUTION: KEEP OUT OF REACH OF CHILDREN

Use Care in Handling. Avoid Contact with Eyes, Skin, and Clothing. Avoid Inhalation of Vapor. In case of Contact with Skin or Clothing, Wash with Soap and Water. If Eyes are Affected, Wash with Water.

NET CONTENTS: 5 GALLONS

EPA Registration No. 4090-131-AB-729

GULF OIL CHEMICALS COMPANY
Division of Gulf Oil Corporation, Houston, Texas 77002 U.S.A.

C1013B-S-LE (A)

HAND SIGNALS

Start The Engine. Move Arm In A Circle At Waist Level, As Though You Were Cranking An Engine.

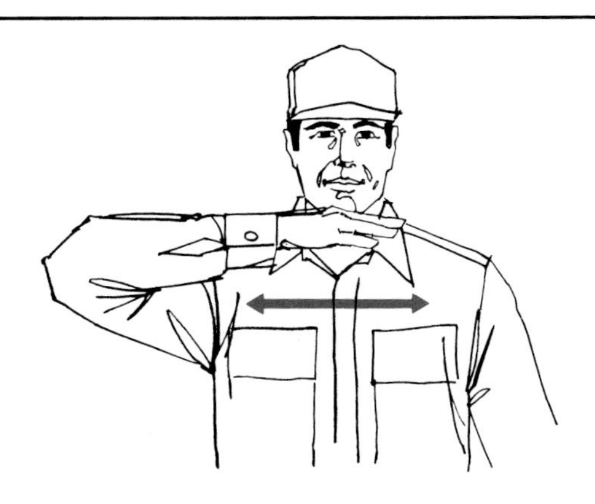

Stop The Engine. Move Your Right Arm Across Your Neck From Left To Right In A "Throat-Cutting" Motion.

Come To Me. (May Mean "Come Help Me" In An Emergency.) Raise Arm Straight Up, Palm To The Front, Move Arm Around In A Large Circle.

Move Toward Me — Follow Me. Look Toward Person Or Vehicle You Want Moved. Hold One Hand In Front Of You, Palm Facing You, And Move Your Forearm Back And Forth.

Move Out — Take Off. Face Desired Direction Of Movement. Extend Arm Straight Out Behind You, Then Swing It Overhead And Forward Until It's Straight Out In Front Of You With The Palm Down.

Fig. 13—Hand Signals

(Continued)

Speed It Up. Clenching Your Fist, Bend Your Arm So Your Hand Is At Shoulder Level. Thrust Arm Rapidly Straight Up And Down Several Times.

Slow It Down. Extend Arm Straight Out To The Side, Palm Down. Keeping Arm Straight, Move It Up And Down Several Times.

Raise Equipment. Point Upward With Forefinger, While Making A Circle At Head Level With Your Hand

Lower Equipment. Point Toward The Ground With The Forefinger Of One Hand, While Moving The Hand In A Circle.

This Far To Go. Put Hands In Front Of Face, Palms Facing Each Other. Move Hands Together Or Farther Apart To Indicate How Far To Go.

Stop. Raise Arm Straight Up, Palm To The Front.

Fig. 14—Hand Signals

These hand signals are in general agreement with the U.S. Army's standard arm and hand signals. Many of them are also used in industry. Learning this new "language" will give you an easy and effective way to communicate when working around agricultural machinery.

MAN-MACHINE COMMUNICATION

Machines can communicate with their operators. Unusual noises may mean the machine is malfunctioning. Frequent glances at the gauges tell the operator how a machine is performing. Listening to and watching the machine's performance tell how well the machine is adjusted, how well it is operating for specific field conditions, or if the machine needs attention. Shape and color coded controls (Fig. 15) also aid in man-machine communications.

The problem with this man-machine relationship is the operator gets tired. Operators strain to stay alert. Operators who watch and listen to their machines, hour after hour, may become so numbed by fatigue they miss an indication of a serious failure. Chapter 2 gives more information on these human limitations.

Fig. 15—Machines And Their Operators Communicate By Shape-Color-Coded Controls

EVERYBODY'S BUSINESS

It is up to everyone associated with agricultural machinery to use safe working practices. All family members and employees can contribute to each others safety. Remember, operators aren't the only ones who get hurt in agricultural machinery accidents.

Here are some things that can be done to help keep others safe:

1. Keep young children away from machines. Don't let them ever get the idea that machines are something to play or ride on.
2. Stay clear of machinery that's being operated by someone else. Don't bother the operator or interfere with his work. If you need him, signal for him to stop (hand straight up) and approach him after he stops.

New operators need to familiarize themselves with the equipment. Family members and employees must refamiliarize themselves with machinery annually. Knowing machinery well can stop accidents and reduce machinery downtime.

1. Read the operator's manual before working with a machine.
2. Make adjustments before starting operation.
3. Go over operating procedures with a skilled operator.
4. Practice operating the machine in an open area.

CONCERN FOR OTHERS

Is it harder to protect other people than it is to protect yourself? Generally, yes.

For example: You need to take a tractor out to one of your fields to bring back a second tractor and two implements. You'll need someone to drive one of the tractors back to the barn. Do you carry that extra driver with you on your tractor or do you arrange some other way for him to get to the field? The second driver should walk to the field or drive out in a car or pickup. Remember riders on tractors are the victims of many fatal accidents.

Dealing with people in situations like this one can be tricky. The right decision is often not the easiest one to make.

YOUR OWN SAFETY

Concern for others isn't enough. Sometimes an operator overlooks personal safety and does things he wouldn't want someone else to do.

Here's an example: Do you *always* disengage all

Fig. 16—Working Around Nature Affects Both Man And Machines

power, shut off the tractor engine, and take the key before unplugging a PTO-driven machine?

Others listen more to what you *do* than to what you *say.* Safety is *doing things right*—not just talking about it. Make safety a habit.

ZEROING IN ON THE ENVIRONMENT

Both human and machine performance is affected by temperature, humidity, soil conditions and terrain.

Prepare for the unusual and expect the unexpected. Mud, rain, and bad weather are inevitable. You need proper equipment and know-how to work in these adverse conditions. You must use good judgment to recognize unsafe conditions such as when a slope is too steep to cross safely with a tractor.

ZEROING IN ON TRACTOR ACCIDENTS

Farm tractors are involved in most of the farm machinery accidents. Most accidents occur in March. Accident rates drop off a little until July, then pick up again (Fig. 17).

Sunday is usually the day for the fewest tractor accidents. Saturday and Wednesday, on the other hand, are peak days for work and for tractor accidents. Tractor accidents can and do occur on any day of the week. Accidents occur when you least expect them.

Many deaths result from tractor overturns (Fig. 18). Most go over sideways. Some go backward. Chances of survival are very good if the tractor is equipped with a rollover protective structure (ROPS) and the operator is wearing a seat belt. The chances of survival in a tractor overturn accident are not very good without a ROPS and seat belt. More information about ROPS and tractor safety can be found in Chapter 5.

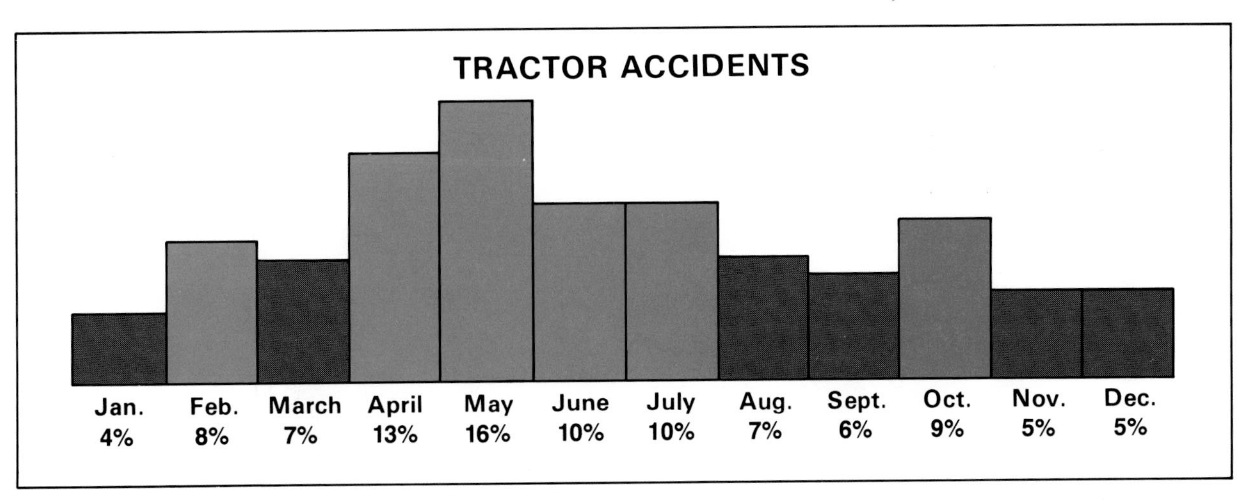

Fig. 17—Most Tractor Accidents Occur Between March and November

Fig. 18—Tractors Can Upset Rearward or Sideways

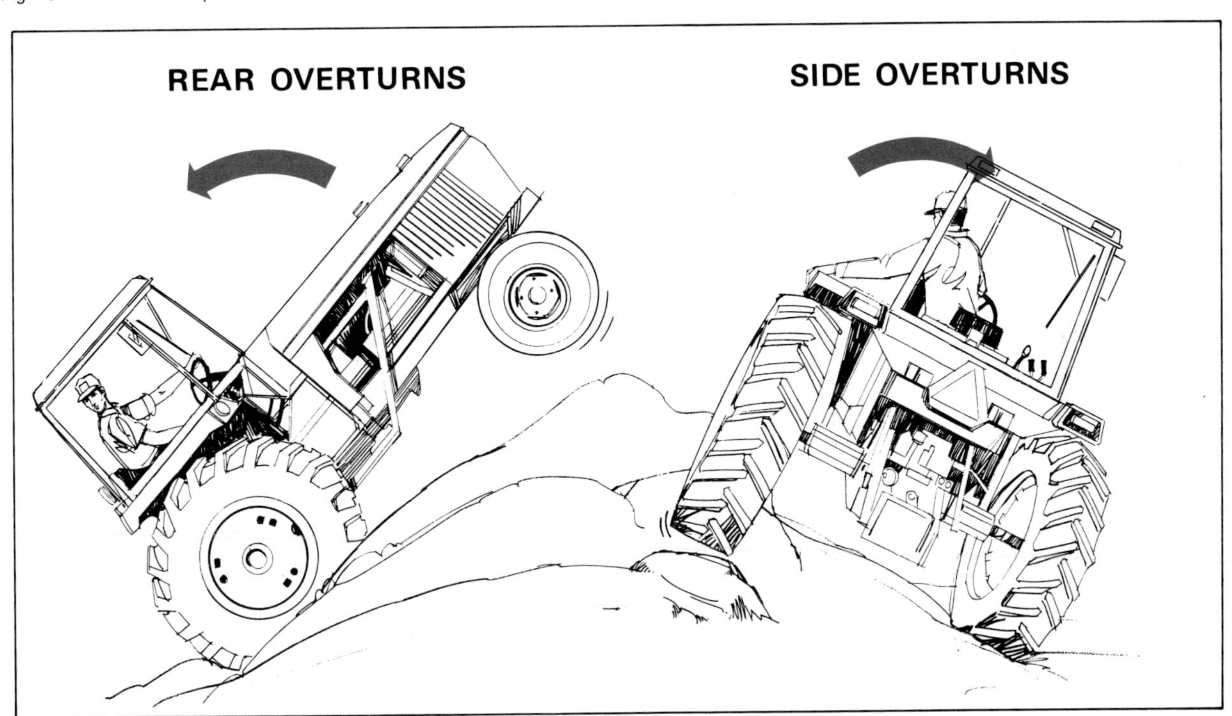

Fig. 19 — Use A Seatbelt If Your Tractor Is Equipped With A ROPS. Do Not Use A Seatbelt If Your Tractor Is Not Equipped With A ROPS.

Keep riders Off Of Your Machines. A Fall From A Tractor Can Be Fatal.

Fatal accidents involving operators or riders falling from tractors are a major cause of tractor-related deaths (Fig. 19).

Who has tractor accidents? Let Figs. 20 and 21 tell the story.

Fig. 20—Tractor Accidents—Who Gets Hurt?

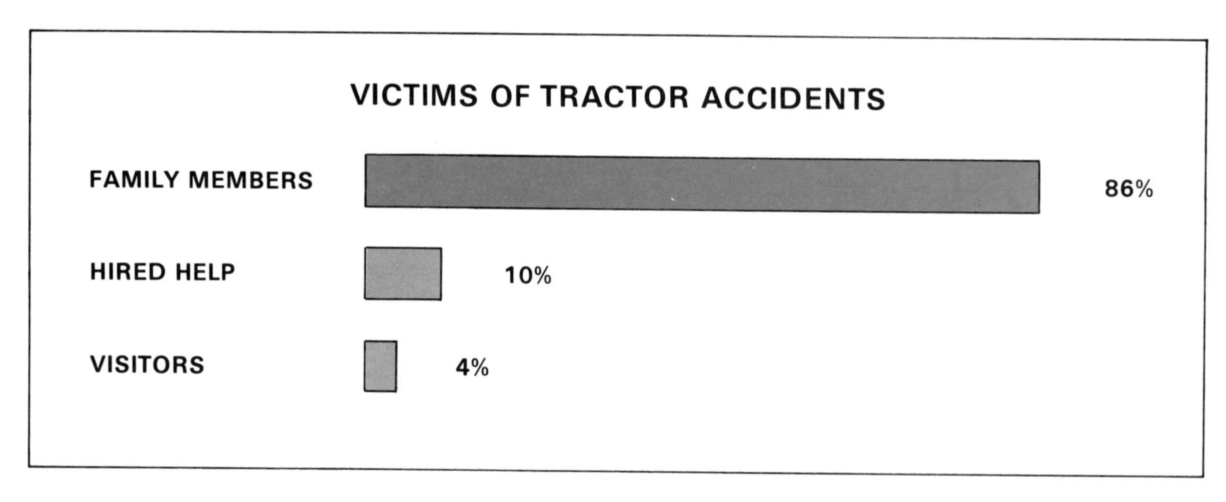

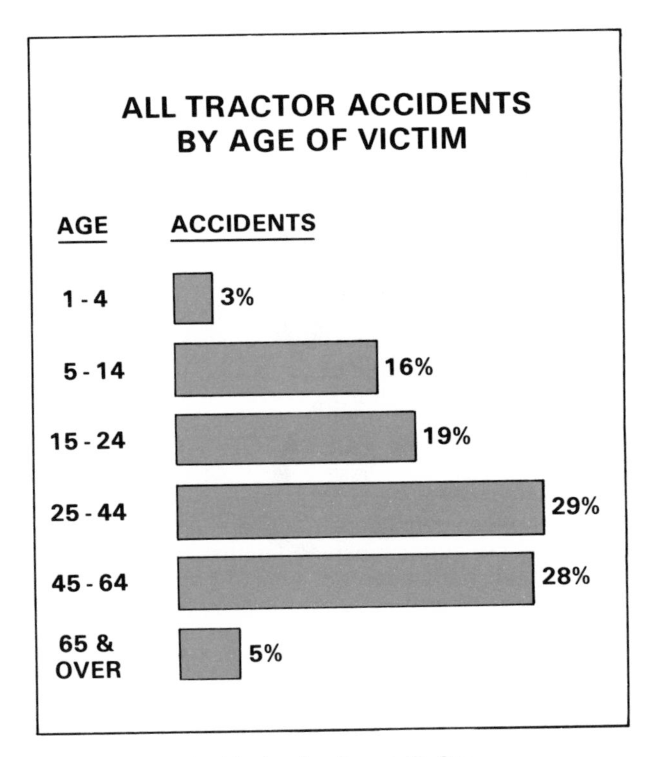

Fig. 21—Tractor Accidents—Age Spares No One

TRACTOR ACCIDENT RATE
(Michigan and Ohio)

OPERATOR AGE	FREQUENCY (Accidents Per Million Hours of Use)
10 - 14	43.0
15 - 24	9.6
25 - 44	4.5
45 - 64	5.6
65	29.7

Fig. 22—Operators Under 15 And Over 64 Have The Highest Accident Frequency Rates

In looking over Fig. 21, it's easy to get the idea that people in the 25-to-64 age group are the worst operators, since they have the most accidents. But Fig. 21 doesn't account for hours of machine use.

According to a study conducted in Michigan and Ohio, tractor operators under 15 and over 64 have seven to ten times more accidents *per hour of machine use* than operators in the 25-to-44 group (Fig. 22).

ZEROING IN ON WAGON ACCIDENTS

Wagons are almost as common as tractors on U.S. farms, and are involved in a significant number of farm machinery accidents.

Half the victims of wagon accidents are injured because they fall off.

Know the hazards when working with farm wagons. Know the places where wagon accidents are most likely to occur (Fig. 23).

Fig. 23—Wagon Accidents—Where Do They Occur?

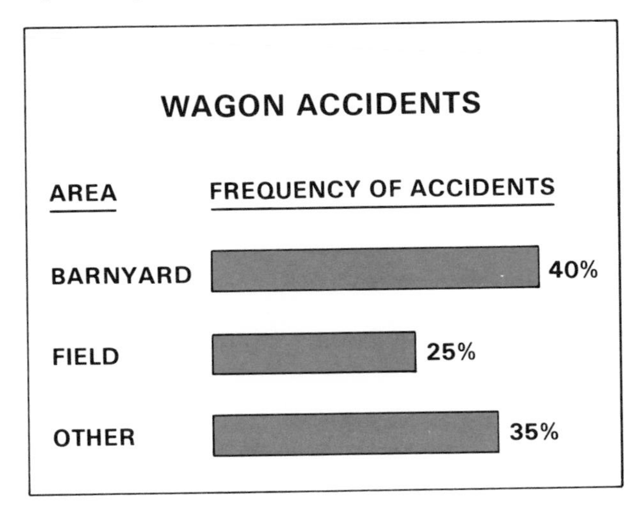

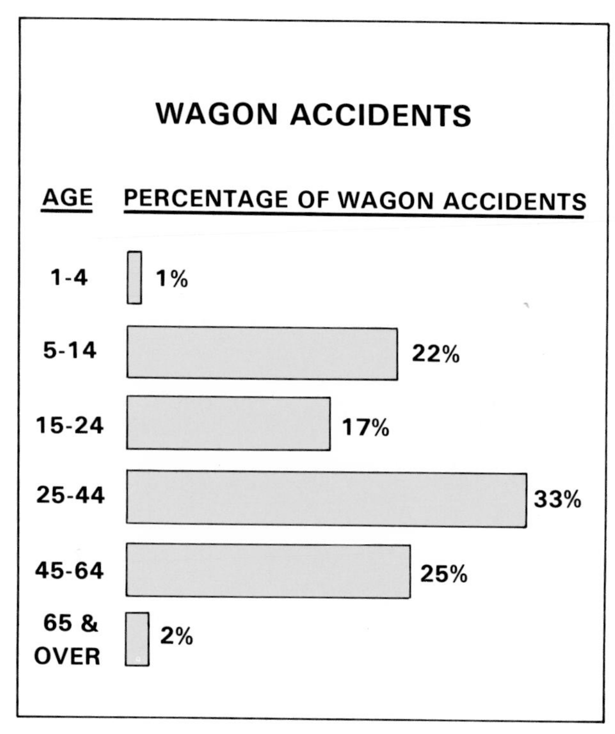

Fig. 24—Wagon Accidents—Everyone's Along On This Ride

Another way of looking at wagon accidents is to see what ages of people are involved (Fig. 24).

These figures show only *numbers* of accidents. They do not account for hours of use. A Michigan-Ohio study indicates that with wagons, as with tractors, people under 15 and over 64 have the most accidents per hours of use. When more states establish accident frequency rates for agricultural machinery, everyone will have a better perspective for their particular region and its farming characteristics.

Check Chapters 8, 9, and 11 for more information on the proper use of wagons.

ZEROING IN ON ELEVATOR ACCIDENTS

Farm elevators have one of the highest accident rates per hour of use. Most people who are injured in elevator accidents suffer crushing injuries or have some part of their body severed. Almost half of the accidents involve getting caught in moving parts, and nearly three-quarters of these injuries involve fingers or hands (Fig. 25). More about elevator hazards and how to avoid them in Chapter 11.

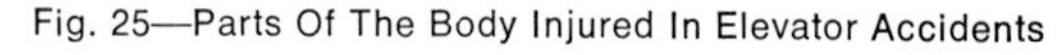

Fig. 25—Parts Of The Body Injured In Elevator Accidents

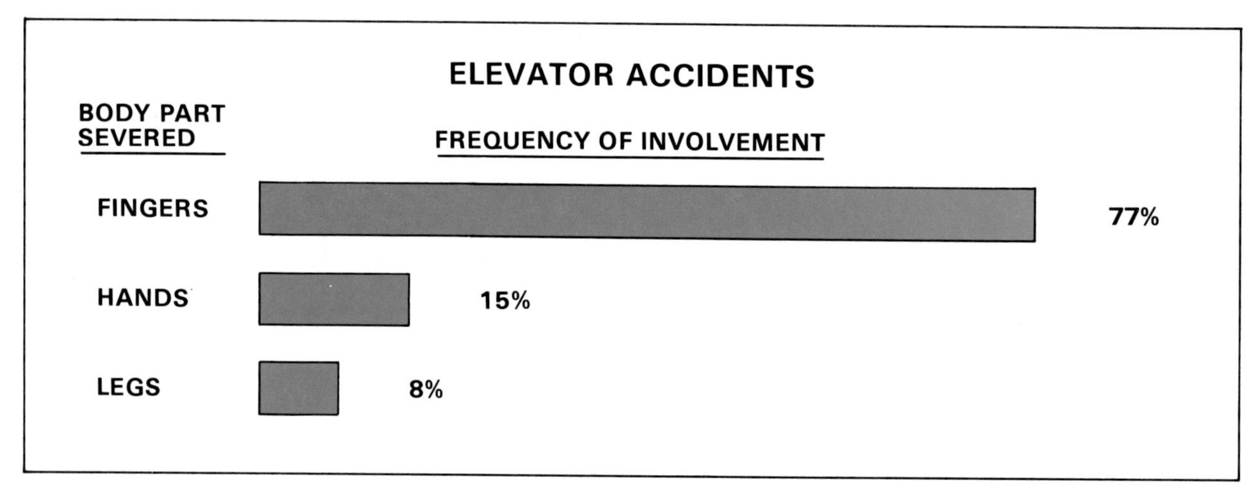

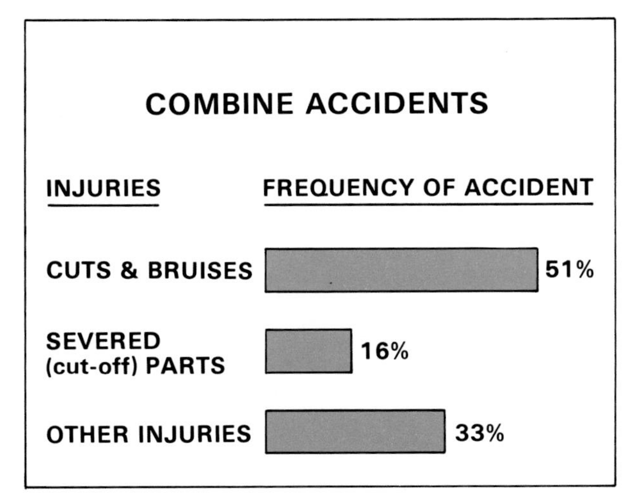

Fig. 26—Types Of Combine-Related Injuries

ZEROING IN ON COMBINE ACCIDENTS

Half of the reported combine-related injuries occur in the field. Fingers are lost in 38 percent of these injuries, and most of these fingers are cut off when caught in moving parts (Fig. 26).

SUMMARY

A good agricultural machinery operator needs a combination of skill, understanding, and knowledge to do his job safely and effectively. There's no substitute for actual experience, but this experience time can be shortened by learning the key points in this chapter:

- *Read and follow the instructions in the operator's manual.*
- *Know your limitations and work within them.*
- *Be your own safety director.*
- *Know the signs and symbols for communicating in agriculture.*
- *Know the common causes of accidents involving tractors and other types of farm machinery.*

CHAPTER QUIZ

1. Give three reasons why safety is important.
2. List four ways that accidents cost money.
3. Which signal word indicates the most serious hazard?
   (a) Danger
   (b) Caution
   (c) Warning
4. Draw stick figures to show the following hand signals used in farm machinery operation:
   (a) This far to go.
   (b) Come to me (Come help me).
   (c) Stop.
   (d) Slow it down.
   (e) Start the engine.
   (f) Raise equipment.
5. Which of the following are true?
   (a) Man can safely work indefinitely under hard conditions.
   (b) Machines always adapt to the demands of the job.
   (c) Communication between man and machine is always visual.
   (d) Operator fatigue is a problem in man-machine relationships.
6. True or false? Both operators and their machines are affected by environmental conditions.
7. Name two protective devices that increase your chances of surviving a tractor upset.
8. How does the accident frequency rate per hour of tractor operators 25 to 64 years old show the accident picture better than just giving the total number of accidents for this group?
9. True or false? Tractor operators under 15 and over 64 have the highest accident frequency rates.
10. What is a common type of wagon-related accident?

2 Target: Human Factors

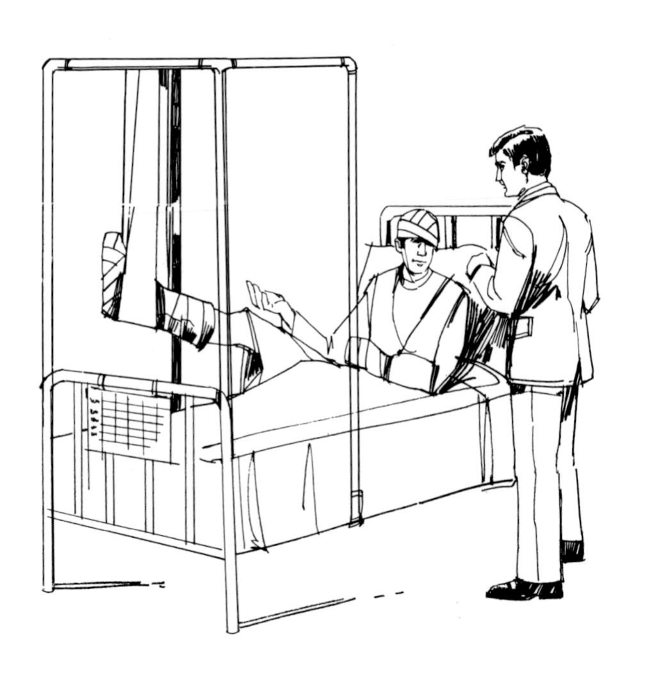

Fig. 1—Human Error Can Cause Accidents And Have Lasting Effects

INTRODUCTION

Human error is caused by many factors: carelessness, fatigue, overload, preoccupation, incompatibility between man and machine, to name a few.

Often human error results in costly damage to equipment and lost time for repairs. If an error causes injury or death the costs are incalculable (Fig. 1).

Fig. 3—Moving A Heavy Ladder Takes A Lot Of Muscle. You May Need Help

HUMAN LIMITATIONS AND CAPABILITIES

If an electrical circuit is overloaded, a fuse will blow or circuit breaker will trip before the system is damaged. Machines have safety devices like slip clutches, shear pins and safety trips to protect them against overloading.

The human body also has many safety devices. When overloaded the body sends signals in the form of pain, increased heart rate, and perspiration. But humans also have a will. That will can cause us to push on

Fig. 2—Physical, Physiological, And Psychological Limitations And Capabilities

after warning signals are given. When overloading continues over extended periods, we begin to make mistakes.

Different people have different limitations. It's important to know your individual limitations.

Human limitations can be classified into three groups (Fig. 2):

- **Physical**
- **Physiological**
- **Psychological**

PHYSICAL

Physical characteristics are: size, weight, and strength. If you recognize your physical limitations and work within them, you'll have fewer accidents than someone who tries to work beyond his limitations. You'll have better control over your job, and you will be able to avoid accidents.

Strength

Some people are stronger than others. One man might be able to move a heavy ladder safely that a weaker person would have difficulty with. Know your strength limits and get help if you need it (Fig. 3).

Muscles make up about 45 percent of the body's weight. When working, they convert the chemical energy of food to mechanical energy. The amount of energy used depends on the size and condition of the muscles, blood supply, heat, respiration, and type of work (Fig. 4). A sprinter works at a rate of 4 to 5 horsepower for about a second at the start of a race. A practical estimation of man's ability to work continuously is about 1/10 to 1/5 horsepower (Fig. 5). It takes about 1/10 horsepower to light a 75 watt electric bulb.

Fig. 4—How Much Energy Is Used In Your Work?

ENERGY CONSUMPTION IN WORK CALORIES

(Total metabolism during work minus the base metabolism)

Activity	Conditions of Work	Travel Speed(Km/hr)	Calories Per Minute
Walking without load	Level, smooth surface	1.86 m.p.h. (3,46)	1.7
		2.48 m.p.h. (3,96)	2.1
	Heavy, deeply plowed soil	1.86 m.p.h. (3,46)	5.2
	Country road, heavy shoes	2.48 m.p.h. (3,96)	3.1
Walking with load carried on the back	Level, firm surface, 22 lb.(10Kg) load	2.48 m.p.h. (3,96)	3.6
	66 lb. load (30Kg)	2.48 m.p.h. (3,96)	5.3
	110 lb. load (50Kg)	2.48 m.p.h. (3,96)	8.1
Climbing	16° gradient, climbing speed 32.72 ft./min(9,97m/mm)		
	without load		8.3
	44 lb. load (20Kg)		10.5
	110 lb. load (50Kg)		16.0
Climbing stairs	30.5° gradient, climbing speed 56.42 ft./min.(18m/min)		
	without load		13.7
	44 lb. load (20 Kg)		18.4
	110 lb. load (50 Kg)		26.3
Pulling cart	2.23 m.p.h. (3,56 Km/hr), level firm surface, pulling energy 25.52		8.5
Working with axe	Two-handed strokes, 35 strokes/min.		
	horizontal stroke		9.5-11.0
	vertical stroke		10.0-11.5
Working with hammer	9.68 lbs. (4,38Kg)weight of hammer, 15 strokes/min.		
	lifting stroke		7.3
	circular stroke		6.7
Working with shovel	Throwing distance 6.56′ (2m)		
	throwing height 3.28′ (1m)	10 liftings/min.	7.8
	throwing height 4.92′ (1.5m)	10 liftings/min.	9.0
	throwing height 6.56′ (2m)	10 liftings/min.	10.0
Sawing wood	2-man cross cut saw, 60 double pulls/min.		9.0
Bricklaying	Normal output .1435 cu. ft./min. (67,72 cm 3/min)		3.0
Digging	Garden spade in clay soil		7.5-8.7
Mowing	Clover		8.3
Milking	Normal milkable cows, 180 pulls/min.		2.2

Fig. 5—High Energy Release Versus Continuous Work

To work safely, avoid muscle fatigue:

1. *Work in a comfortable position.* An awkard position, like working overhead, is far more tiring than one in which the work is done at a more comfortable height.

2. *Work within your limitations.* Don't demand too much of your muscles. A muscle strained to its maximum capacity tires quickly.

3. *Keep moving.* The body movement in dynamic work aids blood circulation, uses a wider variety of muscles, and is less tiring than static work. Dynamic work, such as shoveling, may not fatigue muscles as much as a static job like holding a board while someone nails it to the ceiling.

4. *Take frequent, short rest breaks (10 minutes every 2 hours).* They are more effective in recovering working ability than longer, infrequent breaks.

Fig. 6—Machines Are Faster And Stronger Than You Are—Keep Away From Moving Parts

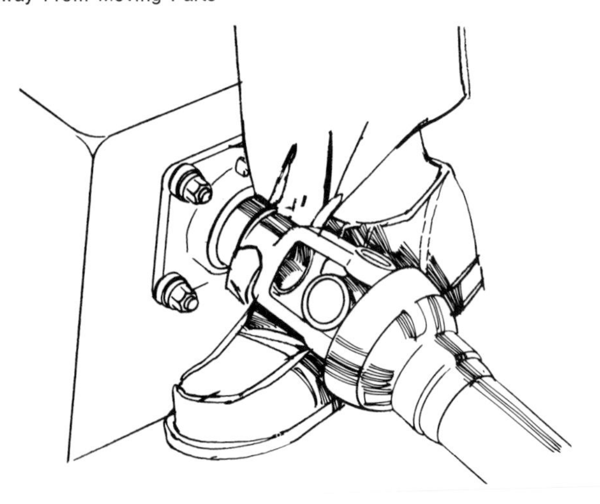

Reaction Time

Reaction time is made up of a series of events beginning with a message to the brain and ending with a body response. For example, tractor driver sees a drainage ditch (the message). The driver's brain analyzes the message. Then the brain tells muscles to move to avoid the ditch. For the brain to receive the message and tell the body to take action takes time. The best human reaction time is *slow* compared to high-speed machinery. Human reaction time to the unexpected is about 1 second.

• If the snapping rolls of a corn picker are moving material at 12 feet per second, stalks would travel 12 feet before a person could let go of them. A man trying to unplug a picker could be pulled into it before he could let go of the stalks. Disengage the PTO, turn off the engine, and take the key before you work on a machine.

• By the time a tractor operator realizes his tractor and loader are turning over and begins lowering the bucket to correct, it will probably be too late. A tractor overturns so quickly that reaction time is usually too slow to stop it. Prevent the problem by keeping the bucket low.

• Gear and belt drives, feed rolls, and PTO shafts turn at high speeds. A person can't react fast enough to pull away once he is tangled (Fig. 6). Keep drives guarded and shut off all power before working near them.

Your reaction time is even slower when you are affected by things like fatigue, medicine, alcohol, and preoccupation. Reaction time may be lengthened in a panic situation. For example, if someone else is caught in a machine you may be shocked so much you can't stop the machine as quickly as normal.

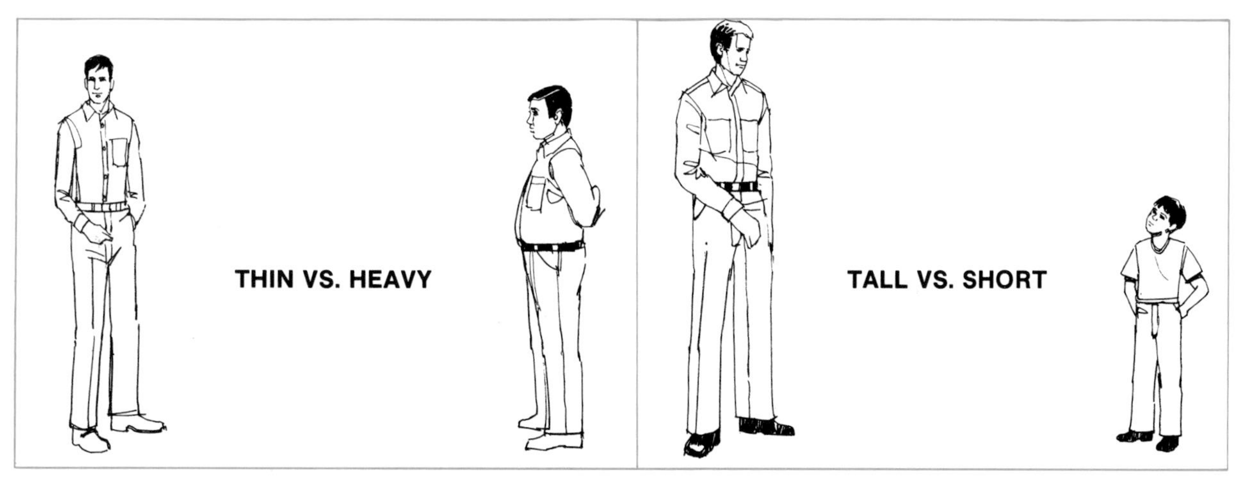

Fig. 7—What Are Your Dimensions?

Think about how you'd react to emergencies *BEFORE you're exposed to them.* Then you can react more quickly in an emergency. Recognize what can cause problems and take steps to avoid the causes.

Body Size

A person's body size often determines the kinds of jobs he can do safely (Fig. 7).

What one person can do easily because of his size could be very difficult or even dangerous for someone else. A tall man can place a heavy box on a high shelf easier than a shorter man.

A lightweight youth can walk across thin ice or stand on a weak stool that would collapse under a heavier person. A tractor seat suspension adjusted for a heavy person will be too stiff for a light person.

Control placement affects operating ability and safety. Controls adjusted for an average-sized person might be beyond the reach of a small person and could cause an accident (Fig. 8). On the other hand, a very large person might be uncomfortably cramped and tire quickly in the same operator station.

What are your size limitations? How high can you reach? Can you properly reach the clutch and brakes on your tractor? How heavy are you? Will a seat, ladder, or other structure support you properly?

Age

Physical capabilities reach their optimum at about 25 to 30 years of age and then decline. Motivation, judgment, and skill usually continue to increase beyond age 30.

Improvement in judgment and skills can offset the decreases in reaction time and muscle strength that occur later in life. However, somewhere between 55 and 70 the typical individual's eyesight, hearing, and

Fig. 8—A Small Person May Not Be Able To Reach Controls

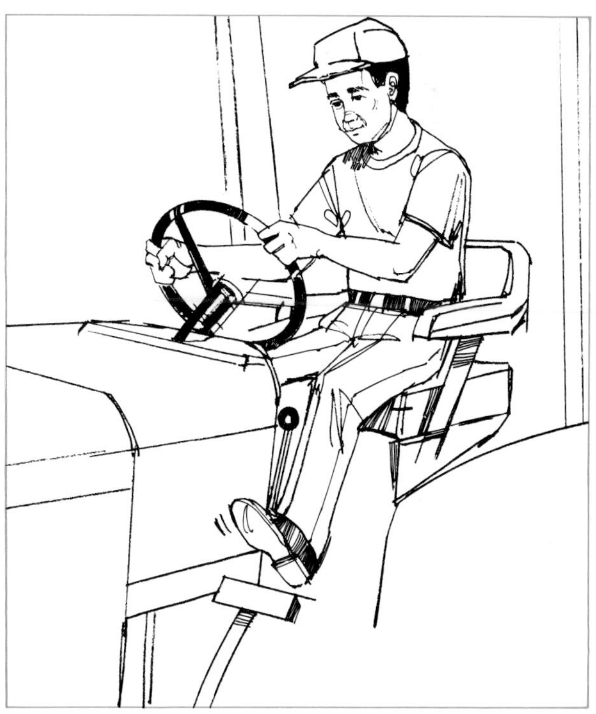

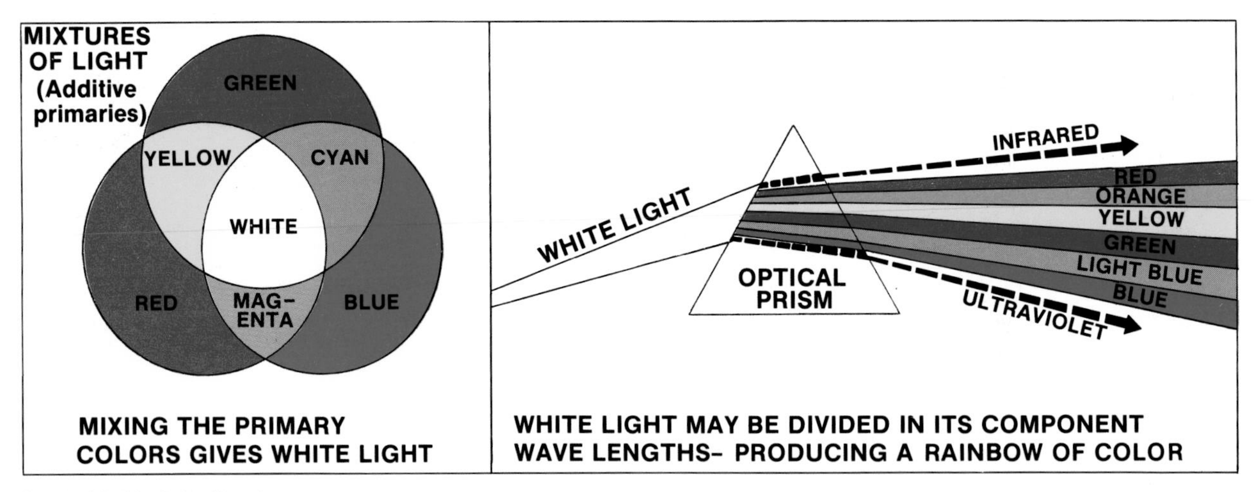

Fig. 9—Visible Color Spectrum

strength deteriorate to the point that his performance is poorer than when he was younger. Respiration ability decreases and cardiovascular disorders are common. People who reach this stage often fail to recognize or admit it and continue to try to work as hard as they did years before. These people are more likely to have accidents than they were in their prime.

Vision

The eye can focus on objects and track them. It is protected by an eyelid, is self-lubricated, and self-cleaned. It automatically adjusts to light conditions and is a sensitive detector of color shades, alignment, and movement.

More than 90 percent of our work is controlled by our eyes, and so are many of our leisure-time activities.

As wonderful as the eye is, it too has physical limitations and needs protection and care. Eyes can adapt to many conditions, but eyes can be strained and even permanently damaged when conditions aren't good. Good vision depends on:

- **Adequate lighting**
- **Size of objects viewed**
- **Color and contrast between the object and its background**
- **Steadiness of the object viewed**
- **Clarity and distinctness of the object**

If eyes are strained because one or more of these factors is poor, headache, fatigue, and weariness soon result.

Like a color TV, the eye is equipped with three primary color receptors; green, red, and blue, and can detect very subtle changes in color hue (Fig. 9).

Fig. 10—Typical Visual Field—Without Head Movement

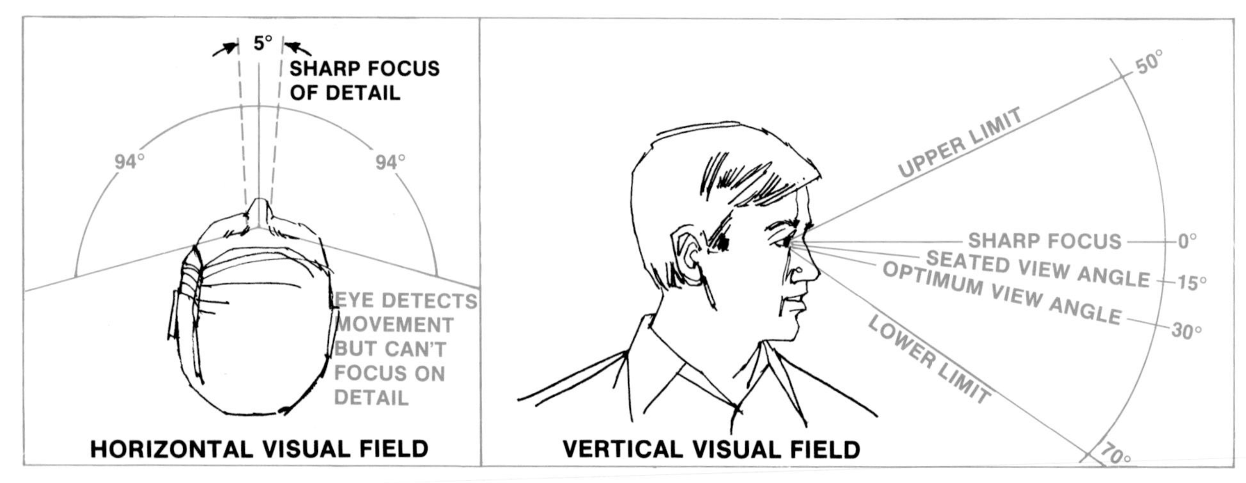

However, defective color vision is common. Total color blindness is rare. It is common for a person to be blind to one color. Have a doctor check your color vision if there's the slightest hint of a problem. Failure to recognize color blindness can cause you to miss a traffic signal or fail to recognize a warning sign.

The normal horizontal field of vision is about 188 degrees (Fig. 10). You can detect objects and movement at the edge of your field of vision, but can't focus detail. The angle of focused vision is quite narrow (about five degrees) and the eyes must be moved constantly to read or to see detail. This means that a tractor operator must turn his head 120 to 180 degrees many times a day to see how a towed implement is doing. If he uses a rear-view mirror, he can watch these implements without the excessive head movement.

People's ability to judge distance varies. A person with wide-set eyes has better binocular vision (stereopsis) and can judge distance better than someone whose eyes are set closely together. A person with only one eye can judge distance only by relative size and experience. Objects tend to look like flat photographs. Imagine trying to judge the distance between the car you're driving and one stopped on the road ahead of you if you could not judge distance.

Since the eye is so important, precautions should be taken to protect it.

Wear eye protection to keep fumes or particles out of your eyes (Fig. 11). Protect your eyes from bright sunlight by wearing tinted glasses. When exposed to ultraviolet rays from arc welding, wear protective goggles with darkened lenses or turn your head away. Closing your eyes doesn't completely prevent invisible ultraviolet rays from passing through your eyelids and damaging the retina. Never look at an electric welding arc without proper eye protection. Not even for an instant!

Hearing

How's your hearing? Is it as good as it was a year ago? How do you know? Hearing loss is less obvious than the losses of other senses. It occurs slowly over the years. You may not even realize your hearing is gradually decreasing because there is no pain.

The normal human ear (Fig. 12) is adapted to wide differences in sound. It can hear the faint peep of a newly hatched chick or the loud blast of a shotgun. However, if your ears are overstimulated by loud noises they lose their sensitivity.

Fig. 11—Keep Eye Protection Handy And Wear It Whenever Fumes or Particles Are Apt To Get Into Your Eyes

Sound is usually measured in *decibels.* Values range from zero to 140 (Fig. 13). Decibles are logarithmic units of pressure and can't be added or subtracted. Doubling the amount of sound pressure energy results in a three-decibel increase (Fig. 14). For example, if two tractors are running side by side and each produces 90 decibels of sound, the combined sound level would be about 93 decibels — not 180. Because the ear also hears logarithmically, the two tractors would *not* sound twice as loud as one.

Many people mistakenly assume that a few decibels increase is not significant when, in fact, every three decibels increase in the sound represents a *doubling* of the sound pressure energy that the ear receives.

The ear can detect changes in sound frequency or pitch. The limit of human hearing is approximately 20 to 20,000 hertz (or cycles per second). The low notes of a tuba or organ would be approximately 20 hertz.

Fig. 12—Cross Section Of The Human Ear

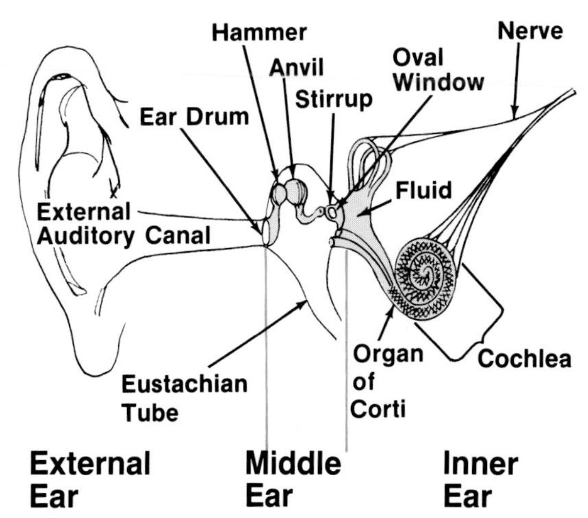

	Decibel Levels of Common Sounds at Typical Distance from Source
0	Acute threshold of hearing
15	Average threshold of hearing
20	Whisper
30	Leaves rustling, very soft music
40	Average residence
60	Normal speech, background music
70	Noisy office, inside auto 60 mph
80	Heavy traffic, window air-cond.
85	Inside acoustically insulated protective tractor cab in field
90	OSHA limit—hearing damage on excess exposure to noise above 90 dB
100	Noisy tractor, power mower, ATV, snowmobile, motorcycle, in subway car
120	Thunder clap, jack-hammer, basketball crowd, amplified rock music
140	Threshold of pain—shot gun, near jet taking off, 50 hp siren (100')

Fig. 13—Decibel Levels Of Common Sounds At Typical Distance From Source

The 20,000 hertz might be reached by a violin. As people age, the higher range usually declines to 12,000 or 15,000 hertz.

Low frequency sounds (below 1,000 hertz) are common in agriculture. Low frequency sounds are measured on the "A" scale of a sound level meter. The "A" scale responds to sound more or less the same as the human ear. The values are given in units of dB(A) i.e.,decibles on the "A" scale. The B and C scales on a sound meter correspond to higher frequencies.

A person's hearing threshold is the lowest sound in decibels that he can hear. Abusing ears with loud sound shifts the hearing threshold upward so a person can hear only louder sounds. Maybe you've experienced this after operating noisy equipment for several hours. When you stop, your ears ring and your hearing doesn't seem normal. Hearing will usually return to normal again overnight. However, repeated exposure will eventually result in a permanent threshold shift (Fig. 15).

Hearing losses occur above the frequency of 1,000 hertz (cps), with maximum loss occurring at the 4,000 hertz level, especially in the 30-to-49 age group. High-frequency losses are most common in people between 30 and 60.

Fig. 14—Doubling The Sound Pressure Energy Gives A Three Decibel Increase

Farmers have greater hearing losses than people in other occupations. This may be partly due to the frequent and continuous operation of noisy equipment (Fig. 16).

Using protective earplugs or muffs in noisy situations helps prevent hearing loss. Protection usually helps a person hear spoken words better since protection reduces high frequency sound coming to the ears.

If you're aware of the sources of noise you can protect yourself from them better. Hearing protection is required when the sound level exceeds 90 decibels on the "A" scale if you're exposed for 8 hours or more. Some experts recommend hearing protection if sound levels exceed 85 dB(A).

Here are some general recommendations to help protect your hearing:

1. *Consider quiet operation when shopping for tractors and farm machinery.* A machine that makes a lot of noise doesn't necessarily do better work or have more power than a quieter unit.

2. *Use ear protection for all noisy jobs.* Operating farm machinery requires ear protection if noise exceeds 90 dB(A) for eight hours. Protection should be used the minute the activity starts.

3. *Keep your equipment well lubricated and maintained.* Replace defective exhaust system parts promptly. Don't use a straight pipe. You'll get ear-damaging sound and no significant increase in power.

4. *Quiet stationary equipment like compressors and feed grinders by building acoustic barriers.*

5. *Limit the length of time that a person hears the noise.*

6. *Stay as far away from noise sources as possible.* Double the distance from the source reduces the sound pressure level to one-fourth.

PHYSIOLOGICAL

Your body has certain physiological characteristics and limitations. Some of these are: muscle tone and strength, your metabolism efficiency, how much food it takes to keep you going, your resistance to illness, and the amount of sleep and rest your body requires.

Physiological limitations vary widely between different people and can change for the same person from day to day. Physiological limits are affected by:

- *Medication intake*
- *Fatigue*
- *Drugs, alcohol, and tobacco*
- *Chemicals*
- *Illness*
- *Environmental conditions, such as temperature, humidity, dust, etc.*

PERMISSIBLE NOISE EXPOSURE

Hours per day that you can safely be exposed to these sound levels.

(As specified by the Walsh-Healey Act for industrial situations)

Duration per day, Hours	*Sound level, Decibels*
8	**90**
6	**92**
4	**95**
3	**97**
2	**100**
1½	**102**
1	**105**
½	**110**
¼ or less	**115**

Fig. 15—Permissible Noise Exposure

Fig. 16—Use A Hearing Protective Device. Continuous Exposure To Excessive Noise Results In Hearing Loss

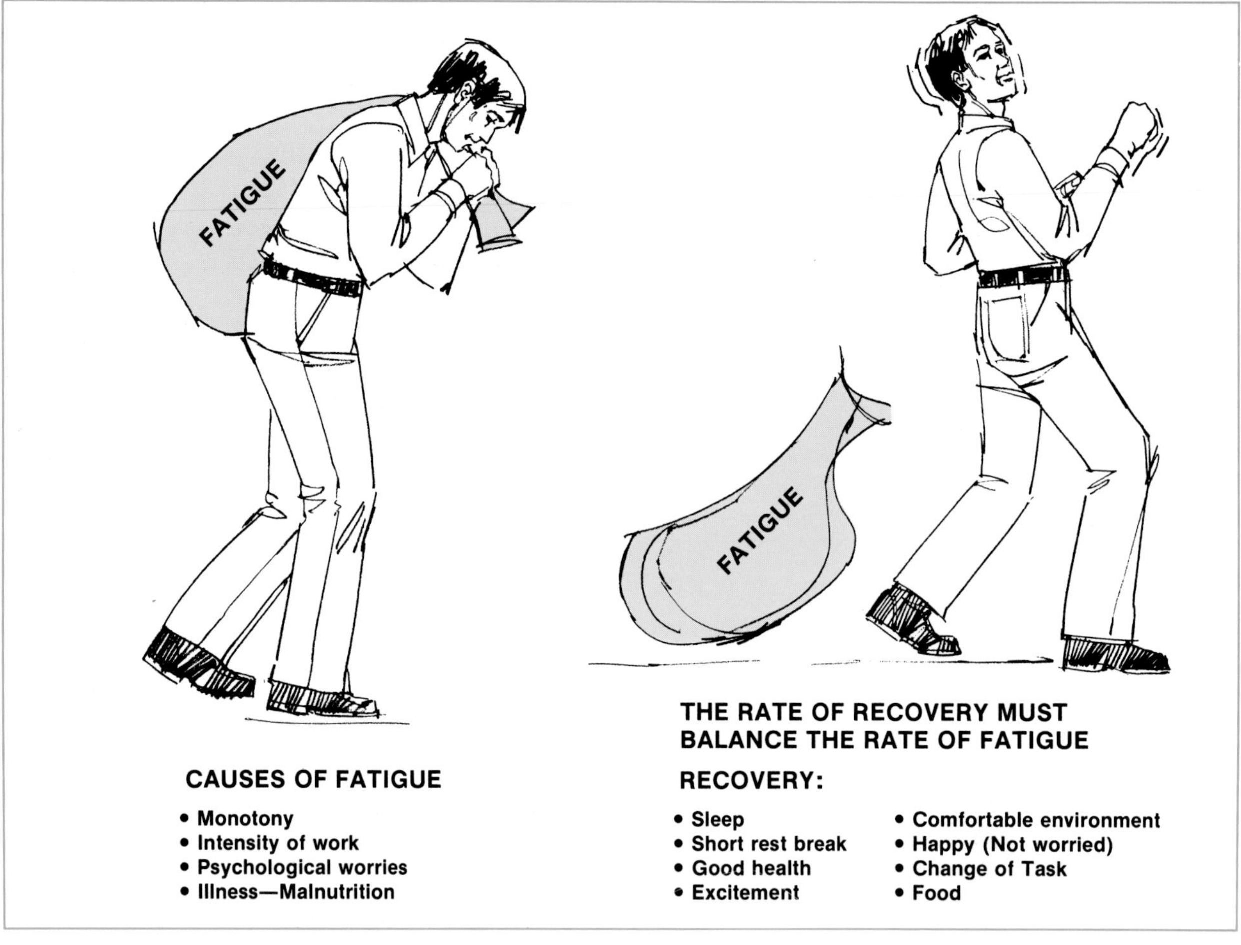

Fig. 17—Fatigue Is Caused By Many Factors

Fatigue

Everyone knows about fatigue. You've probably had the experience of working or exercising to the point that your body could no longer continue and you had to rest. Fatigue may even affect the brain, but you may not recognize it as you do physical fatigue. Many things can cause fatigue (Fig. 17).

When fatigue is due to physical work, typical results are muscle tightness and cramping. Work requires that chemical energy be converted to mechanical energy in the muscles. Blood flow and respiration must increase to supply the muscles with the needed energy and oxygen and to carry off carbon dioxide and chemical waste.

When muscles work at a rate exceeding the ability of the heart and lungs to supply the necessary oxygen and chemical food, aching in the muscles, cramping, tremor, and finally, loss of control may result. You have to stop and rest to recover.

Muscle fatigue is like a safety device that doesn't let a muscle work beyond its capacity. Individual motivation or desire plays a very significant role in determining this limit. Strong motivation will push the limit far beyond the normal one, but there *is* a limit.

If you reach this limit and continue to work, you are more likely to make mistakes. Because of loss of muscle power and control, lessening of attention, slowdown of reactions, and loss of sensitivity, you become accident prone. Exhausted legs may tremble and fail to operate the brakes or clutch promptly. A load that can normally be lifted easily will be far too heavy for you to budge.

To avoid general fatigue and muscle fatigue, rest regularly. Frequent short pauses are more effective than longer rests at wider intervals (Fig. 18). Take rests in the late morning and late afternoon. Eating foods high in sugar is another excellent way to fight fatigue.

Drugs, Alcohol, and Tobacco

The use of drugs, alcohol, and tobacco is usually defended on emotion rather than logic. A person may feel he's an exception and that drugs, stimulants, and depressants don't seriously affect his ability to perform.

Many people begin to smoke, drink, or use drugs to show their independence. In time their self confidence improves, but by that time they've often become hooked on a habit that's expensive and dangerous.

TOBACCO

Smoking reduces work capacity as much as ten percent because of carbon monoxide in the smoker's blood (Fig. 19). Tobacco smoke contains up to four percent carbon monoxide. Carbon monoxide is absorbed by the blood. Carbon monoxide has a greater attraction to the hemoglobin of blood than oxygen does, so it's not expelled during respiration. Blood carrying carbon monoxide can't be used for respiration. And because of the decreased oxygen carrying capacity, a smoker becomes short-winded and has reduced work capacity, especially right after smoking.

Short-windedness can be observed immediately. Long-term effects such as the loss of taste sensitivity, sense of smell, and the development of lung cancer, require a little more time to develop. Medical science has now proven beyond a question that lung cancer and smoking are directly related. Also lung ailments from exposure to dust and toxic gases are more prevalent among smokers because their lungs are already irritated.

Fig. 18—Frequent Short Rests Are More Effective Than Longer Rests

Fig. 19—Smoking Reduces Work Capacity As Much As Five To Ten Percent

Fig. 20—Alcohol In The Blood Affects Judgment And Coordination. The Drinking Driver Of This Tractor Was Killed When His Tractor Went Over A Bridge Abutment

ALCOHOL

Any amount of alcohol in the blood affects human coordination and reflexes (Fig. 20). As the amount of alcohol in the blood goes up, performance goes down (Fig. 21). High alcohol levels affect judgment. Alcohol is a contributing cause in thousands of fatal accidents each year.

Very high levels of alcohol result in complete loss of coordination and rationality. Loss of consciousness is nature's survival switch. The level of alcohol in the blood that causes loss of consciousness is so high that a slight increase in intake can result in death. Passing out usually prevents that fatal last drink.

Experiments in England relating driving skill and alcohol levels showed that bus drivers with high levels of alcohol were often able to steer a bus through a narrow space successfully. But their ability to judge whether a space was wide enough was very poor and they often tried to drive buses through spaces that were too narrow for them. Alcohol affected their judgment ability before their coordination dropped off.

Studies have shown that alcohol physically destroys some of the brain and clouds the thinking process. The evidence that a small amount of alcohol impairs the ability to drive is so overwhelming that a legal intoxication point of .10 percent alcohol in the blood (4 or 5 drinks) has been established in many states. Some Scandinavian countries have gone even lower. In Sweden .08 percent alcohol in the blood constitutes under-the-influence. In Norway a person with .05 percent is legally drunk.

The effects of alcohol on body functions are temporary at first. Toxins can build up in the liver causing hepatitus and eventual permanent damage. Also, dependency can produce terrible problems for the drinker and even more difficult ones for those around the drinker. The danger lies in the loss of judgment and coordination that the drinker doesn't recognize or refuses to admit.

DRUGS AND MEDICINE

Drugs range from aspirin to heroin. They all alter the body and mind. Hard drugs and hallucinatory drugs have direct and devastating effects on the person using them. But there is also a secondary danger that a drugged person will not recognize a dangerous situation.

The narcotic analgesics, heroin, morphine, etc., cause a depression of brain functions. Although skeletal muscle coordination may not deteriorate significantly, the user may be lulled into a false sense of security and ignore signs of impending danger in his surroundings.

The effects of hallucinogens, such as LSD, have the obvious danger that they can distort sensory perception so the user cannot interact with things around him. For example, the drug user may walk off a high building, thinking it only a step down. In addition, LSD may disrupt emotional balance and thought processes. Marijuana has similar affects. It causes a decrease in body functioning ability and muscle coordination making the operation of machinery hazardous.

Sedatives, such as sleeping pills (barbiturates) and tranquilizers (Librium), cause a general depression of brain function. A person may become so sleepy he loses the ability to concentrate and coordination becomes poor. For example, coordination of the foot on the clutch and hand on the gear shift may be disrupted.

A more subtle danger lies in common antihistamine medicines like cold pills, pain killers, allergy remedies, and pep pills. These drugs can cause sleepiness, drowsiness, lowered reflexes, or increased respiration and heart rate. Those that have such an effect carry a warning on the label for the user to avoid driving or operating dangerous machinery. Any drowsiness, dizziness, loss of coordination, or muscle response can be potentially dangerous, depending on your job demands. If you don't know the side effects of a certain drug or medicine you are taking, check with a physician before you operate equipment.

Chemicals

The earth and atmosphere are made of chemicals. The nature of the chemical environment must be considered. The question is not whether a material is harmful, but how much of it can accumulate in the body before harm results. Small doses of some chemicals that show no immediate symptoms may build up over the years to a fatal level.

Chemicals can enter the body by:

- *Oral ingestion through the mouth.*
- *Absorption through the pores of the skin, cuts, or scratches.*
- *Injection under pressure into the skin.*
- *Inhalation (breathing dust or vapors such as aerosol propellants, glues, and spray paints).*

Certain changes can indicate dangerous exposure to chemicals long before clinical signs of poisoning are present. These include changes in:

- *Blood pressure*
- *Urine concentration*
- *Nervous tension*
- *Equilibrium*
- *Work performance*
- *Reaction time*

One indication of harmful exposure to chemicals is your heartbeat not returning to normal as quickly as it should after a routine job is finished.

People vary widely in their reactions. Chemicals may have very little or no effect or they may cause a violent reaction. A modest exposure may irritate a person or impair his performance, while a higher dose could cause permanent damage.

Fig. 21—Reaction Time Is Affected By Fatigue, Illness, Alcohol, And Drugs

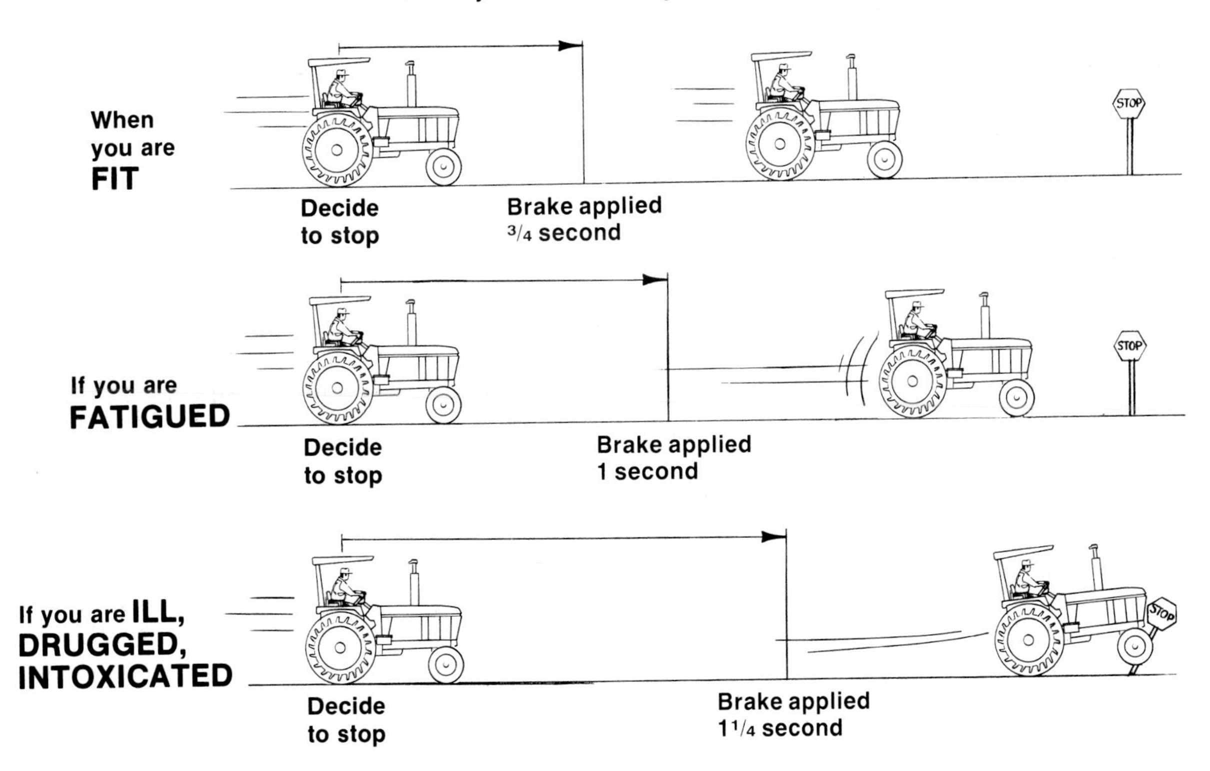

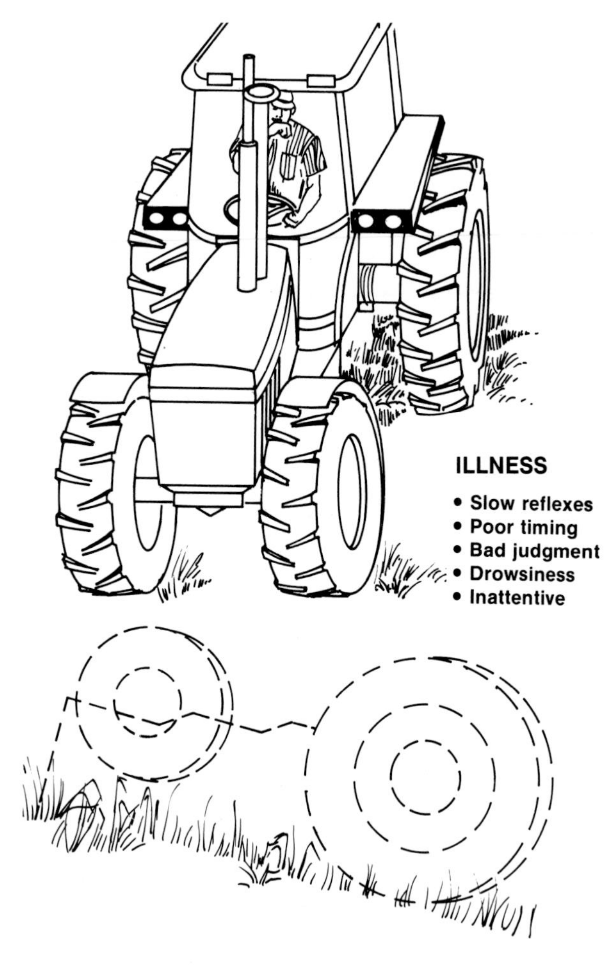

Fig. 22—Don't Work Around Machinery When You're Sick

EYES

Eyes are very sensitive to dust and drift from crop spray. Even small amounts of anhydrous ammonia can cause blindness. Wear unvented or chemical splash goggles when transferring anhydrous ammonia from one tank to another or when unplugging nozzles (see Chapter 7 on "Chemicals" for more details). If you're hit directly in the eyes, flood your eyes with water immediately. A few seconds delay could mean loss of sight. Flush eyes for at least 15 to 20 minutes and then get to a doctor without delay.

SKIN

Skin, particularly on arms and hands, is frequently exposed to chemicals. Exposure to irritating chemicals can cause burning, irritation, itching, hives, rash, and blisters. Avoid contact with chemicals and wear rubber gloves and a face shield. Wash with soap and water if you've been exposed to chemicals.

RESPIRATORY SYSTEM

The respiratory system is sensitive to toxic chemicals and dust and mold from grain and forage material (farmers lung disease). Complications can arise from even the most innocent-appearing substances:

- *Propellant from aerosol cans*
- *Silo gas (NO_2) may not notice smell even in toxic concentrations*
- *Engine exhaust — carbon monoxide (CO)*
- *Crop and fruit sprays*
- *Chemical dusts*
- *Household insect sprays*
- *Sprays for animals*

These are only a few substances that can cause respiratory problems. Safety measures include:

- *Avoid or reduce exposure when possible*
- *Work outside or in a well ventilated room*
- *Use an approved respirator if you can't avoid exposure to the chemical*
- *Avoid drift — stay upwind*

NERVOUS SYSTEM AND VITAL ORGANS

Organic phosphates can be taken in through the mouth or nose, or absorbed through the skin. They can cause nausea, headache, and even death. Other chemicals affect the heart and vital organs. They are usually taken into the body through the mouth or a cut in the skin.

Misuse of chemicals can be disastrous. Keep chemicals away from children and pets. Read the label before you use chemicals so you know how to safely use the chemicals and the antidotes required. Be able to call your doctor or the nearest poison center for help in case of accidental poisoning. See Chapter 7 for more information on safe use of chemicals.

Illness

Illness impairs human performance.

Even minor ailments like headaches and colds can reduce performance (Fig. 22).

A serious illness that requires bed recovery leaves a person weak for several days after he's back on his feet. Don't demand too much of yourself if you're in this situation.

Environmental Conditions

The most common environmental factors that affect farmers are:

- *Temperature and humidity*
- *Vibration*
- *Noise*
- *Dust and Mold*

TEMPERATURE AND HUMIDITY

Extreme temperatures reduce work efficiency (Fig. 23). Below 50°F (10°C) and above 86°F (30°C), strength decreases. Touch sensitivity decreases at low temperatures.

HEAT

Water is very important for the body to adjust to high temperatures. The rate of water intake must equal the rate of water loss by perspiring and body functions to keep body temperature normal. When it's hot, drink plenty of water!

The older and less physically fit you are, the less you'll be able to work in the heat.

Heat leaves the body in several ways (Fig. 24):

- *Transfer of heat from skin to air (convection) (Fig. 25)*
- *Evaporation of perspiration*
- *Exhaling hot air while breathing*
- *Touching a cooler object (conduction)*
- *Radiating from skin to air (radiation)*

Fig. 23—Temperature And Humidity Affect Your Comfort And Work Ability

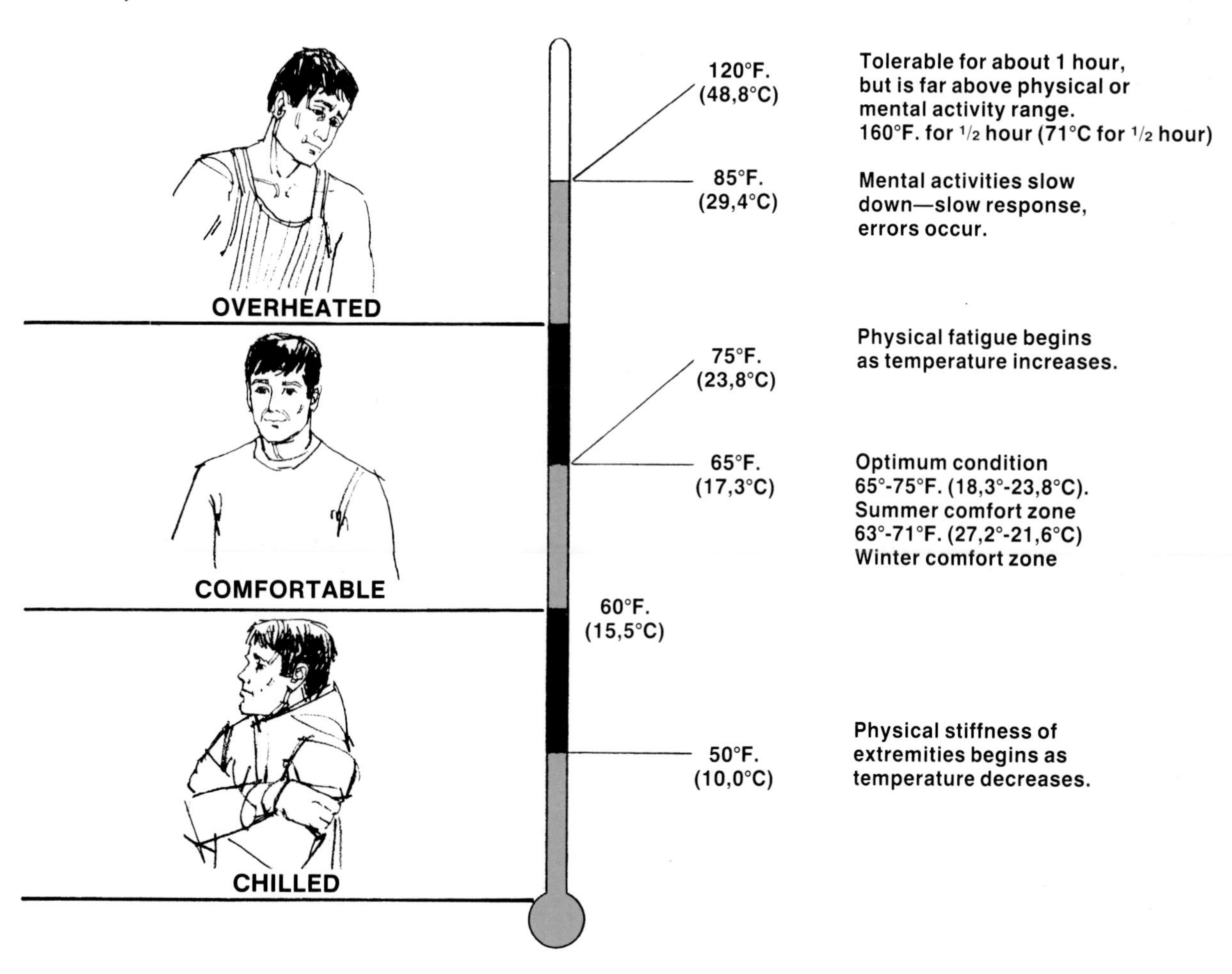

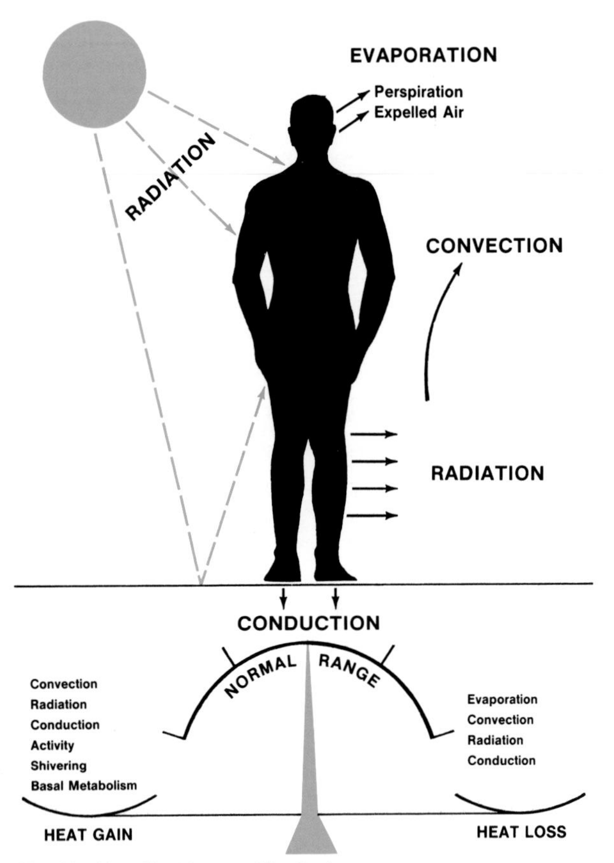

Fig. 24—How Heat Leaves The Body

Fig. 25—Air Temperatures And Velocities Required To Balance Your Body Heat Production

AIR TEMPERATURES REQUIRED

TYPE OF ACTIVITY	BODY HEAT PRODUCED BTU (Kg) PER HOUR	AIR TEMPERATURE NECESSARY (°F) (ASSUMING APPROPRIATE CLOTHING) FOR BODY HEAT BALANCE AT AIR MOVEMENT RATES OF: 20FPM (6m) (indoors)	100 FPM (30,5m/min)	20MPH(33m/min) (outdoors)
At rest	400 (100)	70° (21°c)	75° (24°c)	78° (26°c)
Moderate activity	1000 (252)	58° (14,4°c)	60° (21°c)	63° (17°c)
Vigorous activity	4000 (1008)	28° (-2°c)	30° (-1°c)	35° (2°c)

If both humidity and temperature are high, perspiration can't evaporate. The physiological effect of humidity is insignificant for people doing light work at temperatures below 80°F. (26°C). However, at higher temperatures, body temperature rises and heart rate and blood circulation increase although work level and oxygen consumption may stay constant. You will normally stop work voluntarily when your body temperature exceeds 102°F (95°C).

The *discomfort index* is a formula that relates temperature to humidity (Fig. 26). It's used to determine acceptable working conditions.

When heavy work is done at high temperatures, 90°F (32°C) and up, the sickness rate and the accident rate increase. Some individuals can reach a high level of oxygen intake, maintain a satisfactory heat balance, and work efficiently in the heat. But some individuals cannot. If you can't, don't push yourself beyond your limits. It can be dangerous for you around machinery and harmful to your health.

Some ways to reduce heat exposure are:

- *Shading (sunshield or umbrella)*
- *Increased air speed (fan)*
- *Air conditioning (cooling)*
- *Frequent rests*

Here are some heatwave guidelines:

1. *Slow down* in hot weather. Your body has a greater work load when temperature and humidity are high.

2. *Heed early warnings of heat stress, such as headache, heavy perspiration, high pulse rate, and shallow breathing.* Take a break immediately and get to a cooler place.

3. *Dress for hot weather.* Lightweight, light-colored clothing reflects heat and sunlight and helps you maintain normal body temperature.

4. *Eat carbohydrates in hot weather.* Avoid foods high in fat. Fat needs more oxygen to metabolize and produces higher body heat than carbohydrates. Proteins increase water loss.

5. *Drink plenty of water.* Heatwave weather can wring you out before you know it. Don't dry out!

6. *Get plenty of salt,* unless you're on a salt restricted diet.

7. *Avoid thermal shock.* Get used to warmer weather gradually. Take it easy those first two or three hot days. Your body will probably adjust if you take it slow.

8. *Get out of the heat occasionally.* Physical stress increases with time in hot weather. Try to get out of the heat for at least a few hours each day.

9. *Don't get too much sun.* Sunburn makes the job of keeping cool much more difficult. Sun rays also cause skin cancer. Wear a hat and long sleeved shirt to avoid burning. Don't try to get a suntan while you are working. You would be burning the candle at both ends.

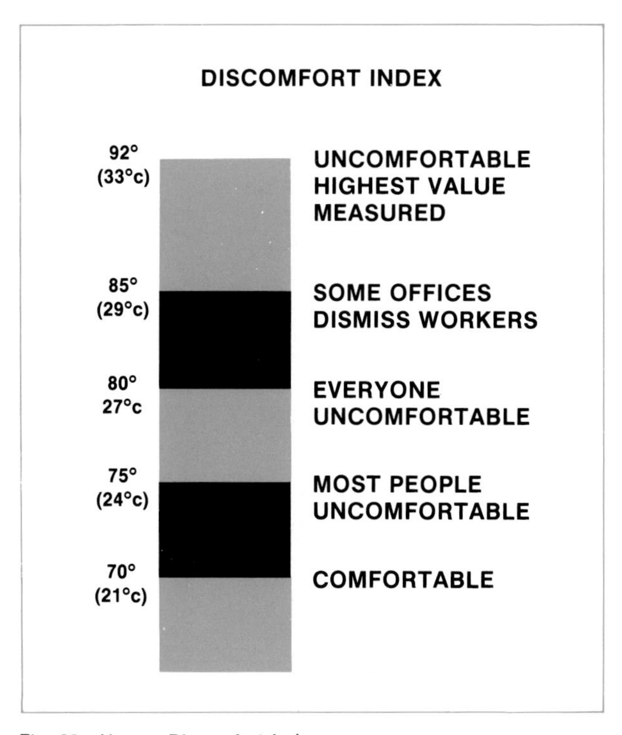

Fig. 26—Human Discomfort Index

COLD

Man can deal with low temperatures much better than high temperatures. Just add clothing. Suitable clothing can protect you against cold. But select clothing carefully for the job you'll be doing.

Heavy, bulky clothes, such as snowmobile suits, are great for snowmobiling, ice fishing, and other inactive sports. However, for active sports or work, they are much too warm. Heavy, bulky clothes can't be ventilated well or removed in layers. They may actually be hazardous if they restrict movement necessary to safely accomplish the task. Also, the body perspires in heavy clothing. Moisture is trapped, and the wearer can easily chill. Clothing for the active person should be layered so pieces can be removed or added depending on body temperature (Fig. 27).

The body supplies the head and vital organs with warm blood first to keep these vital areas warm. Reducing heat loss from the head makes the heat available for other parts of the body. That's why wearing a hat helps keep your hands and feet warm.

A high calorie, high protein diet is required to give a person enough energy to work and keep warm in cold weather.

Fig. 27—Clothing For The Active Person Should Be Layered To Allow Adding Or Removing For Comfort

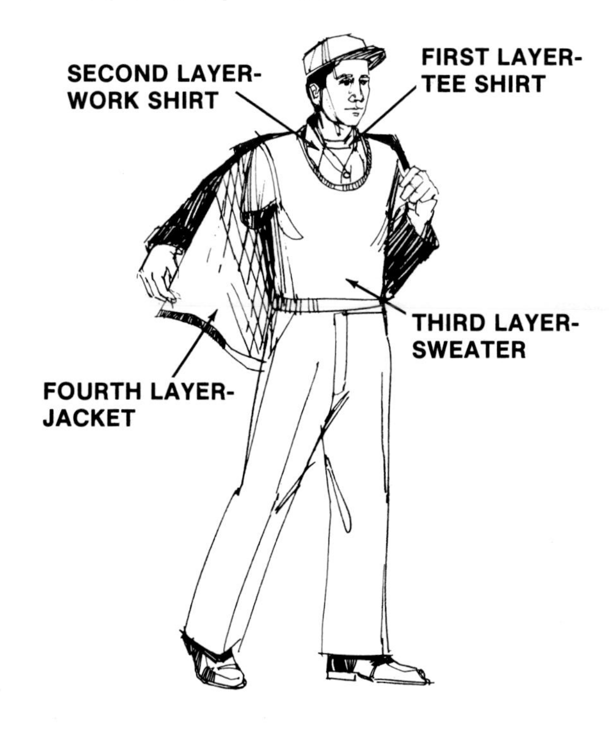

EFFECTS OF VIBRATION ON HUMANS

(Your) Physical Activity	Effect of the Vibration	Vibration Frequency Cycles per Second	Vibration Displacement in inches* (centimeters)
Respiration control	degrades	3.5-6.0	.75 (1,9)
	degrades	4.0-8.0	.14-.61 (0,35-1,54)
Body tremor	increases	40.0	.065 (0,165)
	increases	70.0	.03 (0,07)
Hand tremor	increases	20.0	.015-.035 (0,03-0,08)
	increases	25.0	.035-.055 (0,08-0,13)
	increases	30-300	.02-.20 (0,05-0,508)
	increases	1000	.008 (0,02)
Hand coordination	degrades	2.5-3.5	.50 (1,27)
Foot pressure constancy	degrades	2.5-3.5	.50 (1,27)
Hand reaction time	increases	2.5-3.5	.50 (1,27)
Visual acuity	degrades	1.0-24	.024-.588 (0,60-1,49)
	degrades	35.0	.03-.05 (0,07-0,27)
	degrades	40.0	.065 (0,165)
	degrades	70.0	.03 (0,076)
	degrades	2.5-3.5	.5 (12,7)
Tracking (Steering)	degrades	1.0-50	.05-.18 (0,07-0,45)
	degrades	2.5-3.5	.5 (12,7)
Attention	degrades	2.5-3.5	.5 (12,7)
	degrades	30-300	.02-.20 (0,05-0,508)

*Distance of Movement by Vibration

Fig. 28—Effect Of Vibration On Humans

VIBRATION

The vibration of a massage chair can be soothing. The vibration of a chain saw however can be irritating, even damaging. Equipment that shakes damaging kidneys, spine, and stomach.

Your body deals with vibration by a constant contraction and relaxation of the muscular system. Over a number of hours, this causes a change in the response of the self regulating nervous system. This in turn affects the muscle systems of the intestinal tract and interferes with normal digestion.

Many farmers and truck drivers have been exposed to spine damage caused by vibration. Driving across the furrows on plowed ground can be particularly damaging.

Tractor vibration depends on the type of machine, seat suspension, ground, speed, and power output of the vehicle.

Vibration can affect the ability to steer or match the movement of an object, foot pedal pressure, and vision. Respiratory control, coordination, and concentration are also affected (Fig. 28). If you are particularly sensitive to vibration, don't continue to expose yourself to it.

NOISE

Noise is defined as unwanted sound. A certain sound might be noise to some ears and music to others. For example, a loud, snarling sports car might sound great to its owner, but very unpleasant to his neighbors.

Different people may react differently to the same sound. A sound that soothes one person might keep someone else wide awake, for example, rain on a roof.

The relative loudness of sound also affects people. A dripping faucet, unnoticed in the daytime becomes unbearable at night.

The degree noise bothers people depends on several factors:

- The greater the noise intensity and the higher its pitch, the more people it will annoy.

- Unusual, intermittent noises are more upsetting than familiar constant noises.

- Past experience associated with a given sound may determine the emotional reaction it evokes. Sounds that disturb sleep and are associated with fear are most unpleasant.

- Personal attitudes toward the noise source are important. The sound of a motorcycle engine affects people in different ways, depending on their attitude toward motorcycles.

- The schedule of a person can determine how much annoyance a sound causes. A housewife is much less disturbed by daytime traffic noise than a man who works night and sleeps during the day.

When a particular noise level exceeds background noise by more than three decibels at night or by five decibels during the day, it is classified as intolerable.

Loud noises affect a person. Heart rate increases with increased noise. Energy use increases. This can contribute considerably to fatigue, comfort, and mental ease.

The primary effect of noise is loss of hearing. In a recent report of adult hearing levels (National Center for Health Statistics), hearing impairment was often found among farmers.

When noise exposure goes down, people often:

- *Have less hearing loss*
- *Show fewer signs of stress*
- *Have more energy*
- *Are happier*
- *Seem to have fewer neurotic problems*
- *Have fewer accidents*

PSYCHOLOGICAL

Personal safety and performance depend on psychological factors. People have emotions and moods.

Psychological problems result from:

- *Personal conflict—confusion and uncertainty in a person's mind*
- *Personal tragedy—loss of a friend or relative*
- *Problems in the home, disagreement and friction between people*
- *Vocational problems—difficulties on the job*
- *Financial difficulties*
- *Insecurity*

The effects of these emotional problems show up in a number of ways.

Emotional problems cause accidents. Learn to know if you are upset. Take a break when you feel you are upset.

Fig. 29—An Angry Person Is Dangerous

Temper

An angry person tends to overreact and may take his frustration out on people, animals, or objects that happen to be handy (Fig. 29). A person who is upset is a risk. His judgment is poor, and he may take chances.

Everyone gets angry. It is a normal release. The way to handle anger safely is to stay off farm equipment, and any other potentially hazardous equipment, when you're angry. Wait till you've cooled down.

Anxiety

Anxiety makes you do things you wouldn't normally do. For example, a farmer concerned about getting a crop in before a threatening storm breaks may try to save time by not securing shields on his machinery and thereby setting up an accident situation.

Apathy

A person with a "who cares" attitude toward his job will let details slip by. Sloppiness promotes accidents.

Not many of us would like to be operated on by an apathetic surgeon. There is no place for apathy around farm machinery, either. Attention to detail and concern for the safety of self and others is important.

Preoccupation

- Suppose a man is driving his tractor to the field. His thoughts are occupied by a major business deal. He is preoccupied with the deal. Because he's preoccupied, he is inattentive, and drives into a stump (Fig. 30).

Fig. 30—Preoccupation And Distraction Cause Accidents

• A woman is worried about difficulties her son is having in school. While driving home, she fails to recognize the slow-moving-vehicle emblem on a tractor ahead and runs into it.

• A preoccupied young man overlooks shutting off the power to a plugged corn picker and is caught and mangled while attempting to clear it.

Newspapers are full of accounts of serious accidents that happen because people are preoccupied.

Investigating It Yourself

What makes a person touch a newly painted surface when a sign says "Wet Paint?" Why do people have a tendency to check to see if something is hot or to reach for a moving part? Why does a farmer reach into a dangerous place for a dropped ear of corn, lean against a machine, pat it approvingly, or touch a moving belt? The answer is not simple and involves a deep investigation into the study of psychology. Realize man *has* these tendencies and recognize the potential hazards they create.

Psychological factors, attitudes, emotions and moods, are not usually changed easily or quickly. However, recognize these characteristics and understand the safety hazard and accident potential they possess. If you recognize dangerous traits in your own actions, try to take corrective measures. If you are angry, take a break and cool off before you are involved in an accident. If anxiety or preoccupation continue to bother you, some counseling or dealing directly with the underlying problem may change the situation completely.

If you are aware of psychological problems in someone else, try to keep them away from danger. For example, the person with an apathetic attitude should not be allowed to work where his errors and carelessness could cause a serious accident. Do not permit the "show-off" to operate equipment where he might injure himself, others, or damage the machinery.

The "safe worker" is one who is happy, alert, content, and well adjusted. He works with skill and has concern for his job, for others, and their safety.

PERSONAL PROTECTIVE EQUIPMENT

Many farm injuries can be prevented by personal protective equipment. Devices are available to protect the:

- **Head**
- **Eyes**
- **Ears**
- **Hands**
- **Feet**
- **Body**
- **Respiratory system**

But they *must be used* in order to be effective.

HEAD PROTECTION

The most effective head protection is a hard hat (Fig. 31). It protects against bumps and falling objects. Some hard hats have warm liners for cold weather. Nonconductive hard hats are good for electrical work.

Use hard hats for building work, felling or trimming trees, machinery repair. Lightweight bump caps give some protection. Use them for jobs where dangers to the head are not extreme.

EYE PROTECTION

Eyes are very sensitive. Protect them from impact, chemicals, dust, and chaff. Wear eye protection when spray painting, grinding, drilling, welding, sawing, working in dust or chemicals. There are several types of protective devices for the eyes (Fig. 32).

Fig. 31—Hard Hats Give Protection

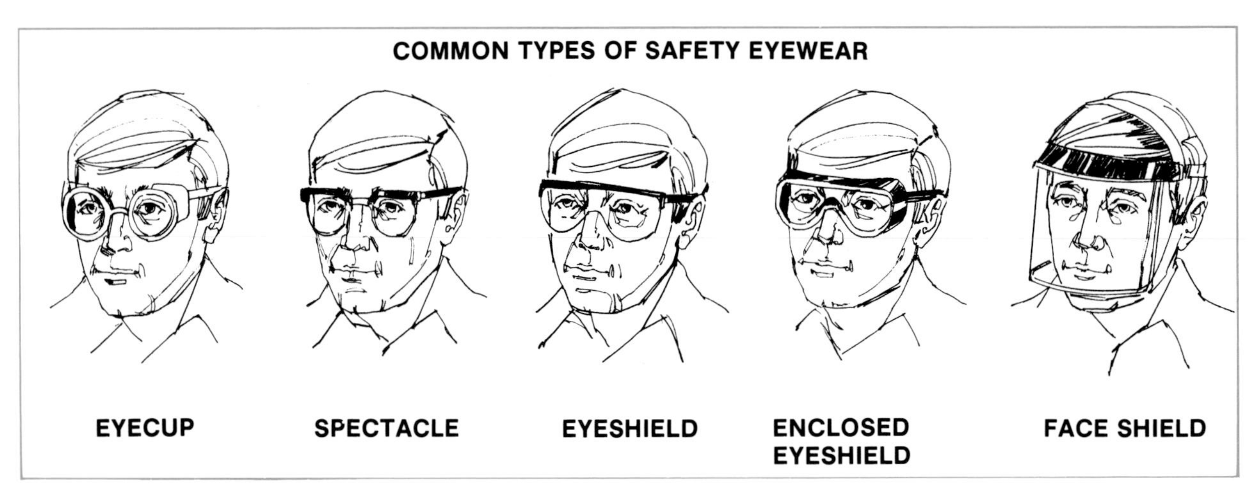

Fig. 32—Common Types Of Safety Eyewear

Safety Glasses

Eyeglasses and sunglasses only give front protection. If you wear glasses, be sure they have impact-resistant lenses (Fig. 33). Safety glasses have heavier lenses and can withstand more shock than ordinary lenses.

Fig. 33—Safety Glasses Have Impact-Resistant Lenses

Fig. 34—Goggles Shield The Eyes From Front And Sides

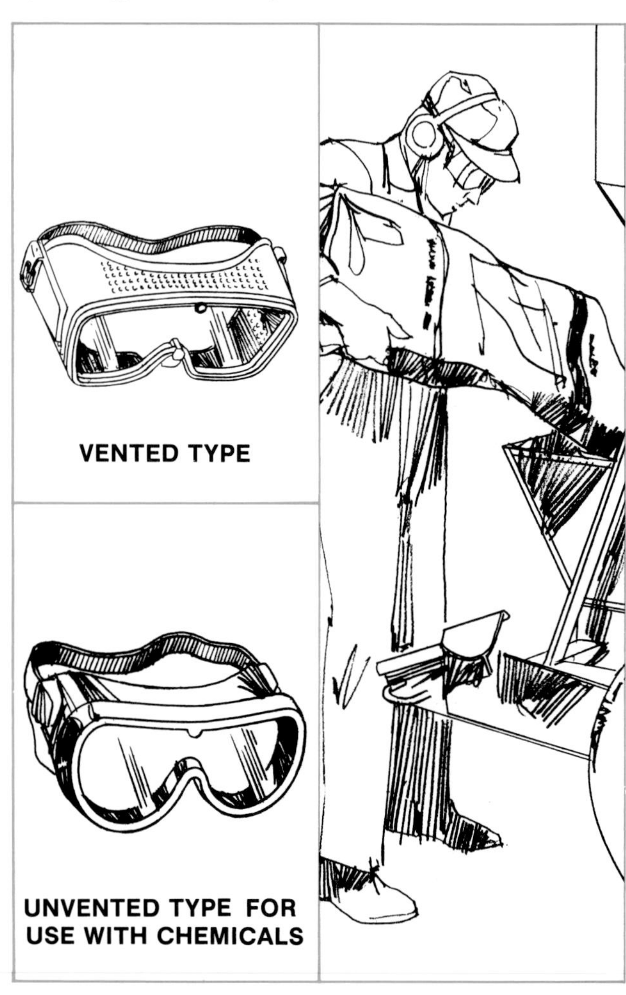

Goggles

Plastic goggles protect eyes from impact from the front and sides. Unvented or chemical splash goggles also offer protection against chemical vapors and liquids (Fig. 34).

Face Shields

Face shields protect the face from splashing, dust, and chaff (Fig. 35). But they offer very little protection against impact.

If you need impact protection, wear safety glasses or goggles under the shield or get a special impact-resistant shield that's fitted to a hard hat.

EAR PROTECTION

Farmers have higher than average hearing loss.

Why? Ear damage can begin at sound levels as low as 85 to 90 decibels, and many farm machines are louder. Wear earmuffs or at least earplugs whenever you're exposed to a continuous noise level of 90 decibels (A) or higher.

Fig. 35—Face Shields Keep Splashes Away From Your Eyes And Face

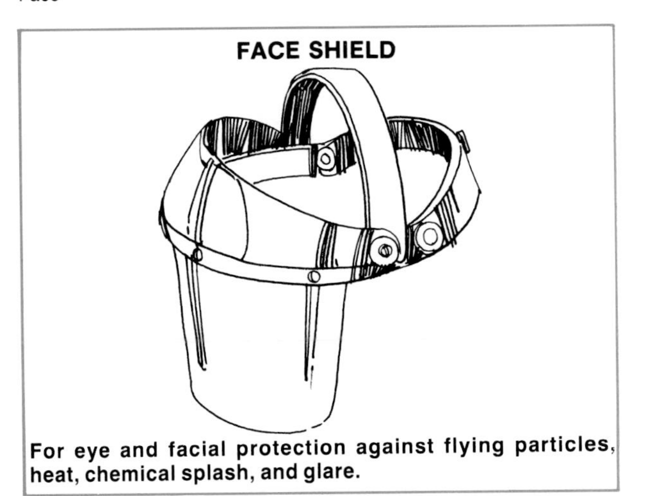

Earplugs

Rubber or plastic earplugs fit into the ear canal and are effective noise suppressors. A snug fit is important, so have them custom-fit for comfort and protection (Fig. 36).

Don't use cotton plugs. They don't block any high frequency sounds and only a few low frequency sounds.

Fig. 36—Use Plastic Or Rubber Earplugs That Fit Snugly

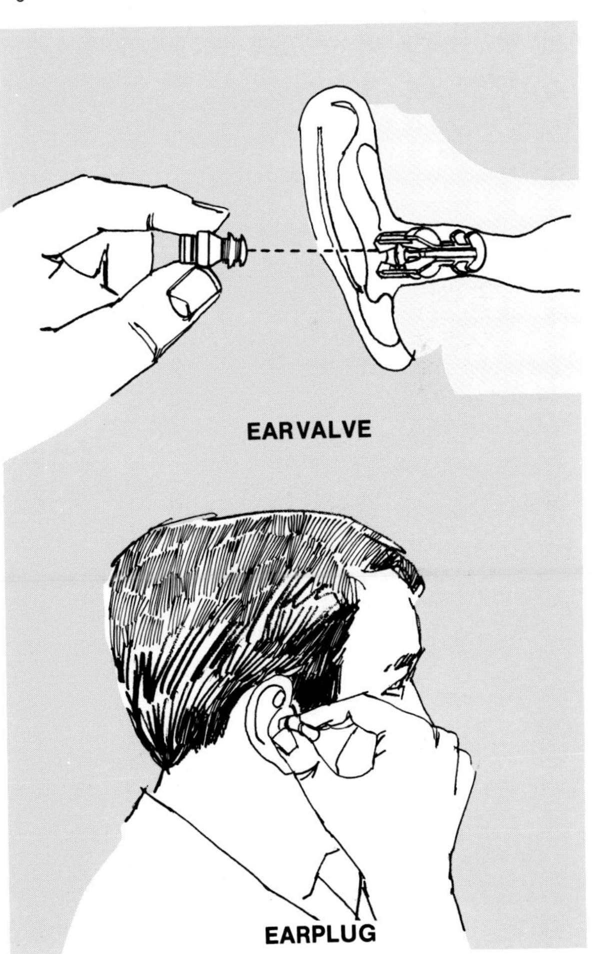

Fig. 37—Accoustical Earmuffs Are More Effective Than Plugs

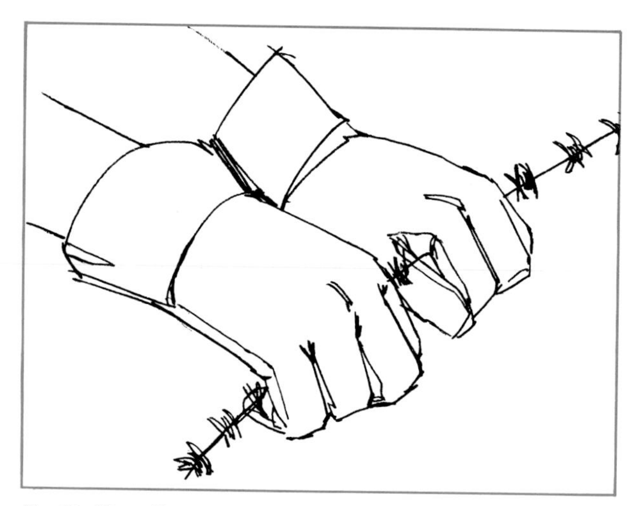

Fig. 38—Wear Gloves

Earmuffs

Ear muffs (Fig. 37) are the most effective protection against noise. They don't contribute to ear infection and discomfort as do ear plugs which fit tightly and carry dirt to the ear canal. Ear muffs block more noise than plugs because they cover the sound conducting bones around ears as well as the ears themselves. They are also comfortable and they keep your ears warm.

Fig. 39 — Wear Safety Boots With Safety Steel-Toes And Metatarcel Support.

HAND PROTECTION

Gloves can't always prevent a finger amputation, but they *can* guard against cuts, abrasions, chemicals, and skin irritation (Fig. 38).

Leather gloves protect hands against rough or sharp objects, and give good gripping power.

Use rubber gloves when working with fertilizers.

Wear unlined neoprene gloves when using pesticides. The chemicals can't penetrate nonporous neoprene. Lining tends to capture and trap chemicals.

Canvas or cotton gloves offer some protection.

Wear gloves that fit. If they're too big, they could easily get caught in moving parts and you could lose a hand.

FOOT PROTECTION

Wear safety shoes. Their steel toes and puncture-proof, skid-resistant soles protect your feet (Fig. 39).

Foot injuries are common if farmers don't wear safety shoes.

BODY PROTECTION

Aprons and padding protect your body. But the most important protective item is your everyday clothing. It should fit comfortably (Fig. 40). Loose clothing near moving parts is hazardous:

1. *Zip up or button your jacket.*

2. *Wear short-sleeved shirts or button the cuffs on long-sleeved shirts.* Avoid rolled-up sleeves.

3. Wear pants or overalls with straight legs that don't drag on the ground.

4. Wear safety shoes or boots with skid-resistant soles.

5. Wear a hat with a wide brim when working in direct sunlight. A bump cap offers comfort and protection from head bumps around equipment.

Fig. 40—Clothing Should Be Snug But Comfortable

RESPIRATORY PROTECTION

There are several types of respiratory protective devices. Each does a certain job. Use the right one.

Dust Filter Respirators

These are soft fiber facepieces that filter air (Fig. 41 A & B). The filter traps dust, chaff, and other particles. If you work where dust contains mold, a dust respirator with a tight seal around nose and mouth is required. There can be severe allergic reaction to mold in hay and forage material.

Dust respirators will not help if you are working with chemicals, toxic gases, or if there is an oxygen deficiency.

Chemical Cartridge Respirators

Cartridge respirators have a partial face mask fitted with one or two replaceable cartridges (Fig. 41C). The cartridges contain an absorbent material (often activated charcoal) that purifies inhaled air and dust filters to trap particles.

Make sure you have the correct cartridge for the material you want to filter out!

Cartridge respirators are effective against all but the most toxic vapors. Use cartridge respirators in open areas where there is no oxygen deficiency. Don't use them in silos or manure pits where there might be a lack of oxygen.

Gas Masks

Gas masks are heavy-duty cartridge respirators (Fig. 41D). They use a chemical filter to remove toxic vapors and particles from the air. They have a greater capacity to protect against high concentrations of toxic gases than chemical respirators. But they too aren't effective when oxygen is lacking.

Supplied-Air Respirators

Supplied-air respirators bring compressed air or outside air in through an air hose (Fig. 42A). They are used either with a blower or with compressed air.

Supplied-air respirators can be used when there is an oxygen deficiency.

Fig. 41—Good Respiratory Protectors — Except Where Oxygen Might Be Deficient

A. RESPIRATOR—FILTER (FOR DUST AND FUMES)

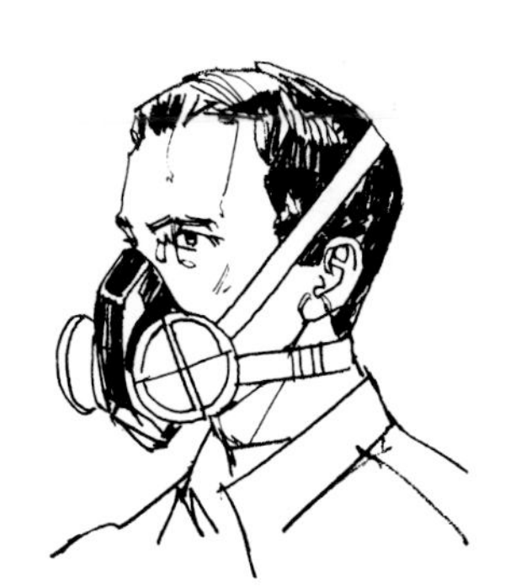

B. FILTER RESPIRATOR

C. RESPIRATOR—CHEMICAL CARTRIDGE

D. GAS MASK

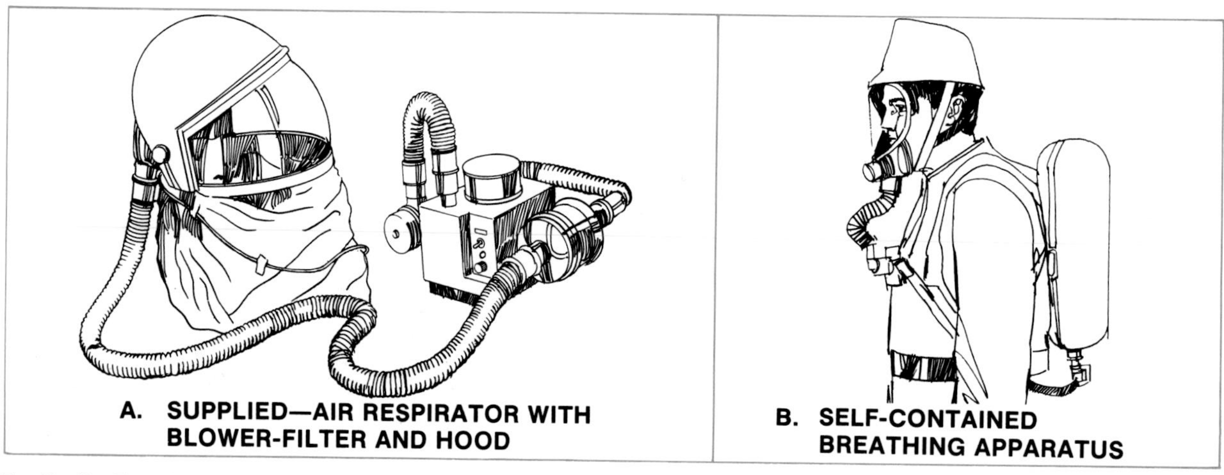

Fig. 42—Use These Respirators When You Must Go Into A Manure Tank Or Silo When There's A Lack Of Oxygen

Self-Contained Breathing Apparatus

A self-contained device has its own air supply in a cylinder (Fig. 42B). The apparatus works in silos, grain bins, manure pits, and other places where gas masks are inadequate because of an oxygen shortage.

LIFTING

The finest safety equipment available is of little value if you ignore it, use it incorrectly, or do not use correct work methods. A back injury resulting from any improper lifting technique is an excellent example.

About 60 percent of us are affected by back disorders sometime during our lives. Back disorders are common, even in the 20-to-30 age group. People engaged in heavy, active work, like dock hands, farmers, and ranchers, are affected more than those who do office work.

Back problems occur when back bone disks are damaged or diseased. The fibrous outer edges become weak and brittle. A sudden heavy load can rupture them, resulting in pain and possible paralysis or muscular problems. The injury is often referred to as a slipped disk.

Back problems usually begin with incorrect lifting practices. A lot of pressure is put on the spine when a person bends his back without bending his knees when lifting (Fig. 43). The poor mechanical position of lifting can cause a force of as much as 1500 pounds (700 kilograms) on the spinal disks when lifting only 110 pounds (50 kilograms). This often causes a ruptured disk. Sometimes the injury is permanent.

Use the proper procedures and body position when lifting to reduce the risk of injury (Fig. 44 and 45).

1. *Protect your hands and feet.* Gloves can protect hands from snags from barbed wire, wood splinters, and sharp edges. Steel-toed safety shoes prevent many painful experiences. Dropping a heavy object on your foot is just one type.

2. *Get a good footing.* Remove obstacles and debris that could cause a fall. Position your feet slightly apart for balance.

3. *Bend your knees.* Get close to the object. Keep your back in a natural position. Do not force your back to be abnormally straight or curved.

4. *Get a good grip.* Grasp the object with a full hand grip instead of your fingertips. It's less likely to slip and fall. Tilt boxes and get one hand under them. Place the other hand diagonally opposite. This principle of gripping diagonally also applies to sacks of fertilizer, grain, etc.

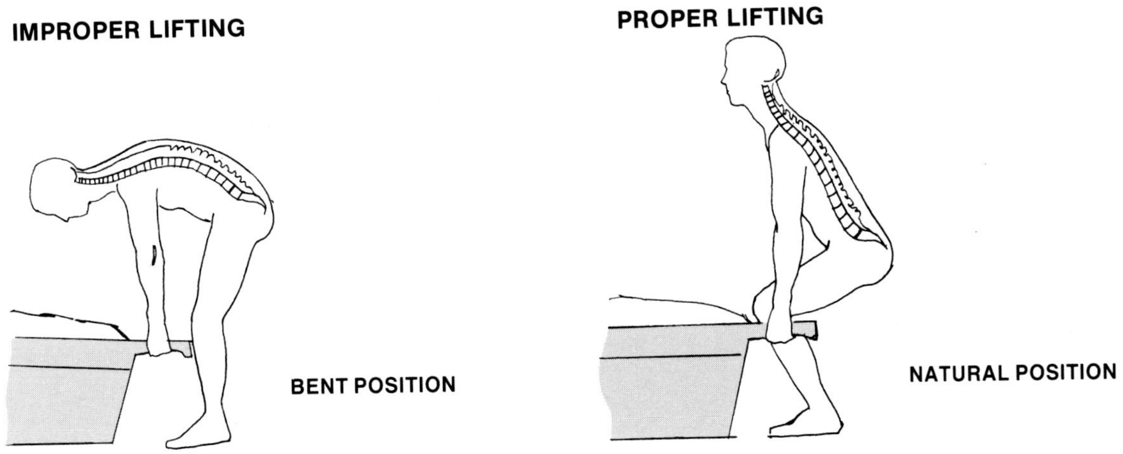

IMPROPER LIFTING

PROPER LIFTING

BENT POSITION

NATURAL POSITION

Fig. 43—Improper Lifting Can Injure The Back

Fig. 44—Practice Personal Protection When Lifting

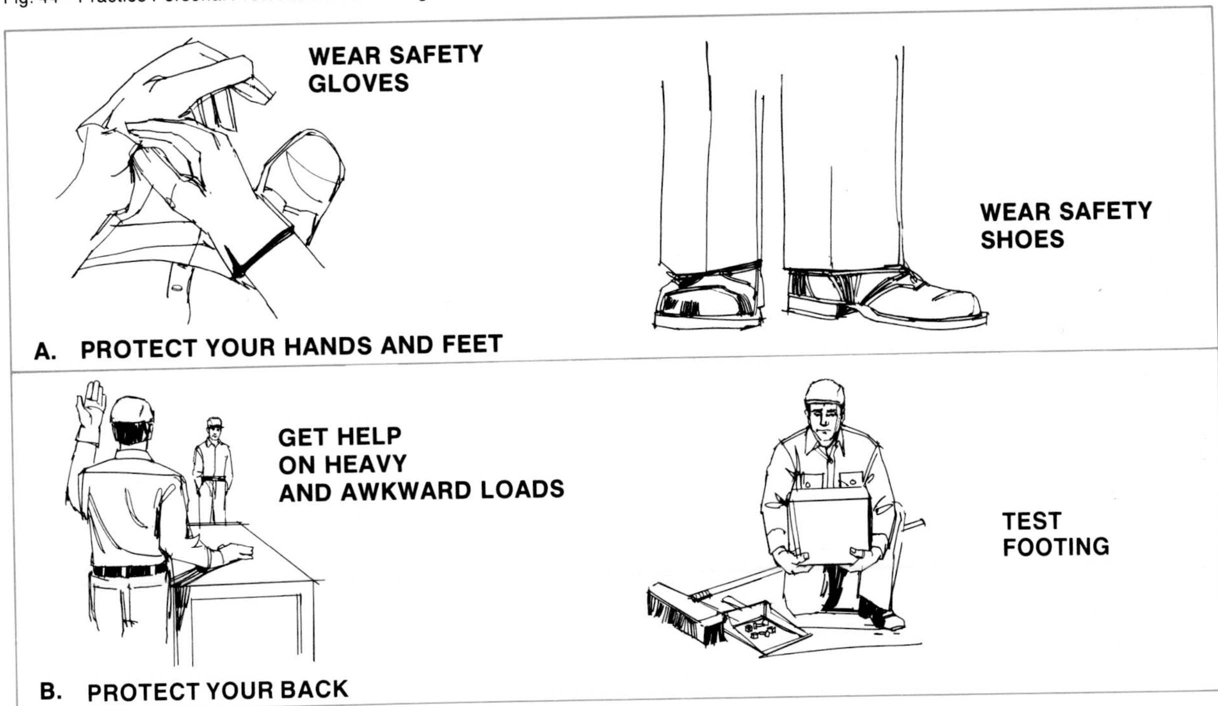

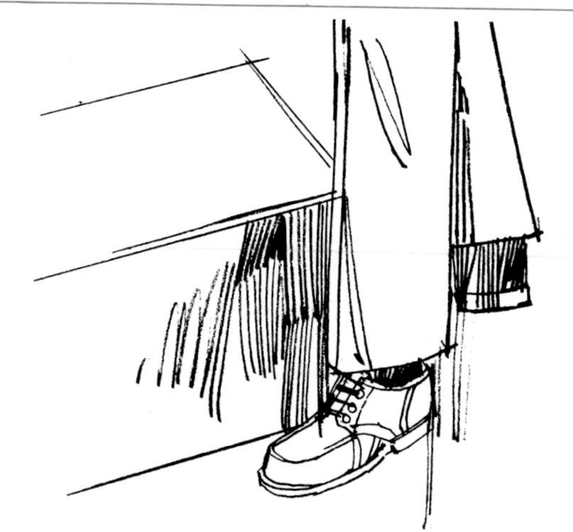

A. STAND CLOSE TO OBJECT TO BE LIFTED

B. BEND THE KNEES

C. TILT THE BOX PLACING ONE HAND UNDER

D. PLACE HAND DIAGONALLY OPPOSITE

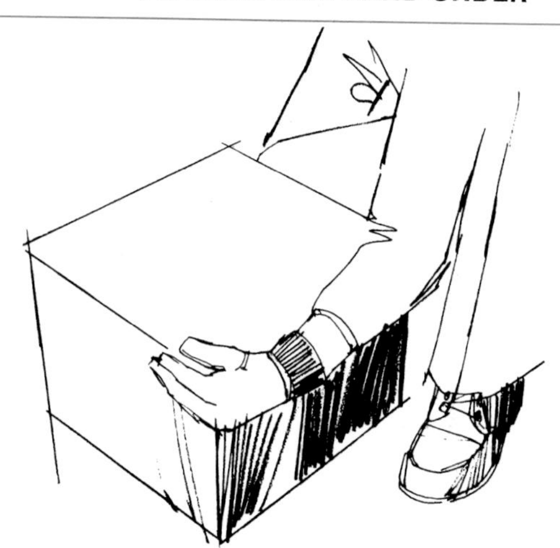

E. GET A FULL HAND GRIP – NOT JUST FINGERTIPS

F. LIFT WITH LEGS, KEEP BACK NATURAL

5. *Lift with your leg muscles.* Lift smoothly, not with a jerk. Keep the load close to your body (Fig. 46).

6. *Use hooks, straps, and pulleys to lift a heavy load from the floor.*

7. *Reverse the procedure to set an object down.* Keep your back in a natural position and bend your knees to lower an object.

Don't try to lift heavy loads without help. Some maximum advisable loads are given in Fig. 47.

A lever and fulcrum can make many heavy lifting jobs easy. Hydraulic jacks may be required. Get the proper help (person or equipment). Don't be laid up for a week or a lifetime with a bad back because you tried to save a few minutes of time.

Carry loads at the center of your body to balance the weight.

MAN-MACHINE SYSTEMS

COMPATIBILITY

A good man-machine system is natural and easy for the man. The man's learning time, adaptation, and conscious movement are minimum.

Controls

Controls are an important element of any man-machine system.

A machine should move in the same direction as its control mechanism. For example, moving a control lever to the right should make the machine move to the right.

It would be very unnatural if a vehicle turned to the left when the steering wheel was turned to the right or if a combine header platform went down when the control lever was moved up.

When selecting a new machine, look for controls and indicators that are easy to use and understand. Be sure there is good compatibility between the man and machine. A compatible machine will be easy to get on

Fig. 45—How To Lift Safely

Fig. 46—Lift Heavy Objects Using Your Leg Muscles Or Use Shop Hoists Or Tractor Lifts

Fig. 47—Recommended Maximum Lifting Loads In Pounds

RECOMMENDED MAXIMUM LIFTING LOADS IN POUNDS (KILOS)	Adults		Youths	
	Men	Women	Boys	Girls
Occasional Lifts	110 (50)	44 (20)	44 (20)	33 (15)
Frequent or Continous Work	39 (18)	26 (12)	24-35 (11-16)	15-24 (7-11)

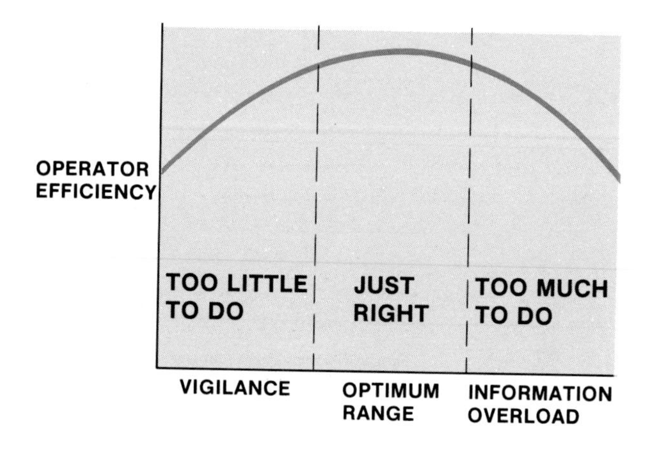

Fig. 48—Too Little Or Too Much To Do—Lowers Operator Efficiency

and off, be easy to see out of, have a comfortable ride, and be convenient and easy to hook up and operate. The direction of control movement will cause the "expected" machine response and minimize operator learning time.

Experience and convention have taught us to expect certain control movements and types of displays. Machines that use these accepted control characteristics are easier and safer for you to use.

Look for conventional controls and displays:

• Handles for controlling liquids are expected to turn clockwise for *off* and counterclockwise for *on.*

• Knobs on electrical equipment are expected to turn clockwise for *on,* or to increase current, and counterclockwise for *off,* or to decrease current. *(NOTE: Liquid flow is regulated in the opposite direction of electricity.)*

• Certain colors are associated with traffic (red, yellow, and green), vehicle operation (flashing yellow or blue), and safety (green).

• Operators expect a steering wheel motion to the right to turn the vehicle right.

• Coolness (engine temperature, etc.) is associated with blue and blue-green colors and warmth is associated with yellows and reds.

• Very loud sounds or sounds repeated in rapid succession and visual displays that move rapidly or are very bright imply urgency.

• Seat heights are expected to be at a certain level.

FATIGUE BREAKDOWN

You perform best when you're challenged by a task, but you shouldn't be overburdened. When a job is too difficult or involves too many controls, gauges, and observations to coordinate, it can overload an operator. He can operate only for short periods at such a task. Breakdown often occurs in the form of mistakes, frustration, and anger.

Bales coming up an elevator faster than a man can stack them is an example of over-load. They may pile up and result in a messed-up bale stack, an angry man, and possibly an accident.

Any system that forces a man to keep up and work much faster than is comfortable will result in an overburdened operator, low-quality performance, and possibly an accident.

Machines have definite upper limits of capacity. If they are exceeded, the machine stops or fails. In contrast, man's upper limit is variable. It depends on environmental conditions, motivation, and physical condition. He can work in an overloaded situation for a while, but eventually he will reach his failure point. On the other hand, if a job is too easy or routine, monotony, boredom, errors, and eventually accidents will result (Fig. 48).

On the other hand, if a job is too easy or routine, monotony, boredom, errors, and eventually accidents will result.

Cultivating corn is a common example. The operator in a long field has little to do. He is not challenged and becomes bored. Some operators even fall asleep and crash at the end of the field.

Mental fatigue as well as physical fatigue contributes to a decline in a person's performance. Long periods of careful thinking, such as repeated calculations or reading, are mentally tiring. The mind begins to slow down and make errors. Short periods of activity are more effective. Take a safety break when you're sleepy, fatigued, or mentally tired.

Factors that increase a person's physical and mental fatigue also increase his work load. When buying new equipment, try to reduce the work load by selecting equipment with features such as:

• *Good visibility*

• *Convenient and easy to operate controls*

• *Comfortable seating and position*

• *Minimum vibration*

• *Shielding from heat of the machine (or cold from exposure)*

- *Quietness*
- *Protection from dust and fumes*
- *Easy to read instruments*
- *Controls that are compatible with the operator and machine responses*
- *Adequate ventilation*
- *Adequate illumination and protection from glare*

Each of these factors affects you. If they cause unnecessary stress, they seriously reduce your work capacity and may result in fatigue breakdown and serious injury to you or someone else.

COMMUNICATIONS

In most man-machine systems, the machine communicates its operating conditions and need for control changes to the operator in several ways. Unless the machine is computer controlled, the operator makes the changes to keep the machine operating correctly.

Machine-to-man communication includes such things as:

- *General observation* — Is the machine on the row? Is the material being fed into the machine at the correct rate?
- *Gauges and visual indicators*—Engine temperature, engine rpm, vehicle velocity, hydraulic or air pressure, malfunction alarm light on oil pressure and generator.
- *Sound* — Overload slows the engine and strains machine parts. Plugged augers, hay pickups, etc. cause slip clutch drives to chatter. Underload may allow excessive engine speed. The experienced operator understands these sound communications from the machine. Some machines are equipped with sound warning alarms such as the malfunction alarm on planters.
- *Feeling* — A machine's vibrations often change with a change in load. You can feel a plow digging into wet or ground. You can feel a plow sliding or tipping.

To improve job performance and reduce fatigue, many control panels are arranged so dials are grouped with indicators pointing in the same direction under normal conditions. The pointer indicating a different direction from the others stands out (Fig. 49). When indicator displays are arranged with no particular agreement in the position of the pointers, each indicator must be looked at individually. This takes more time and conscious effort. When choosing between two different dial arrangements, the fast-scanning, grouped arrangement is preferred.

Fig. 49—Dial Arrangement—Easy Scanning Is Preferred

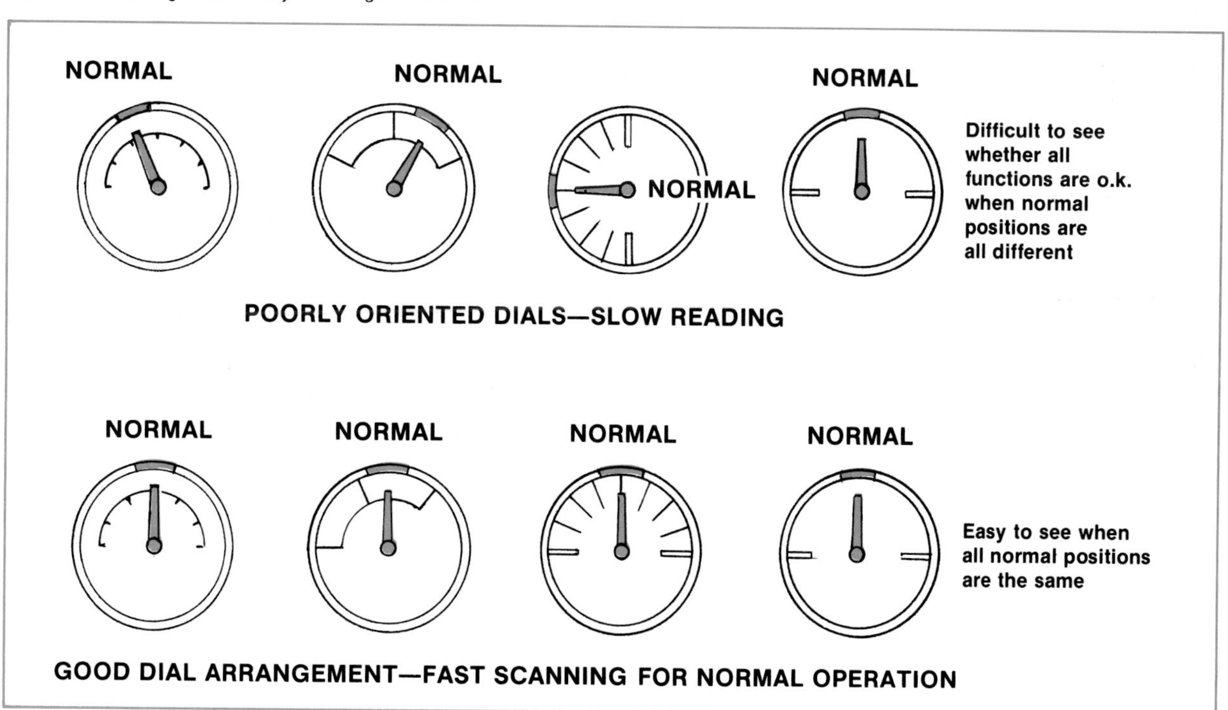

ADAPTABILITY

Simple mechanical machines do not think, learn, or adapt. You must do the adapting when new circumstances develop. The more unnatural the circumstances, the longer you will take to adapt and react. In an emergency you may make an error if your instinctive reaction is different than the required reaction. If dials and gauges are difficult to read and controls are awkward, you will take even longer to adapt and react.

Ask yourself the following questions when evaluating a machine:

- Where are the controls located? Can they be reached easily? Are they spaced far enough apart? Are they labeled and coded?
- What types of controls are used? Are they compatible with the operator and with the machine response?
- Do the controls themselves present a hazard? Is each one clearly identified by sight and touch? Can one be mistaken for another? There should be no confusion about which lever controls a hydraulic cylinder or which knob on a combine controls the header, which the speed, etc.
- Which arm(s) or leg(s) are involved? Are all controls made by one arm with the other limbs doing nothing? A better, less fatiguing arrangement is to use both arms and legs.
- Are the controls standard? Are controls common, or are they unnecessarily "unique" and possibly incompatible with other systems?

OBSOLETE EQUIPMENT

Machines grow old and become obsolete (Fig. 50).

Equipment may not always be worn out before it is obsolete. Hay rakes, for example, have changed from cylinder to parallel bar because parallel bar rakes handle hay more gently.

New farming technology causes obsolescense, not age. Combines have replaced grain binders and threshing machines, herbicides have made checkrow corn planters a thing of the past.

Machines that have been improved over the years are usually safer to operate than older equipment. Some of the improvements that have been made are:

- *Slip clutches and safety devices to prevent damage from overload*
- *Belt and chain drive shields*
- *PTO shaft shields*
- *Neutral start interlock switches to keep engines from starting when a machine is in gear*
- *Seat interlock switches to stop the engine if the operator falls off the seat*
- *Controls that are easier to reach and operate*
- *Rollover protective structures (ROPS) and FOPS (Falling Object Protective Structure). Note: Not all ROPS are strong enough to handle falling objects, and not all FOPS are strong enough to protect the operator in a tractor rollover accident.*
- *Safety signs, such as the bypass start decal, have been added to machines in the last few years.*

If you use older equipment, be especially safety conscious. Reasonable safety can still be achieved if you maintain the equipment properly, follow good safety practices, stay extra alert, and follow the new safety signs that have been developed for your machinery.

Fig. 50 —Machine Designs Change

CHAPTER QUIZ

Select the true answers to the following questions.

1. Effective rest breaks from strenuous work should be:

a) Rather short (1 to 2 minutes) and frequent.

b) Infrequent, but of longer duration (15 to 20 minutes).

c) Not really necessary if a person is in good condition.

2. Two tractors operating side-by-side producing a sound pressure level of 80 decibels each, will provide a combined sound pressure level of:

a) 80 decibels

b) 83 decibels

c) 86 decibels

d) 160 decibels

3. The sound pressure level that begins to produce eardrum pain is:

a) 90 decibels

b) 100 decibels

c) 120 decibels

d) 140 decibels

4. A typical reaction time under good conditions is:

a) .1 second

b) .3 second

c) .5 second

d) 1.0 second

5. True or false? Because young people (15-20) have faster reflexes than older persons (30-50), they have fewer accidents.

6. The body can adjust to working in hot weather by:

a) Going slower

b) Drinking lots of water

c) Dressing cool

d) Taking advantage of shade and cool places

7. One principle of machine compatibility says:

a) A machine should fit the image of the operator.

b) The color and style should be pleasing to the owner.

c) The machine should be the same make as the tractor that operates it.

d) The machine controls must move in ways which correspond to the machine movement and are natural for the operator to use.

8. A man can work continuously at a rate of about:

a) .5 hp

b) .2 hp

c) 2 hp

d) 5 hp

9. The limit of human hearing is approximately:

a) 20-100 hertz

b) 50-1,000 hertz

c) 50-10,000 hertz

d) 20-20,000 hertz

10. Alcohol affects performance because it:

a) Affects judgment, causing errors.

b) Slows up reflexes.

c) Causes poor coordination.

11. Operator efficiency is best when:

a) The operator has little to do and can give full attention to the task.

b) The operator has a moderate task and must stay alert to do everything needed.

c) The operator has the maximum amount he can do and has difficulty keeping up.

12. Psychological factors affect human performance. Those that might cause an accident are:

a) Happiness

b) Anger

c) Contentment

d) Anxiety

e) Apathy

f) Cheerfulness

g) Preoccupation

h) Alertness

i) Personal problems

13. Proper lifting procedure is:

a) Protected hands and feet

b) Good footing

c) Bent knees

d) Good grip

e) Use of leg muscles

f) Use of mechanical aids only for heavy loads

3
Target: Recognizing Common Machine Hazards

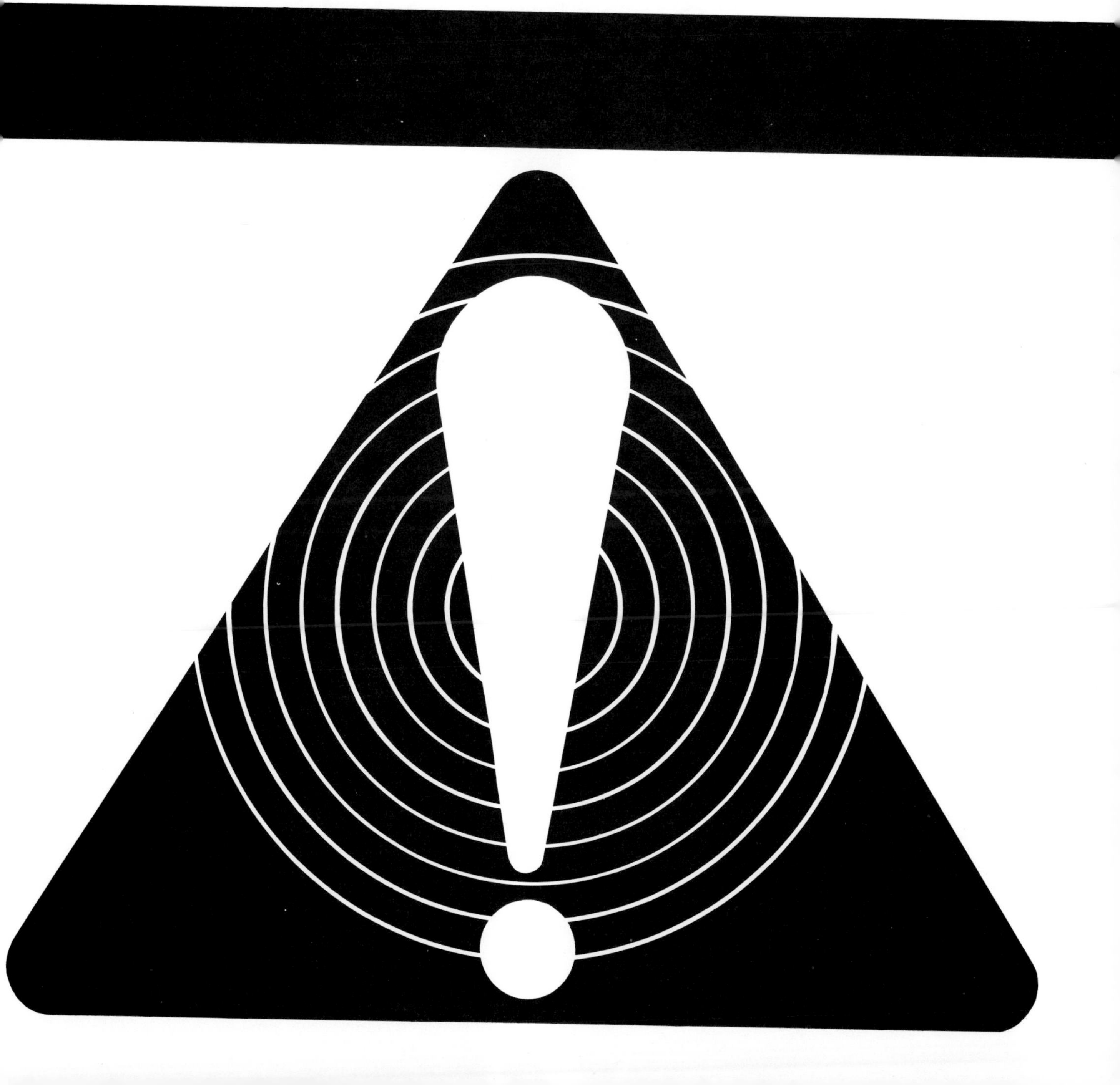

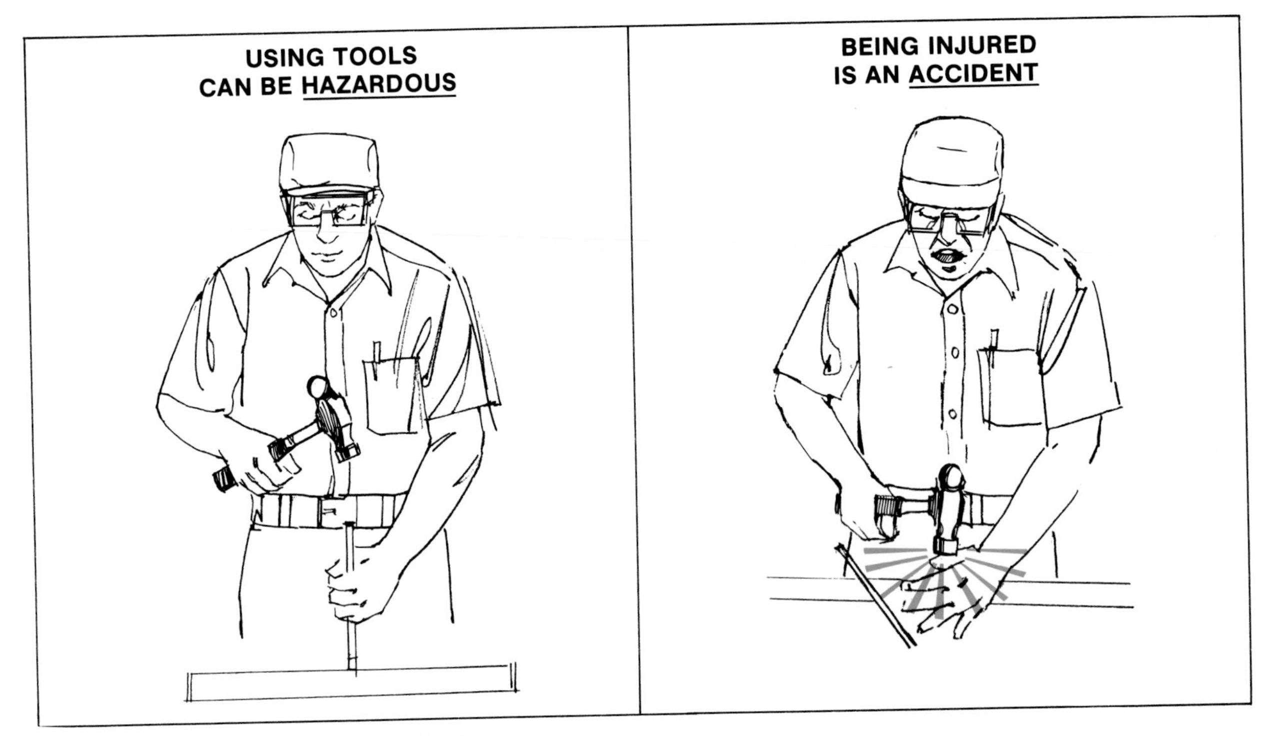

Fig. 1—What Is A Hazard? What Is An Accident?

INTRODUCTION

What is a hazard? A hazard is simply a dangerous object or situation that has the potential to cause injury. When a person comes into contact with a hazard, an accident may occur (Fig. 1). An accident for our purposes is an event that occurs when a person is injured unintentionally as a result of either machine failure or human error.

Let's look more closely at human and machine factors in accidents.

PEOPLE AND HAZARDS

Man's greatest ability is to think and take action according to knowledge, experience, and judgment. In contrast, man's greatest limitation is the tendency to make errors. When an accident occurs due to human error, it is usually because the individual made one of the following mistakes:

- *Forgot something — such as not setting the brake or shifting the transmission into park before dismounting*
- *Took a shortcut — such as trying to start or operate a tractor from the ground*
- *Took a calculated risk — such as stepping over a rotating PTO shaft*
- *Ignored a warning — such as "Disengage power, shut off engine, and take the key before adjusting or lubricating a machine"*
- *Used unsafe practices — such as refueling while smoking*
- *Was preoccupied — such as worrying about losing time while making repairs*
- *Failed to recognize the hazard — this failure results in no corrective or preventive action*

The *environment* may even add more risks which can affect situations. For example, the weather can contribute rain, snow, cold, and heat. In addition, the terrain might be hazardous if it is rough or hilly.

One of the most dangerous mistakes humans make is fail to recognize a hazard. Recognizing hazards and understanding them is the only way you can avoid accidents.

Remember: Many hazards can be avoided if you always disengage the power, shut off the engine, take the key and wait for all parts to stop moving before working on a machine.

MACHINES AND HAZARDS

Machines are designed to work. Because a machine uses power, motion, and energy to do work, it presents a number of potential hazards that are difficult to eliminate completely. For example, the blades on a rotary mower must rotate at high speeds to cut grass and cornstalks. The blades have the potential to injure you if you come in contact with them or if they pick up an object and throw it at you.

Manufacturers build many safety factors into their machines, but you must remember not all potential hazards can be eliminated (Fig. 2). A mower may have a shield over the powershaft. But, if the blades were shielded completely they would not be able to cut the material.

When it is impossible to eliminate or shield a hazard, safety signs are used to warn you. You must recognize the potential hazards and take necessary action to avoid injury. Also, you must recognize and take action when you see a safety sign.

DESIGN SAFETY

Manufacturers eliminate hazards through good design. Good design often eliminates the need for shielding and complex safety training. Here are some of the safety considerations manufacturers have when designing a machine:

1. *Make it easier for the operator to do the safe thing. Minimize fatigue* and maximize convenience, comfort, and safety.

2. *Prevent the operator from coming in contact with a hazard by shielding hazardous machine parts which can not be eliminated.*

3. *Protect the operator.*

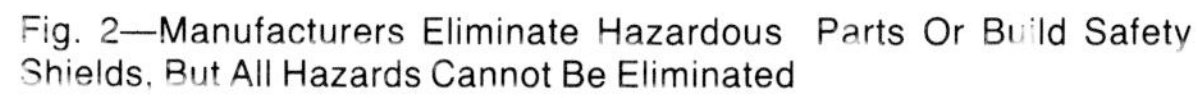
Fig. 2—Manufacturers Eliminate Hazardous Parts Or Build Safety Shields, But All Hazards Cannot Be Eliminated

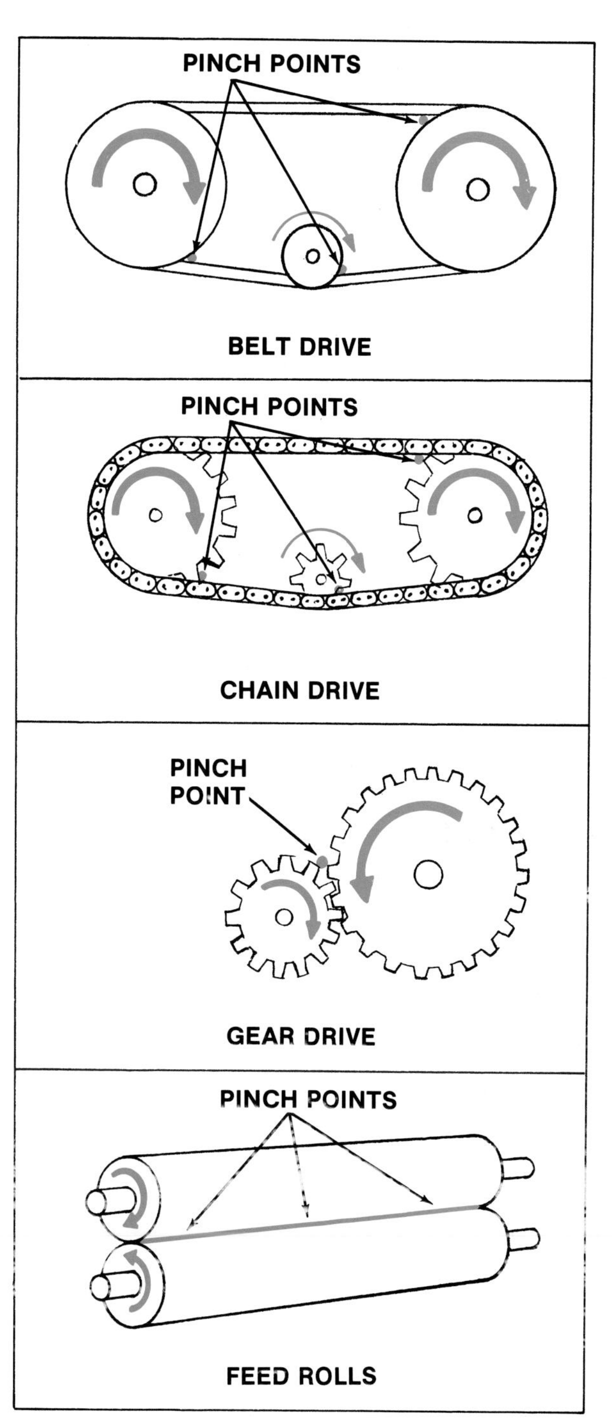

Fig. 3—Pinch Points On Rotating Parts Can Catch Clothing, Hands, Arms, And Feet

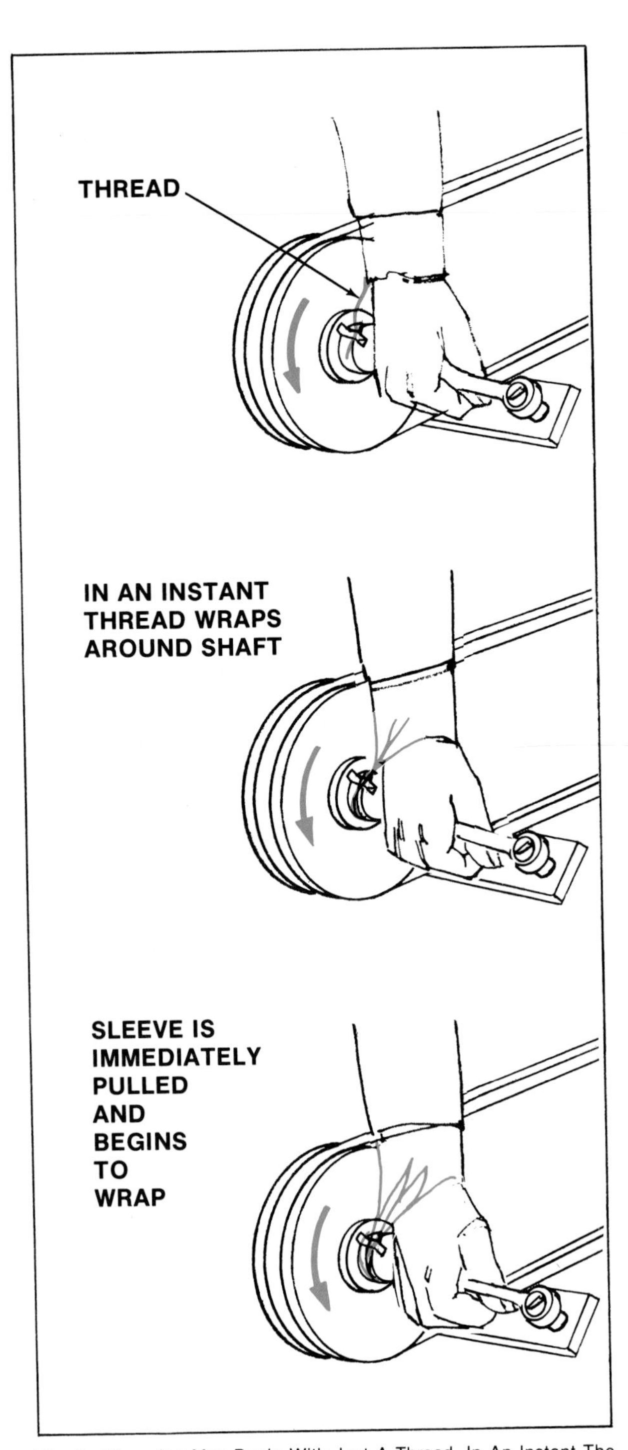

Fig. 4—Wrapping May Begin With Just A Thread. In An Instant The Victim Is Entangled With Little Chance To Escape Injury

4. *Providing warnings in operator's manuals* and warning signs on machines to warn of hazards and explain how to avoid hazards.

COMMON MACHINE HAZARDS

This chapter deals with machine hazards. It will help you recognize hazards and understand why they are hazardous. You will know what to avoid. The information in this chapter will help you develop a safe attitude.

Here are some of the most common machine hazards you should recognize:

- **Pinch points**
- **Wrap points**
- **Shear points**
- **Crush points**
- **Pull-in points**
- **Free-wheeling parts**
- **Thrown objects**
- **Stored energy**
- **Slips and falls**
- **Slow-moving vehicle**
- **Second party**

PINCH POINTS

Pinch points are where two parts move together and at least one of them moves in a circle (Fig. 3). Pinch points are also referred to as mesh points, run-on points, and entry points. There are pinch points in power transmission devices such as belt drives, chain drives, gear drives, and feed rolls (Fig. 3).

You can be caught and drawn into the pinch points by loose clothing. Contact with pinch points may be made if you brush against unshielded rotating parts, or slip and fall against them. Farmers can get tangled in pinch points if they deliberately take chances and reach near rotating parts. Machines operate too fast for a person to withdraw from a pinch point once he is caught.

AVOIDING PINCH POINTS

Manufacturers build shields for pinch points. But some pinch points, such as feeder rollers on corn heads and silage choppers must be open so crops can enter. Always replace shields if you must remove them to repair or adjust a machine. Remember that a portion of the money you paid for the machine went for safety research and design and for the actual hardware involved in shielding. Get your moneys worth and protect yourself.

For pinch points that cannot be shielded, the best protection available is operator awareness. Know the location of the pinch points on your machinery. Avoid them when the machine is operating. And above all, never attempt to service or unclog a machine until you have disengaged all power, shut off the engine, removed the key, and all parts have stopped moving.

WRAP POINTS

Any exposed component that rotates is a potential wrap point. Rotating shafts are usually involved in wrap-point accidents. Often, the wrapping begins with just a thread or frayed piece of cloth, catching on the rotating part (Fig. 4). More fibers wrap around the shaft. There's no escape.

The shaft continues to rotate pulling you into the machine in a split second. *The more you pull the tighter the wrap.* If the clothing would tear away, a person might escape serious injury, but work clothes are usually too rugged to tear away safely. Long hair can also catch and wrap, causing serious, permanent injury.

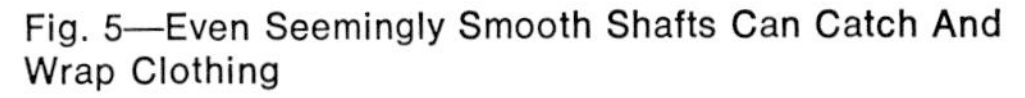

Fig. 5—Even Seemingly Smooth Shafts Can Catch And Wrap Clothing

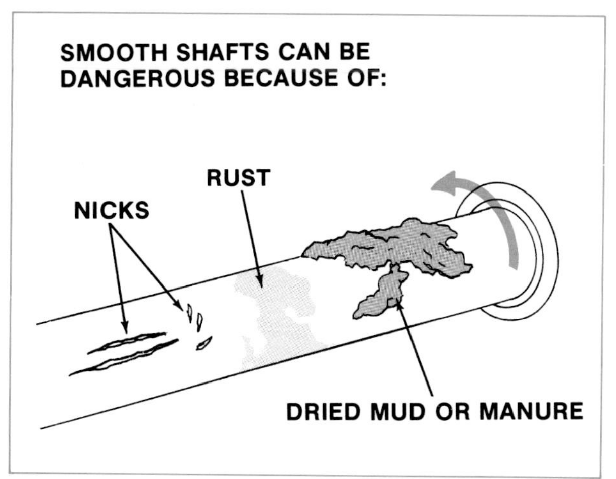

Fig. 6—Shafts That Extend Much Beyond Bearings Or Sprockets Can Be Dangerous

Smooth shafts often appear harmless, but they too can wrap and wind clothing. Rust, nicks, and dried mud or manure make them rough enough to catch clothing (Fig. 5). Even shafts that rotate slowly must be regarded as potential wrapping points.

Ends of shafts which protrude beyond bearings can wrap up clothing (Fig. 6). Splined, square and hexagon-shaped shafts are more likely to wrap than smooth shafts (Fig. 7).

Couplers, universal joints, keys, keyways, pins, and other fastening devices on rotating components will wrap clothing (Fig. 8).

Exposed beater mechanisms will entangle and wrap clothing (Fig. 9).

SHEAR POINTS

Shear points are where the edges of two moving parts move across one another. For example a set of hedge-trimming shears (Fig. 10).

CUTTING POINTS

Cutting points are created by a single object moving rapidly or forcefully enough to cut a relatively soft object. A table knife, a hand-held grass sickle, and a rotary lawn mower blade are examples of single objects that cut because of their sharpness and force (Fig. 11).

Fig. 8—Some Components Are Even More Aggressive Than Some Shafts

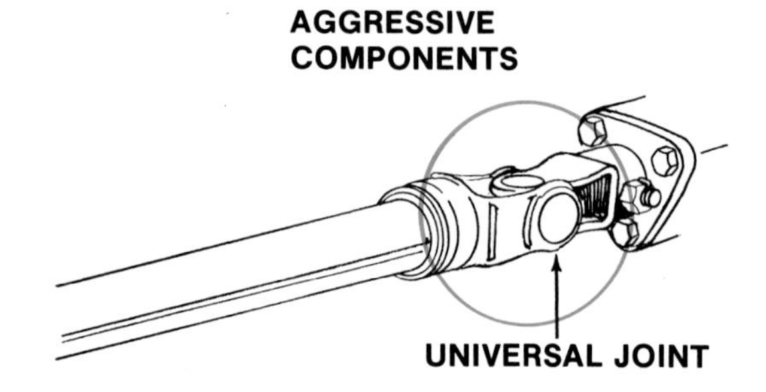

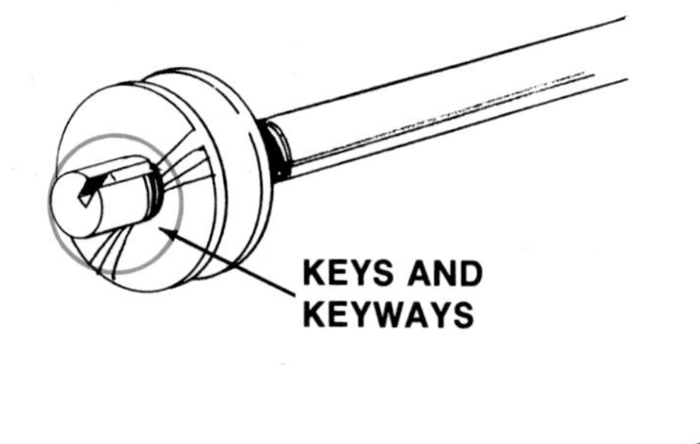

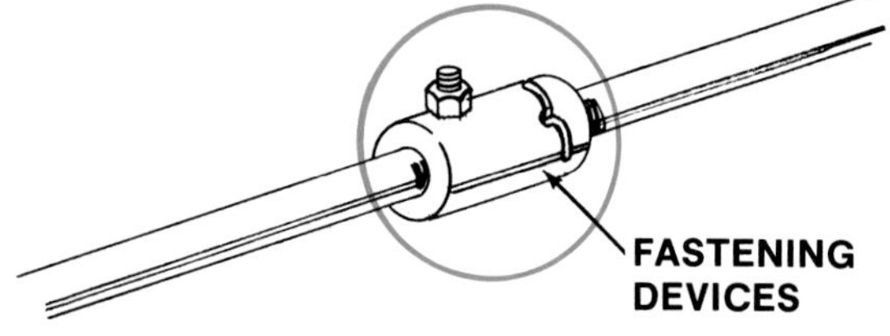

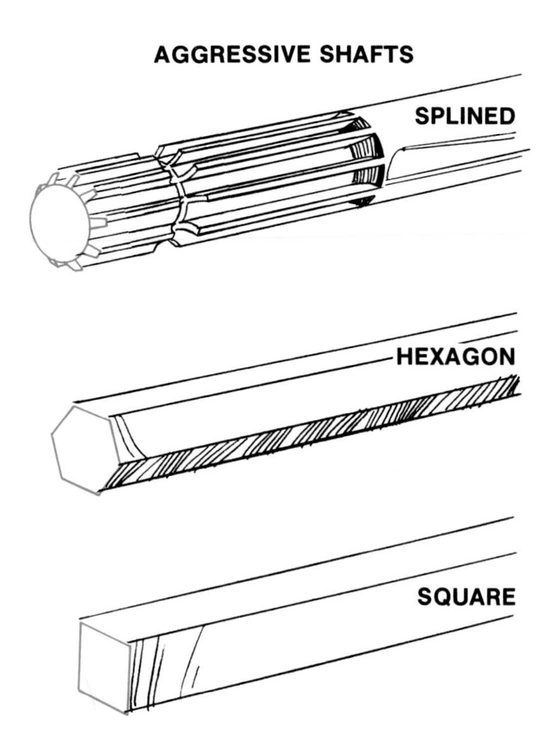

Fig. 7—Splined, Square, And Hexagon-Shaped Shafts Are More Likely To Entangle Than Round Shafts

Fig. 9—Beater Or Feeder Mechanisms Can Tangle Or Wrap Clothing

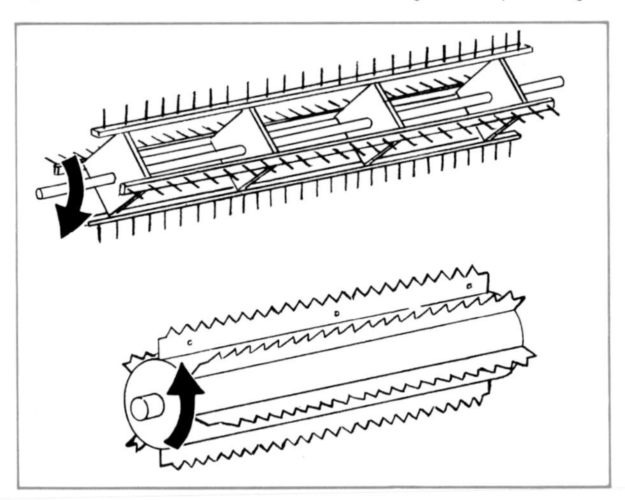

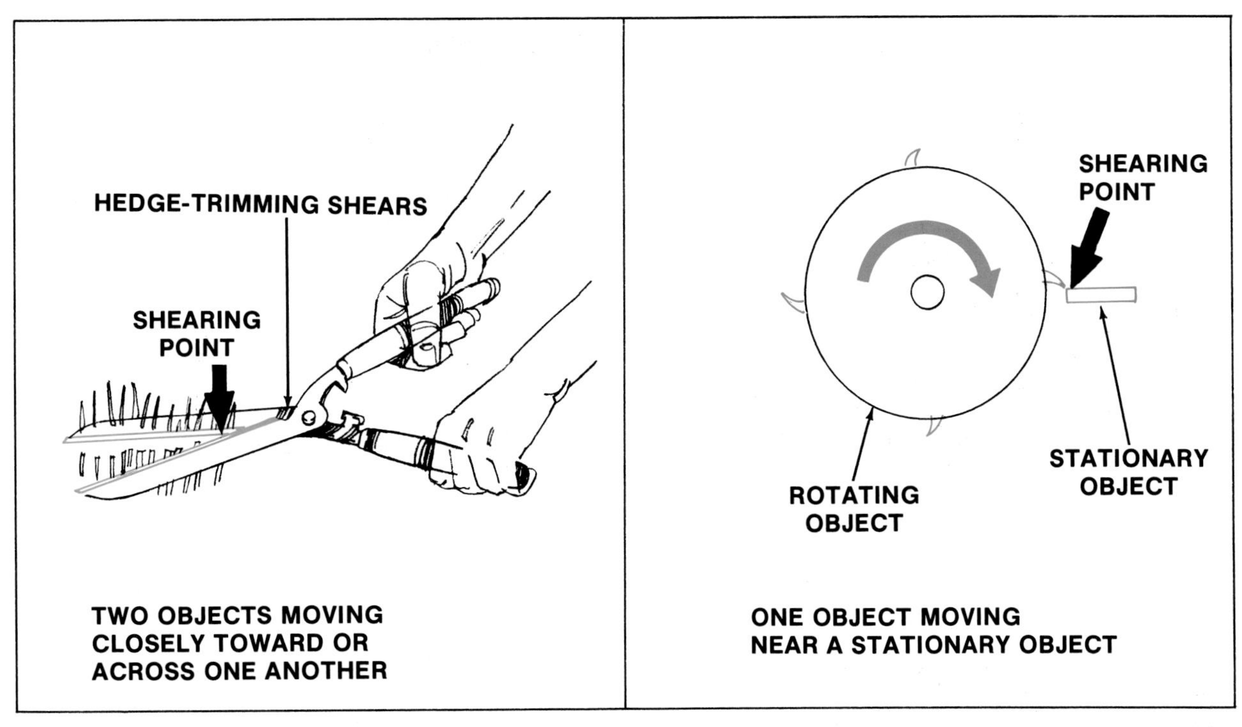

Fig. 10—Shear Points Are Created When Two Parts Move Across One Another

Fig. 11—Cutting Points Are Created By Single Sharp Parts Moving With Enough Force To Cut Softer Material

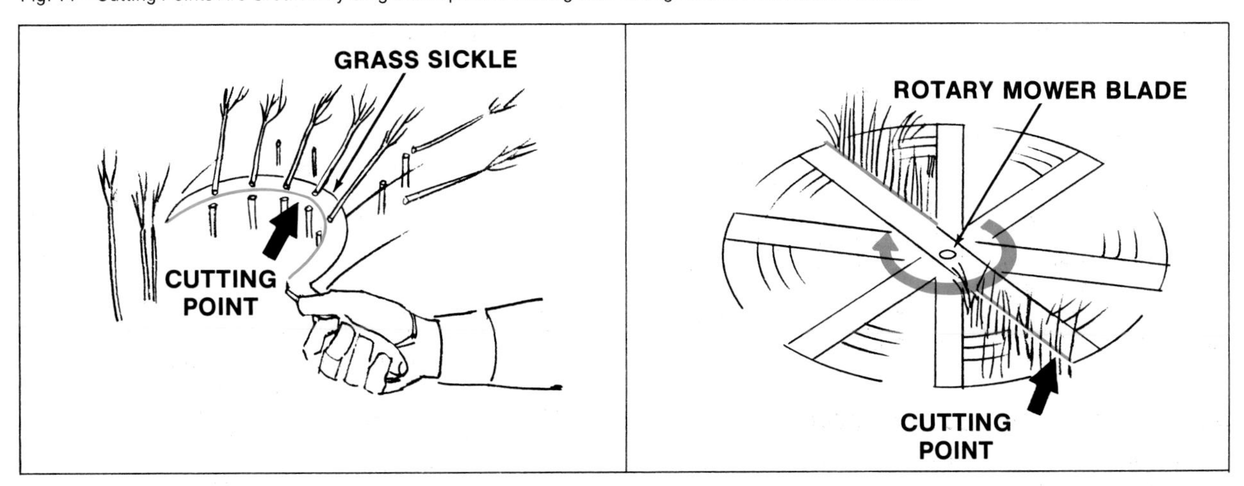

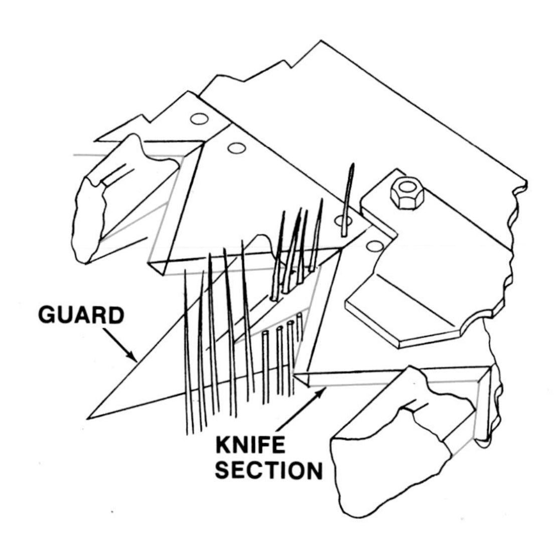

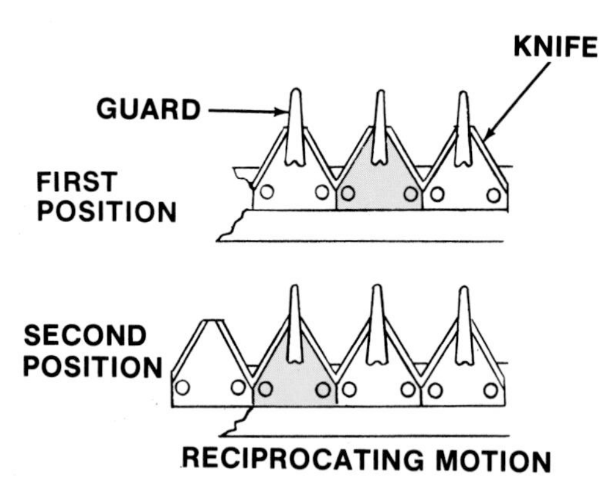

Fig. 12—Some Shearing Parts Move In Reciprocating Motions

DEVICES DESIGNED TO CUT OR SHEAR

Shearing and cutting devices on farm machines cut crops. The shearing and cutting parts may rotate (Fig. 11, right) or reciprocate (Fig. 12). Examples are sickle bar mowers, rotary shredders and cutters, and cutter heads of forage harvesters. Because they must cut crops at several tons per hour, they are aggressive. They do not know the difference between crops and fingers. And because they must cut crops on the move, they usually cannot be guarded to keep out hands and feet.

DEVICES NOT DESIGNED TO CUT OR SHEAR

There are cutting or shear points on devices that are not actually designed to cut or shear. For example, grain augers, chain and paddle-flight conveyors, and hinged implement frame members that move when an implement is raised or lowered (Fig. 13).

AVOID SHEAR POINTS

Because some shear points cannot be guarded, learn to recognize potential hazards and plan activities so you will not be injured.

CRUSH POINTS

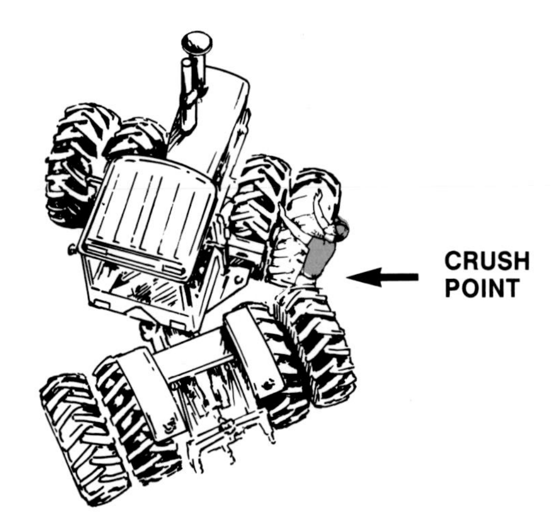

There are crush points between two objects moving toward each other or one object moving toward a stationary object (Fig. 14). Crush accidents often involve a second person. There are several dangerous crush points on hitches (Fig. 15).

A man working under a heavy object (Fig. 16) runs a risk. Hoods and doors can crush or cut if a hand is in the way. Keep in mind that where two objects move closer together, there is a dangerous crush point.

FOUR-WHEEL DRIVE TRACTOR CRUSH POINT

Four-wheel-drive tractors with articulated steering pivot in the middle and can turn and create a crush point at start up. By slightly turning the steering wheel while the tractor is stationary or moving, you can easily create a crush point. Hydrostatic steering systems are extremely sensitive. NEVER stand between the tires when the engine is running. NEVER let anyone stand between the tires when the engine is running.

AVOIDING CRUSH POINTS

Do two things to avoid being crushed. First recognize all potential crush situations. Some of these are mentioned in examples above.

Second, stay clear of the hazards. This statement seems simple and obvious. But many people are killed and injured because they get into a crush point. Be constantly alert. Block all equipment securely if you must work under it. If an implement can roll, block the wheels so it will not roll. When two people hitch implements, make sure both are aware of the danger, and neither stands in a crush point. Know what the other person is doing at all times.

HITCHING PROCEDURE:

The helper on the ground stands clear while the tractor is backed into position. Then the tractor is put into park and the brake set before the helper steps in and hitches up. Then the tractor is inched forward, never backward, to make adjustments.

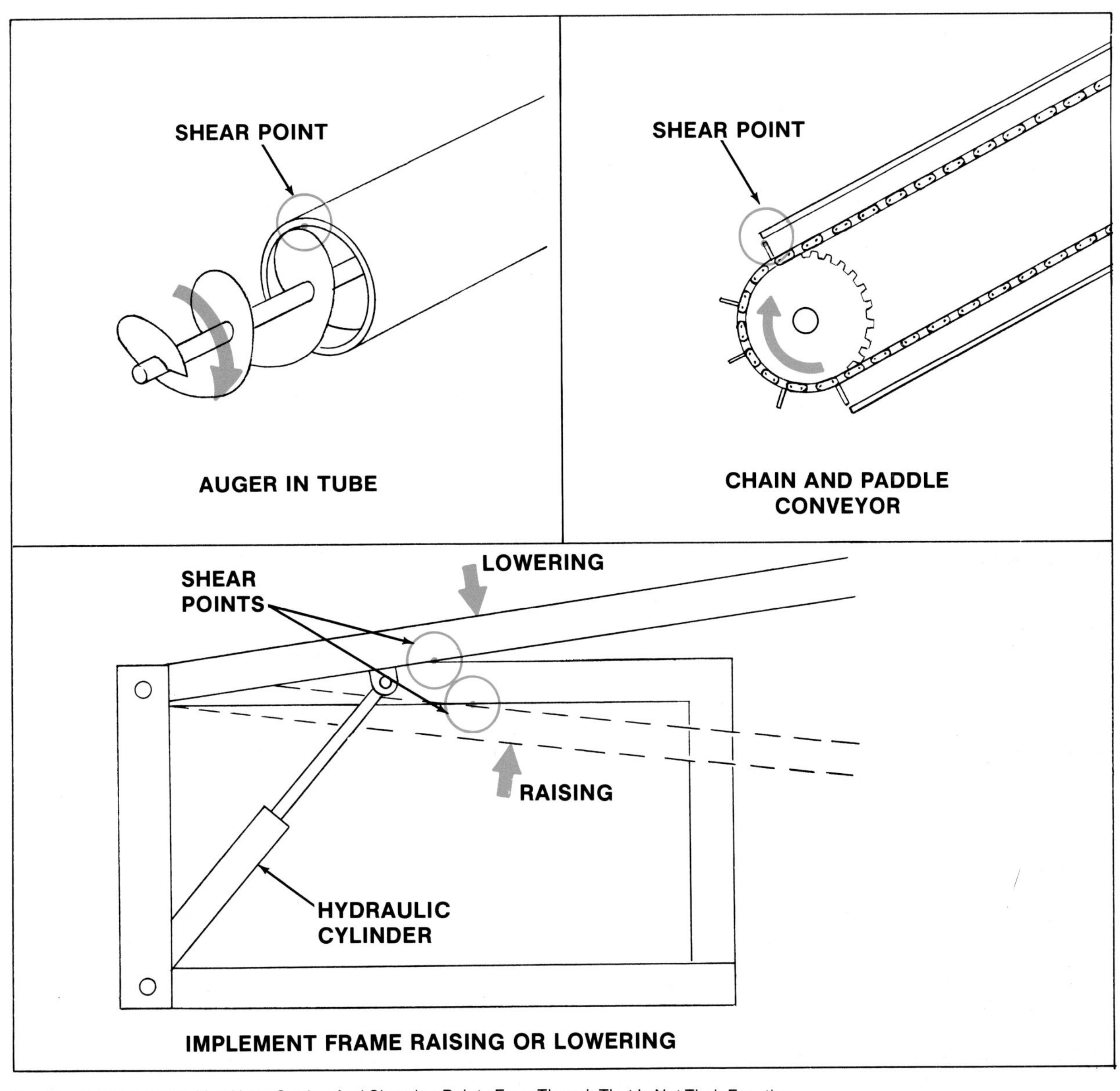

Fig. 13—Some Devices May Have Cutting And Shearing Points Even Though That Is Not Their Function

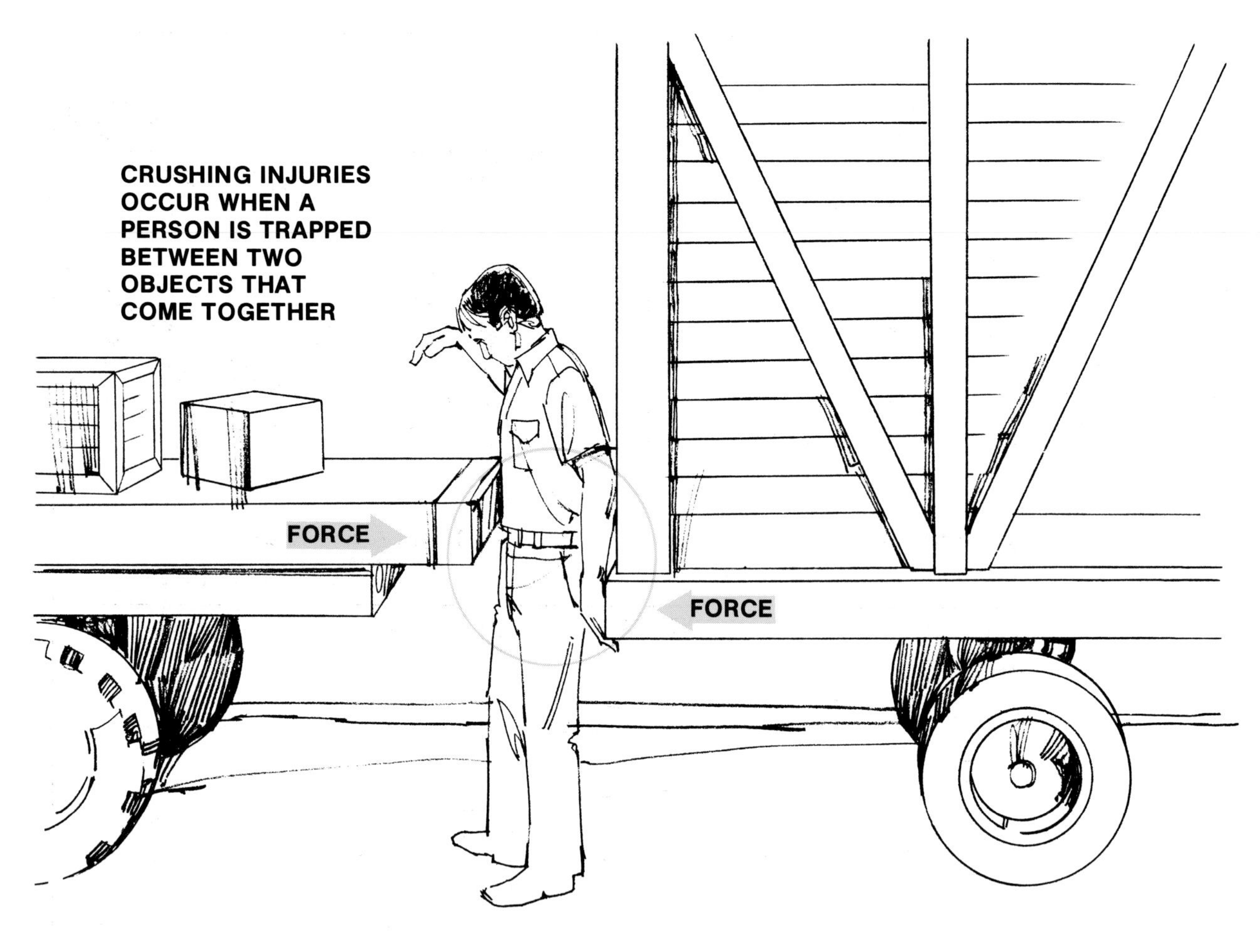

Fig. 14—Crushing Injuries Occur When A Person Is Trapped Between Two Objects Moving Together With Force Or Speed

PULL-IN POINTS

Pull-in accidents often happen when someone attempts to remove a corn stalk, hay, weeds, or something from feed rolls. People may be pulled into the moving parts of the machine and be seriously injured.

The real cause of the accident is attempting to remove material while the machine is running, thinking that the machine will pull in the plugged material and be more quickly cleared. It isn't worth the risk!

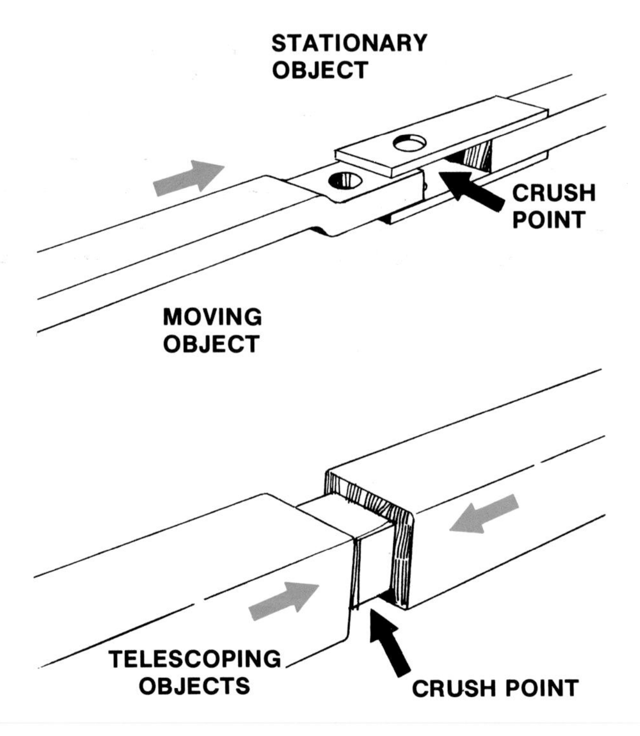

Fig. 15—Some Hitches And Telescoping Shafts Present Crush Points

Another situation that leads to pull-in accidents is attempting to feed materials by hand into feed rolls, grinders, and forage harvesters, to help them along (Fig. 17). Machines are faster and stronger than people. The machine will jerk in your hand and mangle it before you can react. Then the machine will chew and hold on. You can not pull free.

Consider a person attempting to remove or feed corn stalks from corn picker rolls (Fig. 18). Feed rolls move the stalks at about 12 feet per second. If the person grabs the corn stalk about two feet from the rolls and the rolls suddenly take the stalk in at 12 feet per second the person's hand and arm will go in with the stalk. It takes the person about 1 second just to tell himself to release his grip on the corn stalk. In that split second the corn stalk travels 12 feet between the stalk rolls, and that's 10 feet beyond his hand.

Other pull-in hazard situations:

Attempting to remove the stalk, vine or piece of twine wrapped around a rotating shaft. As you pull, it may wrap more tightly and pull you into the rotating shaft instead of you removing the material (Fig. 19).

Attempting to push an obstruction through a machine. For instance kicking or pushing a piece of tree root through a cotton harvester or hay baler with your foot. When the root goes through, your foot follows.

Fig. 16—Improperly Supported Objects Can Provide Extremely Dangerous Crush Points

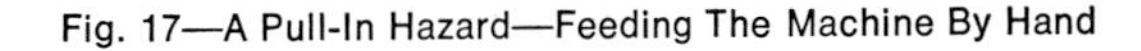

Fig. 17—A Pull-In Hazard—Feeding The Machine By Hand

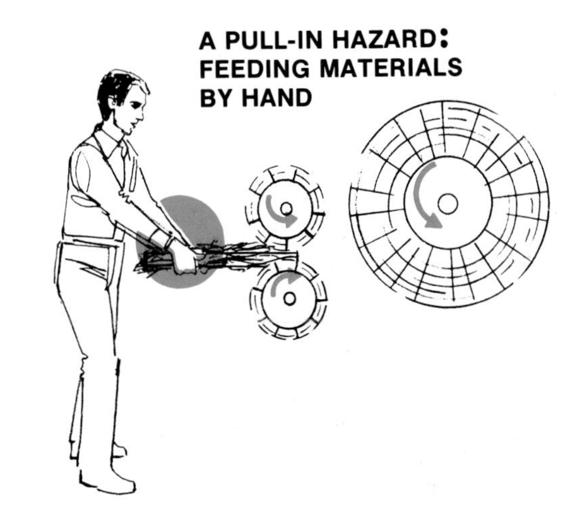

AVOIDING PULL-IN ACCIDENTS

First, recognize the potential hazards. Second, realize that you can't win speed contests with farm machines. Third, remember it is impossible to design machines to do jobs and still be completely safe for people who misuse them.

CLEAN, LUBRICATE, UNPLUG, OR HANDFEED A MACHINE ONLY WHEN IT IS SHUT DOWN. Always disengage power, shut off the engine, remove the key, and wait for all parts to stop moving before performing any of these operations unless the operator's manual instructs you do to otherwise. Some manufacturers have designed in reversers to clear headers. That kind of safety engineered product is the one to buy.

FREEWHEELING PARTS: INTERTIA

The heavier a part is and the faster it moves, the longer it will continue to move after the power is shut off. This is "freewheeling."

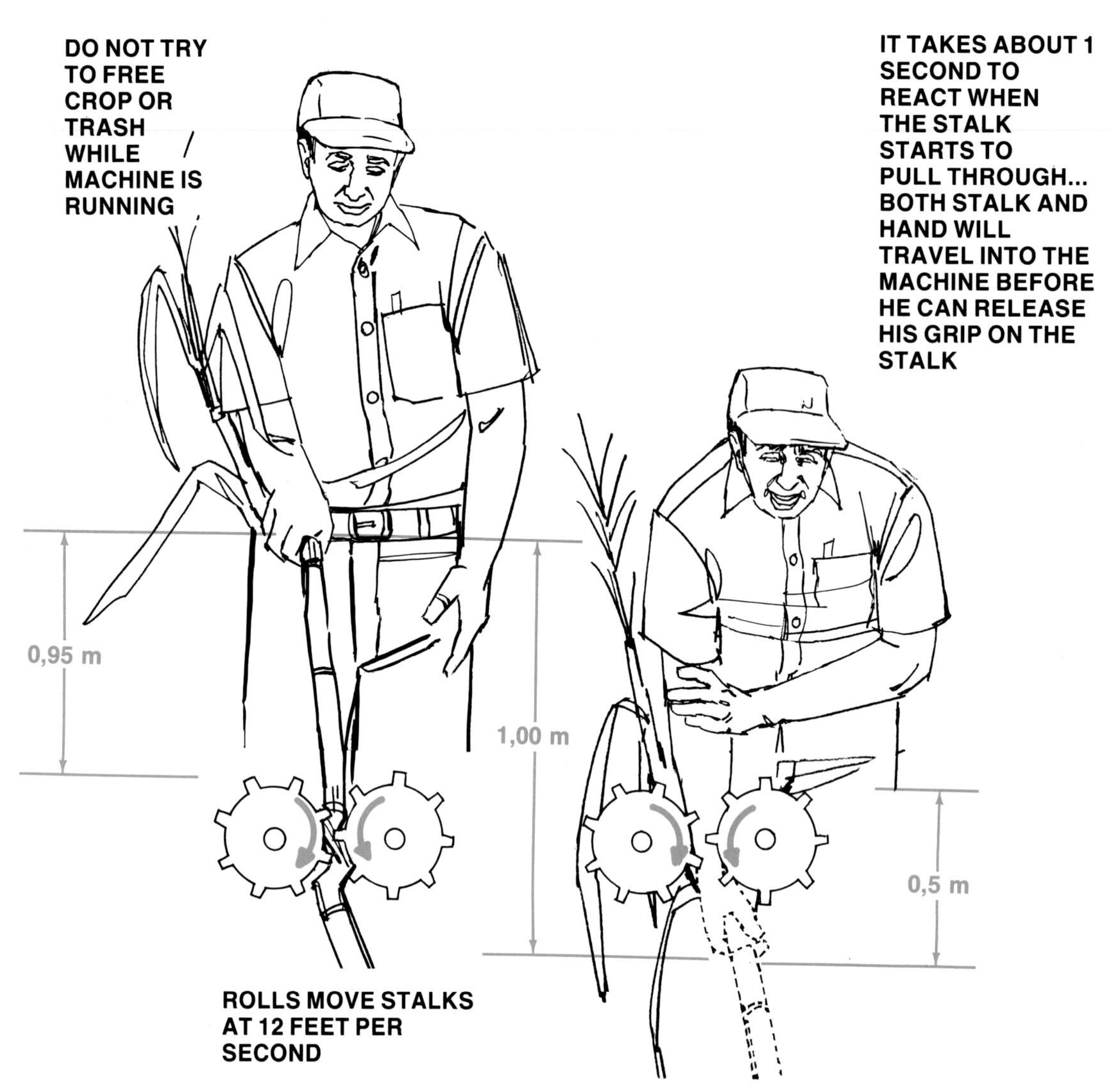

Fig. 18—A Major Pull-In Hazard—Attempting To Unplug A Machine While The Engine Is Running

Many farm machines, especially those which cut or harvest, have components that continue to move after the power is shut off. The parts continue to move because of their own inertia (Fig. 20), or the inertia of other moving parts connected to them (Fig. 21).

FREEWHEELING PARTS

Some freewheeling components on farm machines continue to rotate up to 2 or 2½ minutes after power is disengaged. Though moving slowly, they're still dangerous (Fig. 22). Examples of freewheeling farm machinery components include:

- **Cutter heads of forage harvesters**
- **Hammer mills of feed grinders**
- **Rotary mower blades**
- **Fans and blades on ensilage blowers**
- **Flywheels on balers**

HOW INJURIES OCCUR

Why do people get injured by freewheeling mechanisms? There are several reasons. For instance, they may disengage the power, open an access door, and

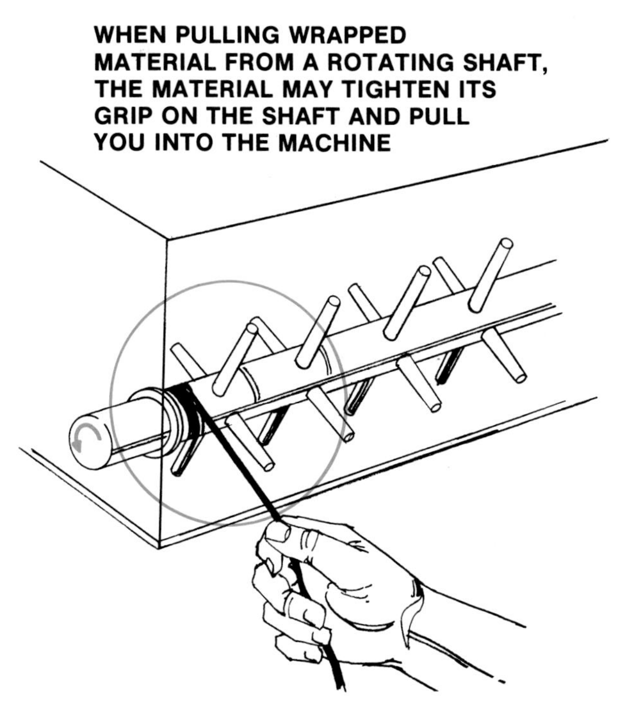

Fig. 19—A Pull In Hazard—Pulling On Wrapped Material To Remove It From A Rotating Shaft

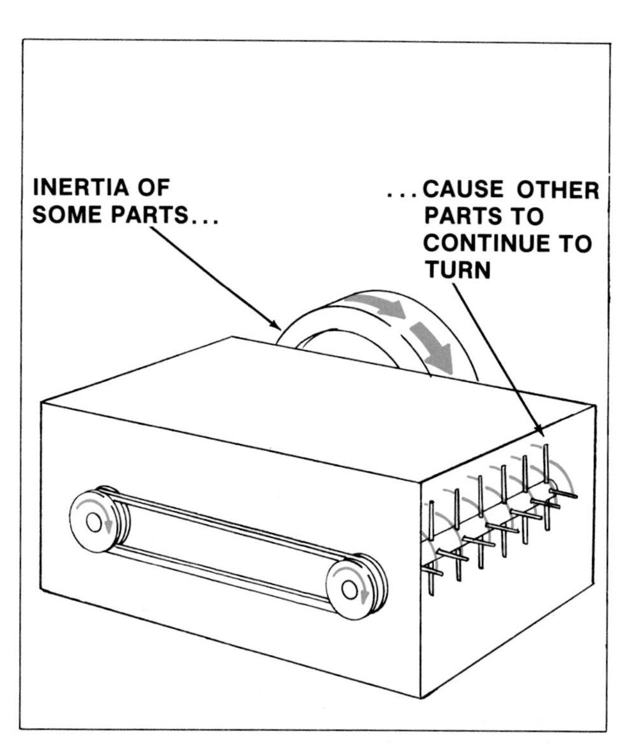

Fig. 21—Some Lightweight Parts May Continue Freewheeling Motion If Connected To Other High Inertia Parts

Fig. 20—Freewheeling Parts Continue To Turn After The Power That Drives Them Is Disengaged

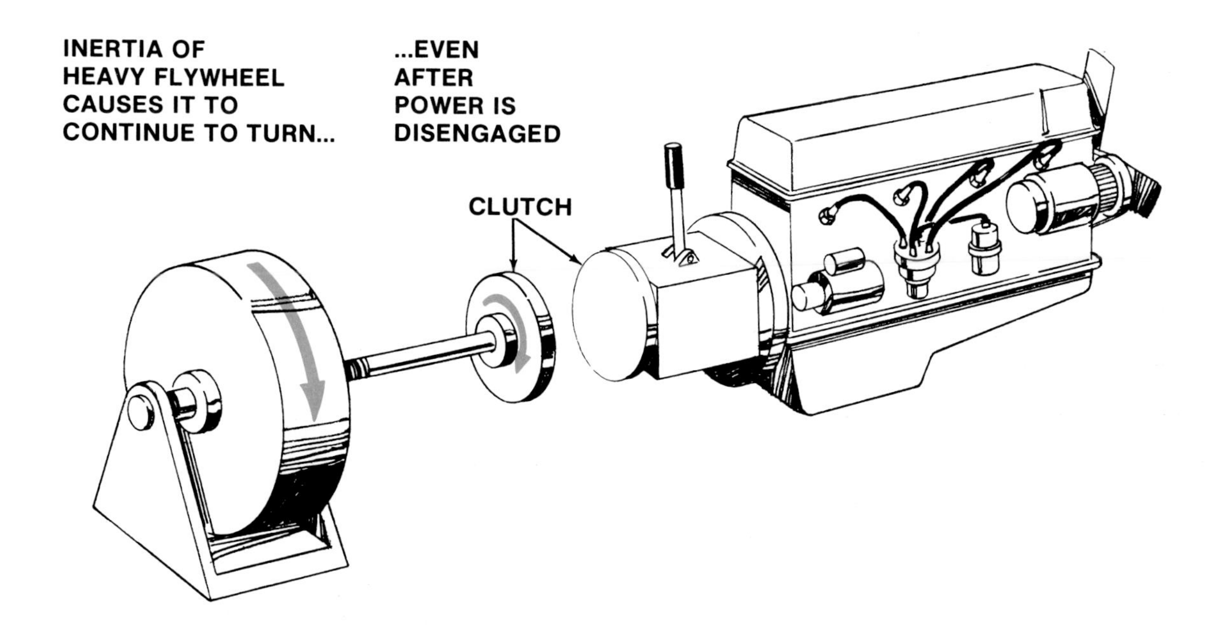

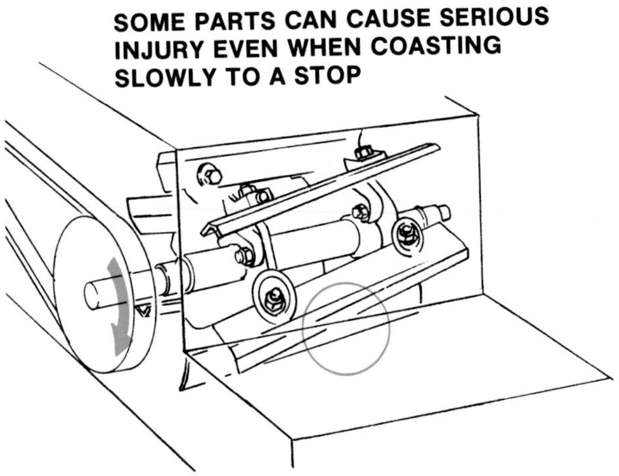

SOME PARTS CAN CAUSE SERIOUS INJURY EVEN WHEN COASTING SLOWLY TO A STOP

Fig. 22—Even Parts That Are Moving Slowly Can Cause Serious Injury

reach in to clean or unclog the machine before the part stops moving. Perhaps they concentrate so much on servicing the machine they forget parts are still moving. Perhaps they get impatient and just can't wait for the last movement to stop. Even though the part may have almost come to a stop it's inertia may still be enough to cause serious injury.

Fatigue can cause people do to things they know are dangerous, and would never do if they were not tired.

AVOIDING FREEWHEELING PARTS

Listen! Almost any freewheeling part makes some sound — often whirring or humming sounds — especially at higher speeds. Also, clutches may make a fairly loud, distinctive clicking or clanking noise until freewheeling parts stop.

Watch for motion! Watch flywheels, pulleys, PTO shafts, and the ends of shafts to see if parts are still moving. Make sure you understand that freewheeling parts, even though moving very slowly, can still cause serious injury. Objects in motion stay in motion until they are stopped by some other force or object. Don't let your hand or foot be the object that stops a freewheeling part.

THROWN OBJECT HAZARDS

A thrown object is a potential personal hazard whether it is a stone from a sling shot, a baseball, or a pea from a peashooter (Fig. 23). If it strikes with enough momentum, it will cause serious injury.

Fig. 23—Any Object Thrown With Enough Speed And Force Can Cause Serious Injury

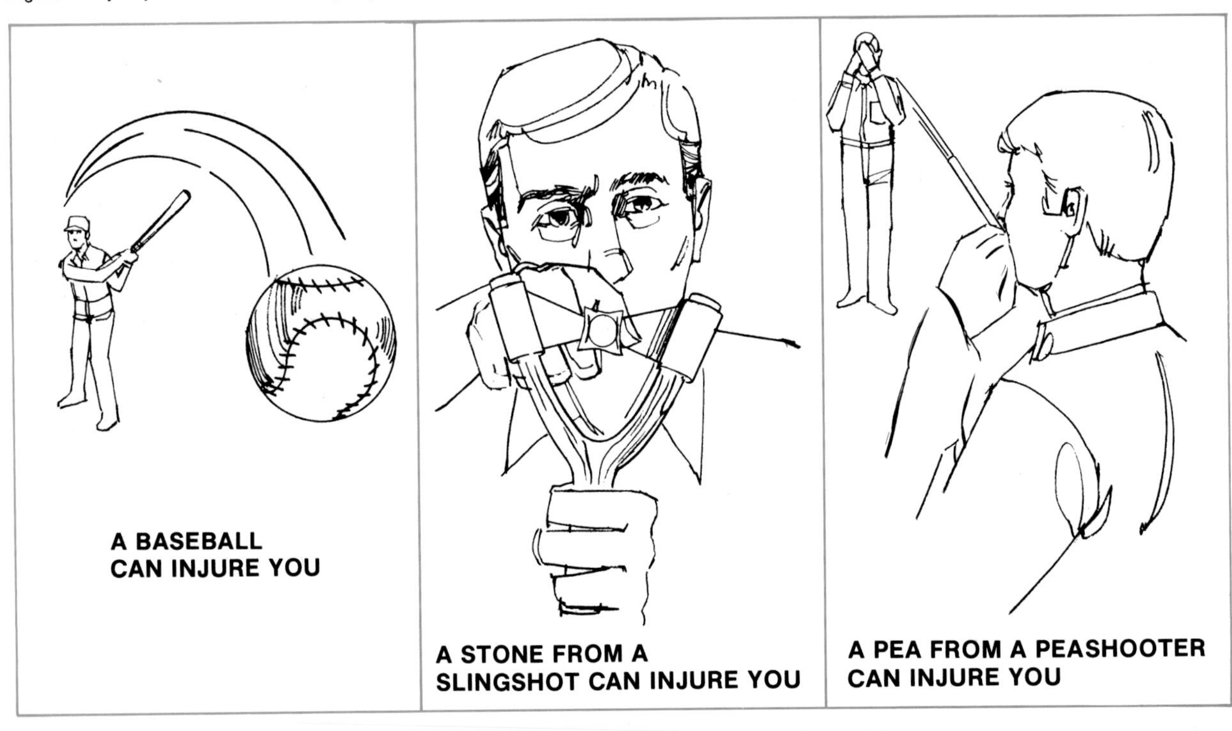

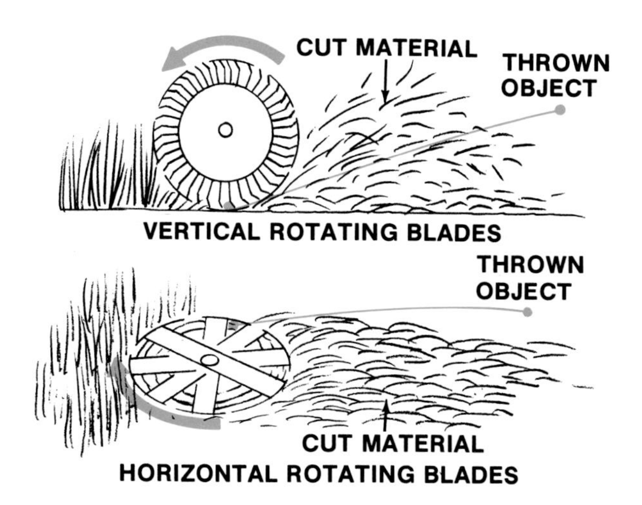

Fig. 24—Stones And Sticks Are Thrown Farther And Harder Than Grass Or Crops

Fig. 25—Particles Of Ground Or Chipped Grain May Be Thrown Hard Enough To Cause Injury

Some farm machines can throw objects great distances with tremendous force (Fig. 24). Recognize machines with a throwing hazard so you can avoid injury.

MACHINES WHICH THROW OBJECTS

Perhaps the greatest potential for injury by thrown objects is around machines that have rapidly rotating parts to cut or chop crops out in the open field. In order to cut or shred, the machine must strike the crop with considerable energy. Examples of such machines are rotary mowers, cutters, and shredders.

There is another type of thrown object hazard around machines which chop or grind crops. Particles of crop may be thrown from the machine (Fig. 25). A straw chopper on a combine or a feed mill may fling pieces of corn cobs and kernels. It may not be thrown as far as a rock thrown by a rotary mower, but you may be near enough to be injured.

AVOIDING THROWN OBJECT HAZARDS

1. *Recognize what machines throw objects.*

2. *Keep the machines properly shielded* to reduce the possibility of thrown objects. Some operations may require removing or adjusting shields or deflectors to handle crops or feeds. Make sure you replace the shielding before you begin the next job.

Follow the manufacturer's instructions for shielding to reduce the hazard of thrown objects.

3. *Know how far and in what direction objects may be thrown,* even with shielding in place (Fig. 26).

4. *Stay a safe distance away from the likely path of thrown objects when you approach a machine.*

5. *When operating a machine which may throw objects, make sure the machine will not discharge near people or animals. Build a shield if necessary.*

STORED ENERGY

Stored energy is energy confined and just waiting to be released. It is completely safe as long as it is confined. But, if it is released unexpectedly, stored energy can cause injury. Learn to recognize potential stored energy hazards and know how to handle them.

A simple slingshot made of a strong rubber band and a wooden handle (Fig. 27) uses stored energy. You slowly stretch the rubber bands to "ready" position. Stretching uses some of your own energy. That energy is "stored" in the rubber bands. It is harmless as long as you hold back the rubber bands. But, when you release the slingshot you release all the energy you used to stretch the rubber bands.

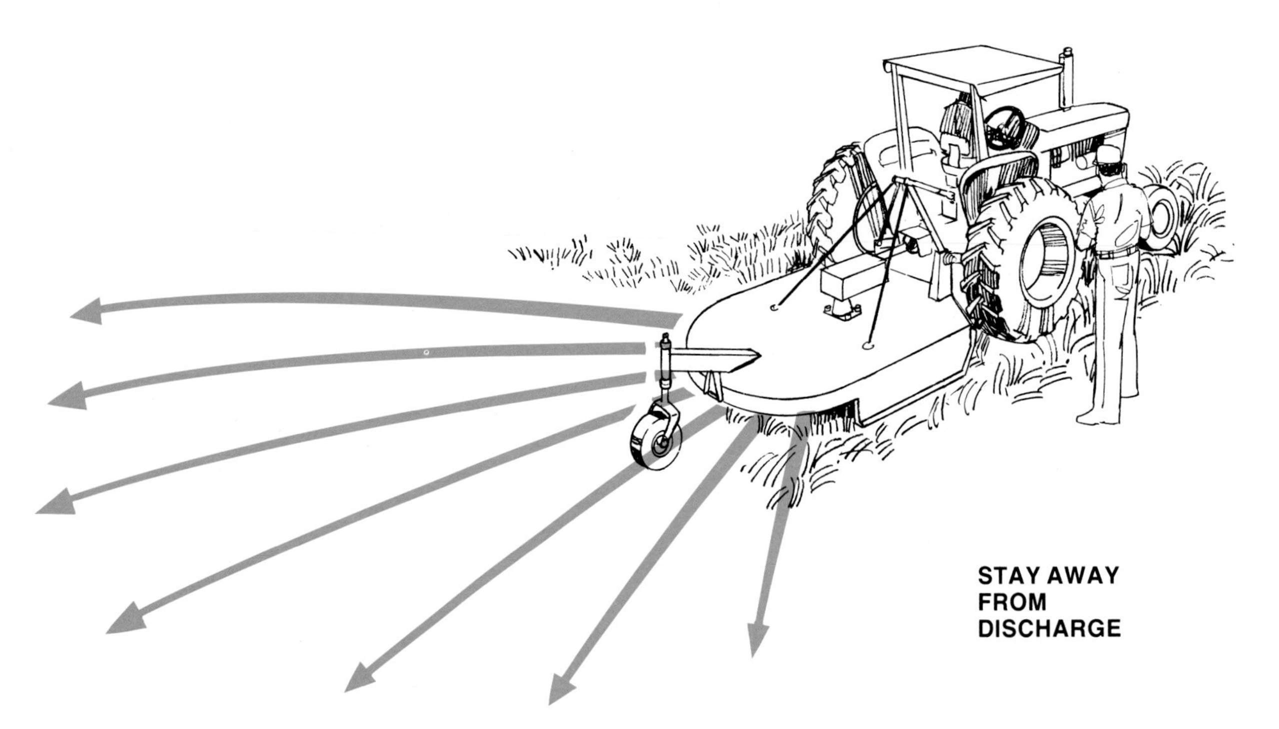

Fig. 26—Know How Far And In What Direction A Machine May Throw Objects. Use Shields As Operation Permits

Stored energy can work for you, or it can be carelessly released and cause injury.

In many farm machines, energy is stored so it can be released at the right time, in the right way for you. Here are some components that store energy. You should recognize them as you use and service farm machinery:

Fig. 27—Energy Is Stored In The Rubber Bands Of The Slingshot, Ready To Be Released

- **Springs**
- **Hydraulic Systems**
- **Compressed Air**
- **Electricity**
- **Raised Loads**
- **Loaded Mechanisms**

SPRINGS

Springs are energy storing devices. They are used to help lift implements, to keep belts tight, and to absorb shock. Springs store energy in tension, like the slingshot, or in compression (Fig. 28).

When you remove any device connected to a spring, be sure you know what can happen. Know what direction the spring will move, and what direction it will move other components when it is disconnected (Fig. 29). A compressed spring can propel an object outward with tremendous force.

Make sure you and others will not be in the path of any part that will move when the spring moves. Plan exactly how far and where each part will move. Use proper tools to assist you in removing or replacing spring-loaded devices. Even small springs can store a lot of energy.

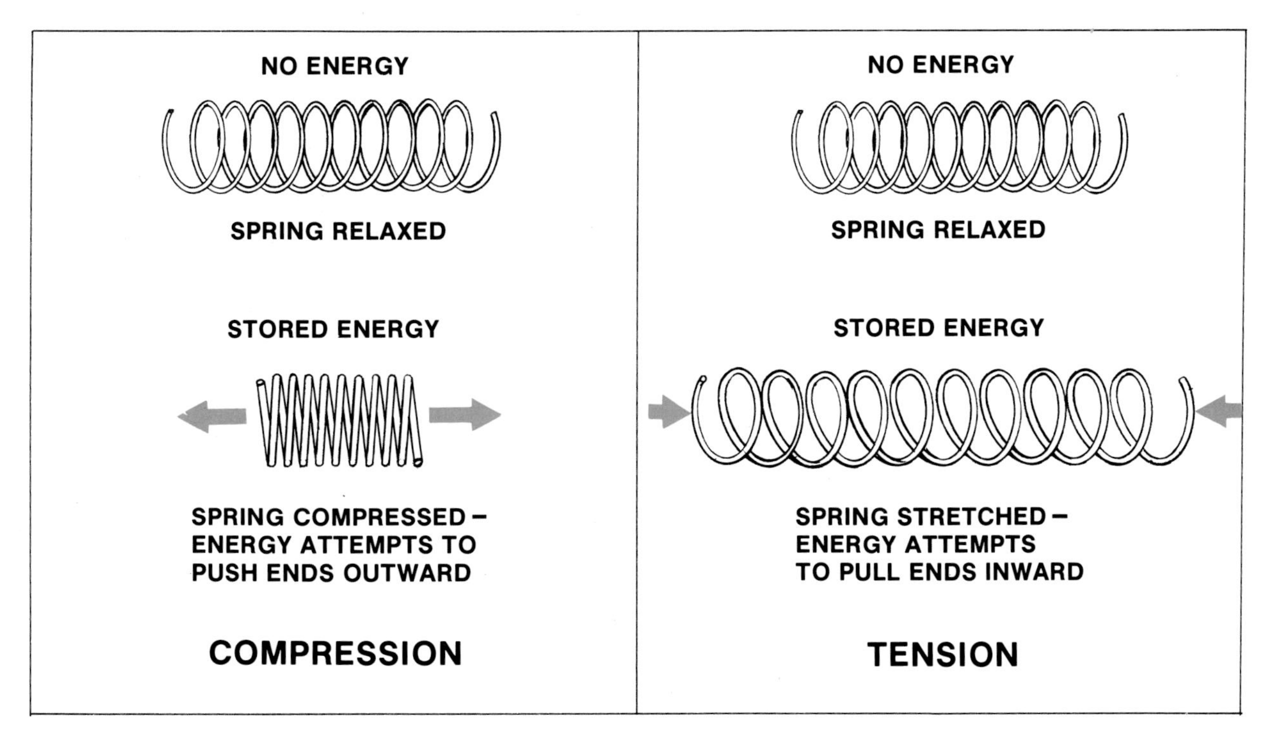

Fig. 28—Energy Stored In Springs

Fig. 29—Be Sure You Know What Can Happen Before Disconnecting Any Part Attached To A Spring

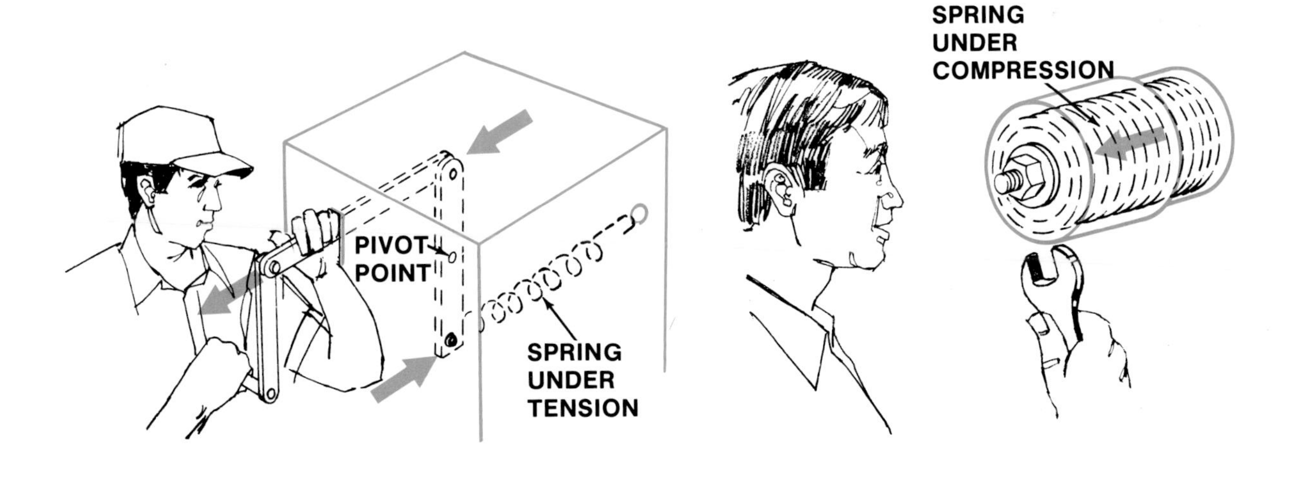

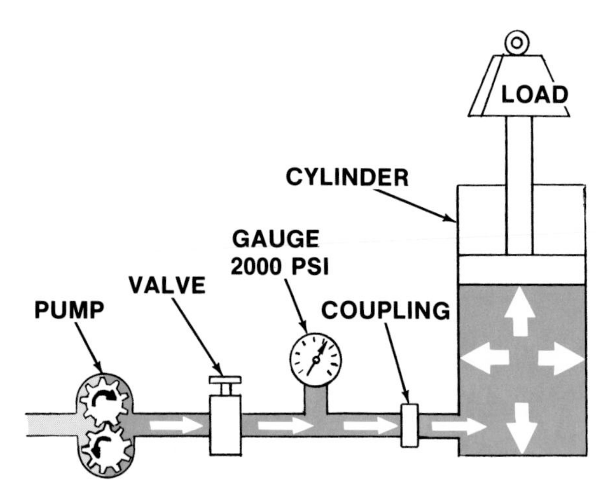

Fig. 30—Hydraulic Fluid Under Pressure Attempts To Escape Or Move To A Point Of Lower Pressure

HYDRAULIC SYSTEMS

Hydraulic systems on farm machines also store energy. Hydraulic systems must confine fluid under high pressure often higher than 2,000 pounds per square inch.

A lot of energy may be stored in a hydraulic system, and because there is often no visible motion, operators do not recognize it as a potential hazard. Carelessly servicing, adjusting, or replacing parts can result in serious injury. Fluid under pressure attempts to escape (Fig. 30). In doing that it can do helpful work, or it can be harmful.

Servicing and Adjusting Systems Under Pressure

Adjusting and removing components when hydraulic fluid is under pressure can be hazardous (Fig. 31). Imagine attempting to remove a faucet from your kitchen sink without relieving the water pressure. You'd get a face full of water! It is much more dangerous with hydraulic systems. Instead of just getting wet from water at 40 psi, you would be seriously injured by oil under 2,000 psi, or more. You could be injured by the hot, high pressure spray of fluid and by the part you are removing when it is thrown at you (Fig. 31).

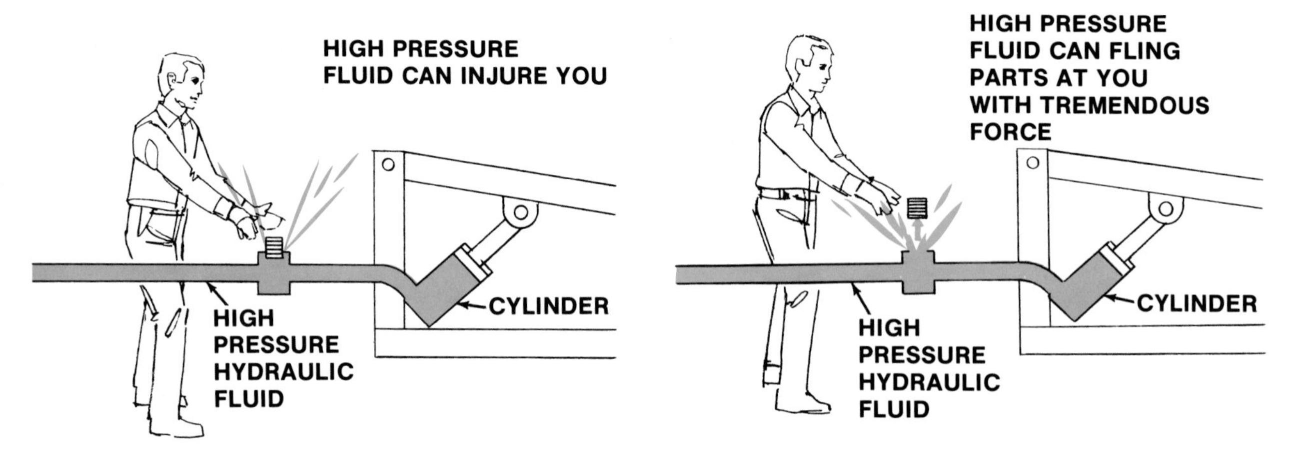

Fig. 31 — Always Relieve Hydraulic Pressure Before Adjusting Hydraulic Fittings. You Could Be Injured By a Hot, High Pressure Spray Of Hydraulic Fluid Or By A Part Flung At You

AVOID HIGH-PRESSURE FLUIDS

Escaping fluid under pressure can penetrate the skin causing serious injury. Relieve pressure before disconnecting hydraulic or other lines. Tighten all connections before applying pressure. Keep hands and body away from pinholes and nozzles which eject fluids under high pressure. Use a piece of cardboard or paper to search for leaks.

If ANY fluid is injected into the skin, it must be surgically removed within a few hours by a doctor familiar with this type injury or gangrene may result.

To avoid this hazard, always relieve the pressure in a hydraulic system before loosening, tightening, removing, or adjusting fittings and components. Keep all hydraulic fittings tight to prevent leaks. But, do not tighten fittings without relieving the pressure. Also, if you over-tighten a coupling, it may crack and release a high-pressure stream of fluid. You will be injured by the fluid and the implement, which will drop to the ground.

Before attempting any service:

1. *Shut off the engine which powers the hydraulic pump.*

2. *Lower implement to the ground.*

3. *Move the hydraulic control lever back and forth* several times to relieve pressure.

4. *Follow instructions in operator's manuals.* Specific procedures for servicing hydraulic systems are very important for your safety.

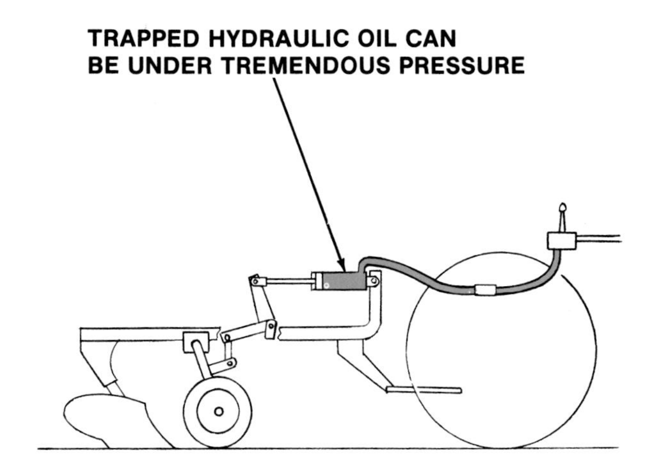

Fig. 32—Even Though The Engine And Hydraulic Pump Are Stopped, Hydraulic Oil Can Be Trapped Under Tremendous Pressure—A Potential Hazard

Trapped Oil

Hydraulic oil can be trapped in the hydraulic system even when the engine and hydraulic pump are stopped (Fig. 32). Trapped oil can be under tremendous pressure, up to 2,000 psi or more. You can be seriously injured by escaping fluid and moving machine parts if you loosen a fitting.

Another hazard with trapped oil (Fig. 33) is heat. Heat from the sun can expand the hydraulic oil and increase pressure. The pressure can blow seals and move parts of an implement or machine.

Fig. 33—Trapped Oil Can Move An Implement And Blow Seals

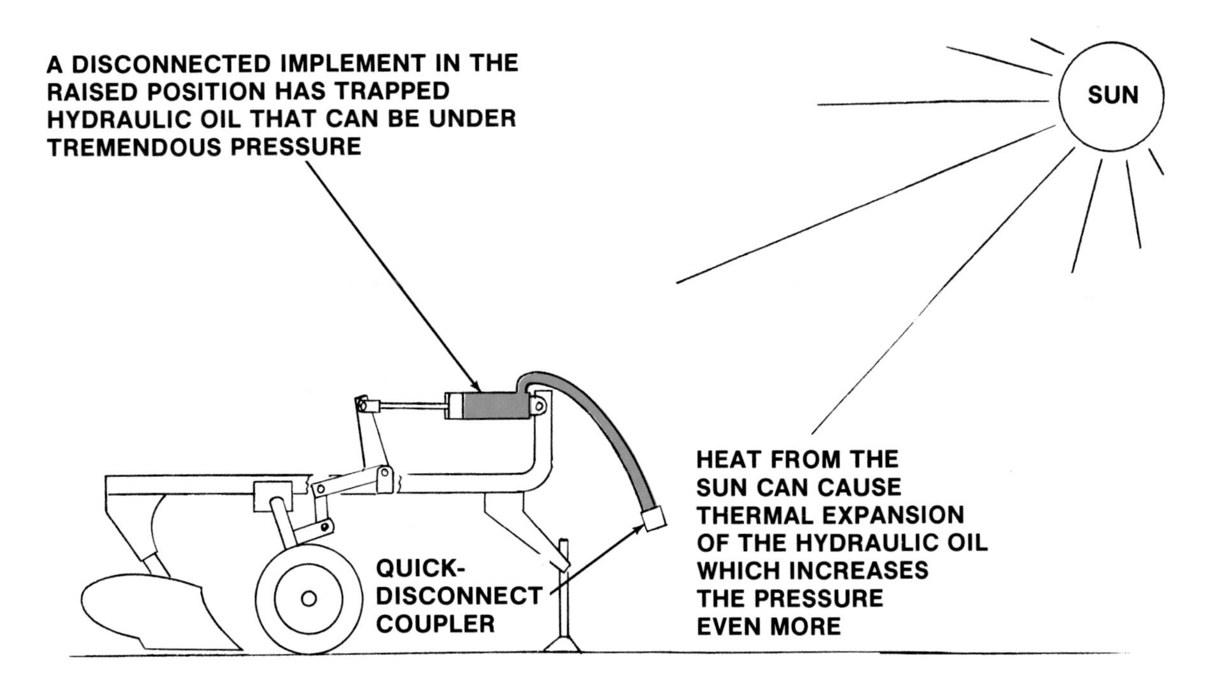

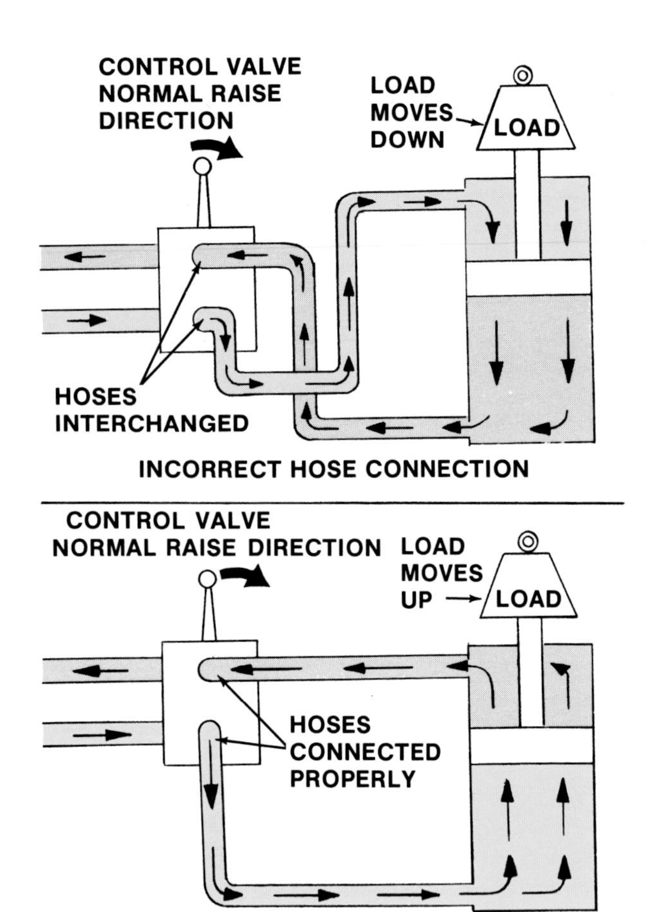

Fig. 34—Don't Interchange Hydraulic Lines

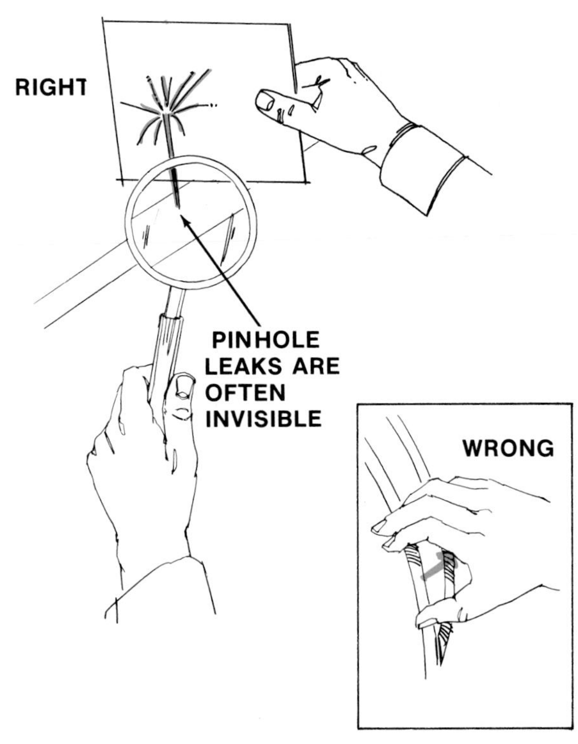

Fig. 36—The Jet Stream Or Mist From A Pinhole Leak In A Hydraulic System Can Penetrate Your Skin—Don't Touch It!

Incorrect Coupling

Crossing hydraulic lines creates hazards. When lines are coupled to the proper part, you get the results you expect. But if lines are crossed, the implement may raise when you expect it to drop (Fig. 34). Serious injury could result. Make sure hydraulic lines are coupled exactly as specified in the machine technical manual. Color code lines with paint or tape. After you attach hydraulic lines try the controls cautiously to see if you get the proper result.

Fig. 35—Don't Connect A High-Pressure Hydraulic Pump To Low-Pressure Systems

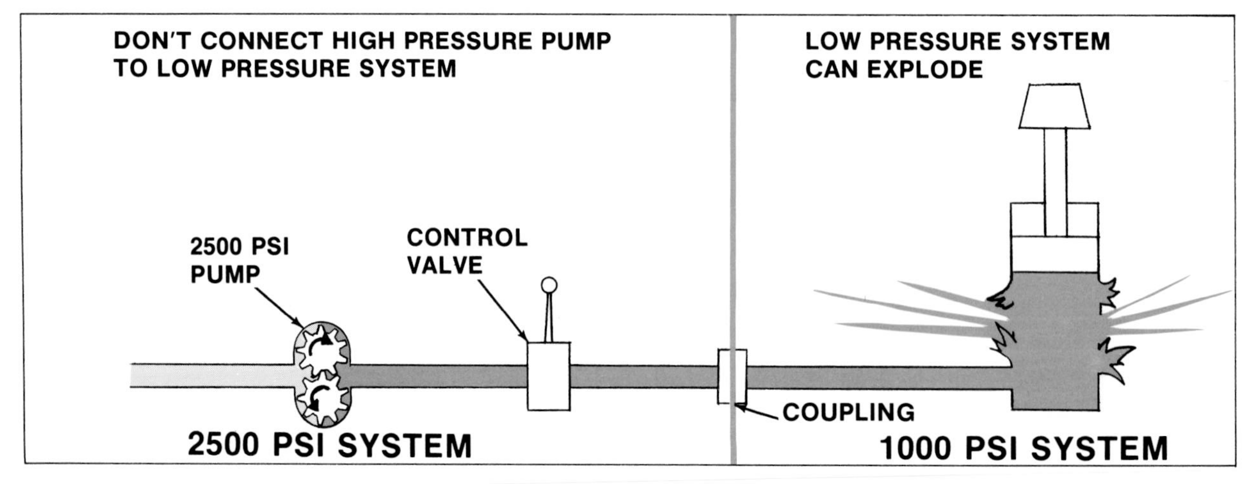

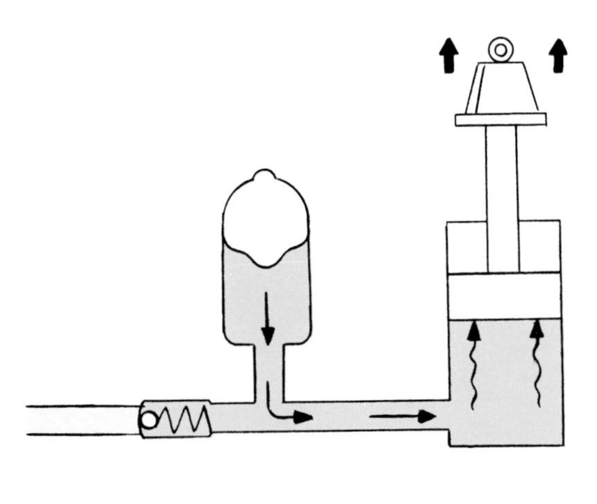

STORE ENERGY

ABSORB SHOCKS

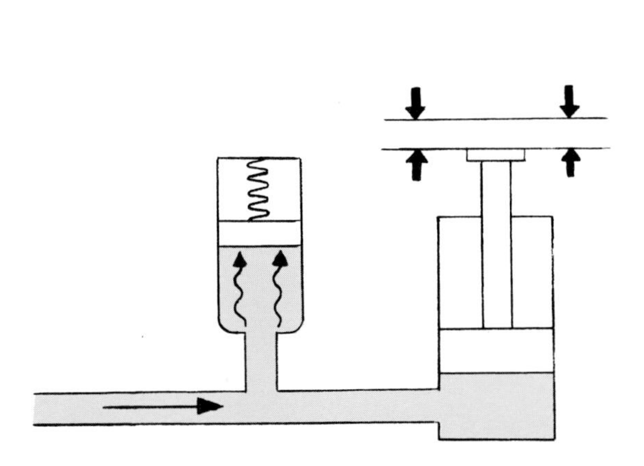

BUILD PRESSURE GRADUALLY

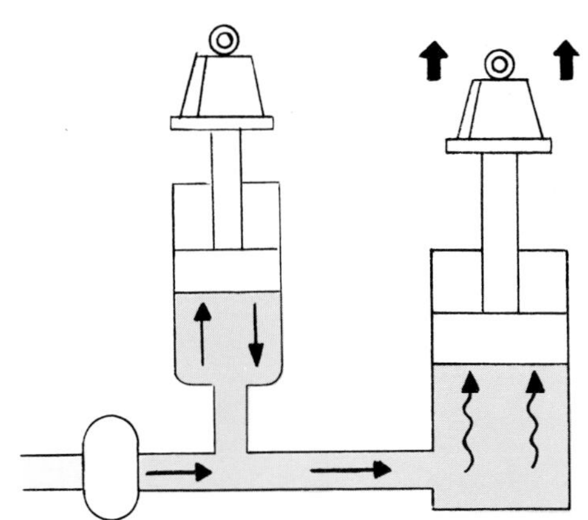

MAINTAIN CONSTANT PRESSURE

Fig. 37—Hydraulic Accumulators Have Four Basic Uses

Do not replace hydraulic pipes with rubber hose. Pipes are usually designed to carry heavier pressure than rubber hose. Hose is usually not as durable as pipes.

Another hazard is coupling a high-pressure pump to a low-pressure system. Attaching a hose from a 2,500 psi hydraulic system to an implement equipped with hoses, cylinders, and fittings designed for 1,000 psi is inviting trouble (Fig. 35). The low pressure system could explode. Never improvise or adapt couplings or fittings on a low-pressure system to attach to a high-pressure pumping system. Always follow the manufacturer's recommendations.

Pinhole Leaks

If liquid, under high pressure, escapes through an extremely small opening it comes out as a fine stream (Fig. 36). The stream is called a *pinhole leak.* Pinhole leaks in hydraulic systems are hard to see and, they can be very dangerous. High-pressure streams from pinhole leaks penetrate human flesh. Hydraulic systems on many farm machines have pressures of 2,000 psi or higher. That's higher than the pressure in hydraulic syringes used to give injections. Injury from pinhole hydraulic leaks comes from the fluid cutting through flesh, and from body reaction to chemicals in the fluid.

AVOID HIGH-PRESSURE FLUIDS

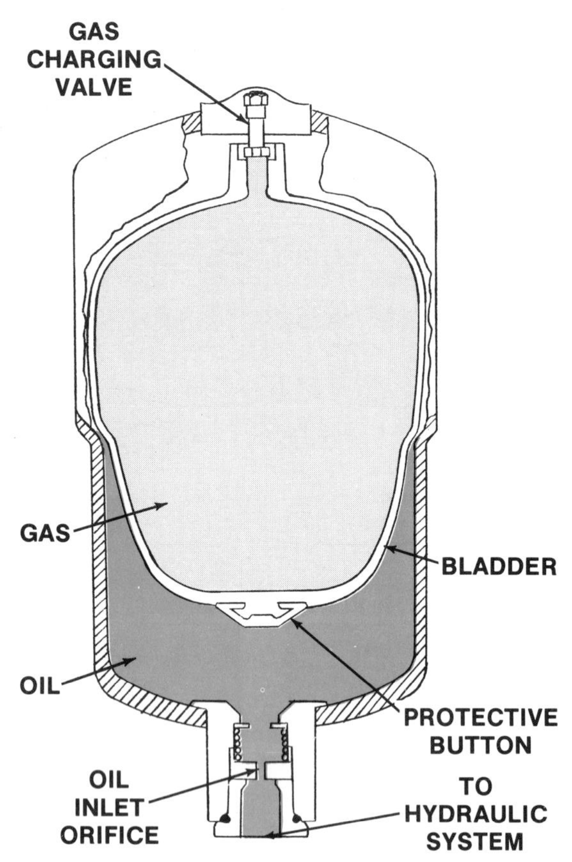

Fig. 38—Construction Of A Typical Hydraulic Accumulator

You may see only the symptoms of pinhole leaks from high pressure systems. There may appear to be only a dripping of fluid, when actually it may be an accumulation of fluid from a high-pressure jet stream so fine it is invisible. Don't touch a wet hose or part with bare or even gloved hands to locate the leak. Pass a piece of cardboard or wood over the suspected area instead (Fig. 36). Wear safety glasses. Then relieve the pressure and replace the defective part.

Diesel fuel injectors are designed to force fuel into engine cylinders under high pressure. Don't touch the jet stream from a diesel injector nozzle. It is just as dangerous as a pinhole leak.

Escaping fluid under pressure can penetrate the skin causing serious injury. Relieve pressure before disconnecting hydraulic or other lines. Tighten all connections before applying pressure. Keep hands and body away from pinholes and nozzles which eject fluids under high pressure. Use a piece of cardboard or paper to search for leaks.

IF ANY fluid is injected into the skin, it must be surgically removed within a few hours by a doctor familiar with this type injury or gangrene may result.

Hydraulic Accumulators

Some hydraulic systems have accumulators to store energy. They may also be used to absorb shock loads and to maintain a constant pressure in the system (Fig. 37). Accumulators that store energy are often boosters for systems with fixed displacement pumps. Accumulators store oil pressure from pumps during slack periods and feed it back when it's needed. Sometimes accumulators are used for backup devices to protect against failure of oil supply. For example, on some large machines an accumulator can be used to maintain pressure for emergency brakes or steering in case the engine stalls.

Accumulators may be pneumatic (gas-loaded), weight loaded, or spring-loaded. They all store energy. That's their job. But because they store energy, they must be respected and properly serviced. Pneumatic accumulators use an inert gas, usually nitrogen, to provide pressure against the oil. (Inert gas is a gas that will not explode.) Air or oxygen will explode if mixed with oil under high pressures. Gas in accumulators is separated from the oil by a flexible bladder (Fig. 38).

Recognize that the accumulators and the entire hydraulic system may have energy stored in it, if the pressure has not been relieved. When the hydraulic pump builds up pressure against the diaphragm or bladder, it compresses the nitrogen in the accumulator (Fig. 39). This energy stored as compressed nitrogen provides pressure against the oil. It may be released when the system calls for pressure or absorb shock if that is its purpose. Even though the pump may be stopped, or an implement has been disconnected from the tractor, energy is stored in the accumulator unless the pressure was relieved before shutdown. The nitrogen is under pressure, so the hydraulic fluid is also under pressure.

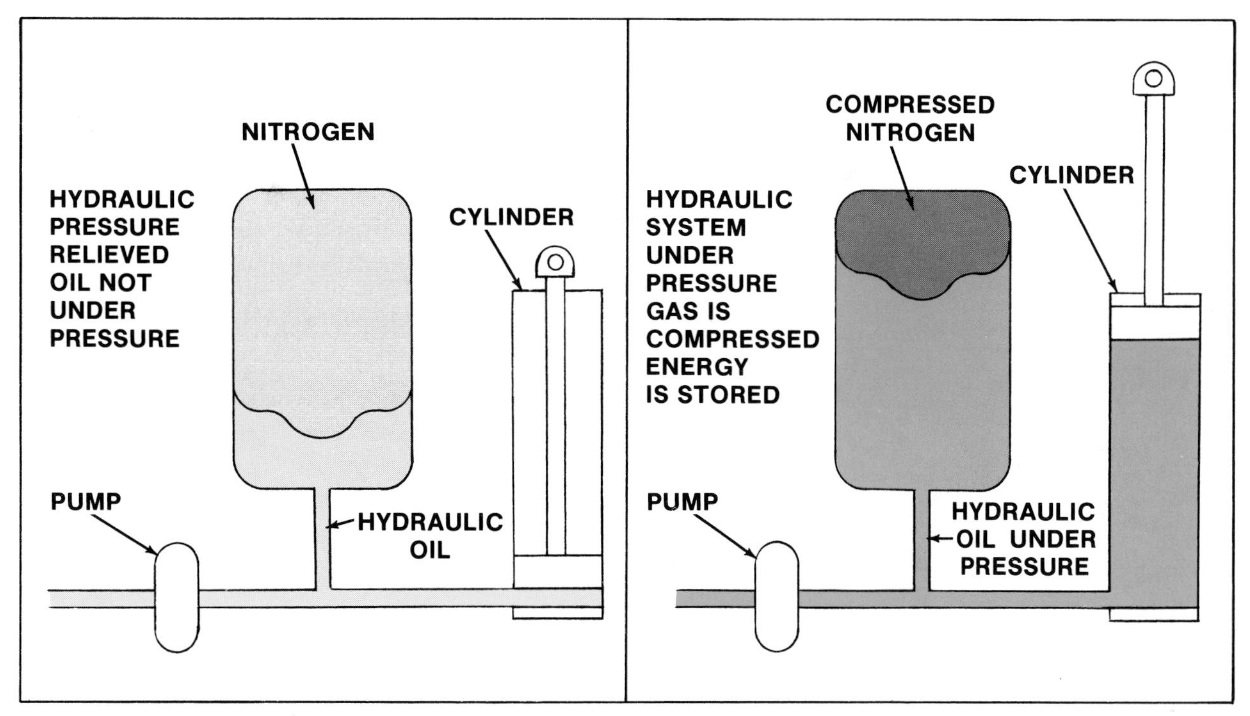

Fig. 39—Hydraulic Accumulators Store Energy

Observe these basic safety considerations for hydraulic accumulators:

1. *Recognize accumulators as sources of stored energy.*

2. *Relieve all hydraulic system pressure before adjusting or servicing any part of an accumulator system.*

3. *Relieve all hydraulic pressure before leaving a machine unattended* for the safety of others as well as for you.

4. *Make sure pneumatic accumulators are properly charged with the proper inert gas (usually nitrogen).* A pneumatic accumulator without gas is a potential "bomb" when charged only with oil.

5. *Read and follow manufacturer's instructions for servicing accumulators.*

Generally, manufacturers recommend only authorized dealers service gas charged accumulators. Read and follow the manufacturer's instructions thoroughly. For more information on hydraulic accumulators, refer to "Fundamentals of Service—Hydraulics," John Deere FOS Series.

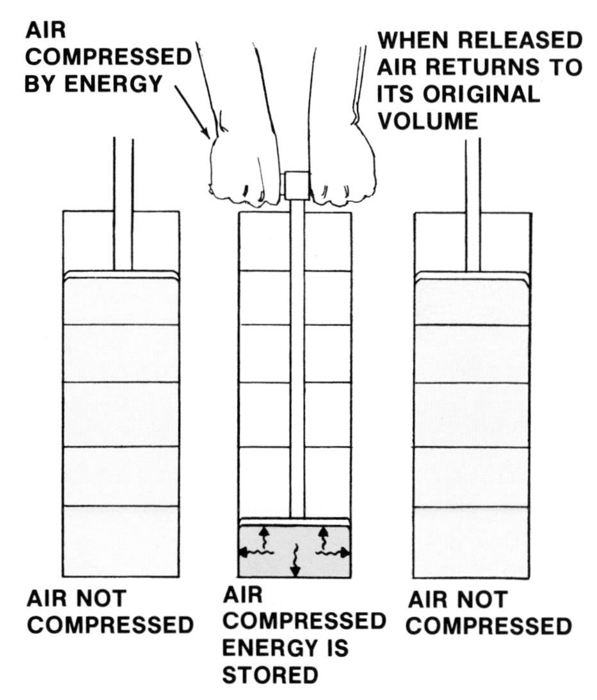

Fig. 40—When Air Is Compressed, Energy Is Stored. The Air Attempts To Return To Its Original Volume

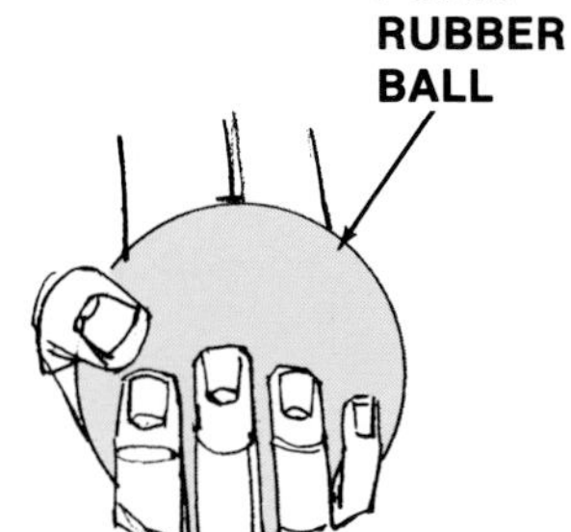

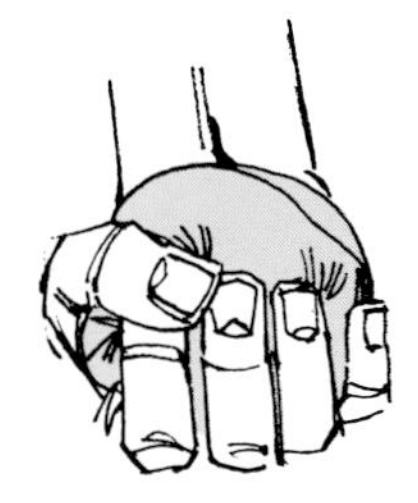

Fig. 41—Air, Like A Compressed Ball, Attempts To Return To Its Original Volume When Released

Fig. 42—Compressed Air Is Stored Energy. Always Stand To One Side When Inflating Tires And Never Over-Inflate Them

COMPRESSED AIR

Compressed air is dangerous. When air is compressed, its volume is reduced (Fig. 40). Energy is stored. It is like squeezing a rubber ball in your hand (Fig. 41). You can squeeze the ball until it is smaller. But it springs back to normal size when you release it. The energy stored by compressed air can be tremendous, compared to squeezing the rubber ball.

Wear safety glasses when using compressed air to clean. Never use more than 30 psi!

Probably the most frequent use of compressed air for farm machinery is for tire inflation. An inflated tire can be dangerous, especially the large ones. When air is compressed into a tire, it attempts to get out.

When that large quantity of compressed air gets an opening to return to its original volume, perhaps through a failure of the tire or its seal, there's a tremendous amount of stored energy released. A 10.00-20 12PR truck tire inflated to 75 psi has 46,510 foot-pounds of energy, enough energy to raise a 3000 pound car 15 feet. A 24.00-49 tire inflated to 75 psi develops 354,260 foot-pounds of energy, which could lift a 134-pound person one-half mile into the air!

You should recognize that compressed air is stored energy. It can be hazardous if not properly controlled and respected. Keep pressure at proper levels. Stand to one side when inflating tires (Fig. 42). Follow manufacturer's recommendations at all times. For more detailed information on tire care, maintenance, and safety refer to "Fundamentals of Service—Tires and Tracks," John Deere FOS manual series.

CAUTION: Every tire and rim or wheel must be handled in a special way. Always use the tire and rim and wheel manufacturer's procedures when you demount and mount a tire.

Information also can be obtained from the following associations:

Rubber Manufacturer's Association
1400 K Street, N.W.
Washington, D.C. 20005

National Wheel and Rim Association
4836 Victor Street
Jacksonville, FL 32207

Further safety information can be obtained from:

U. S. Department of Transportation
National Highway Traffic Safety Administration
400 Seventh St., S.W.
Washington, D.C. 20590

ELECTRICAL ENERGY

One of the most common forms of stored energy is electricity stored in 12 volt batteries. When properly used it makes your work easier. If handled carelessly it can cause serious injury. If you recognize the potential hazards of electricity, you'll be able to avoid serious accidents.

Fires

Every self-propelled machine should have a 5 pound, all-purpose, ABC dry chemical fire extinguisher on board.

Electrical systems can cause fires if not properly maintained. The energy stored in the battery may be tapped to start the engine. But if a bare wire touches a metal part and becomes hot or sparks, it can start a fire in dust, chaff, and leaves. Most machinery fires do not result in personal injury, but every fire is a potential source of injury. Inspect electrical systems. Make sure wires are properly insulated and clean dust, chaff, leaves and oil off wires.

Short-Circuit Starting

If insulation on electrical wires is cracked or worn, a short circuit occurs. Electricity could flow to the cranking motor and start the engine when no one is around. If the positive and negative terminal of a cranking motor are accidentally contacted by another metal object such as a wrench, the current will flow between the two terminals, and accidentally start the engine.

DANGER

Do not short across starter with a screwdriver to start a tractor. You bypass the neutral start switch by doing so. and the tractor can lunge and crush you.

In summary, the energy in an electrical system is waiting to do something. If it does what was planned for it, at the right time, there's no problem. If it does what was planned for it at the wrong time or does the wrong thing injury and property damage result. Inspect electrical systems on all machines. Replace worn wiring, contacts and switches.

Fig. 43—Use The Steps, Ladders And Handholds To Safely Get On And Off Farm Machinery—Don't Jump

SLIPS AND FALLS

Slipping and falling can put a person out of commission for hours, days, or years. Slips and falls can be prevented by recognizing the potential hazards and avoiding them.

ON-OFF ACCIDENTS

Farm machines are equipped with steps, handholds, ladders, and platforms that help you get on and off (Fig. 43). Steps are made so you can have three support points resting on the machine at all times: Two feet and one hand, or two hands and one foot. Use the hand holds and steps, and keep them in good repair. Don't jump off machines.

Recognize what can interfere with steps, handholds, ladders, and platforms. Your good judgment can prevent accidents.

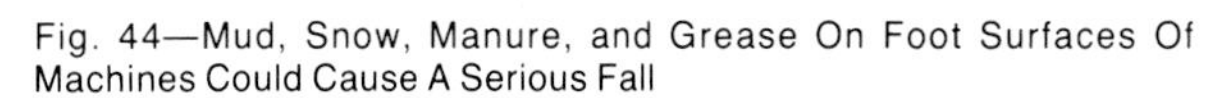

Fig. 44—Mud, Snow, Manure, and Grease On Foot Surfaces Of Machines Could Cause A Serious Fall

SLIPPERY FOOT SURFACES ON MACHINES

Mud, snow, ice, manure, and grease may build up on steps, platforms, and other surfaces. When it does, you can slip and fall (Fig. 44). Or, you could slip and bump a control, causing the machine to lurch into action, injuring yourself or someone else. Also, you could fall into a moving part. Such falls can be fatal.

Take time to clean foot surfaces for your own safety and for others who will use them. Also, wear boots with non-skid soles.

Fig. 45—Keep Machine Platforms Clear

CLUTTERED STEPS AND OPERATOR PLATFORMS

Chains and tools on operator platforms are accidents waiting to happen. You need to be able to move about without having to watch where you place your feet. Slipping on mud or snow may be somewhat excusable, but tripping on something you placed on a step isn't. Machine surfaces intended for your feet should be kept clear for your feet (Fig. 45).

SLIPPERY GROUND SURFACES

When there's snow, ice, or mud on the ground you can't do much to change it. But you can learn to recognize that those slippery surfaces can lead to accidents and try to avoid them.

The danger is usually slipping or tripping and falling against or into a machine that is running. For instance, when grinding feed in a muddy barnyard you could slip and fall into the hopper intake. When you must work under these conditions, the best practice is to slow down, step deliberately, and be on the lookout for slippery surfaces or objects that may cause you to lose your footing (Fig. 46). Also, wear boots with non-skid soles.

Fig. 46—Watch Where You Step

SLOW-MOVING VEHICLES

Everyone knows that speed kills on the highway. Right? Then why consider slow-moving vehicles as major hazards on the highways? A study of slow-moving vehicle hazards may surprise you.

RECOGNIZING THE HAZARDS

Most motorists on public roads are not farmers. In fact the percentage of people who have farming backgrounds or those who even know farmers is becoming smaller each year. So, only a small percentage of the motoring public is likely to give much thought to the unique nature of a farm machine when they see it on the highway. They're traveling comfortably, visiting with passengers in the car while meeting and overtaking other cars and trucks traveling at similar speeds.

When they approach a farm tractor whose speed is only 15 mph, they just don't realize how much slower it is traveling than they are. And even if they think about it, they may not recognize they must react differently to safely handle this situation. After all, they've met and overtaken hundreds or thousands of cars and trucks, but relatively few farm vehicles. They haven't had the practice that provides the reflexes and judgment for proper action.

CLOSURE TIME

Consider a motorist who is traveling 55 mph. He tops a hill, and 400 feet ahead of him is another car traveling 45 mph. The closing speed is only 10 mph. At 10 mph closing speed, the motorist has 27 seconds to realize the speed of the car ahead of him, react, and slow down (Fig. 48A).

Now picture the motorist topping the same hill again. This time 400 feet ahead of him is a tractor traveling 15 mph. Now his closure speed is 40 mph. He has less than 7 seconds to recognize the speed of the tractor, react, and slow down. That's about one-fourth the time he had to slow down for the car (Fig. 48B).

Situations similar to the one above lead to thousands of slow-moving vehicle accidents each year. The motorists aren't able to stop in time. Many of the accidents are fatal.

Many of the rear-end collisions with slow-moving farm machines involve other rural motorists. Every motorist must be alert to avoid collisions.

Fig. 47—Motorists Have Unpredictable Reactions To Slow-Moving Farm Equipment

Fig. 48—The Time A Motorist Has To React To Avoid An Accident Is Much Less For Farm Machinery Than For Automobiles

CLOSURE TIME
27 SECONDS

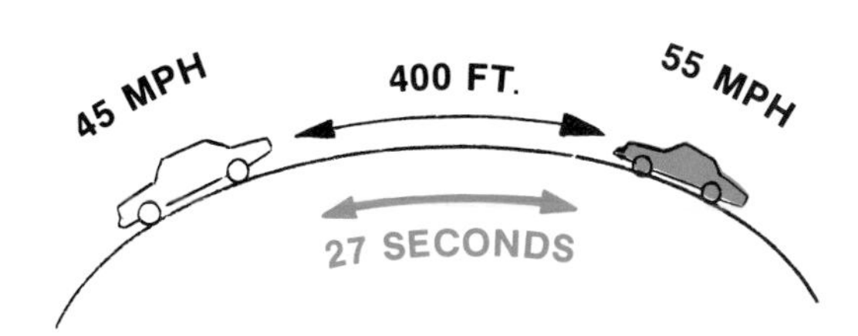

A. TWO CARS: DRIVER HAS 27 SECONDS CLOSURE TIME TO OTHER CAR

CLOSURE TIME
7 SECONDS

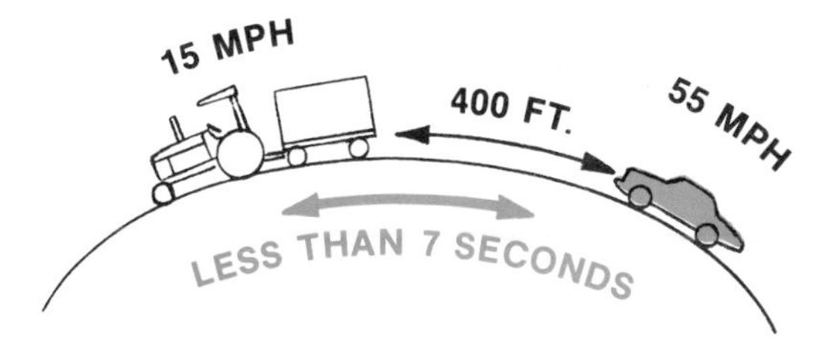

B. TRACTOR AND CAR: DRIVER HAS LESS THAN 7 SECONDS CLOSURE TIME TO TRACTOR

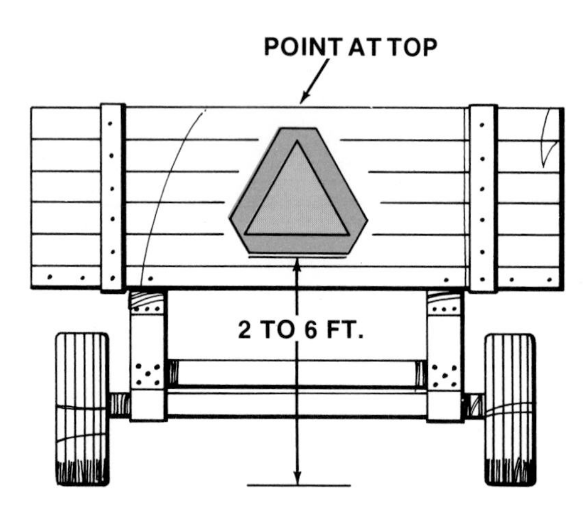

Fig. 49—Use SMV Emblems On Your Farm Machinery To Help Avoid Accidents On Public Roads

Fig. 50—Keep SMV Emblems Clean And Visible

AVOIDING SLOW MOVING VEHICLE ACCIDENTS

As an operator of slow moving farm machinery on public roads, you have a major responsibility for helping motorists avoid hitting you from the rear. It's for your safety and for theirs.

1. *Use a Slow Moving Vehicle (SMV) Emblem.* Identify your farm machinery as slow moving vehicles whenever you travel on a public road, even for short distances. Every vehicle intended to travel 25 mph or less is considered a slow moving vehicle and should be identified with an SMV Emblem (Fig. 49) visible from the rear. The triangular SMV Emblem is the universal symbol to tell everyone the vehicle travels 25 mph or less. It should remind that the closing speed between a car and an SMV is much faster than between cars.

2. *Keep the emblem surface clean and in good repair* for both day and night identification (Fig. 50). When the reflective red border or fluorescent orange center lose their brilliance, replace the emblem. If motorists can't see them clearly, they'll not give you any protection. Mount the emblem securely, and always with a point upward. If a point is downward, the emblem does not look like a universal symbol as intended. It should always give every motorist the same message.

3. *Keep lights and reflectors in good working order for farm machines traveling on public roads.* Operate flashing lights both day and night so you can be recognized as an SMV.

4. *Anticipate problems motorists may have when your machinery is on the road.* Drive with others in mind. Use care, courtesy, and common sense. That may mean pull onto the shoulder and stop so motorists won't be in a tight spot. Do whatever you can to avoid slow moving vehicle collisions even if it's the motorist who is in error. If there's an accident, it really doesn't matter who was "right" if someone is injured or killed. (See Chapter 5 for more details on avoiding transport accidents.)

5. *When you're the motorist, be alert for slow moving vehicles.* Remember that closing speeds are relatively fast. Act quickly.

SECOND PARTY HAZARDS

The rear wheel of a tractor backed over a young farm boy because the father didn't know his son was behind the tractor. A farm operator had his hand tangled in an unshielded V-belt drive when his helper started the combine. A child passenger on a tractor was bounced right out the door of a tractor and run over by a rotary cutter.

These three farm accidents killed a second party. We call them "second party" accidents. They can be divided into two categories: (1) Necessary second party accidents involving other persons who are needed to help with machinery operation, and (2) Unnecessary second party accidents involving other people who are not needed for machinery operation. Both types of accidents can be avoided.

ACCIDENTS TO NECESSARY SECOND PARTIES

Many farm jobs need two men. You may need someone to hold a part in place while you adjust it. Sometimes you need help to get something done sooner. And some jobs are safer if you have help.

Any time there's more than one person involved, there's a chance that what one person does may cause injury to the other. Second party accidents can be avoided if each person knows exactly where the other person is and what he is going to do. For example, when hitching, a helper stands aside while the driver backs up and aligns the hitch. The driver puts the tractor in park or forward gear. Only then does the assistant step in to make the hitch. Final adjustments are made by moving the tractor forward, never backward toward the helper. Clear communication is essential. There are different ways of communicating, and certainly the most effective and convenient method is by voice. When there's too much noise for you to be heard, however, use standard hand signals. (See Chapter 1 for safety hand signals). Whatever communications you use, make sure the other person understands you. Here's one good overall rule to follow when two or more persons are working around machines:

Before starting or moving a machine, tell people to stand where you can see them. Some operators make a habit of sounding the horn before starting the engine.

To avoid injuries to yourself when you are the second party, try to anticipate action and errors of the other guy. Be defensive like a defensive driver. Expect the unexpected. When you're helping hitch an implement to a tractor the driver's foot might slip off the clutch or brake. If you think of that possibility, you'll stand aside while he backs up to the implement. Then when the tractor is in park you will step in to insert the hitch pin (Fig. 51).

If you want to talk to a tractor driver operating a rotary cutter in a stubble field, recognize that the cutter could pick up a rock and throw it at you. Move to a safe place, signal the operator to stop, and then approach the machine.

Accidents to necessary second parties around farm machinery could practically be eliminated if these precautions were followed:

1. Know what other people plan to do.
2. Anticipate the actions and errors of other people.

ACCIDENTS TO UNNECESSARY SECOND PARTIES

The unnecessary second parties in farm machinery accidents shouldn't be there. Usually they just want to watch, or want to be with the person operating the machine. Although some are adults, the highest percentage are children under 15 years of age. Many of them are just riding along for the fun of it when they are injured (Fig. 52). They can interfere with the operator and cause injury to him too. The second party doesn't anticipate what could happen to him because he doesn't understand what is going on. The results are often tragic.

Keep people away from machines. Don't give them rides! Even in tractors with cabs, riders can be bounced around by unexpected bumps, accidentally unlatch the door and fall out.

Look for the unexpected. Check all around your machine before you move it in any direction (Fig. 53).

Fig. 51—Anticipate Actions And Possible Errors By Other Persons

Fig. 52—These Children Are Riding In A Dangerous Place. Do Not Allow It.

CHAPTER QUIZ

1. What are five errors that may result in farm machinery accidents?

2. How can most common machine hazards be avoided?

3. (Fill in the blank.) You must__________common machine hazards in order to avoid them.

4. Which of the following have pinch-point hazards?

a) Belt and pulley drives.

b) Chain and sprocket drives.

c) Gear drives.

d) Feed rolls.

e) Auger in tube.

f) All of the above.

Fig. 53—Make Sure Everyone Is Clear Before Starting Any Machine

5. True or false? "Smooth, rotating shafts are harmless."

6. Why can't a person whose clothing is caught tear away from wrap points?

7. Give two examples of farm machinery devices which have shear points but are not actually designed to cut or shear.

8. How can you prevent second-party crush accidents?

9. True or false? "A person who attempts to pull a 2-foot corn stalk from stalk rolls that rotate 12 feet per second can avoid injury if he releases the stalk as it begins to be pulled in."

10. Give four examples of freewheeling farm machinery components.

11. In general, what machines have the potential to throw objects?

12. What are four parts that store energy?

13. (Fill in the blanks.) You should __________ the pressure of a hydraulic system before attempting to service it.

14. True or false? "The hydraulic system of an implement disconnected from the tractor does not contain high-pressure hydraulic oil."

15. (Fill in the blank.) Hydraulic (gas-loaded) accumulators ____________ tremendous energy under pressure.

16. Where should you stand when inflating a tire?

17. Name two hazards of electrical systems on farm machinery.

18. How can slips and falls be prevented?

19. (Fill in the blank.) Use of an ________ emblem will identify your farm machinery on public roads and help prevent an accident.

20. How can you prevent necessary-second-party hazards best?

21. What two precautions will prevent unnecessary-second-party accidents best?

4
Target: Equipment Service and Maintenance

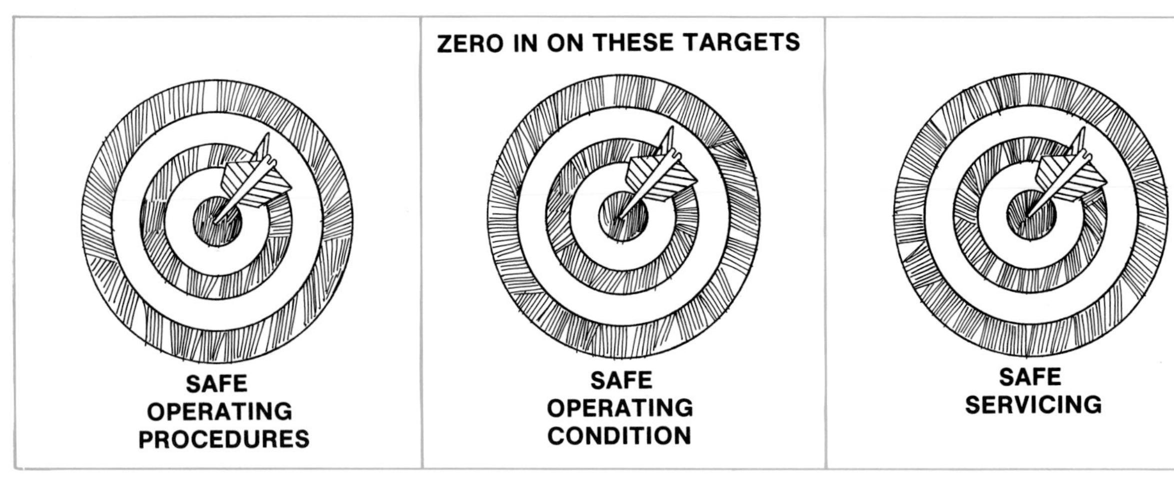

Fig. 1—Hit Each Target For Farm Equipment Safety

INTRODUCTION

The safe farm equipment operator zeros in on three targets (Fig. 1):

- **He follows recommended operating procedures**
- **He keeps equipment in safe operating condition**
- **He recognizes and avoids the hazards during service and maintenance work**

Hit each safety bull's-eye, and *all* your work with farm equipment can be efficient and accident free (Fig. 1).

This chapter aims at keeping equipment in safe operating condition and performing service work safely.

NOTE: You face two hazards when servicing equipment: personal injury and damage to equipment. To prevent personal injury, be aware of the common machine hazards outlined in Chapter 3, and follow this chapter's recommendations. To prevent accidental damage to equipment, you need the information contained in the operator's and service manual provided for the machine. Follow the procedures outlined in these manuals for all service and maintenance work you perform (Fig. 2).

Fig. 2—Follow Recommended Service Procedures To Protect You And Your Equipment

KEEPING EQUIPMENT IN SAFE OPERATING CONDITION

Keep all equipment in top-notch condition. See your dealer for repairs. Accidents are most likely to happen when:

- **Machines are out of adjustment**
- **Worn or broken parts are not replaced**
- **Cutting edges are dull**
- **Shields and other safety devices are not in place or working properly**
- **Safety procedures aren't followed during maintenance**
- **Safety warning signs can not be read**
- **Borrowed or used machinery do not have the proper safety signs in place.**

How can you prevent accidents? Here are three recommendations:

1. Check each machine before you use it.
2. Follow the maintenance and service schedules the manufacturer provides for each machine.
3. Be alert to changes in machine performance and operation.

Here is a closer look.

OPERATION CHECKS BEFORE USE

Make it a habit to check all components and systems that affect the safety and performance of machine operation *before* and *during* each use. These checks are explained in *detail* in other chapters.

Pay special attention to these:

- **Steering**
- **Brakes**
- **Hydraulics**
- **Fuel system**
- **Exhaust system**
- **Warning lights**
- **Tires**
- **Controls**
- **Shields and guards**

When you discover problems, take *immediate* action to make the necessary adjustments or repairs. Follow the procedures outlined in your operator's or service manual, or take the equipment to your dealer for service.

SERVICE AND MAINTENANCE SCHEDULES

Follow the service and maintenance schedule the manufacturer provides for each of your machines (Fig. 3). These schedules, along with the instructions found in your operator's manual, tell you when and how:

- **To lubricate and adjust moving parts to prolong machine life**
- **To replace parts that deteriorate with use and age**
- **To maintain clearances to compensate for wear**
- **To sharpen cutting edges to maintain efficient operation**

Fig. 3—Following The Manufacturer's Service Schedule Helps Keep Each Machine In Safe Operating Condition

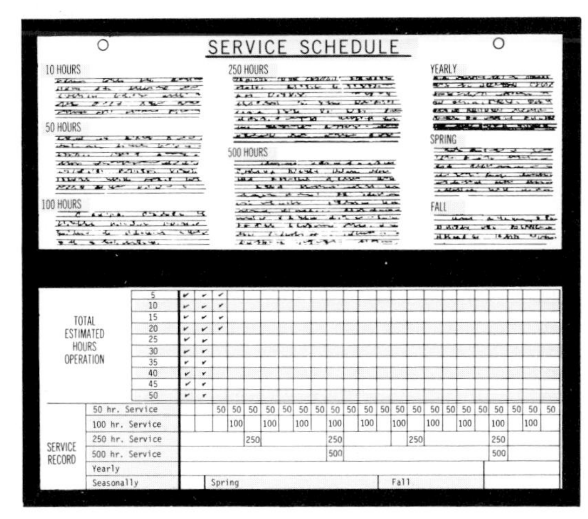

SERVICE SCHEDULE

10 HOURS

50 HOURS

100 HOURS

250 HOURS

500 HOURS

YEARLY

SPRING

FALL

TOTAL ESTIMATED HOURS OPERATION	5	✓	✓	✓																				
	10	✓	✓	✓																				
	15	✓	✓	✓																				
	20	✓	✓	✓																				
	25	✓	✓																					
	30	✓	✓																					
	35	✓	✓																					
	40	✓	✓																					
	45	✓	✓																					
	50	✓	✓																					
SERVICE RECORD	50 hr. Service			50	50	50	50	50	50	50	50	50	50	50	50	50	50	50	50	50	50	50	50	50
	100 hr. Service				100		100		100		100		100		100		100		100		100		100	
	250 hr. Service					250					250					250					250			
	500 hr. Service										500										500			
	Yearly																							
	Seasonally		Spring											Fall										

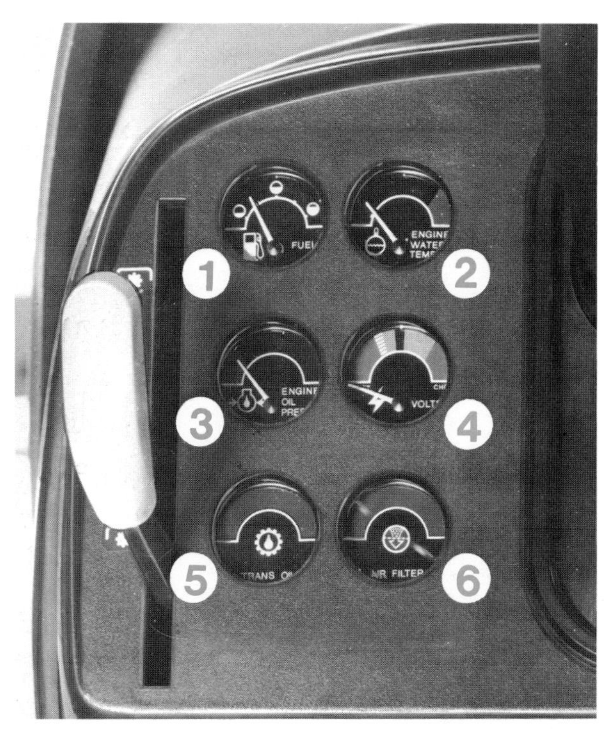

Fig. 4—Watch Your Instrument Panel For Trouble Signs
1. Fuel Gauge
2. Coolant Temperature Gauge
3. Engine Oil Pressure Gauge
4. Electrical Voltage Meter
5. Transmission Oil Temperature Light
6. Air Cleaner Restriction Light

- **To tighten bolts and nuts to protect the safety of the machine and the operator.**

Most operators agree that the small expense it takes in time and money to follow maintenance recommendations pays big dividends by maintaining machine performance and efficiency, prolonging machine life, and keeping the machine's safety devices working properly.

In addition to scheduled maintenance, the machine must be kept clean to operate safely. Pay special attention to steps, handholds, and the working or walking surfaces.

CHANGES IN TRACTOR AND MACHINE PERFORMANCE

A good way to spot problems early is to be alert to changes in the operating characteristics and performance of your equipment. To detect these changes, use your senses of hearing, sight, touch, and smell. Be alert to these warning signals:

- **Unusual noises**
- **Increased vibration**

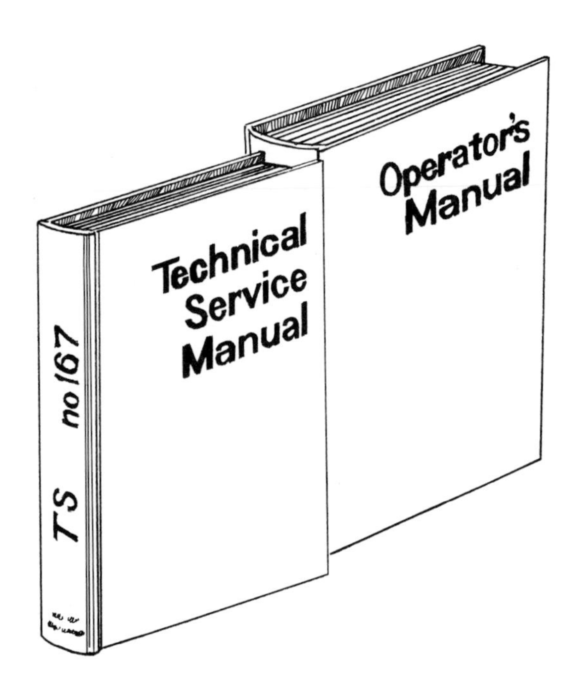

Fig. 5—Technical Information Is Needed To Perform Service Work Properly And Safely

- **Indications that moving parts (bearings, belts, etc.) are too hot**
- **Lack of response to controls**
- **Increased power requirements for engine-driven machines**
- **Changes in operating speed**
- **Changes in engine exhaust characteristics**

Fig. 6—Specialized Tools Are Required For Many Service Jobs

- **Operation of warning lights or horns**
- **Instrument panel gauge readings too high or too low (Fig. 4)**

Stop, investigate, and make necessary repairs or adjustments when you receive any of these warning signals. Many times, safety is involved. Also, you may be able to avoid extensive repairs by catching a minor problem early.

NOTE: Turn off the engine, take the key, shift to park or set the parking brake, and disengage the PTO before dismounting from a tractor, or a self-propelled machine. Do not unplug, clean, adjust, or lubricate any machine that is running.

WHO WILL DO THE JOB?

When service and repairs are needed, you have to decide who will do the job. You can do it yourself, or you can take the work to your farm equipment dealer. Ask yourself these questions to help you decide:

- **Do I have enough time?**
- **Do I know how?**
- **Do I have the necessary parts, tools and equipment?**
- **Can I do the job without getting hurt?**
- **Do I have a safe place to work?**

Let's review each of these factors:

TIME

Try to do your maintenance and repair work during slack periods in your work schedule, and during periods of time when the equipment is not being used. Time is available then to work carefully without the pressures of other jobs.

Keep in mind many service operations will take longer than you think. In many situations, you'll save time in the long run if you go ahead with other jobs and let your farm equipment dealer do your service work for you.

KNOW-HOW

Your operator's manual is full of maintenance, lubrication, and adjustment procedures. It tells you how to service equipment safely. *Always* follow those procedures.

What about jobs not described in your operator's manual? In most cases, the best answer is this: unless you have experience, specialized training, and the technical service manual for the machine, take the work to a qualified dealer (Fig. 5). Service manuals are essential for finding safety recommendations, clearances for close-fitting parts, disassembly and assembly information, and proper adjustment procedures. And, if you don't have previous experience servicing equipment, you may be faced with unexpected problems that you won't be able to solve. Also, you may be exposing yourself to safety hazards unknown to you.

TOOLS AND SERVICE EQUIPMENT

Your farm equipment dealer has a big investment in specialized tools and service equipment (Fig. 6). These are necessary to support equipment safely during repair, and for fitting close tolerances, adjustment, and testing. Without these tools, you may not be able to do the job properly and safely. Before you tackle any job, read through the service procedure to see if you have the necessary tools and equipment.

SAFETY

Time, availability of the proper tools, and "know-how" are essential for safe service work. There are other factors, too.

The safe equipment operator:

- **Maintains a hazard-free shop**
- **Uses service tools and equipment safely**
- **Guards himself against the hazards of service work**

Let's look at each of these factors.

MAINTAINING A HAZARD-FREE SHOP

A safe farm shop meets three important requirements (Fig. 7):

- **It is planned to be as free from hazards as possible.**
- **It is managed to keep it that way.**
- **It is equipped to handle emergency situations.**

SHOP PLANNING

An efficient and hazard-free shop results from careful planning. When a new shop is planned, potential hazards are recognized *in advance,* and are eliminated or minimized as much as possible as the plans for the shop are worked out.

Let's look at some important recommendations for planning a new shop. Even if you don't have the opportunity to plan a new shop, you can use these same recommendations to evaluate and modify, if necessary, the shop facilities you already have.

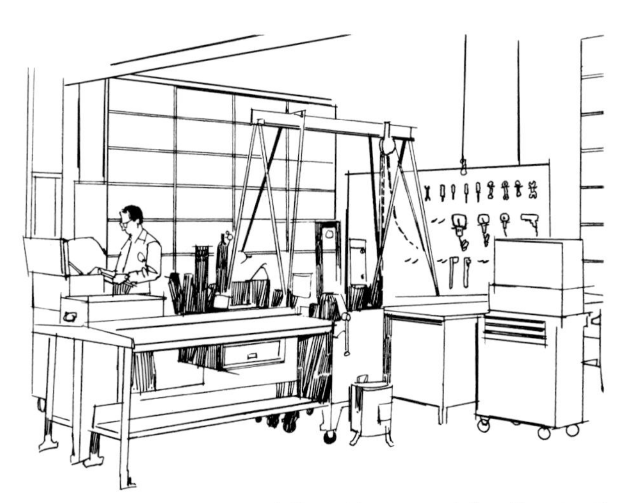

Fig. 7—Planning And Good Shop Management Are Necessary For Safety

LOCATION

Locate the shop in a convenient and accessible location for farm service work, and for the storage of spare parts, tools, and supplies (Fig. 8). Provide adequate drainage to keep the shop floor dry at all times. Give yourself enough space around the building and a service door large enough to maneuver equipment easily and safely. Provide a concrete apron outside for cleaning equipment, for welding, and to provide a solid foundation for hydraulic jacks and support stands. Build a fire proof wall between shop and storage.

Fig. 8—Locate Your Shop For Accessibility And Convenience

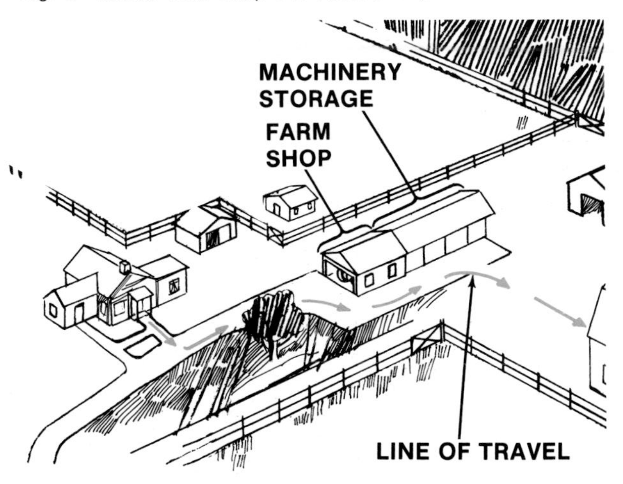

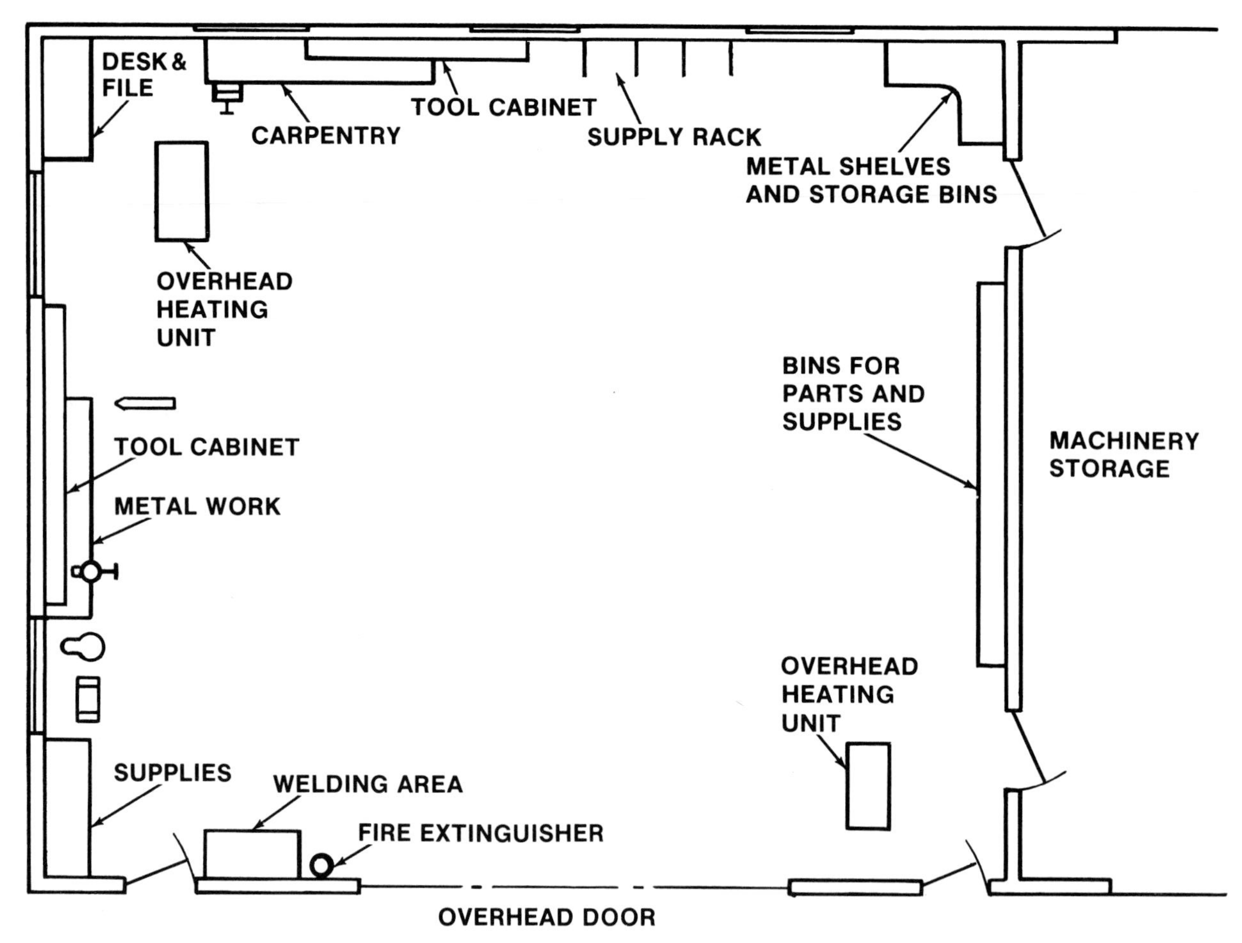

Fig. 9—Draw A Plan To Arrange Your Shop For Convenience And Safety

SIZE AND ARRANGEMENT

To determine size and arrangement make a list of the jobs and activities you'll be doing in the shop, and the sizes of the equipment you'll be working on inside the shop facility. Then sketch floor plans until you are satisfied with the arrangement and location of work and storage areas, benches, and stationary tools (Fig. 9). Keep these points in mind:

1. Provide adequate and uncluttered work space around machines you plan to service.

2. Locate benches, stationary tools, bins, racks, and tool panels in work areas according to use and function.

3. Keep combustible materials out of the welding area, away from heating units, and in proper containers.

4. Locate welding, metalwork, and other dirty work areas away from woodworking and other areas you want to keep clean.

LIGHTING

Provide a sufficient number of windows, skylights, and overhead lights to ensure good general lighting. Place additional lights over benches, stationary power tools, and main work areas (Fig. 10). Use portable lights when necessary to eliminate shadows while servicing equipment. Keep windows clean. Apply light-colored paint to the walls. This will utilize existing light more effectively than darker walls, reduce shadows, and make potential hazards more visible by having contrasts in color.

HEATING AND COOLING

You'll need some source of heat for cold weather work, and in warm climates air conditioning may be needed. Before selecting and installing heating and air conditioning units get the services of a heating and air conditioning specialist. His experience will help you get the most efficient installation. Here are key points to keep in mind when planning:

1. Put units where they will distribute heat and cold efficiently. *Do not* place units where they can be struck by moving equipment (Fig. 11), or near areas where combustible or flammable materials are used or stored.

2. Make sure the installation is safe:

- **Adequate ventilation provided for the units inside**
- **Vented safely outside**
- **Equipped with over-heating shut-off devices**
- **Installed with emergency shut-offs for fuel and electric power.**

VENTILATION

Put enough doors and windows in your shop to ventilate smoke, fumes, and vapors outdoors. Ideally, window area should equal 25 percent of the shop's floor area. Be sure to open windows and doors and work under ventilating hoods when running engines inside, when welding, and when handling chemicals with poisonous dusts or fumes. When working under these conditions, gases, fumes, and toxic substances can rapidly build up to dangerous concentrations.

Use flexible metal tubing to carry engine exhaust fumes outside. If you don't have vent hoods, use exhaust fans to clear smoke and vapors away from welding, cleaning, and painting areas to the outdoors. Hoods equipped with exhaust fans are best for removing fouled air from specific work areas (Fig. 12). Remember that exhaust fans are only efficient if fresh air is available from an open window, or from a ventilator installed to provide an incoming source of fresh air.

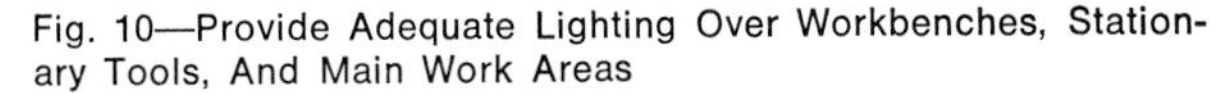

Fig. 10—Provide Adequate Lighting Over Workbenches, Stationary Tools, And Main Work Areas

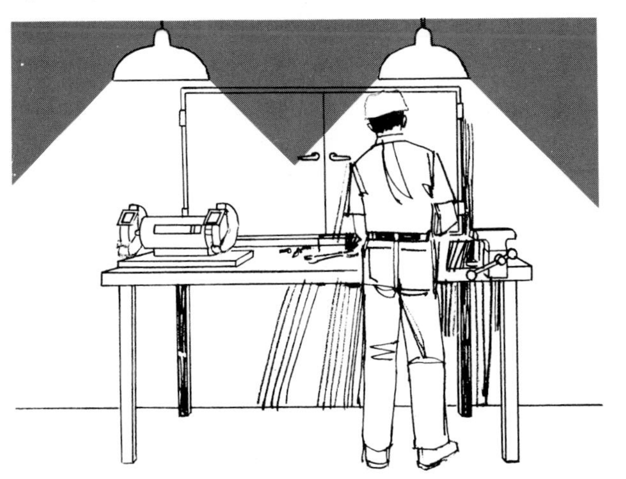

Fig. 12—Adequate Ventilation: Large Doors For Summer, An Exhaust Hood For Winter

WIRING

Get assistance from a qualified electrician when planning the wiring system. In general, the system should meet five requirements:

1. Have adequate capacity to handle lighting, heating, and power tool requirements (Fig. 13).

2. Have enough outlets so extension cords aren't needed.

Fig. 11—Ceiling Unit Heaters Are Good. They Leave Clear Working Space Below

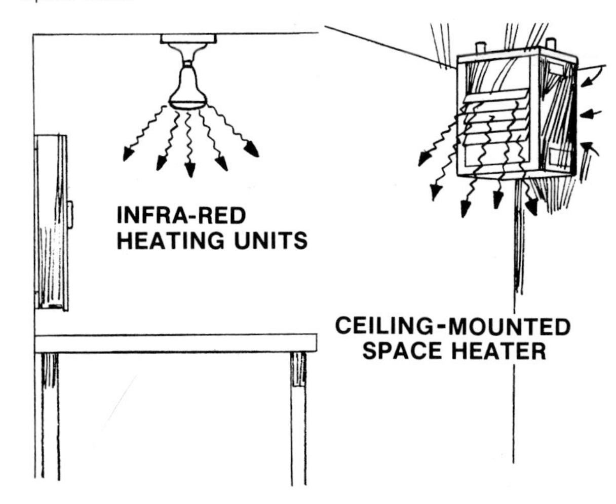

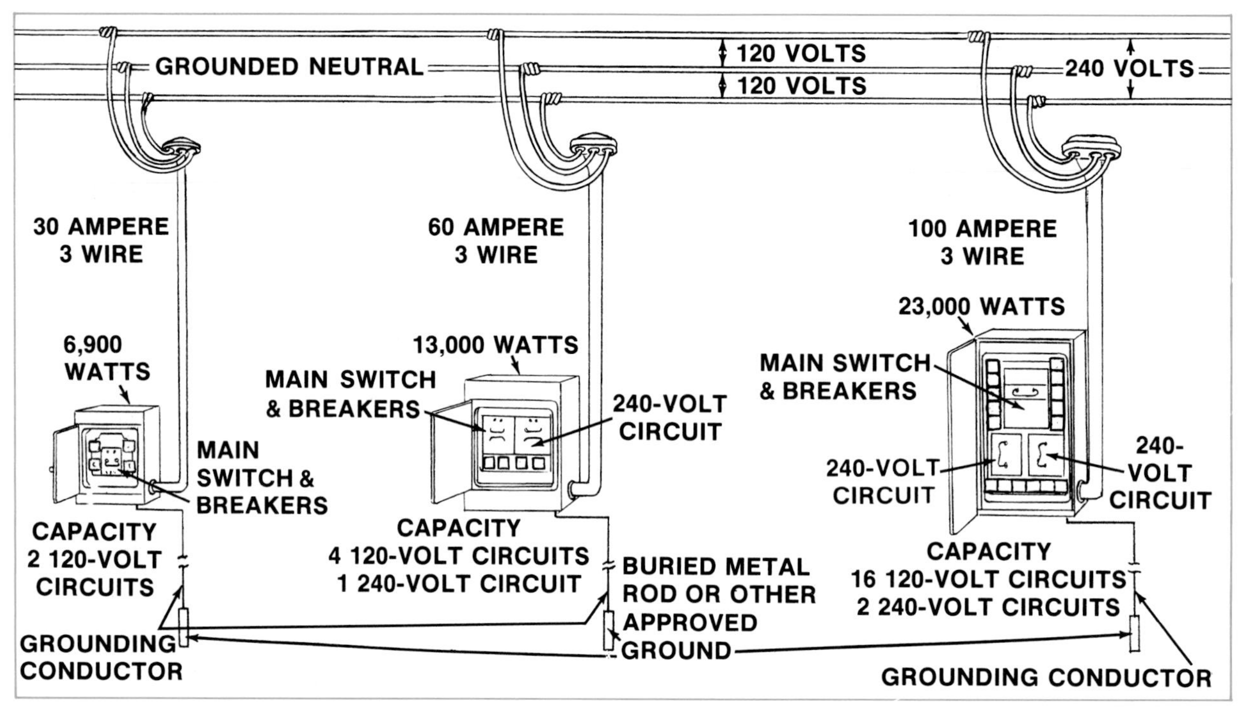

Fig. 13—Install Service Entrance Equipment With Adequate Capacity For Present And Future Needs

3. Have three-wire grounding-type, 120-volt circuits to prevent electric shock while you're using power tools.

4. Have 240-volt circuits for welders and motors over 1/2 horsepower.

5. Be able to expand for future needs.

6. Be sure all circuits are properly grounded and protected.

STORAGE

Develop a system of racks, bins, and tool boards so finding the right tool is easy and quick (Fig. 14). Don't store tools, supplies, or spare parts in aisleways or on the work floor where someone will trip over them. Keep grease, oil, paint, solvents, lumber, and other flammable materials away from heaters and welding areas to prevent fire. Store oil, grease, paint, and solvents in *closed containers* in metal cabinets, or on metal shelves. Stored in this manner they are protected from sparks and flame, and spills don't soak into shelves. Wooden shelves soaked with flammable liquids are fire hazards.

NOTE: Do not store gasoline or other fuels in the shop. If absolutely necessary to keep small quantities in the shop, use approved safety cans, and keep these cans in a ventilated area away from sparks and flame.

Properly store and dispose of hazardous wastes such as battery acid, cleaning solvents, and other harmful chemicals.

Use steel drums for storing trash. Use one drum for noncombustibles and another for materials that will burn. Also, provide a metal container with a self-closing cover for oily rags and oil-soaked filters that sparks, flame, spontaneous combustion, or a careless smoker could set on fire (Fig. 15). Empty trash containers frequently.

Fig. 14—Finding Parts Is Quick And Easy With Labeled Storage Bins

SHOP MANAGEMENT

After a shop has been planned to be as hazard-free as possible, *it must be managed* to keep it that way. Consider these key management procedures:

Make sure someone knows you are working in the shop and will check on you and render aid if you are injured.

1. Keep all tools and service equipment in good condition.

2. Use personal protective equipment, goggles, face shields, gloves.

3. Keep floors and benches clean to reduce fire and tripping hazards.

4. Clean up as you go while doing a job, and clean the area completely after the job is done.

5. Empty trash containers regularly.

6. Keep lighting, wiring, heating, and ventilation systems in good shape.

7. Lock your shop to prevent accidents. A shop is an attraction to a child.

8. Don't let anyone use tools or service equipment unless they've had adequate instruction.

9. Keep guards and other safety devices in place and functioning.

10. Use tools and service equipment for the jobs they were designed to do.

11. Supervise children carefully when they are in the shop.

12. Keep fire extinguishers serviced, and the first aid kit replenished with supplies.

EMERGENCY SITUATIONS

Every farm shop should be equipped to handle emergency situations. The most common types of emergencies are fires and personal injury.

FIRE

Your best protection against fire is prevention. But if a fire starts, you should be ready to take immediate action. Try to judge the situation quickly without panic. If you think you can extinguish the fire easily, do it; if not, immediately call the nearest fire department (always keep its number by the phone).

There are three types of fire: Class A, B, and C (Fig. 16):

Fig. 15—Store Oily Refuse In A Metal Container With a Self-Closing Lid

- **Class A — Combustibles like paper and wood**
- **Class B — Gasoline, diesel fuel, grease and solvents**
- **Class C — Electrical equipment fires**

The equipment you select must be safe and effective for fighting *each* of these fires. Here are the *minimum* recommendations for fire extinguishers for a farm shop.

Fig. 16—Be Prepared To Fight These Types of Fire

CLASS A FIRES	**PAPER, WOOD, CLOTH, EXCELSIOR, RUBBISH, ETC. WHERE QUENCHING AND COOLING EFFECT OF WATER IS NEEDED.**
CLASS B FIRES	**BURNING LIQUIDS (GASOLINE, OILS, PAINTS, ETC.) WHERE SMOTHERING EFFECT IS REQUIRED.**
CLASS C FIRES	**FIRES IN LIVE ELECTRICAL EQUIPMENT (MOTORS, SWITCHES, HEATERS, ETC.) WHERE A NON-CONDUCTING EXTINGUISHING AGENT IS REQUIRED.**

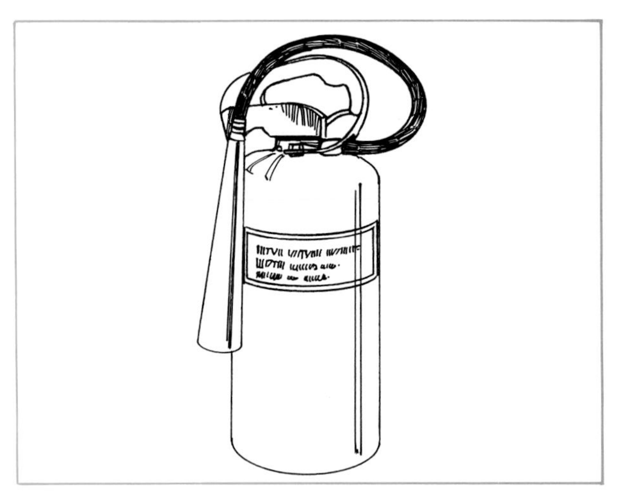

Fig. 17—Equip Your Shop With At Least One All Purpose ABC Dry Chemical Extinguisher

For Class A fires (ordinary combustibles), provide at least one of the following methods for fighting fire:

1. One or more 20-pound multipurpose dry chemical pressurized fire extinguishers specifically approved for fighting Class A fires (Fig. 17). (Not all types of dry chemical extinguishers are approved for Class A fires.)

2. A garden hose attached to one of the water faucets in the shop, kept ready to use, and long enough to reach all areas of the shop (Fig. 18).

3. Several pails submerged in a barrel of water (Fig. 18). These pails can be pulled quickly from the barrel, full of water, and ready to use in an emergency. Add antifreeze if the water could freeze in winter months, but *not* a flammable alcohol antifreeze. Calcium chloride, available from your farm equipment dealer, will not burn and is economical to use.

NOTE: Don't use water on Class B (burning liquids) or C (electrical equipment) fires. Water spreads Class B fires, and water may conduct electricity to give you a severe shock if used on Class C fires.

For Class B fires (burning liquids) and Class C fires (those involving electrical equipment), provide at least one pressurized dry chemical fire extinguisher of 20-pound capacity (Fig. 17), or follow the recommendations of your insurance company and state extension office. Keep extinguishers in a convenient place, and close to, but not in, the fire hazard area. Make sure they are protected from damage and always within easy reach (Fig. 19). Some, but not all, dry chemical extinguishers are effective against Class A fires. When buying extinguishers, select those rated for *all* classes of fires—A, B, and C.

PERSONAL INJURIES

Be prepared to take care of injuries. Advance preparations are essential for coping with serious emergencies and preventing small injuries from becoming serious medical problems. Follow these recommendations:

1. *Learn the basic rules of first aid.* At least one member of your family or work force should take a basic course in first aid, such as those given by the American Red Cross and other agencies. Emergency response time to farms is 4 to 6 times longer than urban places. Farm families should be able to sustain an injured person's life until professional help arrives. Keep a basic first aid book in a handy location known to all family members (Fig. 20).

2. *Know who to call for help.* Keep emergency numbers for doctors, ambulance service, hospital, and fire department near your telephone.

3. *Apply immediate first aid to all injuries.* If an injury appears severe, don't move the victim. Call a doctor and follow his instructions. If the injury is not severe, administer first aid, and take the injured person to a doctor for further treatment.

Fig. 18—Have Water Available To Fight Class A Fires If Type "A" Commercial Extinguishers Are Not Available

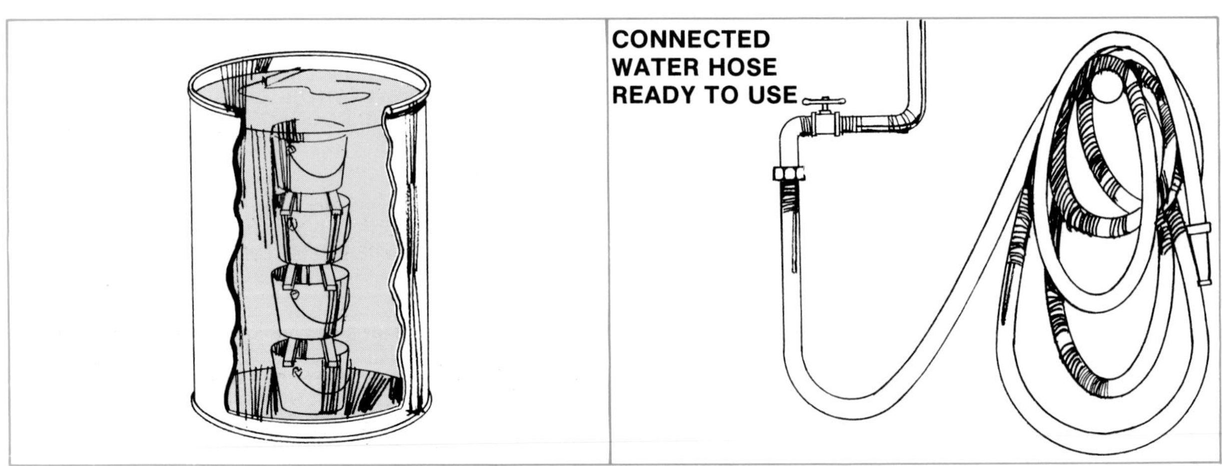

4. *Keep a simple first aid kit in your shop (Fig. 20).* This kit should be backed up with more extensive first aid supplies in your home or office. You may purchase a kit or assemble your own. To assemble a first aid kit:

a) Find a small metal or plastic box that will seal out dust and moisture.

b) Wash the box and rinse it with boiling water.

c) Include these items (and others you may wish to add):

- **Sterile gauze pads, individually packaged, 3 and 4 inches square**
- **Rolls of sterilized gauze, 2 and 3 inches wide**
- **Band-aids in assorted sizes**
- **Sterile absorbent cotton**
- **Roll of adhesive tape**
- **Large triangular bandages made from 40-inch square cotton sheeting, cut diagonally**
- **Scissors with rounded tips**
- **Tweezers**
- **Safety pins**
- **An antiseptic**
- **Cold pack**

NOTE: Antiseptics are not necessary if the victim will soon receive medical care. However, they should be available for use when needed. Ointments should only be used on minor burns. Cold is best!

USING SERVICE TOOLS AND EQUIPMENT SAFELY

Don't take the use of hand tools and service equipment for granted. You're more likely to be injured when servicing equipment than when you're operating it. Small hand tools can inflict great injury.

Consider the safe use of these service tools:

- **Hand tools**
- **Power tools**
- **Welding equipment**
- **Hoists and jacks**
- **Cleaning equipment**

HAND TOOLS

You can avoid hand tool injuries if you follow four basic rules:

1. Select the right tool for the job.

Fig. 19—Always Keep Extinguishers Where They Are Easy To Reach

2. Use it in the right way.
3. Keep it in good condition.
4. Store it safely when it's not in use.

To follow all of these rules, you need more information than can be given here. Study several good publications, available from book stores or libraries, that describe the proper use and care of hand tools, and follow the recommendations given for their use. Let's look at some of the important principles relating to personal safety for the hand tools most frequently used for service work.

Fig. 20—Keep A First Aid Book And A First Aid Kit Available For Immediate Use

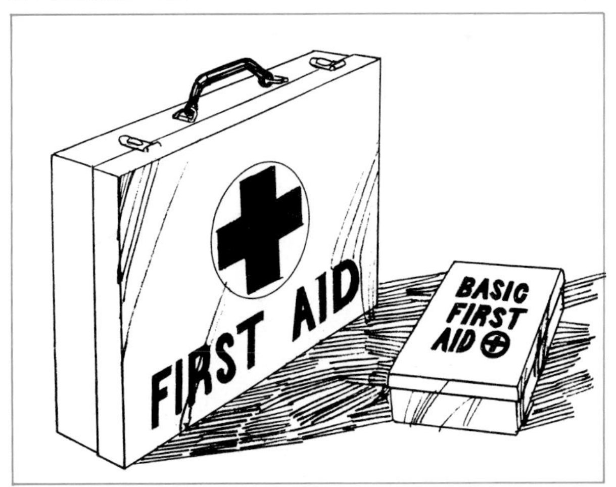

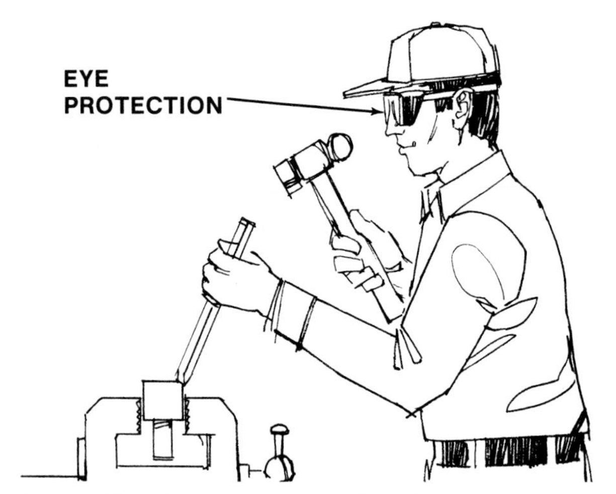

Fig. 21—Wear Eye Protection And Hold Chisels And Punches Near The Head Of The Tool

Fig. 22—Grind Mushroom Heads From Chisels And Punches

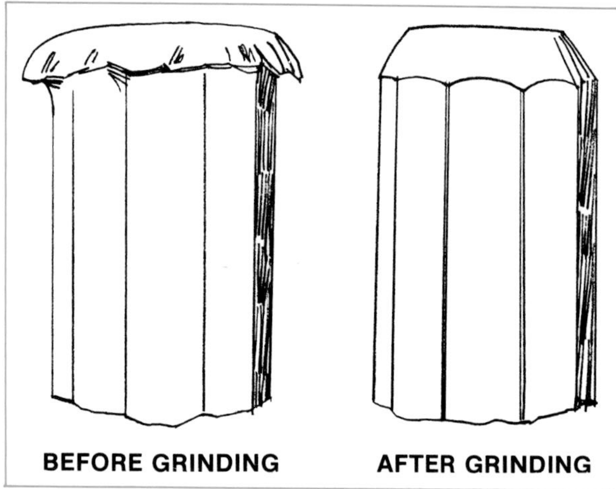

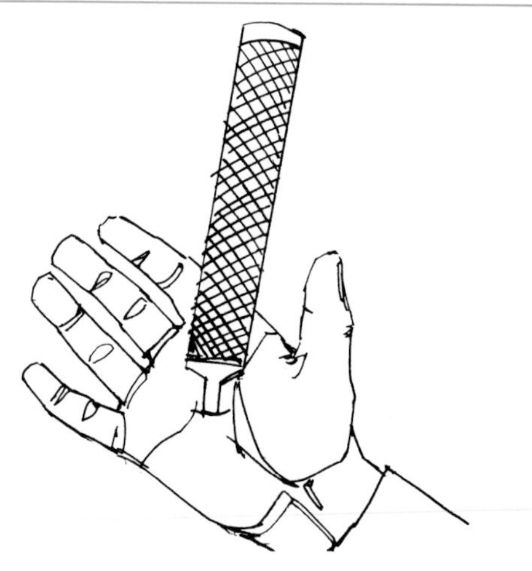

Fig. 23—Files Without Handles Can Pierce Your Hand Or Wrist

Chisels And Punches

1. *Wear eye protection (Fig. 21).* The hardened face of the hammer or the end of the tool may chip or shatter and send metal fragments flying through the air.

2. *Grind off mushroom heads (Fig. 22).* The sharp edges can tear your skin if the tool slips. And, when the tool is struck, chips could break off the mushroomed head and fly into your eyes. Keep a smooth bevel ground on the heads of all punches and chisels.

3. *Don't use chisels and punches for prying.* They are hard and brittle, and excessive force could break them with a snap.

4. *Hold the tool steadily but loosely.* The best place to hold it is just below the head. If you miss and strike your hand, your hand will not be caught between the hammer and the work piece.

5. *Select the proper sized tool for the job.* Heavy pounding on tools too small for the job increases the risk of injury from tool breakage. Tools too large for the job may not be safe either. The full cutting edge of a chisel, for example, should be used. Using only a section of the cutting edge of a larger chisel could result in breaking off the corners from the cutting edge.

Files

Keep a handle on every file (Fig. 23). This will keep the tang from piercing your palm or wrist if the file should slip or catch.

Knives

1. *Keep blades sharp.* The greater the force you have to apply, the less control you have over the cutting action of the knife. The safest knife usually has the sharpest edge. Keep all of your knives uniformly sharp.

2. *Cut away from the body.* Your hands and fingers should always be behind the cutting edge. Keep knife handles clean and dry to keep your hand from slipping onto the blade.

3. *Never pry with a knife.* Blades are hardened and can break with a snap.

4. *Store knives safely.* Keep knives in their own box or scabbard when not in use. An unguarded blade could cut you severely.

Screwdrivers

1. *Use screwdrivers only for driving screws.* Using them for punches or prybars breaks handles, bends shanks, and dulls and twists the tips. This makes them unfit to tighten or loosen screws safely.

2. *Sharpen screwdrivers properly.* File or grind worn or damaged tips to fit the slot of the screw (Fig. 24). A sharp, square-edged tip won't slip as easily as a dull one, and less pressure will be required to hold the tip in the slot. Keep an assortment of screwdrivers on hand to fit different sizes of screws.

3. *Don't hold parts in your hand (Fig. 25).* Put the work on a bench or in a vise to avoid the possibility of piercing your hand with the screwdriver tip.

4. *Use screwdrivers with insulated handles for electrical work.* If the blade or a rivet extends through the handle, an electrical circuit could give you a serious shock.

Hammers

1. *Wear eye protection.* Always wear goggles when striking hardened tools and hardened metal surfaces. This will protect your eyes from flying chips. Whenever possible, use soft-faced hammers (plastic, wood, or rawhide) when striking hardened surfaces.

2. *Check the fit and condition of the handle (Fig. 26).* Keep handles tightly wedged in hammerheads to prevent injury to yourself and others nearby. Replace cracked or splintered handles. And don't use the handle for prying or bumping. Handles are easily damaged and broken this way.

3. *Select the right size for the job.* A light hammer bounces off the work. One that's too heavy is hard to control.

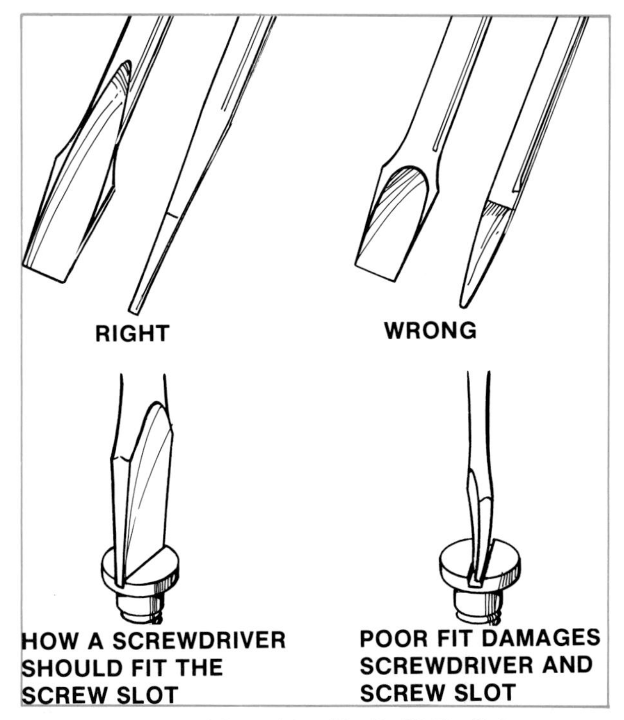

Fig. 24—File Or Grind Screwdriver Tips To Fit The Slot Of The Screw

Fig. 25—Working On A Bench Will Keep The Screwdriver From Piercing Your Hand

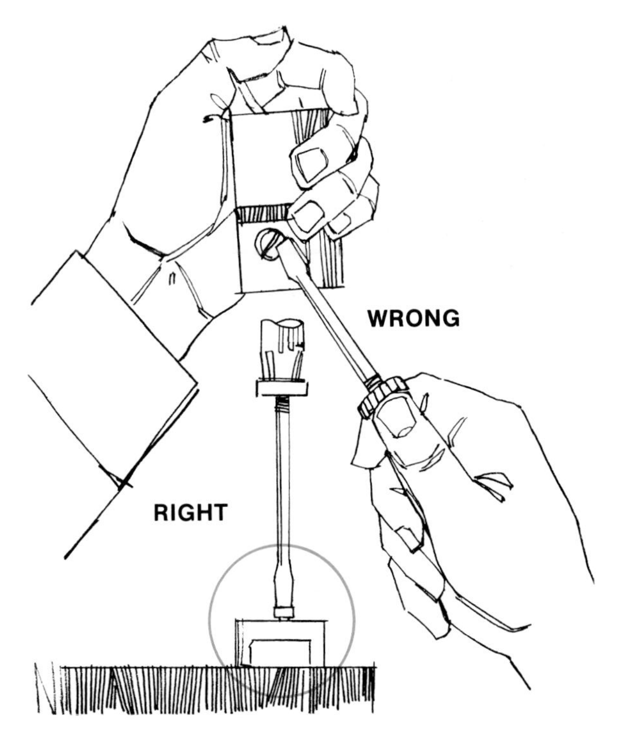

Fig. 26—Keep Handles Tighly Wedged To Prevent Injury To Yourself Or Others Nearby

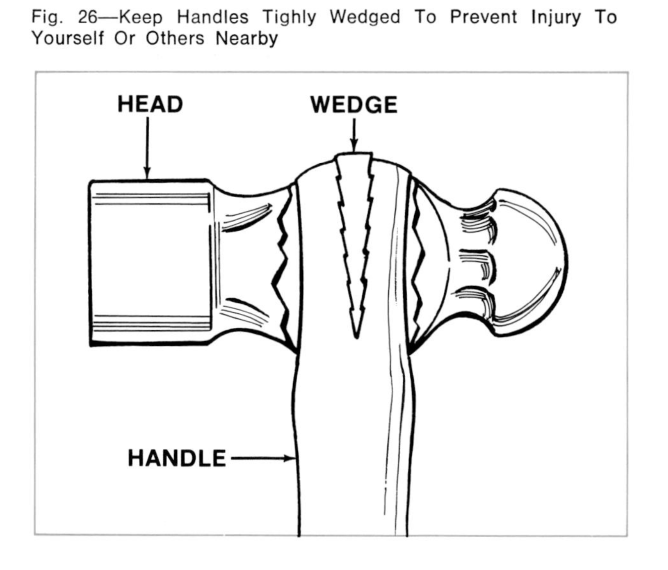

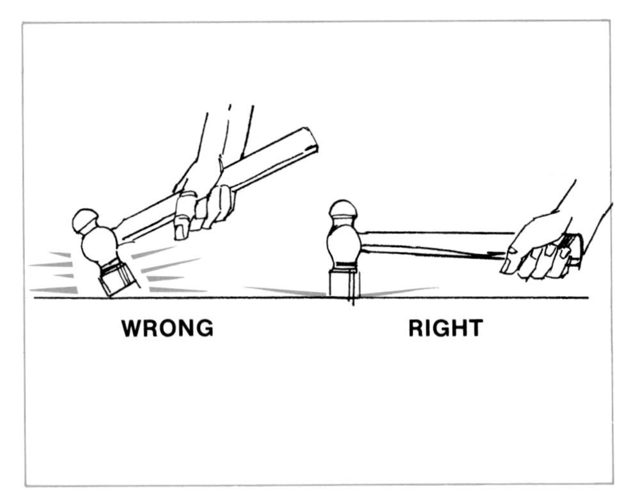

Fig. 27—Grip The Hammer Near The End And Strike Squarely With The Surface

4. *Grip the handle close to the end (Fig. 27).* This increases leverage for harder, less tiresome blows. It also reduces the possibility of crushing your fingers between the handle and the projecting parts and edges of the work piece if you should miss.

5. *Prevent injuries to others.* Swing in a direction that won't let your hammer strike someone if it slips from your hand. Keep the handle dry and free of grease and oil.

6. *Keep the hammer face parallel with your work (Fig. 27).* Force is then distributed over the entire hammer face, reducing the tendency of the edges of the hammerhead to chip, or slip off the object being struck.

Wrenches

1. *Use wrenches that fit (Fig. 28).* Wrenches that slip damage bolt heads and nuts, skin knuckles, and lead to falls. Don't try to make wrenches fit by using shims.

Fig. 28—Use Wrenches That Fit

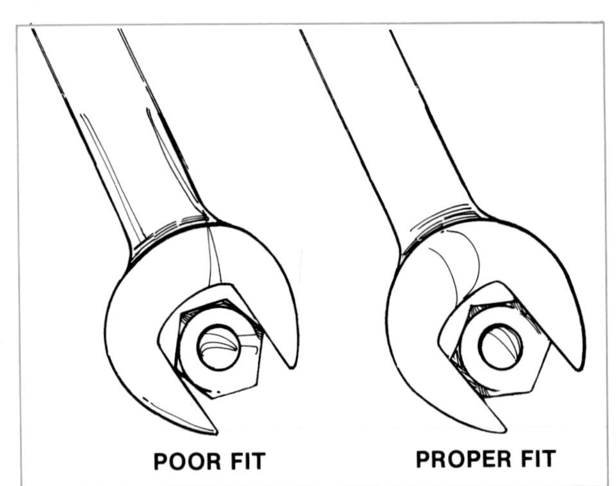

Do not use metric wrenches on English sized bolts and vice versa.

2. *Don't extend the length of a wrench.* Do not use pipe to increase the leverage of the wrench. The handle was made long enough for the maximum safe force to be applied. Excessive force may break the wrench or bolt unexpectedly, or the wrench may slip rounding off the corners of the bolthead or nut. Skinned knuckles, a fall, a broken wrench, may result. Don't hammer on wrenches unless they're designed for this type of use.

3. *Pull on the wrench (Fig. 29).* This isn't always possible, but if you push, you take this risk: if the wrench slips, or if the nut suddenly breaks loose, you may skin your knuckles or cut yourself on a sharp edge. Use the open palm of your hand to push on a wrench when you can't pull it toward you.

4. *Replace damaged wrenches.* Straightening a bent wrench weakens it. Cracked and worn wrenches are too dangerous to use, as they could break or slip at any time.

5. *Keep the open jaws of adjustable wrenches facing you.* Have the open jaws toward you when placing adjustable wrenches on bolt heads and nuts. Then pull on the wrench. This forces the movable jaw onto the nut, reduces its tendency to slip, and places most of the pressure on the solid, stronger jaw (Fig. 30). Adjust these wrenches to fit bolt heads and nuts to a snug fit.

6. *Use pipe wrenches only for pipe or round stock.* If you tighten bolt heads and nuts, their sharp corners may slip and break the hardened teeth in the jaws. Then, the wrench may slip when used on pipe making the wrench unsafe to use.

Pliers And Cutters

1. *Do not use pliers as a wrench.* They do not hold the work securely and can damage bolt heads and nuts.

2. *Guard against eye injuries when cutting with pliers or cutters.* Short and long ends of wire often fly or whip through the air when cut. Wear eye protection, or cup your hand over the pliers to guard your eyes.

3. *Wear eye protection when cutting with bolt cutters.* Chips of metal may break away from the cutting edges and be flung into your eyes. To help prevent chipping or breaking the cutting edges, observe these precautions:

- **Select a cutter big enough for the job.**
- **Keep the blades at right angles to the stock.**
- **Don't rock the cutter to get a faster cut.**
- **Adjust the cutters to maintain a small clearance between the blades.** This prevents the hardened blades from striking each other when the handles are closed.

POWER TOOLS

Reducing the physical effort required and speed are the main reasons that workers use power tools. In your attempt to get a job done *quickly,* however, take the necessary precautions to get it done *safely:*

1. Read the operator's manual and observe all precautions.

2. Protect yourself from electric shock (refer to next section).

3. Keep guards and shields in place.

4. Keep the work area clean.

5. Give your job full attention.

6. Let each tool work at its own speed without forcing it.

7. Wear snug-fitting clothes to prevent entanglement of clothing.

8. Wear eye protection when recommended.

9. Maintain secure footing and balance at all times.

10. Keep your tools clean and sharp.

11. Be sure the switch is *off* before plugging in the power cord.

12. Keep your fingers away from the switch when carrying a portable tool.

13. Turn the switch off immediately if the tool stalls or jams.

14. Use portable tools only in areas completely free of flammable vapors and liquids. Sparks could cause a fire or explosion.

15. Before making adjustments or changing bits or cutters, disconnect the power cord or you could accidently touch the switch and be injured when the tool starts.

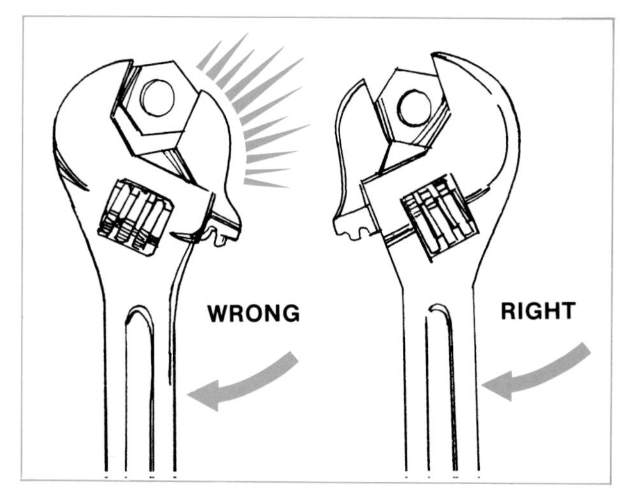

Fig. 30—Apply Most Of The Force To The Solid Jaw By Keeping The Open Jaw Toward You

16. Repair or replace damaged extension cords and plugs.

17. Use clamps or a vise to hold your work.

18. Use power tools only for their intended functions.

19. Remove adjusting keys and wrenches before operating a tool.

20. Use three-wire cords if a grounding conductor is needed for shock protection.

21. Store idle tools safely to prevent damage to the tool and cord, and to prevent unauthorized use.

22. Provide enough light so you can see what you're doing.

Fig. 29—Pull On The Wrench, Or Push With The Open Palm Of Your Hand To Avoid Injury

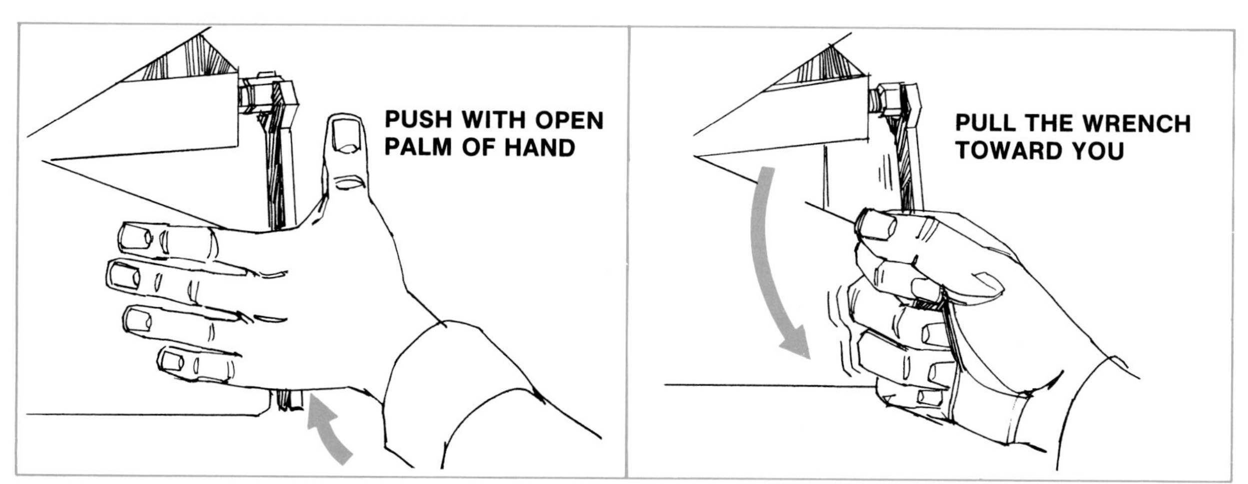

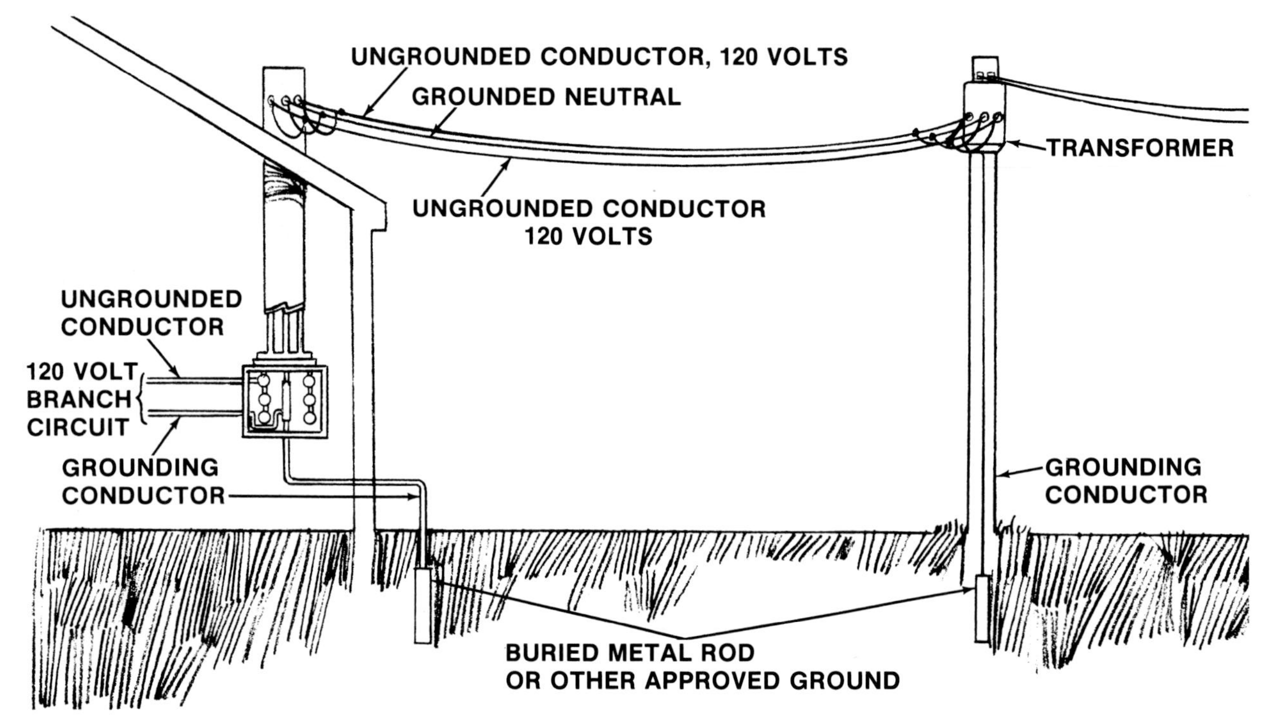

Fig. 31—240-Volt Electrical Service Showing A Two-Conductor Branch Circuit

ELECTRIC SHOCK

Three conductors supply electrical power to your farm shop if it's equipped with 240-volt single-phase service (Fig. 31). One of these conductors is the *grounded neutral.* It is connected to a metal rod buried in the ground at the transformer, and again at the service entrance box that provides the branch circuits for the shop. The grounded neutral may or may not be insulated. If it is, the insulation is always white if the circuit has been wired according to the specifications of the National Electric Code of the United States.

The other two conductors are *ungrounded.* Each carries a voltage of 120 volts with respect to the grounded neutral. Because the grounded neutral is connected directly to ground, each of the ungrounded conductors has an electrical pressure of 120 volts above the earth's "neutral" charge. Ungrounded conductors may be covered with insulation of any color except white or green. (White insulation identifies grounded neutral conductors. Green insulation identifies grounding conductors.)

To illustrate how a person can receive an electric shock when using a power tool, let's look at the operation of a portable drill (Fig. 32). The drill is connected to a 120-volt branch circuit consisting of two conductors: an ungrounded conductor and the grounded neutral.

When the drill is turned on, current flows from the ungrounded conductor, through the insulated conductors in the drill, back through the grounded neutral, then to ground. If the insulation in the drill is in good condition, all of the current entering the drill will return through the grounded neutral. But if the drill's insulation is defective, current may also flow from the drill housing through the body of the operator to ground (Fig. 33). This causes electric shock.

The amount of current that flows through the operator's body is determined by the condition of the insulation in the tool, and the electrical resistance of the skin of the operator's body. His resistance is lowest when his skin is damp from perspiration, or when he's working in a damp location. These situations are dangerous. A current of only .006 ampere can electrocute a healthy man in less than a second.

In new power tools, the insulation is usually adequate to prevent current leakage from the insulated conductors within the tool to its frame or housing. Age and abuse, however, can deteriorate insulation, and so a shock hazard always exists when a power tool is used.

There are three ways to prevent electric shock:

- *Using three-conductor, grounding-type circuits*
- *Using tools equipped with double insulation*
- *Installing ground-fault interrupters*

Here is a closer look at these.

Three-Conductor, Grounding-Type Circuits

Grounding-type, 120-volt circuits use three conductors. The third conductor is called the *grounding conductor.* It's connected to the grounded neutral

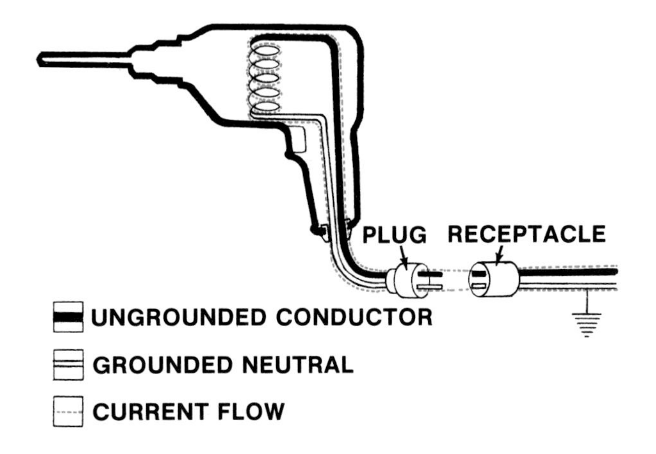

Fig. 32—Safe Current Flow Through An Electric Drill

Fig. 33—Two-wire, 120-Volt Circuits Do Not Provide Shock Protection From Defective Or Inadequate Insulation

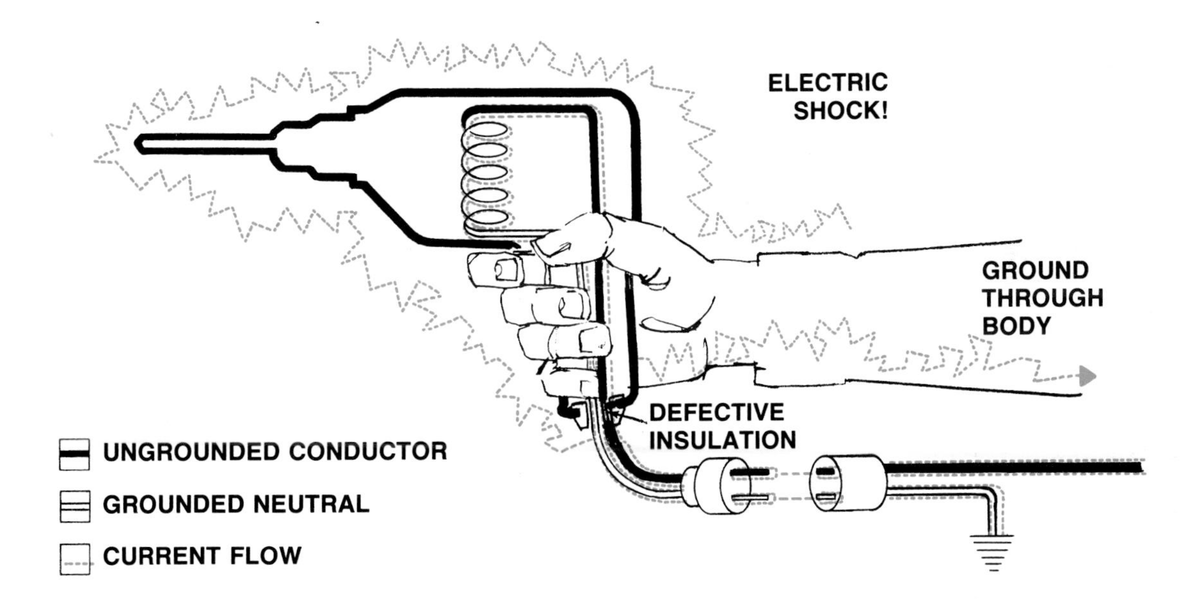

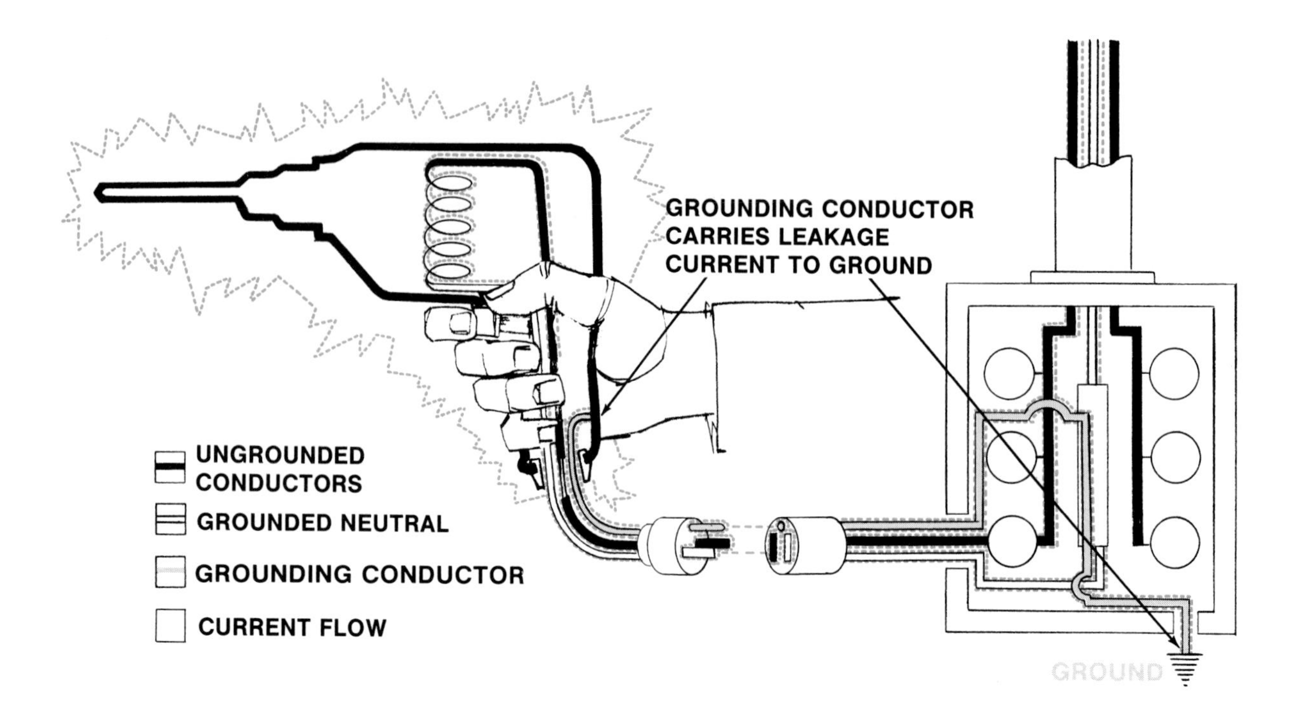

Fig. 34—The Grounding Conductor Carries Current Leakage Safely To Ground

in the shop's service entrance box, and to the center terminals of the grounding-type outlets installed in the circuit (Fig. 34).

Power tools designed for use on grounding-type circuits are equipped with three-wire cords. At one end, the grounding conductor is connected to the tool housing. The other end is connected to the center blade of a three-prong plug. (The insulation on this conductor is green.) When the tool is plugged in, the grounding conductor in the power cord and the branch circuit provides a continuous electrical path from the tool housing directly to ground. Should any current leakage occur from the insulated conductors, the grounding conductor carries the current directly and safely to ground (Fig. 34).

Double-Insulated Tools

Because many power tool users don't have three-conductor, grounding-type circuits, manufacturers make power tools equipped with two layers of insulation. If one layer becomes defective, the second layer provides the necessary protection from shock.

Tools equipped with double insulation can be safely used on two-conductor circuits, since there's no need for the grounding conductor. These tools can be identified by the words "Double Insulation" or the symbol ⧈ marked permanently on the tool housing or nameplate (Fig. 35). Double-insulated tools are frequently encased in a plastic covering.

Double-insulated tools provide good protection from electrical shock if dry. But, there is a danger of electrical shock if you are working in damp conditions. Three wire grounded tools provide almost complete protection.

Fig. 35—Double-Insulated Tools Are Clearly Marked On Case Or Nameplate

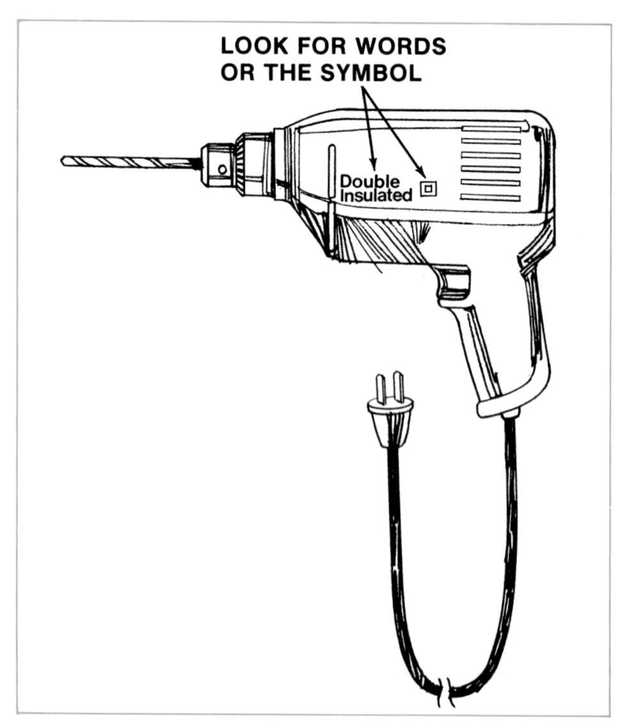

Ground-Fault Interrupters

All of the electrical current that enters a power tool on an ungrounded conductor should flow back again from the tool through the grounded neutral to ground. If it doesn't, a *ground fault* exists. This means that defective insulation is allowing some of the current to flow to ground by some other path. If this path is the operator's body, there is the danger of severe shock or electrocution.

A ground-fault interrupter is a device that compares the amount of current flowing to a power tool in the ungrounded conductor with the amount returning in the grounded neutral. If the ground-fault interrupter (often called a GFI) senses a difference as low as .005 ampere, the GFI snaps off the current by opening the circuit. In this way, the GFI protects the operator from shock.

Two types of GFIs are available. One type is permanently wired into the branch circuit at the service entrance box. The other is a portable unit that plugs into standard 120-volt outlets (Fig. 36). The plug of the power tool is then plugged into the GFI. Installation of GFIs in the service entrance panel is a job for a qualified electrician.

Protect yourself from electrical shock by following these recommendations:

1. *Select a shock protection system.* If you have two-conductor circuits and a variety of tools—some with two-wire cords and plugs and some with three-wire cords and plugs—you have four alternatives:

- Have an electrician install ground-fault interrupters permanently in each of the shop branch circuits
- Plug in a portable ground-fault interrupter when individual power tools are used
- Convert your two-conductor circuits to grounding-type circuits
- Replace your present tools with new ones equipped with double insulation

Get the advice of a competent electrician to help you decide which alternative is best or most economical for your farm.

2. *Purchase tools designed to prevent shock.* Look for tools which carry the approval label of a recognized inspection and approval agency. The label "UL Listed" indicates that the tool has been safety-approved by the Underwriters Laboratory. A UL label on a cord means only the cord has been tested. The PTI "Safety Seal" indicates approval by the Power Tool Institute. Approved tools are equipped with three-wire, grounding-type cords and plugs or with double insulation (Fig. 37). Buy either type if your shop has grounding-type circuits. If you have two-conductor circuits, with or without GFI protection, buy double-insulated tools.

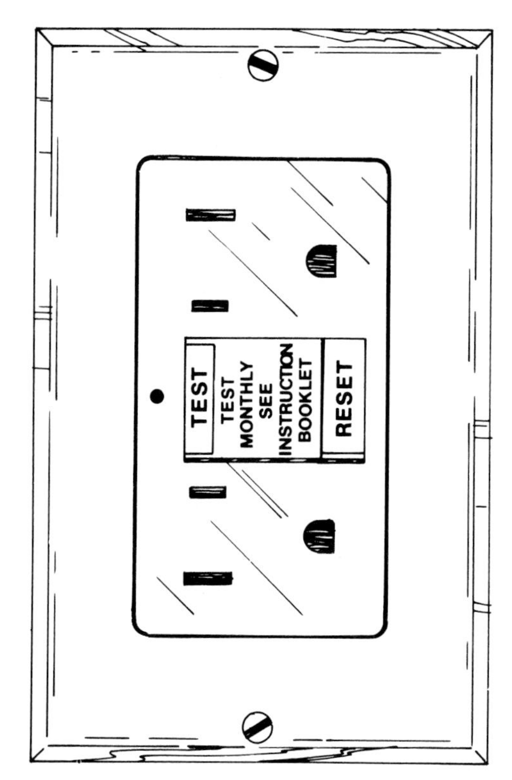

Fig. 36 — The Ground Fault Interrupter Protects The Electric Circuit. The Power Tool Plugs Into the GFI. Portable and In-Outlet Ground Fault Interrupter Plugs Also Are Available.

NOTE: DO NOT assume that a 3-prong plug is grounded. It may not be.

Fig. 37—Look For These Symbols When You Buy Power Tools

3. *Avoid the use of grounding adapters (Fig. 38).* You can buy adapters for plugging grounding-type plugs into two-conductor circuits. These are not recommended, and their use is prohibited in Canada by the Canadian Electrical Code. They are dangerous because two-conductor circuits don't have a grounding conductor to connect to the "pigtail" of the adapter. If you *must* use an adapter, have a competent electrician install a separate grounding conductor to the outlet to adequately ground the adapter.

4. *Inspect extension cords regularly.* They have the same shock hazard as power tools. Keep them away from sharp objects, heat, oil, and solvents that can damage insulation. Do not patch a damaged cord — shorten it or get a new one. And use extension cords of adequate capacity. Undersized cords cause a loss of voltage (electrical pressure). Loss of voltage within the cord also makes the extension cord heat, possibly causing a fire. To determine the capacity needed for each power tool, check the nameplate for its ampere rating, and then refer to Fig. 39 to determine the recommended conductor size. Use extension cords for temporary connections only. Extension cords deteriorate, and are unsafe in permanent installations.

5. *Don't abuse tools.* This may destroy the insulation on the conductors inside the tool. Don't drop power tools, throw them around, or pick them up by pulling on the power cord. Avoid overheating. And when tools become hot from continuous use or from temporary overloads, stop and let them cool.

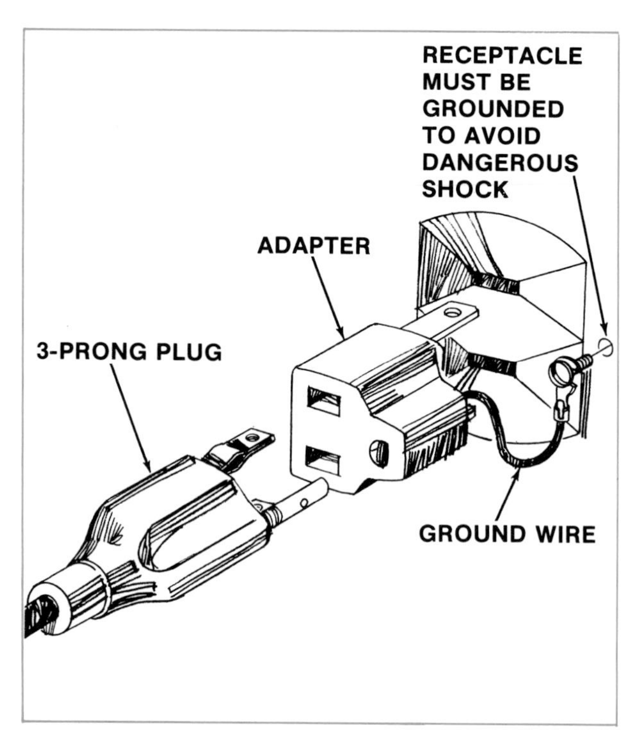

Fig. 38—If Adapters Must Be Used, Install A Separate Grounding Conductor

DRILL PRESSES

Three types of accidents are common with drill presses:

Fig. 39—Use Extension Cords Of Adequate Capacity To Prevent Overheating And A Possible Fire

	EXTENSION CORD SIZES					
Ampere rating (on nameplate)	0 to 2.0	2.10 to 3.4	3.5 to 5.0	5.10 to 7.0	7.10 to 12.0	12.1 to 16.0
Ext. Cable length	Wire Size (American Wire Gauge)					
25 ft.	18	18	18	18	16	14
50 ft.	18	18	18	16	14	12
75 ft.	18	18	16	14	12	10
100 ft.	18	16	14	12	10	—
150 ft.	16	14	12	12	—	—
200 ft.	16	14	12	10	—	—

Fig. 40—Wear Eye Protection And Hold Small Work Pieces In A Drill Press Vise

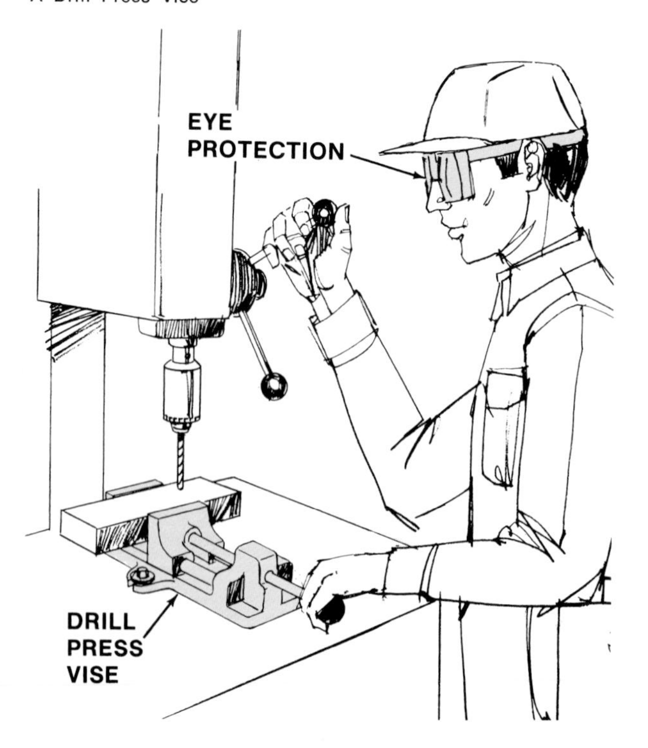

• *A drill breaks and metal fragments are thrown into the operator's eyes*

• *Clothing is caught by a revolving drill or chuck*

• *A work piece is caught and spun around with the drill, tearing skin from the operator's hands*

Avoid injuries by following these suggestions:

1. *Wear eye protection (Fig. 40).* Drilled-out chips or fragments from a broken drill could be flung through the air and injure your eyes.

2. *Prevent drill breakage.* Use sharp, straight drills. Discard bent drills, and sharpen those that are dull or chipped. Mark the locations of holes to be drilled with a center punch to keep the drill from wandering when starting to drill (Fig. 41). Support the work piece so it can't move or tip after you start drilling. Don't force the drill. If it doesn't cut, it is dull. Or, you may need to drill a pilot hole before using a larger drill. Finally, relieve the pressure when the drill starts to cut through. At this time, the drill may catch on the work pieces, breaking the drill, or spinning the work piece if it's not held securely.

3. *Remove the key from the chuck before switching on the motor.* Wear close-fitting clothes, and keep your sleeves buttoned. Loose clothing can get caught by a revolving chuck or drill.

4. *Avoid injuries from work pieces.* Clamp small pieces to the table or hold them with a vise, wrench, or pliers. Turn the motor off immediately if the work piece starts to spin. Don't try to catch it—it could gash your fingers or hands. Remove metal chips and spirals with a wooden stick or small brush. Don't let spirals spin around with a revolving drill (Fig. 42).

Fig. 41—Mark The Location Of Holes With A Center Punch To Keep The Drill From Wandering

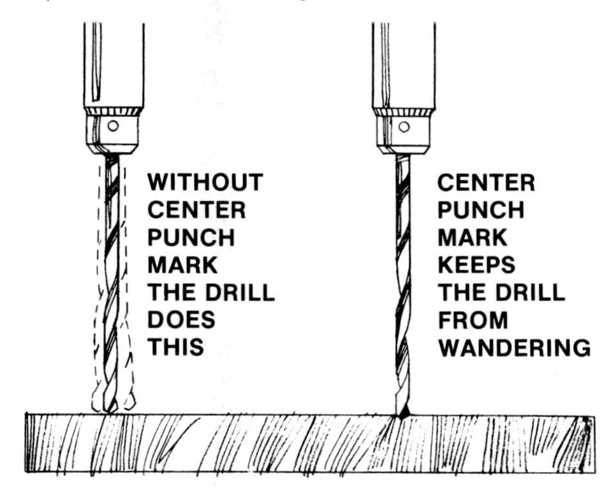

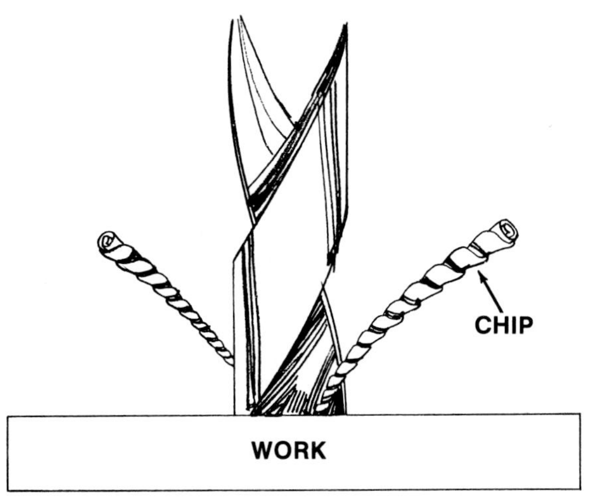

Fig. 42—Dont Let Spirals Spin Around A Revolving Drill. Remove Them With A Wooden Stick Or Small Brush

PORTABLE DRILLS

In addition to the precautions for drill presses, keep these in mind when using portable drills:

1. *Keep a firm grip on the drill with both hands.* Drills of ½-inch capacity or larger could throw you off balance into a fall.

2. *Pull the plug when changing drills (Fig. 43).* If the

Fig. 43—Pull The Plug Before Changing Any Bit Or Cutter In A Power Tool

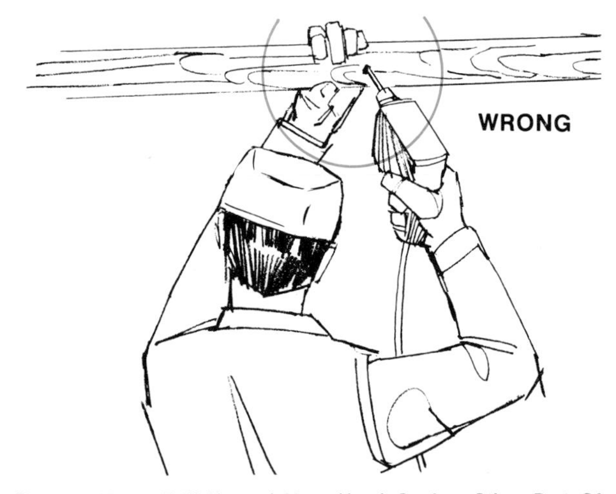

Fig. 44—Never Drill Toward Your Hand Or Any Other Part Of Your Body

switch is accidentally turned on, the chuck, key, or drill could tear skin from your fingers or hands.

3. *Don't lock the switch in the on position.* Use the lock only when the portable drill is mounted in a stand.

4. *Never hold small work pieces in your hand.* The drill is very likely to catch and spin them around. There is also the danger of drilling *through* a work piece into your hand. Never put your hand behind the work *in line with the drill* (Fig. 44). The drill could go through the work and severely puncture your hand.

5. *Wear eye protection so flying debris can't enter your eyes.*

STATIONARY GRINDERS

There are three common hazards associated with using grinders:

- *If a grinding wheel explodes at high speeds, shattered pieces of it could fly into your face.*
- *If your hands touch the wheel, you will lose skin and flesh.*
- *If the work piece gets very hot, your fingers could be burned.*
- *Flying particles can damage your eyes if you don't wear safety glasses.*
- *The tool can jam between the tool rest and the wheel if the tool rest is not against the wheel properly.*

Protect yourself from these dangers:

1. *Always wear eye protection (Fig. 45).* Protect your eyes even if your grinder is equipped with a shatter-proof eye shield.

2. *Keep shields in place.* The eye shield and the wheel shield are both needed to protect you from wheel fragments if the wheel breaks or shatters at high speed.

3. *Check for a defective wheel before installing a new one.* Tap grinding wheel gently with a light metal object. A clear ring indicates a sound wheel. "No ring" indicates a defective wheel and it should not be used.

4. *Use compression washers and flanges on each side of the wheel (Fig. 46).* Make sure the size of the arbor hole in the wheel matches the diameter of the grinder shaft. If not, obtain and install bushings of proper size.

5. *Make sure that the speed of your grinder doesn't exceed the recommended speed for the wheel.* Grinder speed can be determined from the motor nameplate. The maximum recommended speed for the wheel is indicated on the label glued to the side of the wheel.

6. *Set the tool rests slightly above center and 1/8 inch from the face of the grinding wheel (Fig. 46).* This position will help prevent thin work pieces and your fingers from getting wedged between the tool rest and the grinding wheel.

7. *When starting the grinder, stand to one side of the wheel, turn on the switch, and let it run for a full minute before doing any grinding.* Then grind with a light pressure gradually until the wheel warms up. Cold wheels may shatter.

Fig. 45—Wear Eye Protection When Grinding

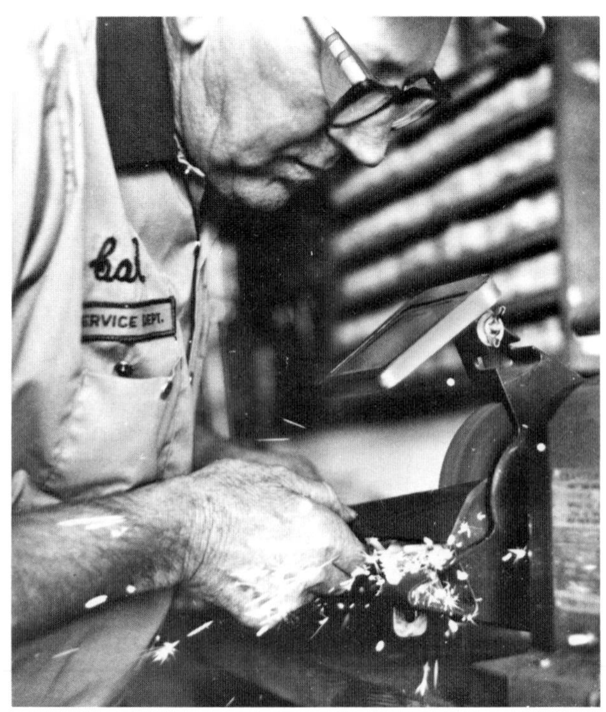

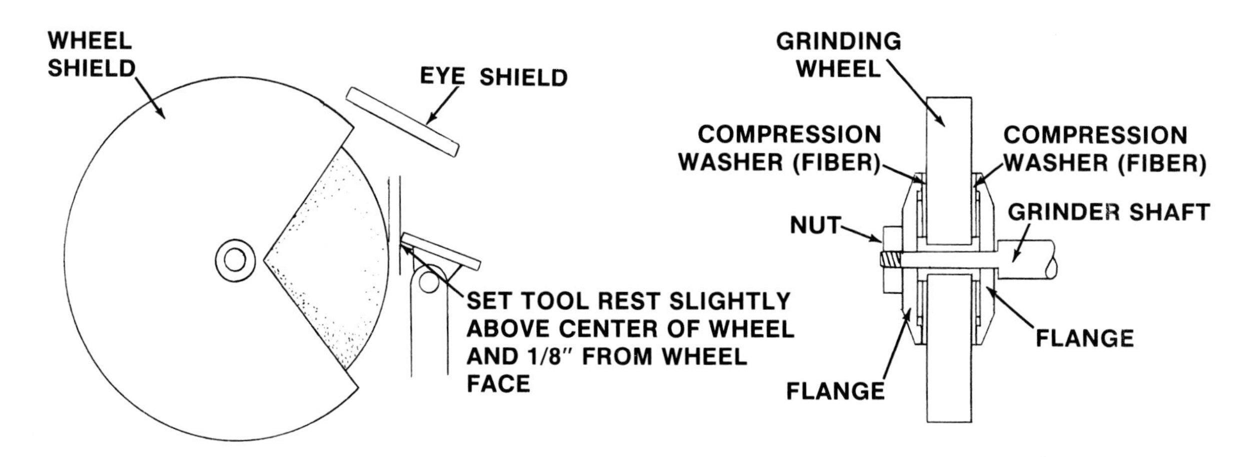

Fig. 46—Install Grinding Wheels And Adjust Tool Rests Properly. Keep All Shields In Place

8. *Grind only on the face of the wheel.* Side pressure may break the wheel if it's not specifically designed for side-pressure grinding.

9. *Protect your fingers and hands.* Never adjust the tool rests when the wheel is turning. Hold small pieces to be ground with pliers or a locking wrench. Position work pieces on the tool rest to prevent them from getting wedged between the tool rest and wheel.

10. *Grind with moderate pressure.* Forcing the work against the wheel heats the work piece quickly, wears the grinding wheel out of round, and increases the chance that your fingers may slip onto the wheel. Grind with only moderate pressure, and dip the work piece in water frequently to keep it cool. To eliminate the need for applying the work piece against the wheel with more than moderate pressure, keep the wheel sharp and true by dressing it when needed.

PORTABLE GRINDERS

Portable grinders are difficult to handle because of their size and weight. Extra care is needed to avoid injury and to protect the grinding wheel from damage. When using portable grinders, observe these precautions in addition to those listed for stationary grinders:

1. *Hold the grinder firmly with both hands.* And hold the grinder or position yourself to keep the stream of sparks and dust directed away from your body and away from persons who may be nearby (Fig. 47).

2. *Before starting to grind, make sure that everyone within range is wearing eye protection.*

3. *Let the grinder come to a complete stop before laying it down.*

4. *Guard against blows to the wheel, either from dropping the grinder, from other shop tools, or by engaging the wheel too quickly or abruptly to the work.*

POWER BRUSHES

To prevent accidents when using wire wheels, use these safe practices:

1. *Follow grinding safety rules.* These are most important: providing eye protection, using flanges to mount the brush, and setting the tool rest properly if one is used.

Fig. 47—Hold The Portable Grinder Firmly With Both Hands

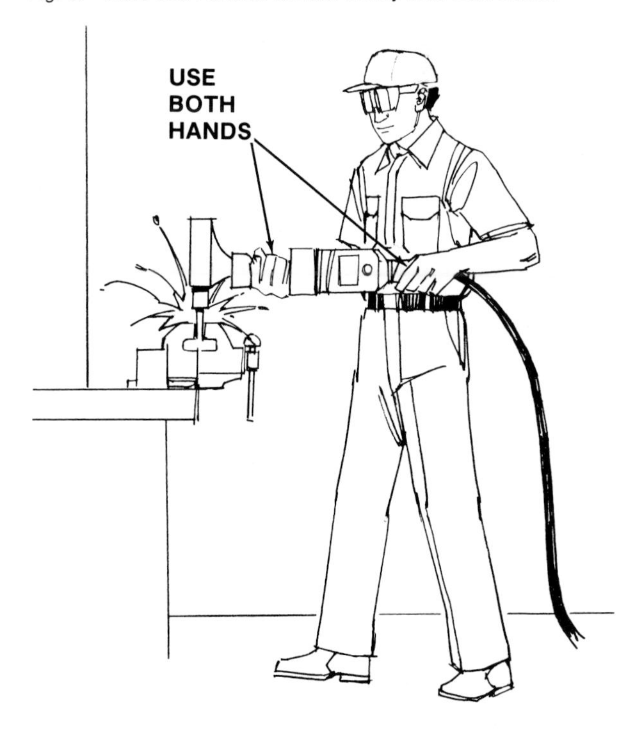

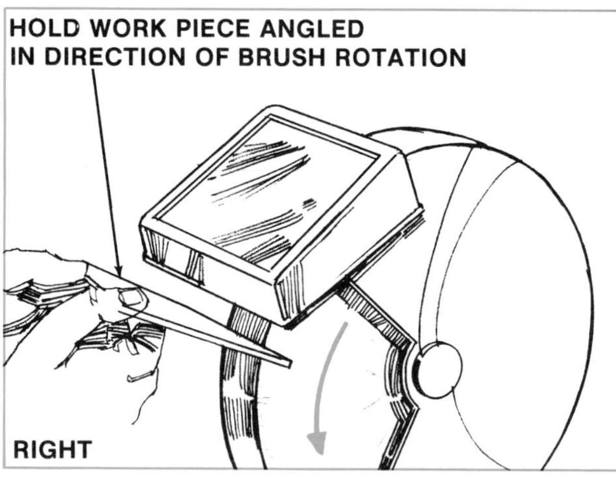

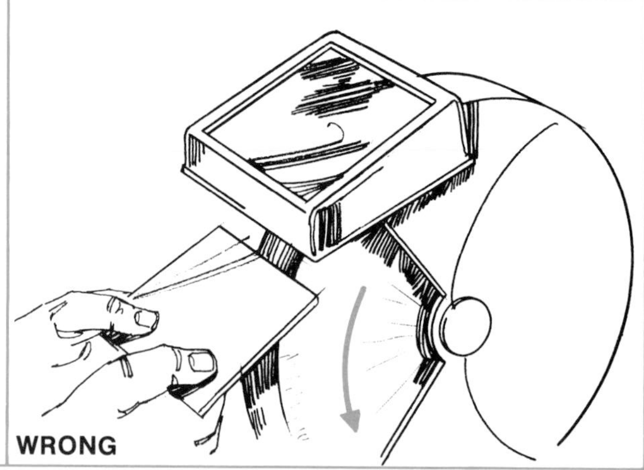

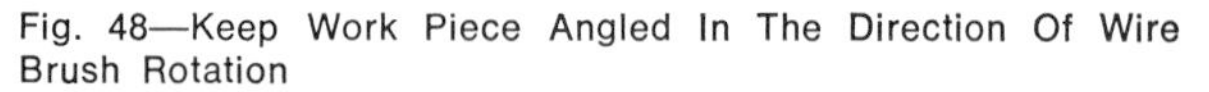
Fig. 48—Keep Work Piece Angled In The Direction Of Wire Brush Rotation

2. *Hold the work piece at the proper angle.* Hold it with both hands at or below the horizontal center of the brush, and angled as shown in Fig. 48. Don't push the edge of the work piece upward against the direction of wheel rotation. If you do, the wheel could jerk the piece out of your hands and cause an injury.

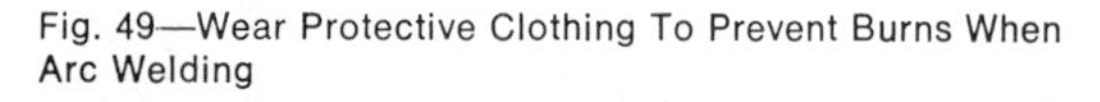
Fig. 49—Wear Protective Clothing To Prevent Burns When Arc Welding

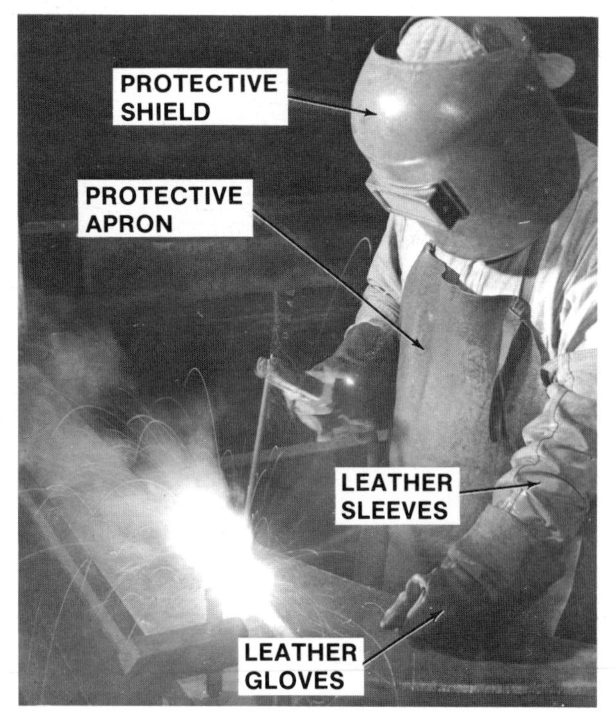

3. *Let the tips do the work.* Forcing the work piece against the brush increases wire breakage and the chances of snagging the work. Force doesn't make the wheel clean faster—it merely bends the wires.

4. *Hold small pieces with pliers or a locking wrench.* This will save your skin if the work piece catches or your hands slip against the brush.

ARC WELDERS

The hazards of arc welding include intense heat, the brilliance of the arc, fumes, and working with a powerful welding current. Protect yourself from these hazards in the following ways:

1. *Wear a helmet.* Your eyes, face, and neck need protection from the burning rays of the arc and from the splatter of molten metal and slag. To protect your vision, make sure your helmet has a colored lens with at least a No. 10 shade. Lighter shades, indicated by lower numbers, will not protect your eyes from the harmful rays. Never strike an arc before your helmet is in place. And never look at the arc from any distance with naked eyes while another person is welding.

NOTE: A No. 10 shade will protect your eyes adequately when welding with 200 amperes or less. Use darker shades when using higher currents. Refer to your welding instruction manual.

2. *Protect yourself from burns (Fig. 49).* Always wear leather or heat-resistant gloves. Wear high-topped shoes to prevent leg and ankle burns. Button your shirt and collar, and turn down cuffs. Wear a skull cap under your helmet to protect your scalp and hair from welding splatter and hot fragments of flux when chipping the weld.

3. *Protect others.* Warn others nearby when you are ready to strike an arc. Be sure their helmets are in place if they intend to watch (Fig. 50). When finished, don't leave pieces of hot metal or exposed and hot finished welds where they can be accidentally touched by others. Dispose of hot electrode

stubs in a metal container—never drop them on the floor.

4. *Prevent fires and explosions.* Before welding, clear away all combustible materials. Never weld on barrels, tanks, or other containers that once held flammable materials such as fertilizers. Vapors remaining in the container may explode. Let a professional welder tackle these jobs.

5. *Provide ventilation.* Clear away welding smoke by opening doors and windows, or by switching on an exhaust fan. Avoid breathing welding fumes, especially from metals that produce toxic fumes (like zinc used to produce galvanized steel). Avoid breathing welding fumes from containers that have been cleaned with chlorinated hydrocarbons.

6. *Protect your eyes at all times and face when chipping slag.* Never chip slag when your eyes or those of others nearby are not protected by goggles, an eyeshield, or the clear lens of a welding helmet (Fig. 51). Fragments of hot slag burn. If they hit the eye, medical attention will be required to remove them and blindness may result. Remember that the risk of permanent eye injury is so great that *you should never chip* slag from a weld without protecting your eyes.

Fig. 51—Protect Your Eyes While Chipping Slag

Fig. 50—Never Expose Naked Eyes To The Arc At Any Distance

Fig. 52—Provide An Outlet For Your Welder Near The Large Service Door So You Can Weld Outside

7. *Guard against severe shock or electrocution.* Weld in a dry location. Don't change electrodes with bare, sweaty hands. Wear dry gloves. Do not weld in a damp location. Remember that water, unless chemically pure, is a good conductor of electricity, and that it increases the conductivity of your skin. To avoid a shock when welding, stay dry.

8. *Keep your equipment in good condition.* Protect welding cables from damage and keep all connections tight. Inspect them frequently for damaged insulation, frayed conductors, loose connections, and broken electrode holders or grounding clamps. Keep them in good repair.

9. *Treat eye and skin burns promptly.* Don't let any arc flash your eyes receive go unattended. If you receive a flash, get medical help right away. If you are fortunate, no lasting harm may have been done, but medical help can relieve the pain that usually appears several hours after the flash has been received.

10. *Weld outside (Fig. 52).* If possible, provide an outlet for your welder near the large service door of your shop. The cost will be repaid in safety and convenience. Fire hazards are minimized, there is better ventilation, and long welding cables are not needed to reach large pieces of equipment that can't be brought inside.

Always read your welding equipment operation manual, and follow directions.

Fig. 53—Keep Cylinders Chained In An Upright Position

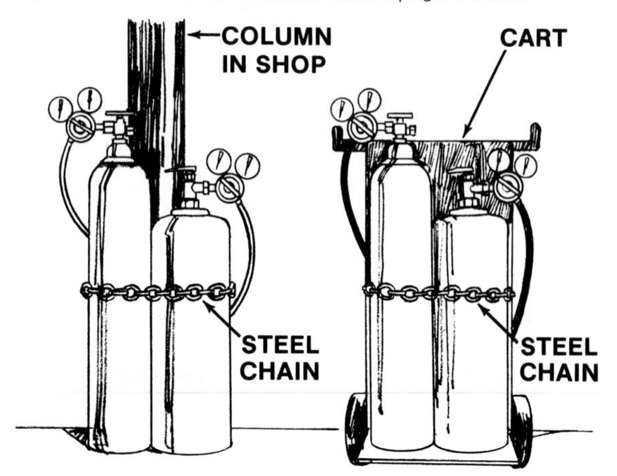

OXYACETYLENE WELDERS — BASIC SAFETY INFORMATION

Follow operating recommendations carefully. Emergency situations resulting from misuse can be painful, costly, and disasterous.

1. *Safeguard the fuel supply.* Keep cylinders chained in an upright position (Fig. 53). Don't let them drop or fall. When stored, keep the caps in place to protect the valves from damage. If the valve assembly of a tank full of oxygen was broken off, for example, by tipping the tank over accidentally, the oxygen released could result in a fire or explosion.

If a valve leaks, move the cylinder outside, keep all sources of fire away, and notify your supplier. Don't use cylinders as rollers, supports, or for any other purpose besides supplying oxygen or acetylene to your welding outfit. Never use the oxygen tank as a source of "compressed air." An oxygen-enriched atmosphere is extremely flammable.

2. *Prevent equipment fires.*

a) Never use oil to lubricate any part of your welding equipment. And never weld or handle and adjust equipment with greasy hands. Oxygen and oil or grease is a perfect combination for a spontaneous fire. Never use compressed air to blow talc or clean out a new air hose. Compressed air is treated with a lubricant.

b) Test for leaks. Use a bucket of water to test (non-petroleum base soap) for leaks in the hose, and a brush and soapy water for leaky connections (Fig. 54). Keep hoses protected from hot and molten metal and your welding flame. Repair hose leaks by cutting the hose and inserting a splice. Don't try to repair a leaky hose with tape.

c) Select gas pressures according to the size of the tip you're using. The operator's manual will give you this information. Remember that 15 psi is the maximum safe pressure for acetylene. Normally, the pressure should never be adjusted this high. If there is a steady buildup of regulator pressure when the torch valves are closed, close the cylinder valve and remove the regulator for repair.

d) Open the oxygen cylinder valve to prevent leaks at the valve stem. The withdrawal rate from a cylinder of acetylene must not exceed one-seventh of the acetylene cylinder capacity. For example, a cylinder with 140 cubic feet of capacity should not release more than 20 cubic feet of acetylene per hour. Open the acetylene valve completely.

e) In case of fire in the equipment, turn off the air cylinder valve first, and then the acetylene valve. Do not relight until the welding equipment has been inspected for damage and the cause of fire determined.

3. *Wear goggles* (Fig. 55). Use the specified goggles listed in your welding operation manual. When in doubt, use the darkest lense possible that allows you to clearly see the outline of your work.

4. *Be sure the regulators are closed before opening the cylinder valves.* If the regulators are not closed, and if the cylinder valves or the regulator diaphrams are damaged, gas pressure may shatter the glass on the regulator when you open the cylinder valves. Protect yourself from the possibility of being injured by shattered glass by following this procedure:

a) Be sure the regulators are closed by turning each pressure adjusting screw counterclockwise until it fits loosely in the regulator.

b) Stand to one side away from the face of the regulator.

Fig. 55—Stand To One Side When Adjusting Regulators. Note Goggles Ready For Use

c) Open each cylinder valve slowly. See step 2 item d.

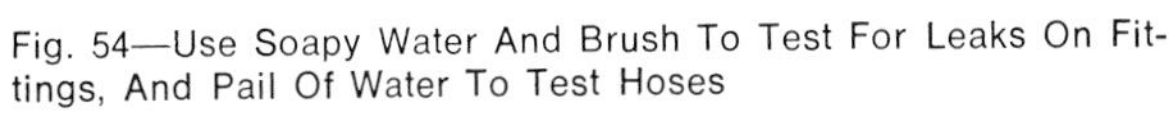
Fig. 54—Use Soapy Water And Brush To Test For Leaks On Fittings, And Pail Of Water To Test Hoses

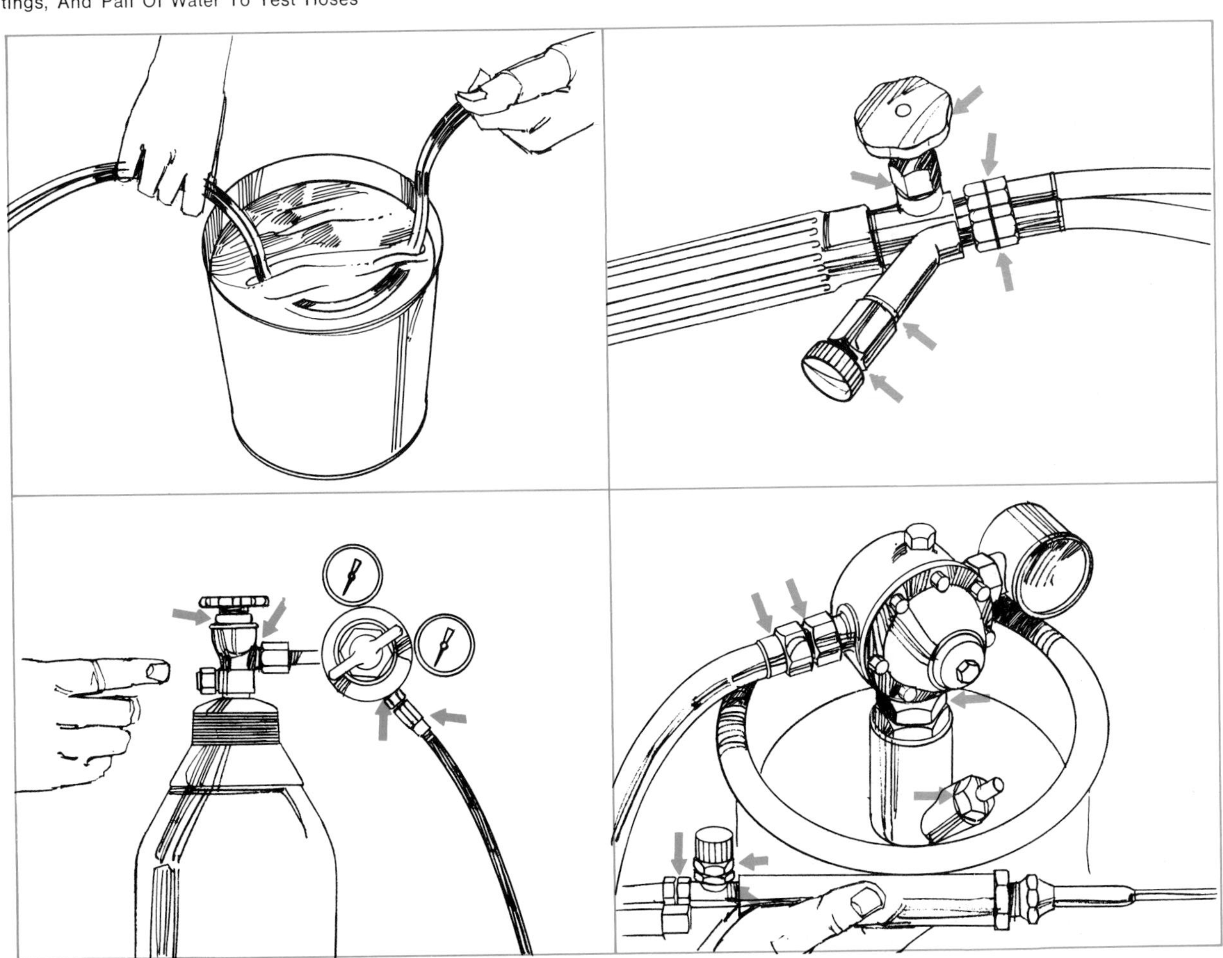

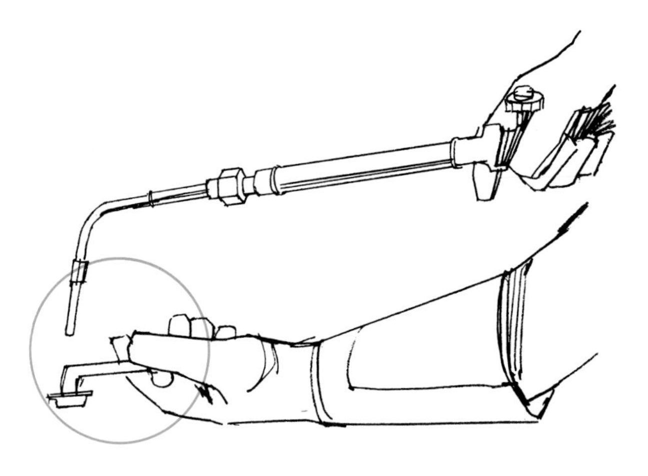

Fig. 56—Light The Torch Safely

5. *Prevent burns.* Wear heat-resistant gloves to protect you from hot metal and the heat of the flame. Wear tight fitting dark colored fire-resistant clothes that are free of oil and grease. Wear a long-sleeved shirt, and high top boots or shoes with hard soles. Light the torch with the tip facing downward, away from your face and legs. Open the acetylene valve no more than a quarter turn to light the torch. Always use a friction-type lighter (Fig. 56). Lighting from a hot surface could result in a large flare-up, and use of matches could burn your hands.

6. *Prevent fires.* Weld in an area cleared of all combustibles. When cutting, remember that the shower of hot, molten metal can ignite materials some distance away.

7. *Shut down the equipment properly.* Close the cylinder valves and release the gas pressure from the hoses by opening and closing each torch valve one at a time. Then loosen the tension of each regulator pressure adjusting screw by turning it counterclockwise until it fits loosely in the regulator. If necessary, move your equipment to a safe location for storage.

8. *Make sure your acetylene and oxygen supply lines are equipped with valves to help prevent flashbacks.* See your welder supplier for more information.

9. *Never return a completely empty cylinder to your acetylene supplier.*

10. *Use ear protection.*

NOTE: These are only basic safety procedures. Completely read and follow your operation manuals before you begin.

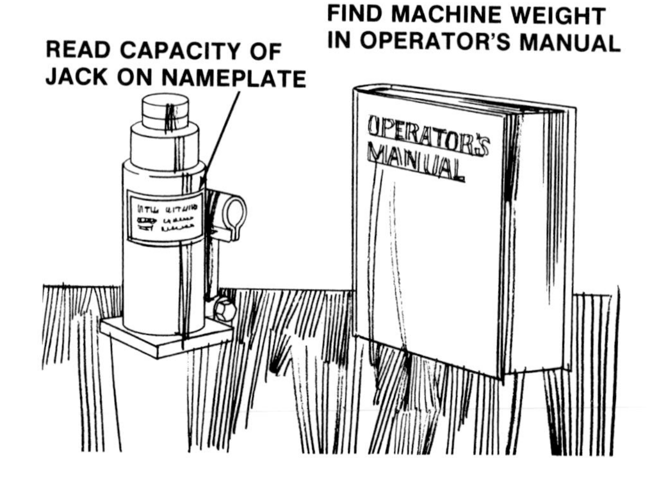

Fig. 57—Check Jack Capacity And Weight Of Machine Before Lifting

JACKS

Serious crushing accidents can result from the improper use of jacks. Each of the following recommendations is important. Be sure to observe them:

1. *Don't overload the jack.* Make sure any jack you use is strong enough to carry the load. If there is any doubt, check the rated capacity of the jack against the weight to be raised (Fig. 57). Jacks will lift some overload, but there is always the danger of immediate failure. Do not overload them.

2. *Lubricate jacks frequently, but only at points specified in the operating instructions.* Some points may be designed to run dry.

3. *Keep reservoirs of hydraulic jacks full of the recommended hydraulic jack oil.* Don't use other fluids or dirty crankcase oil as these may damage seals and valves within the jack. Take leaky jacks out of service and get them repaired.

4. *Handle jacks carefully.* Dropping or throwing them around may distort or crack the metal, and the jack may fail under load.

5. *Position the jack properly.* Select a point on the machine strong enough to carry the lifted weight. The lift point should be flat and level with the floor or ground supporting the base of the jack. Position the jack so the lift will be straight up and down.

6. *If working on the ground, place a heavy block under the base of the jack.* Select one long enough and wide enough to keep it from sinking, shifting, or tipping over when weight is applied. If the jack will not lift high enough, place additional blocking *under the jack* (Fig. 58). Don't put extenders between the jack saddle and the load.

7. *Stabilize the equipment.* If the machine is self-propelled, place the transmission in gear or in park position, and set the brakes. Block at least one wheel that will remain on the ground. When lifting pull-type equipment, hitch it to a tractor drawbar to keep it in place.

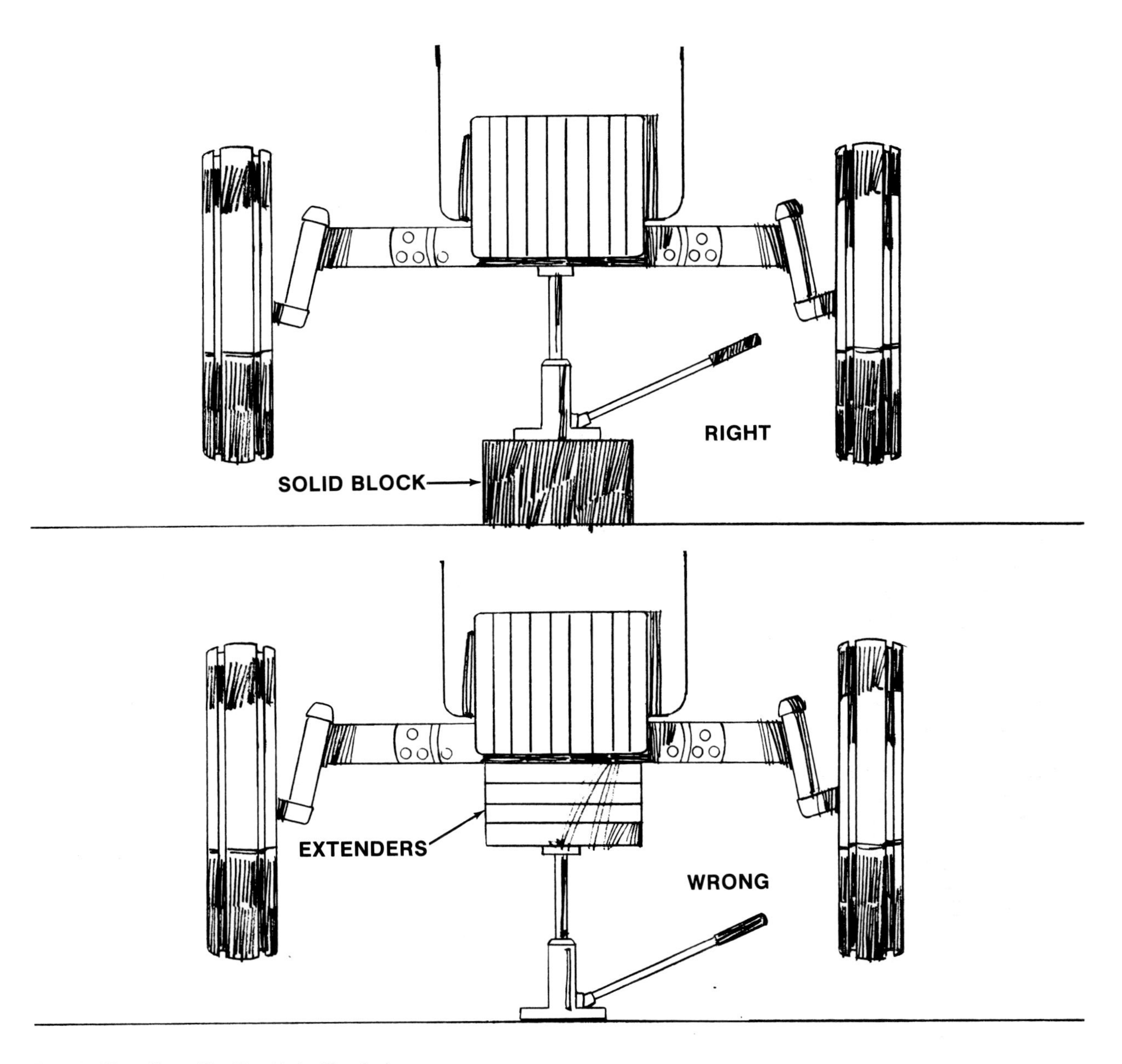

Fig. 58—Place Heavy Blocking Under The Jack

Fig. 59—Avoid This Happening! Reset The Jack If It Starts To Lean, And Block The Equipment More Securely

8. *Recheck the position of the jack after it has started to lift.* If it starts to lean because the equipment rolls or shifts, lower the jack and reset it, and block the equipment more securely. If you do not, and continue to lift, the jack will push the equipment forward or backward, forcing it to roll off the jack and fall (Fig. 59).

9. *Keep equipment level.* Don't lift one side of a machine so high that there's danger of the jack saddle sliding on the equipment frame, permitting the equipment to fall (Fig. 60). Prevent this type of accident by observing these precautions:

- *Lift no higher than necessary.*
- *Place a wooden shim between the equipment frame and the jack saddle to eliminate the metal-to-metal contact.*
- *Keep equipment level by jacking or blocking each side alternately, or at the same time.*

10. *Beware of jack handles.* The handles of some mechanically operated jacks can pop up and kick when the load is lifted or lowered. Stand to one side when jacking to avoid being struck by the handle (Fig. 61). Hold the handle firmly to prevent kicking, and do not release the handle from your hands when it is in a position where it can kick. Never straddle a jack handle with your legs. And remove the handle when it's not being used.

11. *Support the load with blocks or stands.* Never allow raised equipment to remain supported by jacks alone. Jacks can fail and tip permitting the equipment to fall unexpectedly. Place solid blocks or stands under the equipment *immediately.* Two types of stands frequently used in high-lift applications are shown in Fig. 62. An adjustable stand is used with a hydraulic jack in Fig. 62B. The support pipe serves as an extender for the jack. When the equipment has been lifted to the desired height, a locking pin inserted through the support pipe converts the adjustable stand to a rigid one. Use stand rated for the specific job.

Fig. 60—Keep Equipment As Level As Possible Or The Jack May Slip

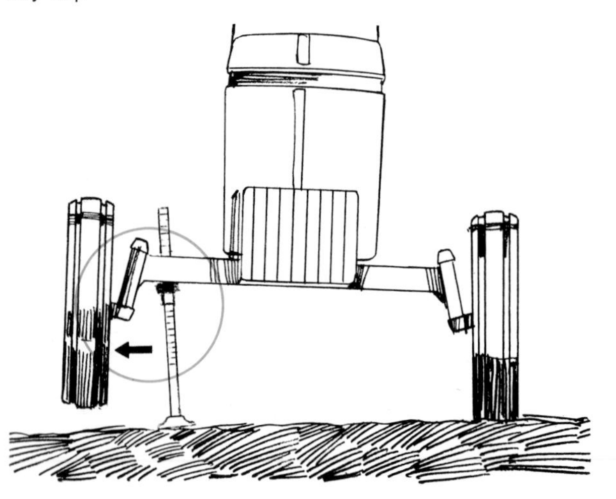

Fig. 61—Some Jack Handles May Kick. Stay Clear. Remove The Handle When Not Jacking

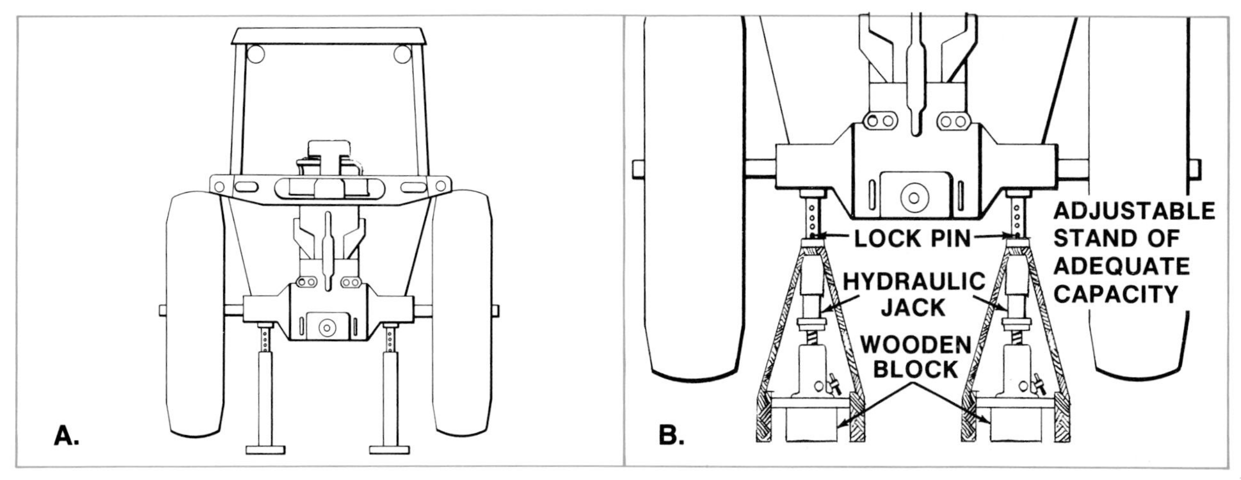

Fig. 62—Support Equipment Immediately With Stands Or Blocks After Jacking To The Desired Height

HOISTS

Overloading the hoist and failing to rig lifting chains so they will not slip are the two most common causes of hoist-related accidents and injuries. When using hoisting equipment:

1. *Use a chain hoist (Fig. 63).* Chain is durable and has great strength for its diameter and weight. Don't use block and tackle and natural or synthetic fiber rope for lifting implements and machines. The size of fiber rope usually found on farms is not big enough to carry the weight of heavy equipment, and you can never be sure how strong a piece of rope is. It weakens with age and abuse. Knots reduce its strength by one-half, and sharp bends made by looping the rope around some equipment frames break the fibers internally, making the rope unsafe for any use.

2. *Inspect chains often.* Look for bent links, cracks, gouges, and extreme wear (Fig. 64). Don't repair chain by tying knots or by fastening with bolts and nuts. Obtain and use approved repair links available from hardware stores and farm equipment dealers. Ideally, damaged chain should be returned to the manufacturer for repair.

Fig. 64—This Chain Shows Extreme Wear And Should Be Replaced

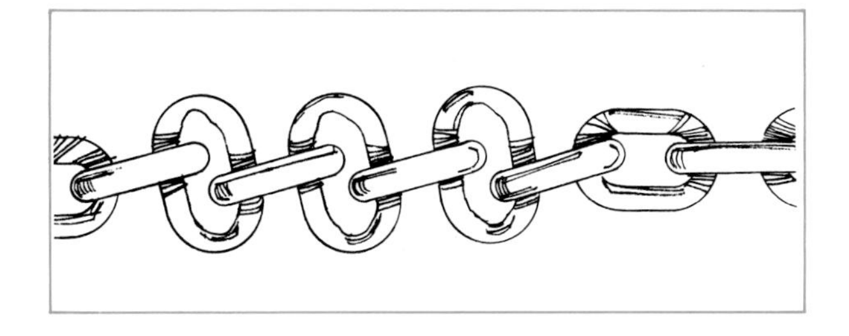

3. *Check the condition of the hooks.* Replace those that are bent, sprung, or cracked. If in doubt about the condition of a hook, check all dimensions of the questionable hook with a new one. If you find any differences in shape or size, replace the hook.

4. *Know the capacity of your hoist and don't exceed it.* Remember also that the beam or A-frame that carries the hoist must be strong enough to support the load. If in doubt about its strength, have a qualified person check it.

Fig. 63—Use A Chain Hoist. Do Not Use Rope

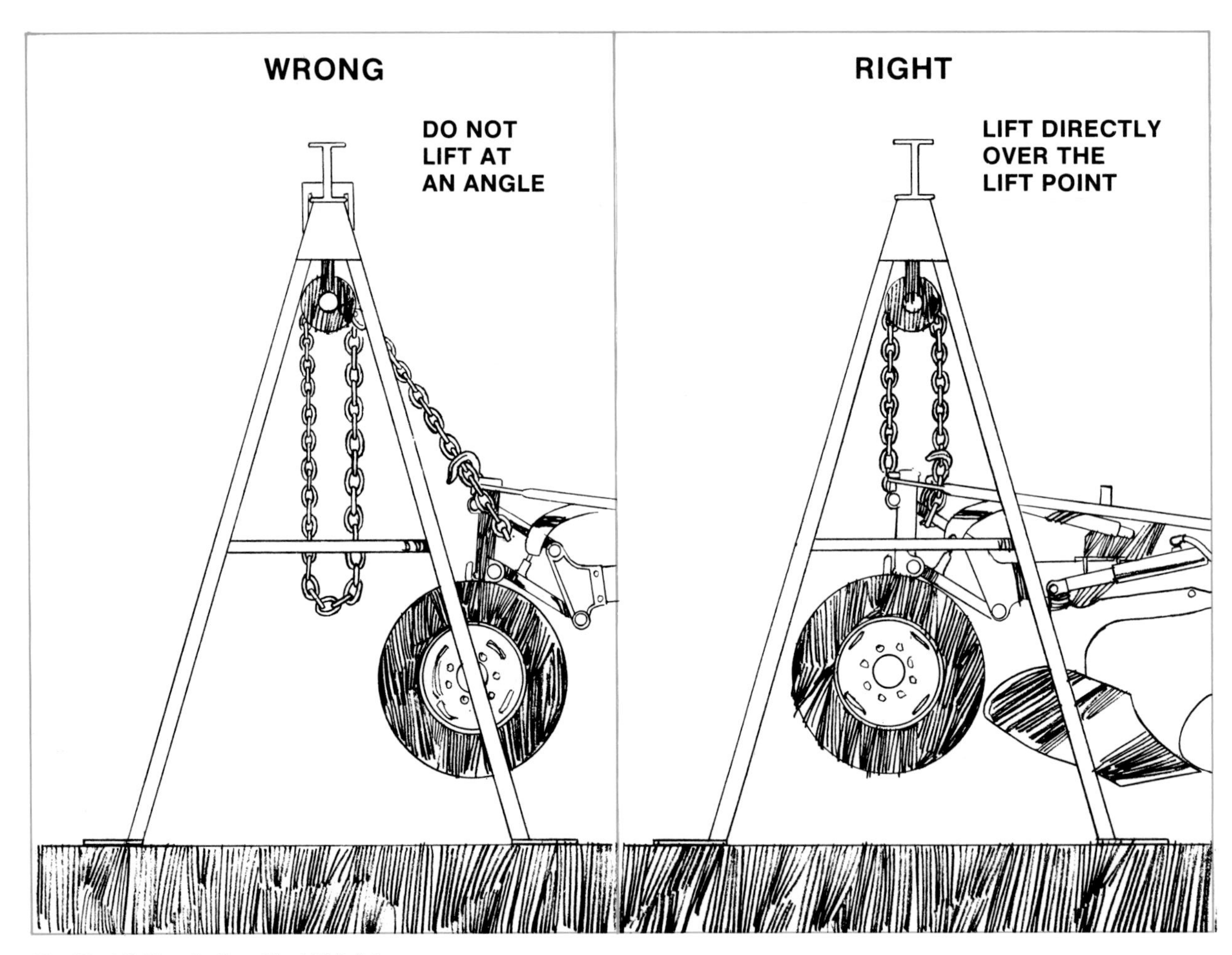

Fig. 65—Lift Directly Over The Lift Point

Fig. 66—Keep The Pull Point From Slipping

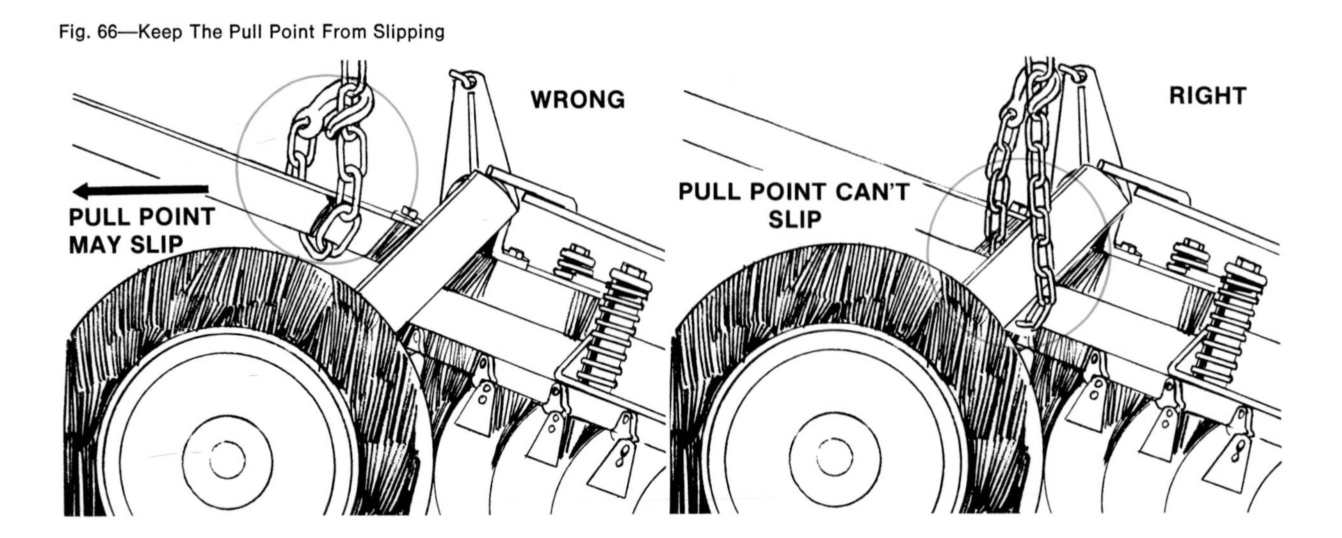

5. *Select a suitable lift point.* Hook to a heavy section of the frame to prevent breaking parts, or bending the frame out of alignment. Attach the hook directly over the point of lift on the implement or machine as some hoists, particularly A-frames, are not designed to withstand a sideways pull (Fig. 65). Attach the chain (or rig the lifting chains) to prevent the pull point from slipping (Fig. 66). If lifting an entire machine or implement, lift from a point where it will balance and not tip.

6. *Protect yourself from injury while the equipment is being raised.* Keep your hands away from pinch points when the lift chain tightens. Don't stand on or within the frame of any implement you are lifting. Make sure the implement doesn't swing and strike you as it leaves the ground. Have stands or blocks ready to use. *Never get under equipment carried by a hoist unless it is securely supported on blocks or stands.*

CHEMICALS AND CLEANING EQUIPMENT

In service operations, cleaning is needed in order to:

- **Keep dirt from entering the machine when parts are disassembled**
- **Inspect parts for wear and damage**
- **Install and adjust parts properly during reassembly**

We have already discussed the safe use of power brushes. Let's turn now to the safe use of solvents, steam cleaners, and high-pressure washers.

SOLVENTS

Most solvents are toxic, caustic, and flammable. Be careful, therefore, to keep them from being taken internally, from burning the skin and eyes, and from catching on fire. Read the manufacturer's instructions and precautions before using any commercial cleaner or solvent.

Whenever possible, use a commercially available solvent (Fig. 67). Many types are available for general-purpose cleaning and for specific cleaning jobs. Always read and follow the manufacturer's instructions to get the best results and to be able to handle each product safely.

Never clean with gasoline. It vaporizes at a rate sufficient to form a flammable mixture with air at temperatures as low as 50° F below zero. It is *always* unsafe to use. If you don't use a commercial solvent, use diesel fuel or kerosene. These *will* burn, but not as easily or as explosively as gasoline (Fig. 68).

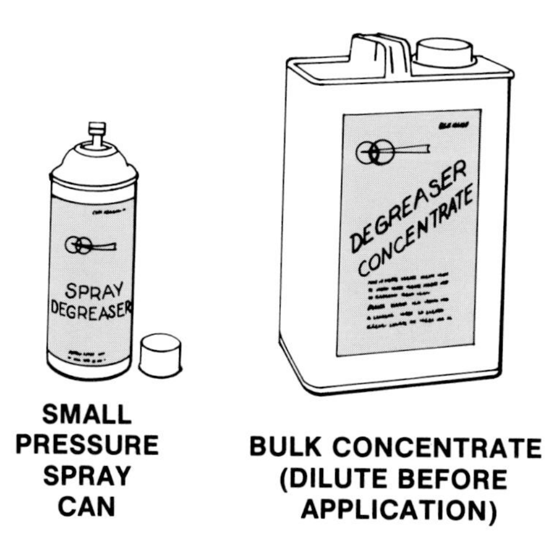

Fig. 67—Use Commercial Cleaning Solvents—Not Gasoline

Follow these general safety practices when using chemical cleaners:

1. *Follow the manufacturer's instructions.* Cleaning agents intended for the same purpose may be quite different in chemical composition. Read the label

Fig. 68—One Gallon Of Gasoline Mixed With The Right Amount Of Air Can Have The Explosive Force Of 83 Pounds of Dynamite

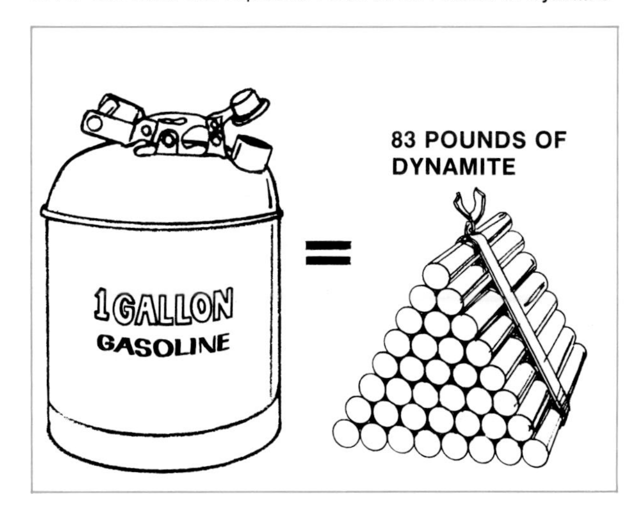

Fig. 69—Use Cleaning Solvents Only In A Ventilated Area, Outdoors If Possible

on each container you buy and follow instructions carefully.

2. *Work in a well-ventilated area (Fig. 69).* Do the cleaning outdoors if possible. If you can't, provide ventilation by opening doors and windows.

3. *Keep solvents away from sparks and flame.* Don't let anyone smoke in the immediate area. Don't use solvents near heaters, sparks or open flames. Some solvent tanks and tubs should be "grounded." Contact your solvent supplier for instructions on how to ground your solvent tank or tub.

4. *Never heat solvents unless instructed to do so.* And don't mix them, because one might vaporize more readily and act as a fuse to ignite the other.

5. *Wipe up spills promptly.* Keep soaked rags in closed metal containers and dispose of them promptly.

6. *Store solvents in their original containers or in sealed metal cans properly labelled.* In a sealed container, the solvent will be kept clean from further contamination, toxic fumes will be controlled, the fire hazard will be reduced, and there will be less likelihood of spillage. Never use open pans (Fig. 70).

7. *If a commercially made parts washer is used, close the lid when you're finished cleaning.* Never destroy the fusable link that closes the lid automatically in case of fire.

8. *Protect your skin and eyes.* Wear a face shield and rubber gloves when working with strong, concentrated cleaning solutions (Fig. 71). Some are caustic, and they can destroy the natural oils of the skin and burn it severly. Always read and follow the manufacturer's instructions carefully. Even when gloves are not called for, avoid long exposures to solvents, and wash your hands when the job is done. Wear an apron if it's needed to keep your clothing dry.

9. *Avoid accidental poisoning.* Wash your hands and arms before eating or smoking. Keep all solutions in labeled containers. Never use empty containers, no matter how thoroughly cleaned, for carrying food or beverages. Keep poison containers sealed even when they are empty. Keep them out of the reach of children.

10. *Be prepared for emergencies.* Keep fire-fighting equipment near your cleaning area (Fig. 17, page 94). Save the original containers for all solvents until they are completely used. In case of accidental poisoning, follow the instructions provided by the manufacturer immediately, and when going for medical aid, take the container with you so the chemical can be quickly identified.

Fig. 70—Store Solvents In Sealed And Labeled Cans, Never In Open Pans

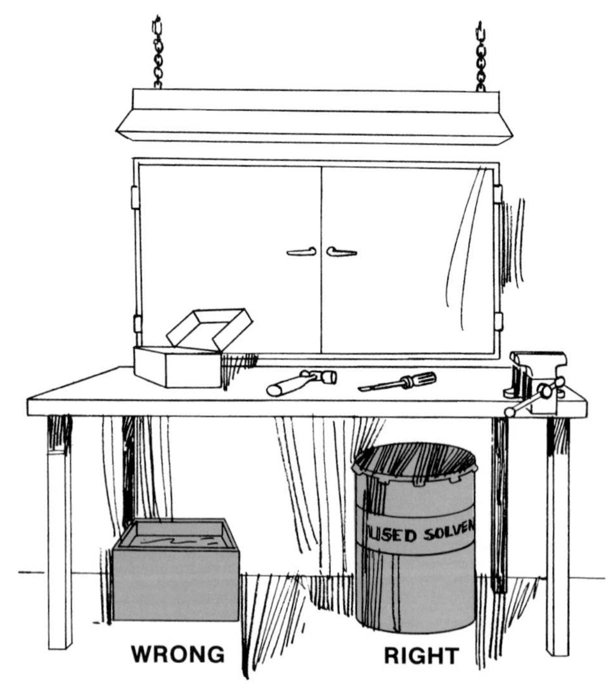

NOTE: If a solvent is splashed into your eyes, flush them thoroughly with water, keeping your eyelids open. Then get medical attention immediately. Some solvents may cause skin irritation or dizziness. If that happens, stop. Wash thoroughly with mild soap and water, and get some fresh air. If possible, use solvents in a location where a clean water source is available and ready for immediate use if needed.

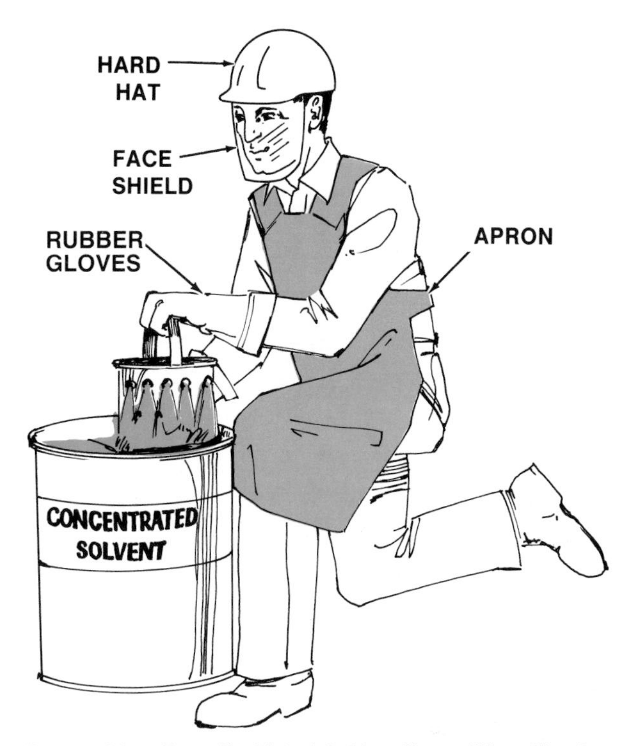

Fig. 71—Wear Face Shield And Rubber Gloves When Cleaning With Cleaning Solutions

STEAM CLEANERS

A number of cleaners are on the market. Operating instructions vary with type and model. Before using any steam cleaner, become thoroughly familiar with its operation, and read its instruction manual carefully.

NOTE: Do not steam carburetors, electrical components, diesel injection pumps, or air-conditioning lines as damage to the components may result. Avoid oversteaming hydraulic lines, wiring, and bearing seals. And don't use cleaning compounds on painted surfaces where appearance is important. Many cleaning compounds will dull the paint.

Certain safe practices apply to the operation of all steam cleaning machines:

1. *Work in a ventilated area.* Clean equipment outdoors when possible (Fig. 72). Otherwise, provide adequate ventilation for your indoor cleaning area. Steam must be ventilated away for good vision, and to prevent damage and rust to your shop facilities and tools caused by condensing moisture.

2. *Protect yourself and others from burns.* Wearing gloves is not essential when using a well-insulated gun, but gloves are recommended if you light the burner by hand. When cleaning, warn others to stay away. Watch for unexpected bystanders when swinging the gun around. Always hold the gun securely and never lay it down until the cleaner has been shut down. When shutting it down, turn off the burner and keep the water running until all steam disappears from the nozzle. Then you can be sure that the equipment will not burn or scald anyone.

3. *Keep the cleaner in good condition.* Avoid damage to the steam gun and hose. Remember that tape does not make safe repairs. Keep fuel connections tight. If your cleaner does not ignite readily, burn cleanly, or maintain steam pressure within safe limits, have a qualified repairman service it.

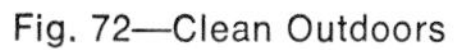

Fig. 72—Clean Outdoors

HIGH-PRESSURE WASHERS

Cleaning with high-pressure washers is becoming common on farms. Take these precautions when using this equipment:

1. *Wear eye protection.* The high pressure stream may cause dirt to be flung into your eyes. In addition, your eyes need protection in case the high

Fig. 73 — Follow the Safety Instructions for Your High Pressure Washer. Read Your Operator's Manual Carefully, Use Only the Chemicals Recommended By the High Pressure Washer Manufacturer.

Fig. 74—Disengage Power And Turn Off The Engine Before Servicing Any Machine

pressure stream should become directed toward your face.

2. *Read the installation instructions.* In addition to an adequate water supply, you'll need a grounding-type circuit or one protected with a ground-fault interrupter. The electrical circuit must have sufficient capacity to handle the load of the motor, and it should be equipped with a safety switch that can be locked in the *off* position. If you need a new circuit for your washer, have a qualified electrician install one for you.

3. *Use extension cords of adequate capacity.* Conductors must be large enough to prevent excessive voltage drop. And three-wire cords must be used if the circuit is not protected with a ground-fault interrupter. Refer to your owner's manual for specific recommendations.

4. *Always direct the spray away from the body.* The pressurized stream can penetrate skin. If the water is heated, it can burn. When cleaning, warn others to stay away, and watch for unexpected onlookers when moving the spray wand around. Hold the wand securely and don't put it down unless the machine is turned off. Even if the wand is equipped with a shut-off valve, the valve could fail. Then, the high pressure discharge would violently whip the wand around and possibly cause a serious injury. If injured by the spray, see a doctor at once.

5. *Protect yourself from splatter and throwback (Fig. 73).* Always keep yourself out of line of the sludge and water thrown back from the surface you're cleaning. Hold the gun at an angle to the surface, and stand to one side when cleaning corners from which the sludge is thrown straight backward.

COMPLETING SERVICE WORK

There are three requirements for safe service and maintenance work. We've outlined the first two: maintaining a hazard-free shop and using service tools and equipment safely. There is one more: using safe work procedures when servicing the equipment itself.

Let's look first at general recommendations that apply to *all* implements and machines. Then we will discuss precautions to observe when working on specific components and systems. Because we can't cover all components and systems, be sure to have operator's and service manuals on hand for all jobs you intend to do, and read them before tackling any job.

GENERAL RECOMMENDATIONS

1. *Disengage the power and stop the machine before servicing (Fig. 74).* This is the most basic rule of service and maintenance safety. Don't try to save time by keeping the machine running. You may be tempted to do this, thinking that, with care, you can do the job safely. *Don't.*

A slip, a caught piece of clothing, or one false move could cut, chew, twist, or pull you into a machine before you realize what has happened. *Do not clean, unplug, lubricate, adjust, or repair any machine while it's running, unless it's specifically recommended in the service or operator's manual. Lock out the ignition and put a warning sign over the ignition that tells everyone that you are working on the machine.*

2. *Support equipment on blocks or stands.* Don't rely on jacks, hoists, or hydraulic cylinders. They are designed for lifting only, and they can fail without warning. Never take the chance of being crushed by a machine or implement which could fall unexpectedly. Instead, protect yourself by following one or more of these procedures:

- *Lower all equipment to the ground.*
- *Use support stands if provided for the equipment.*
- *Engage safety locks if the hydraulic cylinders are so equipped.*
- *Support the equipment on metal stands or solid blocks (Fig. 75).*

3. *Follow all recommendations outlined in instruction manuals and sheets.* Keep these materials in your shop for future use.

4. *Use containers to store removed parts.* Cans, small boxes, or metal pans work well. Having a place to put these parts will help you work in an orderly manner in an organized work area. Keep larger parts out of the way so you won't be stumbling over them.

5. *Provide ventilation.* Engine exhaust fumes contain carbon monoxide, a colorless, odorless, and deadly gas. Never run engines in buildings unless there's ample ventilation to carry exhaust fumes away (Fig. 76). If doors and windows don't provide sufficient ventilation, use, *in addition,* flexible metal tubing to help carry fumes outside. If you plan to run engines in your shop frequently, install an exhaust system equipped with a suction fan to pull the fumes through the tubing and safely outdoors.

6. *Protect yourself from sharp edges and protruding parts (Fig. 77).* Knives, disk and coulter blades, sheet metal, and other sharp-edged parts can skin knuckles and cut severely. Wear gloves. Cover sharp edges with tape, rags, or wooden guards. Grasp all sharp-edged parts away from the cutting edge and hold firmly enough to keep them from slipping in your hand. Never push or pull toward a sharp edge without protecting yourself in some way.

7. *Clean up spilled oil, grease, or fuel.* In addition to removing a fire hazard, you won't be tracking it all over the shop where it could cause a slip or fall.

8. *Make sure coolant level is O.K. in an engine before plugging in an engine block heater.* A fire could result if coolant is low.

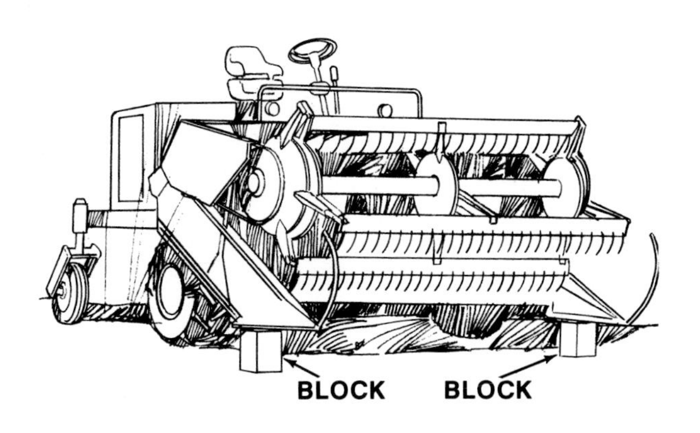

Fig. 75—Support Equipment On Adequate Blocks Or Stands

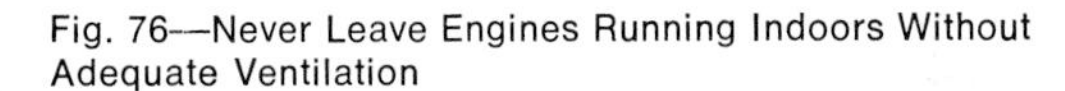

Fig. 76—Never Leave Engines Running Indoors Without Adequate Ventilation

Fig. 77—Protect Yourself From Sharp Edges And Protruding Parts

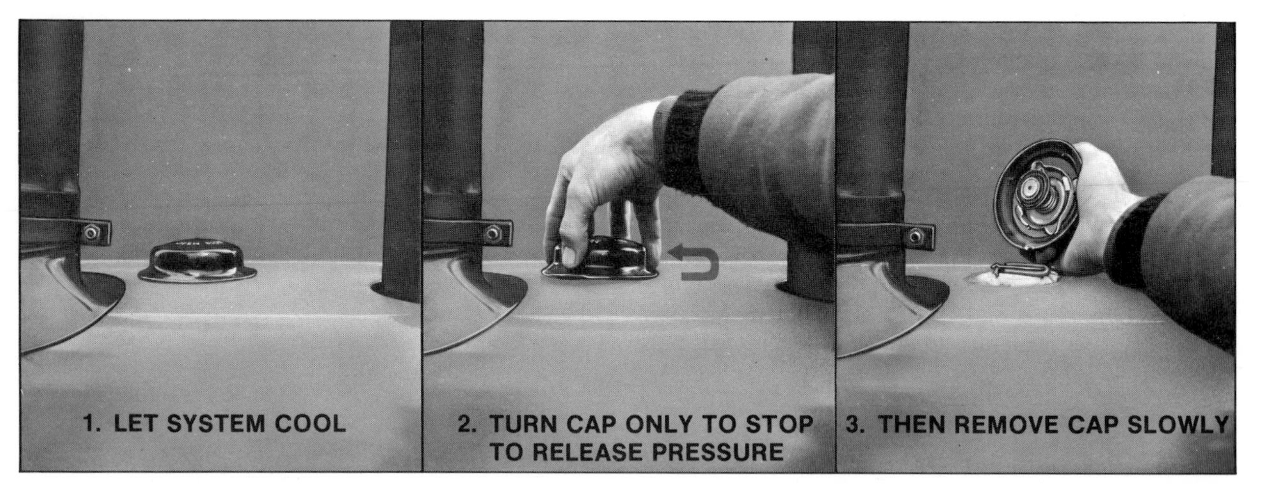

Fig. 78—If Possible, Let The System Cool Before Removing The Radiator Cap. Then Remove Slowly.

9. *Use jacks and hoists to handle heavy components.* Struggling with awkward and heavy parts may injure your back, or you might drop them, crushing your hands or feet. Before removing heavy parts, estimate their weight and be prepared to handle them safely.

10. *Protect your helper.* Work as a team. Warn your helper of your intention to do anything that might affect his safety. Make sure that he also warns you. Turning a shaft, for example, might pinch has fingers on the other side of the machine. *Never start an engine, engage power, or raise or lower a machine without warning your helper in advance.* Make sure you know where he is, and that he knows where you are.

Fig. 79—Keep Electrolyte At The Proper Level To Reduce The Volume Of Accumulated Hydrogen Gas

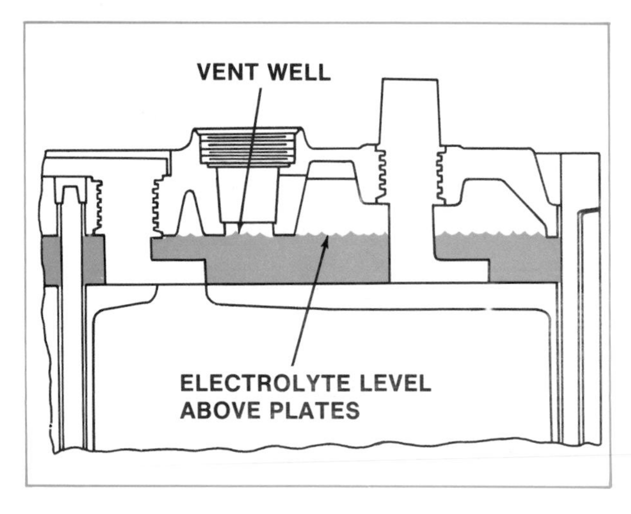

11. *Make sure the job is done right.* Your safety and the reliability of your equipment depend upon how well you do your job. Follow the service procedures outlined in your operator's and service manuals. Observe all precautions. As you work, check each step as you complete it to make sure it's right. Finally, turn all moving parts over several times by hand if possible before engaging power. Then gradually engage power to the machine.

COMPONENTS AND SYSTEMS

Cooling System

Engines are equipped with pressure radiator caps. These caps maintain pressure of 6 to 7 psi in the cooling system, which raises the boiling point of the coolant to about 230°F. If you remove the radiator cap when the coolant is hot, steam and violently boiling coolant may gush out on your hands and body.

When checking the coolant level:

1. *Let the engine and radiator cool* (Fig. 78).

2. *Protect yourself if you must remove the cap from a hot radiator.* Fold several thicknesses of a large rag and place it over the cap. Stand to one side and turn your face away. Then, before removing the cap, loosen it slightly and wait for all pressure to escape.

When handling antifreeze, keep these points in mind. Both types commonly used, alcohol-based antifreeze and ethylene glycol (permanent type), are poisonous. And alcohol-based antifreeze will burn. For these reasons:

a) Keep alcohol-based antifreeze away from sparks and flame.

b) Keep both types of antifreeze away from children.

c) Never put food or beverages in empty antifreeze containers.

Electrical System

Avoid these hazards when servicing an electrical system:

- **Battery explosions**
- **Being burned by the battery electrolyte**
- **Electrical shock**
- **Bypass start hazard**

NOTE: When servicing the electrical system, always follow the steps outlined in your operator's or service manual to avoid damaging the electrical system. Lock out or disconnect the electrical power source from the electrical system before you begin your service procedure.

BATTERY EXPLOSIONS

When charging and discharging, a lead-acid storage battery generates hydrogen and oxygen gas. Hydrogen will burn, and is very explosive in the presence of oxygen. A spark or flame near the battery could ignite these gases, rupturing the battery case and throwing acid all over.

To prevent battery explosions:

1. *Maintain the electrolyte at the recommended level.* Check this level frequently. When properly maintained, less space will be available in the battery for gases to accumulate (Fig. 79). Refer to your operator's manual. Put only water or battery electrolyte in the battery.

2. *Use a flashlight to check the electrolyte level.* Never use a match or lighter, because these could set off an explosion.

3. *Do not short across the battery terminals.* If the battery isn't completely dead, the resulting spark may set off an explosion if hydrogen gas is present.

4. *Remove and replace battery clamps in the right order (Fig. 80).* This is very important. If your wrench touches the ungrounded battery post and the tractor chassis at the same time, the heavy flow of current will produce a dangerous spark. To prevent this, follow these rules:

a)Battery removal: disconnect the grounded battery clamp first. Note: Some systems may be positive ground. Make sure you know which post is grounded.

b) *Battery installation:* connect the grounded battery clamp *last.*

Fig. 80—Remove And Replace Battery Clamps In The Correct Order

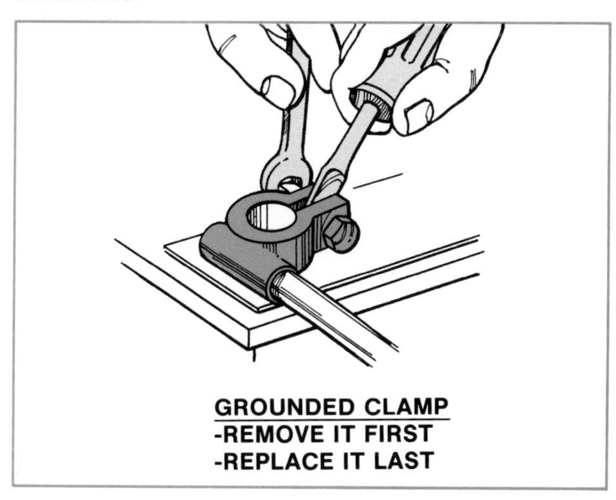

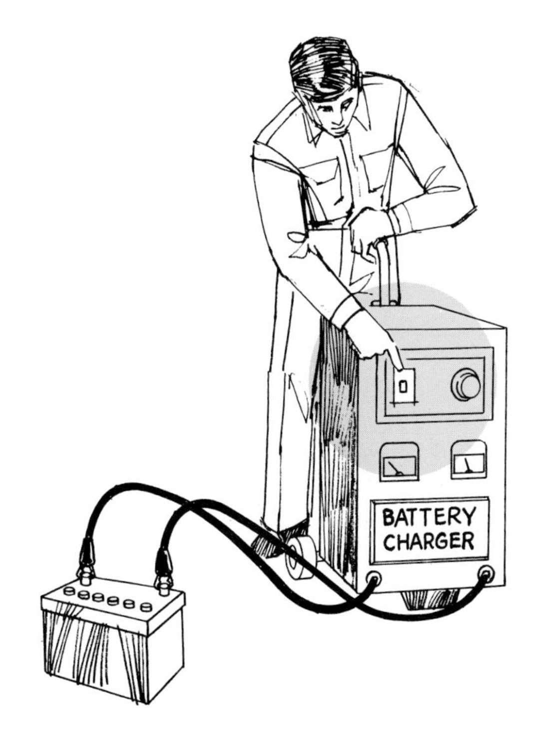

Fig. 81—Turn Charger Off Before Disconnecting Clamps From Battery Posts

5. *Prevent sparks from battery charger leads.* Turn the battery charger off or pull the power cord before connecting or disconnecting charger leads to battery posts (Fig. 81). If you don't, the current flowing in the leads will spark at the battery posts. These sparks could ignite the explosive hydrogen gas which is always present when a battery is being charged.

CONNECTING A BOOSTER BATTERY

Improper jump-starting of a dead battery can be dangerous. Follow these procedures when jump starting from a booster battery.

1. Remove cell caps (if so equipped).

2. Check for a frozen battery. Never attempt to jump start a battery with ice in the cells.

3. Be sure that booster battery and dead battery are of the same voltage.

4. Turn off accessories and ignition in both vehicles.

5. Place gearshift of both vehicles in neutral or park and set parking brake. Make sure vehicles do not touch each other.

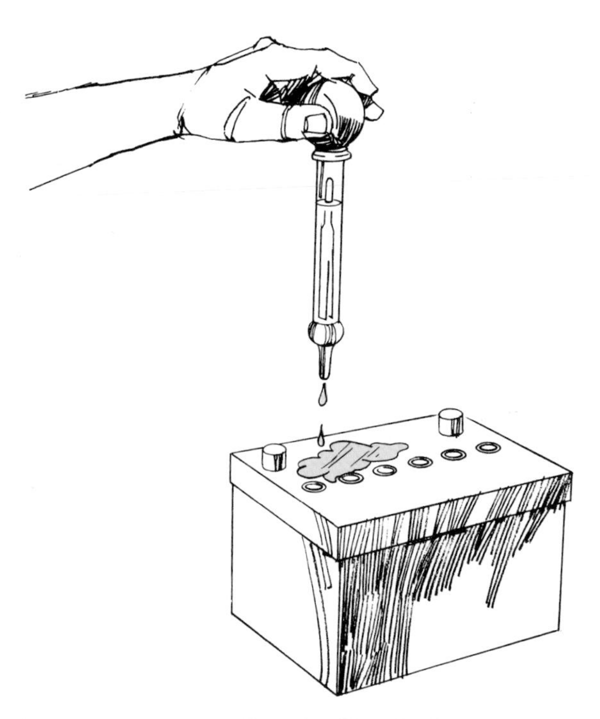

Fig. 82—Avoid Dripping Electrolyte When Reading Specific Gravity

Fig. 83—Never Touch Wires In The Secondary Circuit Or Spark Plug Terminals While The Ignition Switch Is Turned On

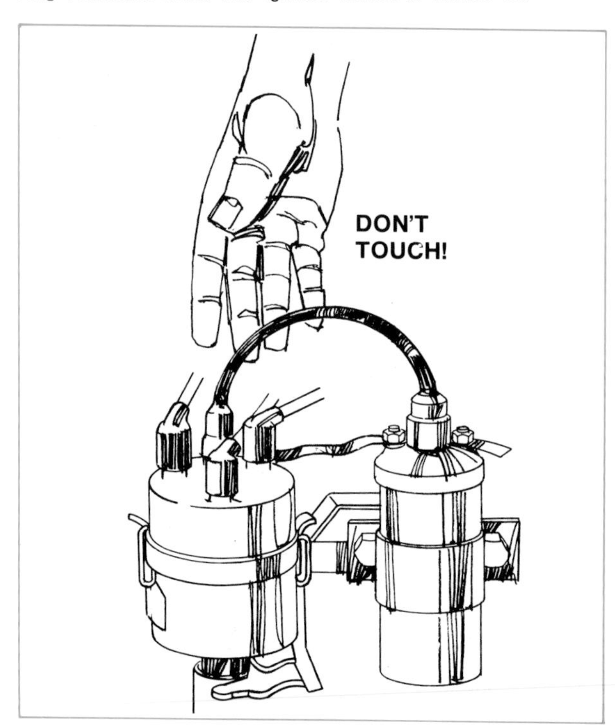

6. Remove vent caps from both batteries (if so equipped). Add electrolyte if low. Cover the vent holes with a damp cloth, or if caps are safety vent type, replace the caps before attaching jumper cables.

7. Attach one end of one jumper cable to the booster battery positive terminal. Attach other end of the same cable to the positive terminal of the dead battery. Make sure of good, metal-to-metal contact between cable ends and terminals.

8. Attach one end of the other cable to the booster battery negative terminal. Make sure of good, metal-to-metal contact between the cable end and the battery terminal.

CAUTION: Never allow ends of the two cables to touch while attached to batteries.

9. Connect other end of second cable to engine block or vehicle metal frame *below* dead battery and as far away from dead battery as possible. That way, if a spark should occur at this connection, it would not ignite hydrogen gas that may be present above dead battery.

10. Try to start vehicle with dead battery. Do not engage the starter for more than 30 seconds or starter may overheat and booster battery will be drained of power. If vehicle with dead battery will not start, start the other vehicle and let it run for a few minutes with cables attached. Try to start second vehicle again.

11. Remove cables in exactly the reverse order from installation. Replace vent caps.

ACID BURNS

Battery electrolyte is approximately 36 percent full-strength sulfuric acid and 64 percent water. Even though it's diluted, it is strong enough to burn skin, eat holes in clothing, and cause blindness if splashed into eyes. Fill new batteries with electrolyte in a well ventilated area, wear eye protection and rubber gloves, and avoid breathing any fumes from the battery when the electrolyte is added. Avoid spilling or dripping electrolyte when using a hydrometer to check specific gravity readings (Fig. 82).

If you spill acid on yourself, flush your skin immediately with lots of water. Apply baking soda or lime to help neutralize the acid. If acid gets in your eyes, flush them *right away* with large amounts of water, and see a doctor at once.

ELECTRICAL SHOCK

The voltage in the secondary circuit of an ignition system may exceed 25,000 volts. For this reason, don't touch spark plug terminals, spark plug cables, or the coil-to-distributor high-tension cable when the ignition switch is turned on or the engine is running (Fig. 83). The cable insulation should protect you, but it could be defective.

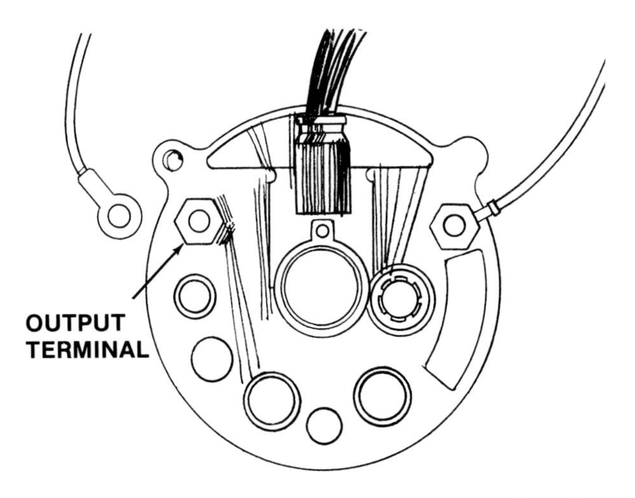

Fig. 84—Don't Run The Engine When The Wire Has Been Disconnected From The Alternator Or Generator Output Terminal

Fig. 85—Relieve All Hydraulic System Pressure Before Starting Service Work

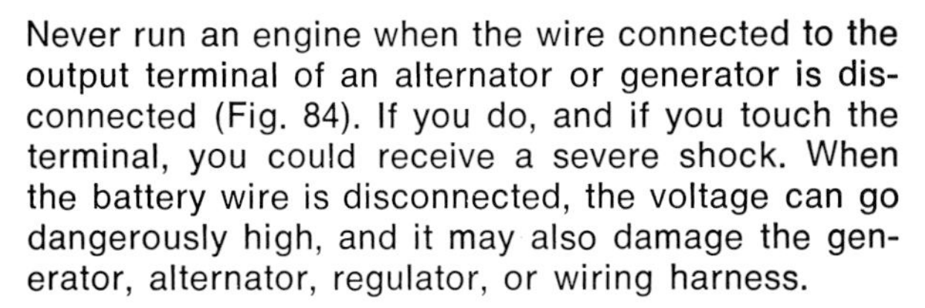

Never run an engine when the wire connected to the output terminal of an alternator or generator is disconnected (Fig. 84). If you do, and if you touch the terminal, you could receive a severe shock. When the battery wire is disconnected, the voltage can go dangerously high, and it may also damage the generator, alternator, regulator, or wiring harness.

Hydraulic System

Damaged or improperly maintained hydraulic systems are hazardous in four ways:

- *High-pressure oil could escape and hit you in the eyes or penetrate your skin.*
- *High-temperature oil could burn you badly.*
- *A raised implement could fall and crush you.*
- *A loosened component could be flung off as a projectile.*

To prevent these hazards, protect the system from damage and maintain it. Keep the hydraulic system in good repair.

1. *Follow all maintenance procedures outlined in your operator's manual.* Change oil and filters on schedule. Keep dirt, moisture, and air out of the system by following the maintenance recommendations in your operator's manual. Maintain the recommended level of oil in the reservoir by adding clean oil when necessary. Keep the fins of oil coolers clean.

2. *Relieve system pressure before beginning maintenance work (Fig. 85).* Lower equipment to the ground, turn off the engine, and move the hydraulic levers back and forth several times. This will protect you from being crushed by falling equipment, and from being struck by high-pressure oil when disconnecting lines, hoses, and fittings. When disconnecting

Fig. 87—Tighten Fittings With Two Wrenches, But Only Enough To Prevent Leaks

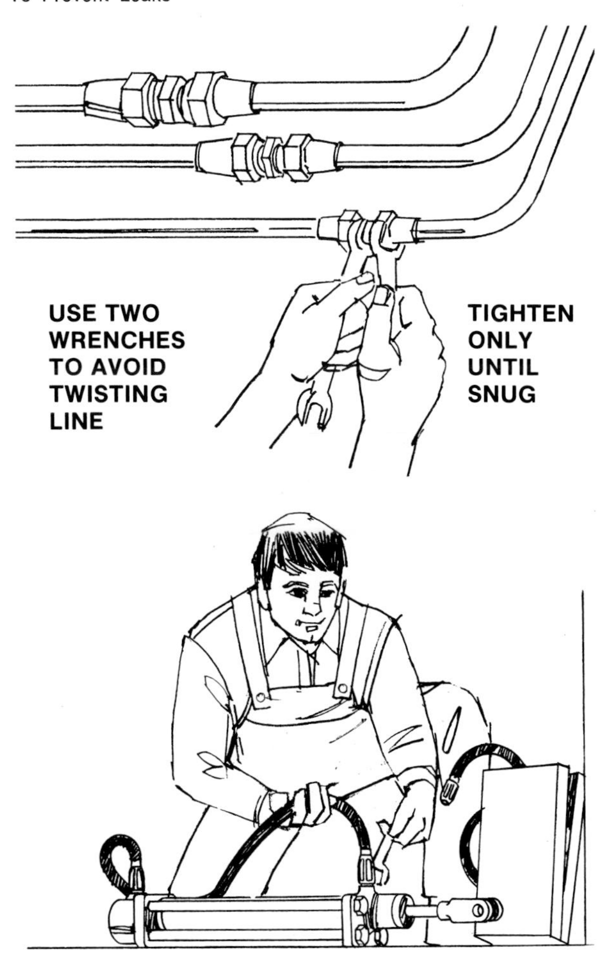

Fig. 86—Replace Damaged Hoses And Fittings

hydraulic fittings, loosen them slowly to relieve residual pressure. Never service the hydraulic system while the engine is running unless the service instructions indicate that you should.

NOTE: Equipment which has a hydraulic accumulator for controls, such as power steering, power brakes and implements with remote hydraulic accumulators, require special precautions because these systems store more reserve pressure and take longer to relieve. Refer to your operator's manual for proper procedures for relieving hydraulic pressure in these systems.

Fig. 88 — Avoid High Pressure Fluids

3. *Avoid damage to lines and hoses.* Keep sharp edges, sharp points, and heavy objects from striking or rubbing against lines and hoses. Rig hoses in clamps, avoiding twists and sharp bends. Allow slack for the changing distances between moving parts when installing new hoses. Replace damaged lines that could rupture under load. Damaged hoses are particularly hazardous (Fig. 86). Inspect hoses and lines frequently.

4. *Maintain fittings.* Keep fittings tight. Never reassemble fittings that are dirty, chipped, or distorted. Fittings must be clean and in *perfect* condition. Keep O-rings and other seals in place. They, too, should be in *perfect* condition. Keep an assortment of O-rings in your shop so they are available when you need them. When tightening fittings, use two wrenches to avoid twisting the lines, but do not overtighten. Tighten the fittings just enough to prevent leaks (Fig. 87).

5. *Prevent burns.* Let the hydraulic system cool before changing hydraulic filters or removing fittings or lines. Hot oil can burn you severely.

Fig. 89—Take All Major Fuel System Work To Your Dealer

6. *Replace parts with identical or approved replacements.* Use replacements that have the same strength and capacity. Parts may look alike, but they may not be the same.

7. *Avoid the hazards of pinhole leaks.* Hydraulic fluid under pressure can penetrate the skin and cause an infection or a severe reaction. When checking for pinhole leaks, wear eye protection. Let oil spray on a piece of cardboard to protect yourself from injury (Fig. 88).

CAUTION: Escaping fluid under pressure can penetrate the skin causing serious injury (Fig. 88, left). Relieve pressure before disconnecting fuel or other lines. Tighten all connections before applying pressure. Keep hands and body away from pinholes and nozzles which eject fluids under high pressure. Use a piece of cardboard or paper to search for leaks (Fig. 88, right). Do not use your hand.

If ANY fluid is injected into the skin, it must be surgically removed within a few hours by a doctor familiar with this type injury or gangrene may result.

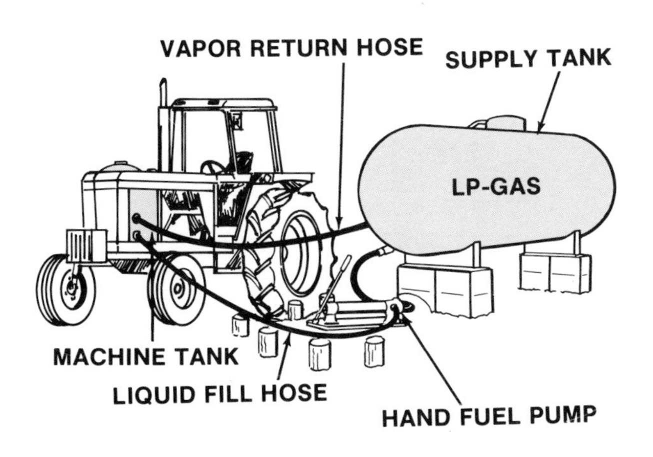

Fig. 90—Fueling With LP-Gas

Fuel system

Have all major fuel system service work done by your farm equipment dealer (Fig. 89).

Unless you're qualified and have special service tools, never attempt to service diesel injection pumps or nozzles. In nearly all cases, when this type of service work is attempted by individuals without specialized training and service equipment, serious damage to the fuel system results. In addition, inexperienced personnel run the risk of injury from high pressure oil.

Most components of LP-gas fuel systems should also be serviced by specialists. The danger of fire and explosion is too great in a farm shop.

Gasoline fuel systems may be serviced and adjusted by operators who have repair experience. However, service work is best left to service personnel.

When performing routine maintenance jobs, remember these general safety precautions.

Fig. 92—Always Install The Air Filter Before Starting The Engine

Fig. 91—Use An Approved Cleaner For Dry Elements

1. *Fuel safety.* Fire hazard is greatest with LP-gas and gasoline (Fig. 90). Diesel fuel and kerosene are less hazardous, unless spilled on a hot engine manifold. Read and follow the precautions in Chapter 5 when fueling tractors.

2. *Use an approved cleaner for dry element air filters (Fig. 91).* Get a package of filter element cleaner from your dealer, and follow the instructions that come with the package. Don't use gasoline, fuel oil or solvents. Gasoline is always too dangerous to use for cleaning. Fuel oil and solvents may destroy the fibers of the filter or leave an oily film.

Fig. 93—Your Dealer Should Service The LP-Gas Converter

Fig. 94—On-The-Farm Service Should Include Only The Fuel Tank, Sediment Bowl, And Filters

3. *Install the air filter before starting the engine (Fig. 92).* The air filter is needed to protect the engine from dust and dirt drawn in with the air. And if the engine backfires, the filter is needed to contain flame in the engine.

4. *Clean up spilled fuels.* Turn off the fuel supply at the fuel tank when changing fuel filters and when cleaning sediment bowls. Promptly clean up fuel to reduce the fire hazard.

Fig. 95—When Bleeding Injection Lines At The Injectors, Loosen One Turn Only

5. *Never adjust a carburetor when the tractor is in motion.* Hook a load to the PTO instead. If you can't, hitch up and operate the tractor with a drawbar load between trial settings. Repeat your trial adjustments until the correct setting is obtained.

WORKING WITH LP-GAS FUEL

Read your operator's manual. Learn the procedures needed to work with LP equipment safely. If certain procedures are not explained in detail in the operator's manual, the manufacturer expects a dealer to do that service (Fig. 93).

Restrict your maintenance operations to cleaning the fuel strainer and *replacing* valves and gauges. Never service the fuel tank except to install new valves or gauges. Don't repair valves or gauges or attempt to repair the converter.

If you use LP-gas equipment, follow these precautions carefully:

1. *Refuel safely.* Observe all precautions outlined in your operator's manual and in Chapter 5.

2. *Do all service work outdoors.* LP-gas is heavier than air. LP-gas will displace air and suffocate you. Moving air is needed to carry the gas away.

3. *Don't place your hands in the path of the vapor stream when releasing pressure.* The gas is so cold it will freeze flesh.

4. *Empty the fuel tank before removing any line, valve, or gauge.* If you can't run the engine to empty the tank, consult your fuel supplier for an approved method.

5. *Don't store a tractor with a leaking LP-gas system in a building.* The accumulated gas could be ignited easily and there is the suffocation danger (see above).

6. *Watch for leaks.* The appearance of frost at any point in the fuel system when the engine is not running indicates a leak. To test for leaks, use soapy water (nonpetroleum base soap) and a brush. Never use an open flame!

7. *Don't use equipment that isn't approved for use with LP-gas.* All equipment, including hose lines, should meet the safety standards of a recognized authority, such as the American Gas Association or the Underwriters Laboratory, Inc.

WORKING WITH DIESEL FUEL

Since injection pumps and nozzles require special tools and equipment, restrict your service work to the low pressure side of the system: the fuel filters, sediment bowl, and fuel tank (Fig. 94).

Air is left in the system when fuel lines and filters are drained. For this reason, you'll need to bleed the air out of the high pressure lines. Normally, you can bleed half of the injection lines, and the others will bleed themselves after the engine has started. When bleeding the lines, loosen the injection line nuts only one turn to avoid excessive spray (Fig. 95). Use two wrenches to avoid bending or twisting the lines. After cranking the engine with the starter, to remove the air, tighten the connections until they are snug and free of leaks.

Before running the engine, be sure all fuel connections are tight. Diesel fuel escaping under high pressure can penetrate the skin (see CAUTION on page 126).

When using an ether starting aid in cold weather, remember ehter is *extremely flammable.* Keep sparks and flames away. Store ether in a cool, protected place. Prevent accidental discharge. Keep the safety cap on the pressurized container. Do not burn empty cans or attempt to crush them. They will blow up!

Power Transmission System

Repair and adjustment of transmissions and other major drive line components should be performed by your dealer. These jobs require specialized training and service equipment. Put your efforts into maintaining and servicing the drive mechanisms that don't require major teardowns, specialized training, or special tools.

To get these jobs done safely:

1. *Observe general safety recommendations for doing service work.* These are discussed earlier in this chapter. All of them apply when servicing drive line and power transmission components. The first precaution is especially important: *Never clean, unplug, lubricate, adjust, or repair any machine while it is running* (Fig. 74, page 120).

2. *Make your work accessible.* Remove shields, guards, and other components to get access to the parts you want to service. Use good judgment. Your work will proceed more easily and safely if you have room to use your tools. Always replace guards and shields that were removed.

3. *Avoid pinch points.* Turning a shaft can pinch your fingers if they are caught between meshing gears, a belt and pulley, or a sprocket and chain (Fig. 96). A common error is to grasp and pull on a V-belt. As the belt moves and rotates the pulley, the mechanic's fingers are pinched between them.

Fig. 96—Don't Turn The Machine Over By Hand If Your Fingers Are At One Of These Pinch Points

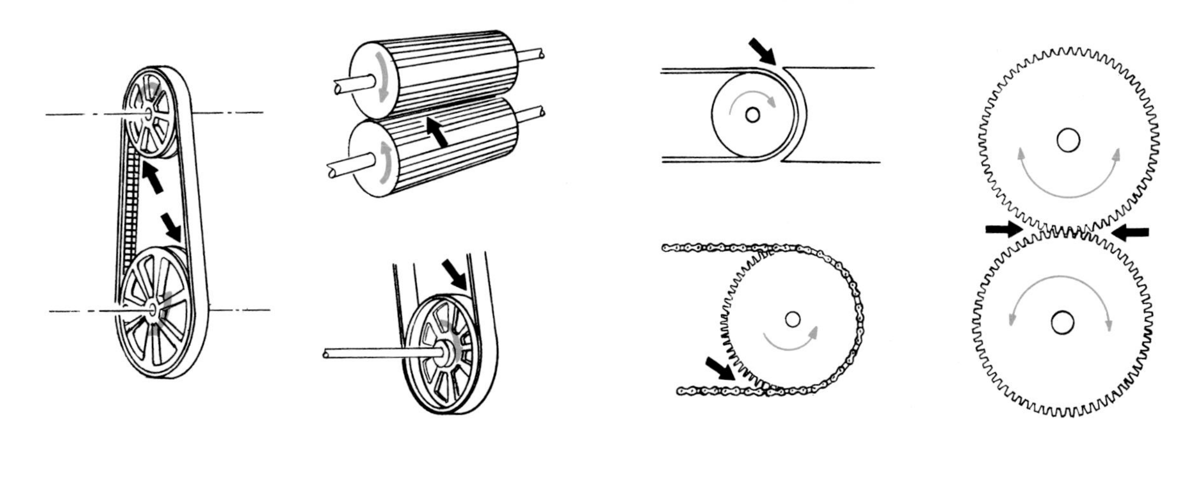

Fig. 97—Wear Goggles When Pulling Bearings Or Striking Bearing Parts

4. *Pull bearings cautiously.* Removing bearings can be dangerous. Hardened steel bearings can shatter under pressure. Pieces fly outward with deadly force. Always wear eye protection when pulling bearings or striking bearings (Fig. 97). Shield the bearing if you can, and keep others away. Always apply force to the tight bearing ring. Exerting force on the free ring may crack the bearing causing pieces to fly with dangerous force.

5. *Don't spin bearings with compressed air.* The rollers or balls in a bearing are held to the inner ring with a bearing cage. If the cage is spun with compressed air, centrifugal force could cause a ball or roller to fly outward with enough force to cause an injury.

6. *Clean with a commercial solvent, fuel oil, or kerosene.* Remember that these are flammable liquids, and keep sparks and flame away. Don't clean with gasoline. It's too hazardous. Refer to page 117 for additional precautions when using solvents.

7. *Keep rotating edges smooth.* File or grind sharp edges smooth. When installing sprockets, pulleys, and gears drive the keys all the way in past the end of the shaft. Clip and bend the ends of cotter keys. These procedures will help prevent cuts from sharp edges, and keep clothing and gloves from being caught in revolving parts (Fig. 98). Always replace guards and shields that were removed.

Tires

Unless you have the experience and the proper tools for mounting and removing tires, it's better to have your tire work done by a farm equipment dealer or a tire service company.

A professional is essential for tires mounted on split rims. Changing split rims is hazardous, especially if proper equipment is not available. If a tire explodes, a lock ring blows off, or a hydraulic bead breaker slips, you could be struck with instant, killing force. Leave split rims to professionals.

If you mount and remove tires on farm equipment, follow the step-by-step instructions given in a tire repair manual, such as John Deere's Fundamentals of Service: Tires and Tracks. The book will make your job much easier and safer.

If you mount and remove tires on farm equipment, follow the step-by-step instructions given in a tire repair manual. These types of books will make your job much easier and safer. See page 76 for address of tire and rim organizations.

Fig. 98—Install Keys Properly To Keep From Tearing Skin Or Clothing

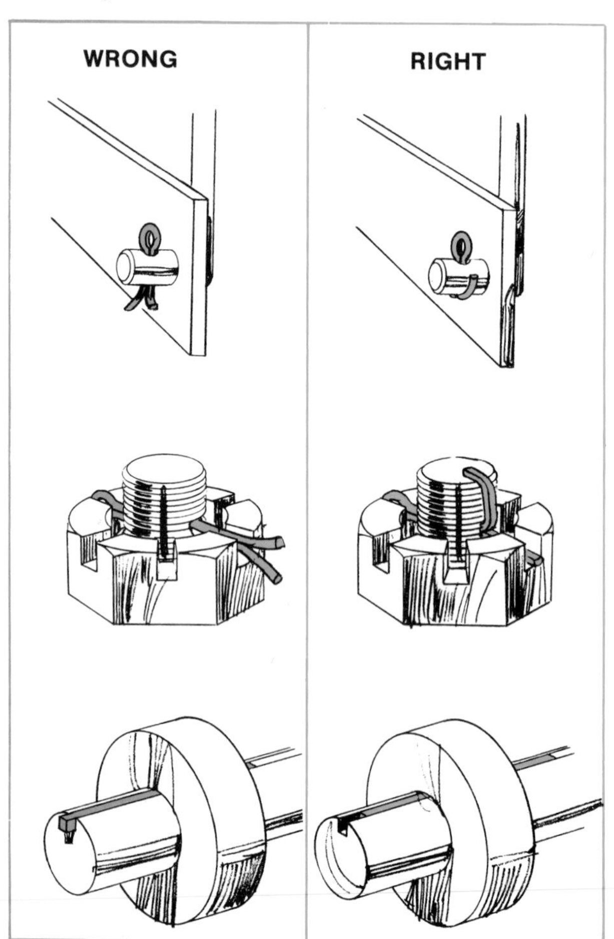

Fig. 99—Apply Rubber Lubricant To Help Slip Tire Bead Over The Rim

Fig. 100—Take Small Bites With Irons To Avoid Hard Prying

If you mount and remove large traction tires, observe these basic safe practices:

1. *Support the machine safely.* Before jacking the equipment off the ground, take steps to keep it from rolling and falling. If the machine is self-propelled, put the transmission in gear or in park position and set the parking brake. When changing tires on pulled equipment, block the wheels. It is usually safest to hitch pull and semi-mounted equipment to a tractor hitch to keep the raised equipment in place.

2. *Use a rubber lubricant.* After tire beads have been broken from the rim, apply a recommended rubber lubricant to the bead you need to remove (Fig. 99). Lubricant will make tire irons easier and safer to use because less force is needed to slip the bead over the rim. To remove a tire, apply lubrication to both beads. Likewise, when mounting tires, use a lubricant to help slip the beads over the rim, and seat the beads against the flanges of the rim.

3. *Handle tire irons safely.* When prying with tire irons, keep these points in mind:

- Never let go of an iron when prying. The iron may spring back with terrific force.

- Don't try to remove large sections of the tire from the rim with each bite. Instead, take smaller bites to avoid hard prying (Fig. 100).

- Keep your balance. Tire irons may slip, forcing you to fall.

4. *Get help when handling large tractor tires (Fig. 101).* Big, heavy tires can tip and fall injuring your back, straining muscles, or breaking bones. When handling large tires, get help. Stand on each side of the tire as shown in Fig. 101. If you have a hoist, lift truck, or loader-equipped tractor, use it to handle tires.

Large tires can fall over on you and fatally injure you. use hoists, block and tackle, and blocks for larger tires and liquid filled tires.

5. *Inspect tires and rims before mounting.* If necessary, clean rust, corrosion, and old rubber off the rim with a chisel or wire brush. Also, inspect the rim for cracks and other damage. Check the tire casing and tube to make sure they're in serviceable condition, and clean and dry. Moisture trapped inside a tire can deteriorate the cord fabric and cause a premature failure.

CAUTION: Do not try to fit a North American-made tire on a European-made rim even when the tire and rim are the same size. Similarly, do not fit a European tire on a rim made in North America. See your tire and rim dealer.

Fig. 101—Get Help With Tires And Wheels

Fig. 102—Make Sure Tire Is Supported Before Putting Hands Between Tire And Rim

tube to make sure they're in serviceable condition, and clean and dry. Moisture trapped inside a tire can deteriorate the cord fabric and cause a premature failure.

6. *Watch for pinch points.* Whenever you're working with a partially mounted tire, there is the danger of the tire slipping and crushing your hands or fingers between the tire and the rim. Block tires securely to avoid being pinched or crushed, especially when installing or removing the tube (Fig. 102). And never place your fingers between the tire bead and the rim when inflating a tire. The beads usually "pop out" against the flange with crushing force.

7. *Stand to the side when inflating tires.* Use an extension hose with a clip-on chuck (Fig. 103). Then you will not need to hold the chuck on the valve stem when you are filling the tire with air. If the beads do not seat at the highest

Fig. 103—Stand To The Side. Never Stand Over Or In Front Of A Tire While Inflating It. Use A Clip-On Chuck

recommended pressure given by the tire dealer, deflate the tire reposition the tire slightly on the rim, lubricate the beads again, and reinflate. *Never exceed* the highest recommended pressure given by the tire dealer when seating beads. Higher pressure could break the bead or even the rim with explosive force. Never fill tires with flammable gas.

8. *Store tires so they can't fall over and hurt someone.* Tie them to the wall or lay them down flat.

9. *Always follow tire and rim manufacturer handling procedures when working on tires and rims.* Multi-piece rims and liquid filled tires can present special problems and hazards if they are not handled properly.

CHAPTER QUIZ

1. What are the three "targets" of farm equipment safety discussed in this chapter?

2. What are the hazards when farm equipment is serviced?

3. Give an example of a change in machine performance that can be detected by:

a) Sight

b) Hearing

c) Feeling

d) Smell

4. What factors should an operator consider when deciding whether to service a machine himself, or take the work to a qualified service shop?

5. Classify each of the following fires by type: Class A, B, or C.

a) Wood shavings

b) Liquid solvent

c) Electrical motor

d) Grease

e) Straw

6. List four basic rules for the safe use of hand tools.

7. Which of the following statements are true?

a) Keep a smooth bevel ground on the heads of chisels and punches.

b) Handles are placed on files for comfort and ease of use.

c) Eye protection is recommended when striking any hardened metal tool or surface.

d) Cold chisels and punches may be used safely as pry bars.

8. Explain why the face of a hammer should be kept parallel with the surface struck.

9. Name two quick checks that should be made before plugging in a power tool.

10. Explain how a three-wire, grounding-type circuit protects a power tool user from electric shock.

11. To which terminal of a three-blade plug or receptacle should the green insulated wire of an extension cord be attached?

12. True or false? Grinder tool rests should be set slightly below the center of the wheel and at least a quarter-inch away from the face of the wheel.

13. How would you test a grinding wheel for defects before mounting it on the grinder?

14. Why can't the eye protection used for oxyacetylene welding be used for arc welding?

15. True or false? The amount of pain felt immediately after receiving an arc flash indicates the extent of eye injury.

16. Select the correct answer. To turn oxygen and acetylene regulators off, adjusting handles should be turned:

a) Clockwise until tight.

b) Counterclockwise until loose.

17. Explain why oil is never used to lubricate oxyacetylene welding equipment.

18. In what situation will a hydraulic jack actually push an implement off the jack?

19. Why should blocks or stands be placed under equipment immediately after it has been jacked up?

20. Give two reasons why natural or synthetic fiber rope may not be safe for hoisting heavy implements and machines.

21. True or false? Gasoline does not vaporize enough in cold winter months to form an explosive or flammable mixture with air.

22. Where should rags soaked with oil or solvent be kept until disposal?

23. What immediate action should be taken if any chemical is accidentally splashed into your eyes?

24. How can you protect yourself from being crushed under a machine if a jack or hydraulic cylinder should fail?

25. What procedure is necessary to safely remove a cap from a hot radiator?

26. Fill in the blanks with "first" or "last." When removing a battery, disconnect the grounded clamp ________. When replacing a battery, connect the grounded clamp ________.

27. What are four potential safety hazards when servicing the hydraulic system?

28. What trouble is indicated if frost forms at a connection in an LP-gas fuel system?

29. How many turns should the high-pressure lines of a diesel fuel system be turned when bleeding the lines of air?

30. What is the main hazard when using an ether starting aid?

31. Which are recommended practices?

a) Wear goggles when pulling bearings.

b) Apply force to the tight bearing ring when removing a bearing.

c) Don't spin bearings with compressed air.

d) Clean bearings with gasoline.

e) Strike bearing parts only with a hardened steel hammer.

32. True or false? Install cotter keys by bending the ends so they fit closely against the shaft or nut.

33. Where is the safest position to stand when inflating a tire?

34. What is the maximum safe tire pressure when forcing the tire beads of a newly mounted tire against the rim flanges with compressed air?

5
Target: Tractors and Self-Propelled Machines

Fig. 1—Safe Tractor Operation

INTRODUCTION

Very few operators receive awards for operating their tractors safely. Safe operators just enjoy many hours of rewarding, productive work. They escape the pain and expense of serious injuries. And they don't become accident victims.

How can you be a safe operator? First, decide that an accident isn't going to ruin your chances of living a happy and productive life. Become dead set against accidents. Get the knowledge and skill you need to perform safely. Then, *always* use this knowledge and skill to protect *yourself and others.*

This chapter describes what you need to know for general safe operation. Specific recommendations are covered in other chapters.

NOTE: This chapter deals with safety. For operating procedures refer to Fundamentals Of Machine Operation—Tractors, Combines, and other manuals in this series and to the operator's manual provided with your machine.

TRACTORS

The safe operator assumes six basic responsibilities. He meets these responsibilities by using the knowledge and skill he acquires. The safe tractor operator:

- **Performs proper maintenance according to the operator's manual**
- **Uses his tractor only for the jobs it was designed to perform**
- **Makes preoperation checks**
- **Removes risk of fire or explosion when refueling**
- **Follows recommended procedures when starting and stopping his tractor**
- **Takes special care to avoid accidents during operation**
- **Observes special precautions when towing implements and machines**

Let's examine these in detail.

USING TRACTORS FOR THEIR INTENDED FUNCTIONS

Tractors perform a wide variety of jobs in farming operations (Fig. 2). They can function as:

- *Load movers that pull heavy implements for tillage operations*
- *Transport units that move equipment from place to place*
- *Implement carriers that carry mounted equipment*
- *Remote power sources that power other machines through their power takeoff shaft or hydraulic system*

To handle such a wide variety of jobs, tractors have many different features. These include:

- *Adjustable wheel tread spacing*
- *Quick steering response*
- *High-power engines*
- *Multi-speed transmissions*
- *High chassis clearance*
- *Unequal weight distribution between front and rear wheels*
- *Individual brakes for each rear wheel*
- *Adjustable drawbar hitches*
- *Hydraulically controlled hitches*
- *Power-assisted controls*
- *Provisions for adding weights and dual wheels*
- *Hydraulic circuits to power or lift implements and machines*
- *PTO shafts to operate trailed and mounted equipment*
- *Differential locks*
- *Provisions for mounting equipment on the tractor chassis*

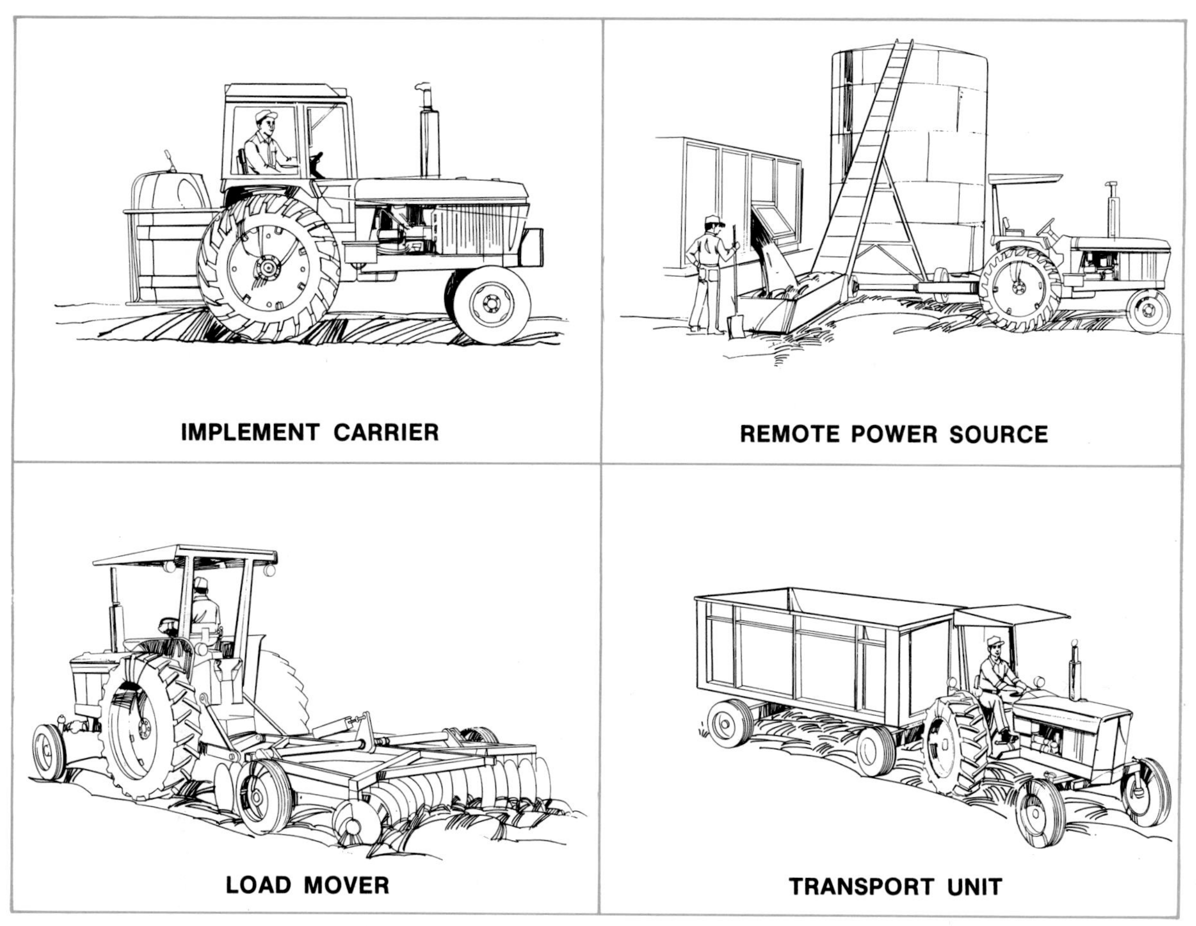

Fig. 2—Use Tractors For Their Intended Functions Only

Features like these let a machine perform a wide variety of jobs in farming operations and increase the tractor's efficiency and ease of operation. They also make tractor operation safer. The better a machine is designed for its job, the fewer operating hazards there will be.

Features designed into tractors can become operating hazards if the operator misuses or ignores recommended operating procedures. Turning quickly and braking only one rear wheel, for example, can tip over a tractor. Engaging a gear quickly when starting a heavy load can tip a tractor over backwards. And failing to disengage the power and shut off the engine before cleaning, servicing, or adjusting PTO-driven machines frequently results in terrible injuries. To benefit from the features your tractor provides, and to prevent any of these features from contributing to an accident, always:

Fig. 3—A Tractor Is Not A Recreational Vehicle

1. Use your tractor only for the specific jobs it was designed to perform. It is *not* a recreational vehicle (Fig. 3).
2. Follow all operating precautions and recommendations outlined in your operator's manual (Fig. 4).

MAKING PREOPERATION CHECKS

Check your tractor like a pilot checks his plane (Fig. 5). Go down a checklist.

Check these:

- **Tires**
- **Shields**
- **Platform and steps**
- **Fuel and hydraulic lines**
- **Visibility from within operator enclosures**
- **Brakes**
- **Steering**
- **Lighting equipment**
- **Seat adjustment**
- **Neutral-start safety switches**

Fig. 4—The Most Important Safety Manual—The Operator's Manual

Fig. 5—Safety Begins With A Preoperation Check

Fig. 6—This Tire Is Hazardous

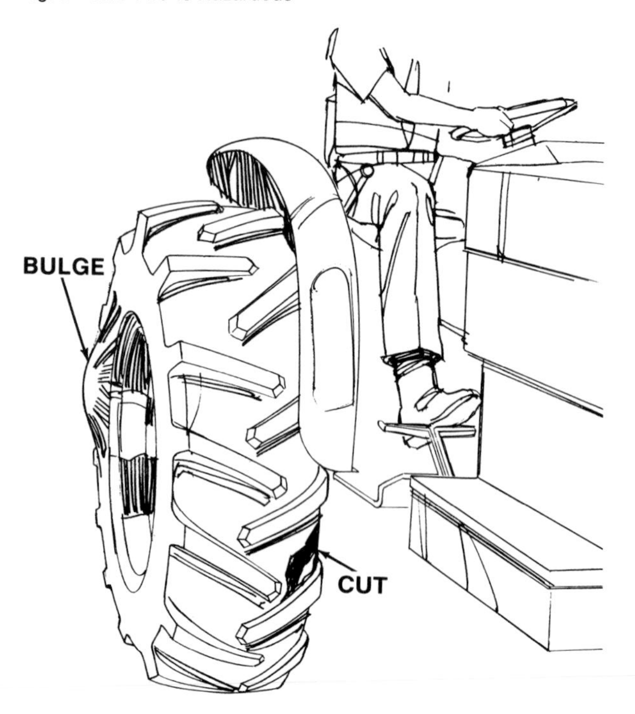

Before checking your tractor examine yourself. Loose, frayed clothing can catch on moving parts and jerk you into machinery. Check your safety boots. Are they in good condition? Do they have slip-resistant soles? Are laces tied tight?

Tires

Look for cuts in the tread and sidewalls. Damaged tires blow out and throw your tractor out of control (Fig. 6).

Check tire pressure. Underinflated tires develop internal damage. Overinflated tires make the front wheels bounce on rough ground and damage the tire internally.

NOTE: Slow down on rough ground. Never let the front wheels bounce. You need to maintain complete steering control.

Shields

Make sure all shields are in place. They protect you from moving parts and pinch points. Always make sure the tractor PTO shaft is shielded, and that all shields provided for PTO-operated equipment are in place.

Refer to page 161 for more information on shielding the tractor PTO shaft. Chapter 3 gives details on shielding moving machine parts.

Platform And Steps

Clean off slippery mud, grease, and any crop residue that may have accumulated on the steps and operator's platform of your tractor (Fig. 7). Remember that chains and tools carried on the platform may interfere with pedal operation or cause a slip or fall from the tractor. Remove them.

Fuel And Hydraulic Lines

Leaky fuel lines and connections are fire hazards and waste fuel. Loss of pressure or fluid from hydraulic lines can cause loss of power steering, power brakes, and control of the three-point hitch. Keep hydraulic lines in good repair.

Hydraulic oil and diesel fuel released under high pressure are especially hazardous. Many hydraulic and diesel fuel systems develop pressures over 2,000 psi, which is three times the pressure needed for oil to penetrate skin. Pinhole leaks are often invisible, but especially dangerous and can be located only with a magnifying glass or piece of cardboard (Fig. 8). They can penetrate the skin. Don't touch them.

CAUTION: Escaping fluid under pressure can penetrate the skin causing serious injury (Fig. 28, left). Relieve pressure before disconnecting fuel or other lines. Tighten all connections before applying pressure. Keep hands and body away from pinholes and nozzles which eject fluids under high pressure. Use a piece of cardboard or paper to search for leaks (Fig. 28, right) Do not use your hand.

If ANY fluid is injected into the skin, it must be surgically removed within a few hours by a doctor familiar with this type injury or gangrene may result.

Fig. 7—A Safe Platform Is Clean And Free Of Tools And Debris

Fig. 8 — Avoid High Pressure Fluids

Fig. 9—With Windows Cleaned, Mirrors Adjusted, And Wipers Functioning, This Tractor Is Safer To Operate

To prevent injury from escaping high-pressure oil and to avoid loss of power steering, power brakes, and control of the three-point hitch, replace defective lines, fittings, and seals. Keep all connections tight. When making replacements, make sure all pressure is released from the system before loosening any connection or fitting. Diesel fuel system pressure is released when the engine stops. To release hydraulic system pressure, lower equipment to the ground, turn off the engine, and move all hydraulic controls back and forth in each direction a few times. Refer to Chapter 4 for service procedures and additional precautions.

NOTE: On some tractors, you must adjust the load to a position of low hydraulic pressure before you stop the engine. Refer to your operator's manual for the proper procedure.

Visibility

Safe tractor operation requires good visibility in all directions. Check for clean windshield or wiper operation and mirror adjustment on tractors equipped with operator enclosures. If the windows need cleaning, take time to complete this chore (Fig. 9).

Brakes

Test your tractor's brakes before starting tractor movement (Fig. 10), and again at slow travel speed. If you have power brakes, start the engine, press each pedal, and observe its operation. Brakes should have a solid feel, and the distance of pedal travel should not exceed that specified in your operator's manual. Then lock the brake pedals together, drive ahead slowly, declutch, and press on the brakes. If brakes aren't equalized tractor will swerve. If the brakes do not pass these tests, check your operator's manual for adjustment procedures, or have a service shop adjust the brakes for you.

Steering

Turn the steering wheel in each direction and note the amount of rotation before the wheels start to turn. Steering should be quick and responsive without excessive play in the steering linkage. Units equipped with power steering should respond when only finger pressure is applied to the steering wheel.

Lighting Equipment

Lights are required during daytime periods of low visi-

Fig. 10—Brakes Should Have A Solid Feel

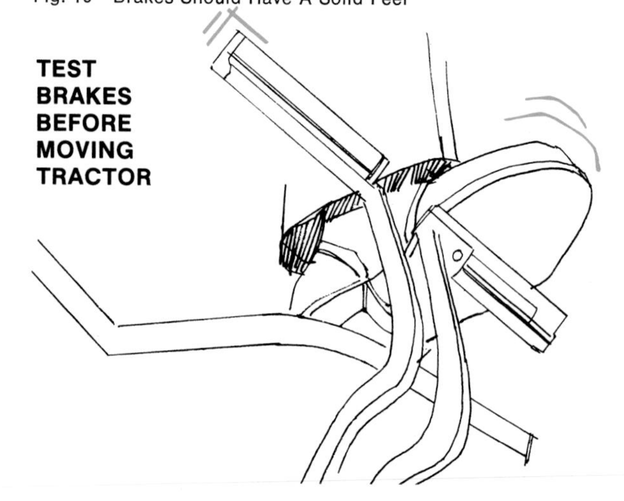

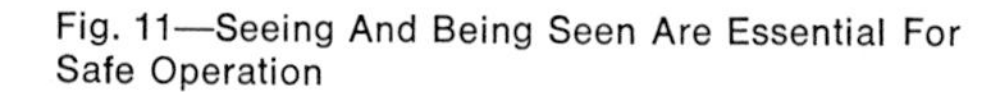

Fig. 11—Seeing And Being Seen Are Essential For Safe Operation

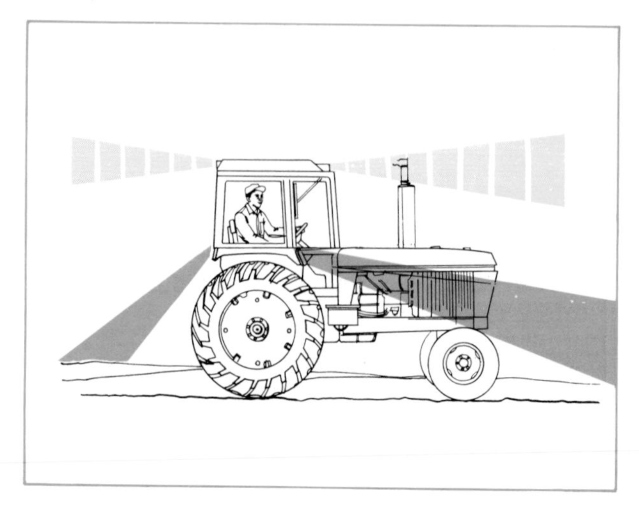

bility, after sunset, and when traveling on public roads (Fig. 11). If you anticipate the need for lights, check their operation before you start. Check them periodically for proper adjustment by following the procedures outlined in your operator's manual. Keep your flashing lights on night and day when driving on public roads so people can see and recognize you easier. Refer to page 168 for lighting requirements when traveling on public roads.

Seat Adjustment

Check the position of the seat if another person has been operating your machine. Your arms and forearms should form a 90-degree angle when your hands are on the steering wheel (Fig. 12). Your legs should remain slightly angled after depressing the foot pedal.

Many seats can be adjusted to the operator's weight to cushion him from bumps and jolts.

Neutral-Start Switches

Neutral-start switches keep the engine from starting when the transmission is engaged or when the clutch is engaged. Check them periodically to make sure they're working properly.

Neutral-start switches can be located so that starting is only possible when:

- *The clutch or inching pedal is depressed*
- *The shift lever is in neutral or park position*
- *Any combination of the above*

These switches should prevent engine cranking when the rear wheels are engaged with the engine. If they don't, have your dealer adjust or replace them. Here's how to check their operation:

1. Determine from your operator's manual which controls are connected to the starting switches.
2. Drive to an open field.
3. Try to start the engine with the controls *in each position* that should *prevent* the engine from starting:
   - *Without depressing the clutch or inching pedal*
   - *With range shift lever in each position other than neutral or park*

Fig. 12—Proper Seat Adjustment Keeps The Operator Comfortable And Within Easy Reach Of Controls

REFUELING

Prevent fires and explosions when refueling. The greatest danger occurs when handling gasoline or LP-gas. These fuels vaporize at temperatures as low as 0°F (22°C), and mix with surrounding air to form explosive mixtures at relatively low (6-percent) fuel-vapor concentrations.

Diesel fuel is safer to handle. It vaporizes at approximately 110°F (43°C). But to be completely safe, observe the same precautions when handling diesel fuel that you do for gasoline.

Safety begins with the proper installation and use of fuel storage equipment. Review the recommendations presented in Chapter 4.

PREVENT MACHINE RUNAWAY

Avoid possible injury or death from machinery runaway.

Do not start engine by shorting across starter terminals. Machine will start in gear if normal circuitry is by passed.

NEVER start engine while standing on ground. Start engine only from operator's seat, with transmission in neutral or park.

Fig. 13 — Never Bypass The Neutral Start Switch

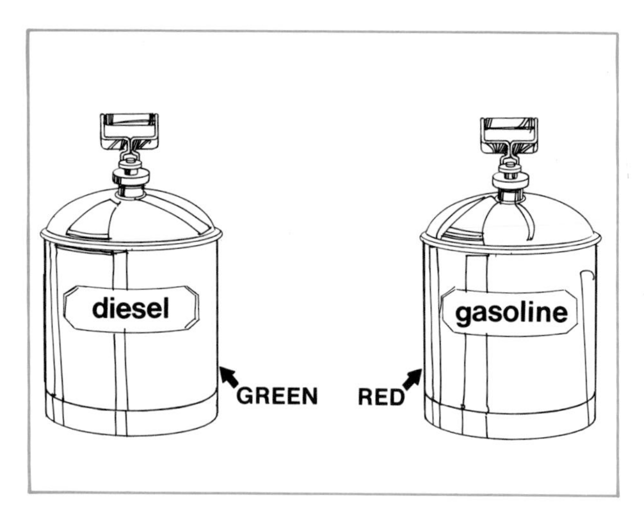

Fig. 14—Carry Fuel In Cans Labeled And Painted To Identify The Type Of Fuel

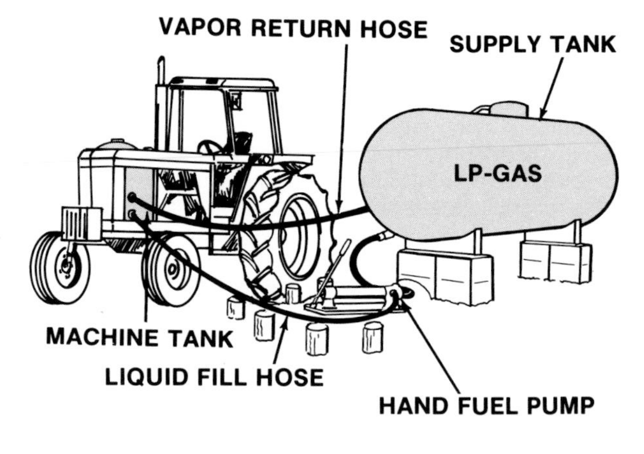

Fig. 15—Refuel With LP-Gas Outdoors And With Vapor-Return Line Attached

Gasoline And Diesel Fuel

1. *Identify fuel containers.* Don't run the risk of someone confusing gasoline with diesel fuel. Gasoline has almost no lubricating qualities, and its use in a diesel tractor would quickly seize and damage the injection pump. Tractor fuels are flammable and explosive, so people should be able to identify them quickly and positively. Label your containers, and paint gasoline containers *red* and diesel fuel containers *green.*

2. *Carry gasoline and diesel fuel in approved safety cans (Fig. 14).* These cans are made from heavy-gauge metal or tough polyethylene. They are equipped with caps that seal completely and with pressure-relief valves that open when vapor pressure within a can exceeds 3 to 5 psi. In addition, safety cans are frequently equipped with flash-arresting screens in the filler opening and pouring spout. These screens reduce the possibility of fire or explosion. Identify approved safety cans by reading the label. Only those approved by the Underwriters Laboratory, Inc. or the Associated Factory Mutual Fire Insurance Companies, or those meeting United States government standards are approved safety cans.

   Never carry fuel in glass or plastic containers designed for household or farm chemicals. These containers are fragile and do not identify by their shape or color the fact that they may contain flammable, explosive fuel. Thin-gauge metal economy gasoline cans often found in discount stores are not recommended.

3. *Let hot engines cool before refueling.* Wait at least five minutes before refueling a hot engine, and longer if possible. If you can't wait for a hot engine to cool at the end of a day's work, refuel it the next morning or after supper. Refueling after supper is preferred. Moisture may condense in the fuel tank during the night if you wait until morning to refuel. If you spill fuel, wait for it to evaporate before starting the engine.

4. *Stay away from flame.* Welding or smoking in the area of your fueling operation could touch off a flash fire or explosion. Turn off the engine and all electrical switches. Do not refuel your tractor near fires or sparks. Gasoline fumes are heavier than air. They float down. A hot engine or spark can ignite the fumes.

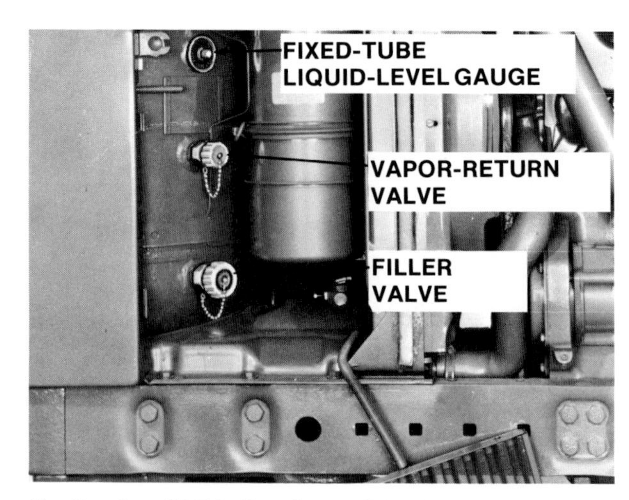

Fig. 16—Open LP Tube-Type Gauges Only Momentarily, Then Close Them

LP-Gas Fuel

1. *Refuel outdoors.* Never refuel in or near buildings or over pits. There is too much risk of the heavy LP-gas collecting and increasing the possibility of an explosion.
2. *Always connect the vapor-return line.* The vapor-return line equalizes the pressure between the tractor and supply tanks and permits the tractor tank to fill without loss of fuel vapor to open air (Fig. 15).
3. *Never fill an LP-gas tractor tank more than 3/4 full.* Overfilling can cause excessive pressure buildup in the tractor tank, and this will open the relief valve. Gas will then escape from the tank, wasting fuel and creating hazardous explosive conditions. LP-gas tanks are equipped with either automatic fuel tank gauges and manually operated foxed tube-type gauges. To operate a fixed tube-type gauge, open it *momentarily* while the tank is being filled and then close it (Fig. 16). When the tank is filled, the mist that comes from the outlet changes to a spray of liquid fuel. Do not *overfill. Never permit the gauge to remain open.* Doing so is extremely hazardous.
4. *Don't refuel an LP-gas tractor when it's hot,* when the engine is running, or when the ignition switch is on.
5. *Don't smoke during the refueling operation.* Wait until all of the colorless LP-gas vapors have ventilated away.

STARTING

After completing your preoperation safety checks, take the following precautions when starting the engine:

Fig. 17 — Provide Ventilation Before Starting

1. *Provide adequate ventilation if starting the tractor indoors (Fig. 17).* Carbon monoxide, a colorless, odorless, and deadly gas, is present in the exhaust gases of all engines. Be sure you have plenty of good, fresh air when starting up indoors. Open doors and windows. Don't take a chance.
2. *Start the tractor from the operator's platform.* With some tractors, the only alternative is to stand on the side of the tractor between the front and rear wheels. *Don't do it!.* If the transmission is in gear when the engine starts, you could be crushed (Fig. 18).

If you use ether, follow the instructions printed on the can and those in your operator's manual. These instructions are designed to prevent fire and damage to your engine. Ether is extremely flammable. Keep sparks and flame away when using it. To prevent accidental discharge when storing the pressurized can, keep the cap on the container, and store it in a cool, protected place.

PREVENT MACHINE RUNAWAY

Avoid possible injury or death from machinery runaway.

Do not start engine by shorting across starter terminals. Machine will start in gear if normal circuitry is by passed.

NEVER start engine while standing on ground. Start engine only from operator's seat, with transmission in neutral or park.

Fig. 18 — Do Not Start The Engine While You Are Standing On The Ground.

3. *Be sure the PTO is disengaged.* The power takeoff shaft may have been shifted into gear after you last parked the tractor. If engaged and connected to a machine, the machine might start when the engine starts, creating a definite hazard.

4. *Shift to neutral or park and disengage the clutch before starting the engine (Fig. 19).* A tractor started in gear will lurch forward or backward, possibly causing serious personal injury or property damage. Most tractors and self-propelled machines are equipped with neutral-start switches that make it impossible to start the engine when the transmission is in gear. But make it a habit to disengage the clutch or inching pedal before starting, even if starting switches are provided. They can fail!

5. *When hitching to implements, line up the hitch from the operator's seat (Fig. 20). Never allow anyone to stand between your tractor and your implement when you are hitching your implement to your tractor.*

6. *Look ahead and to the rear before moving the tractor.* Make sure no people or obstructions are ahead or behind. Sometimes machines are accidentally started in the wrong direction. Small children's actions are unpredictable. Sound the horn to warn your tractor is about to move.

7. *Engage power to the drive wheels slowly.* Engaging the clutch suddenly, or quickly shifting a hydraulic transmission to high speed, could tip the tractor over backwards, especially when starting up a slope. Be especially careful when starting a heavy load and when the rear wheels are frozen to the ground. If they're frozen, break them loose by starting out slowly in reverse.

STOPPING AND PARKING

Your ability to stop your tractor and park it safely is just as important as being able to get it underway safely.

Tractor upsets, collisions, runaway tractors, and people being crushed under machines and implements can happen when operators ignore safety recommendations.

To avoid these accidents, special precautions are necessary when stopping and parking tractors. Here are some important ones:

1. *Signal before stopping, turning, or slowing down.* If other drivers are not warned in advance of your intentions to change rate or direction of travel on the highway, they may run into your equipment, or be forced to swerve into a dangerous situation themselves. Avoid this risk by giving the proper signal before you slow down, turn, or stop.

   Some newer tractors are equipped with turn signal lights. Use these if your tractor has them. If not, use the standard arm signals for motor vehicle drivers on public roads (Fig. 21). Remember to signal well in advance and for the distance of travel required by the laws in your state. And remember that towed equipment may obstruct the vision of drivers behind you. If it does, they won't be able to see the signals you give.

Fig. 19—Before Starting Engine, Shift To Neutral Or Park And Disengage The Clutch

2. *On highways, pull over to the right-hand shoulder before stopping if possible.* This is usually the safest procedure. Before doing so, however, make sure the shoulder is wide enough and sound enough to support your tractor and equipment.

3. *Slow down before braking.* Use engine braking to reduce speed before applying the brakes and disengaging the clutch. This will help keep your tractor under control. Latch brake pedals together for highway use.

Fig. 20—Line Up The Hitch From The Operator's Seat

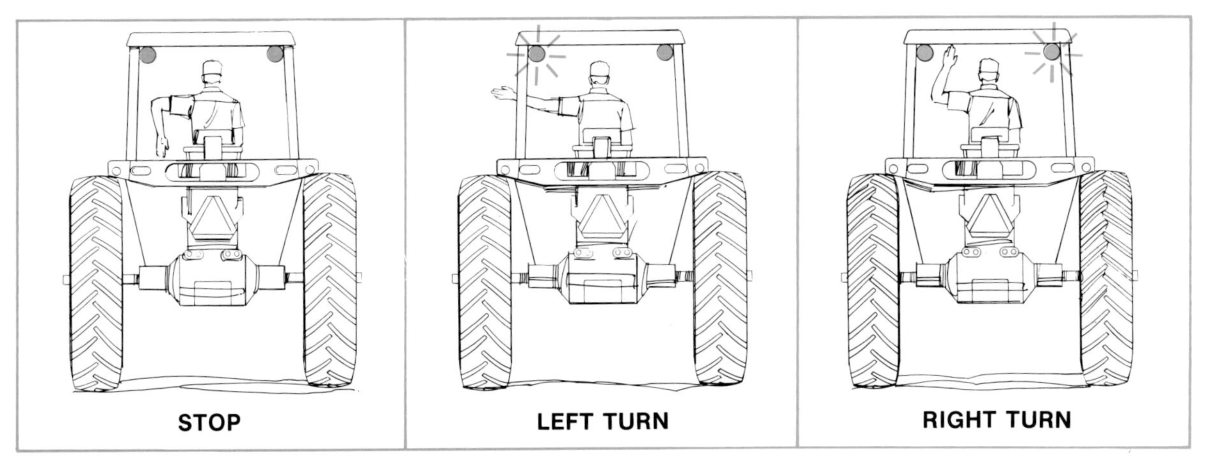

Fig. 21—Signals Prevent Accidents

4. *Pump the brakes when stopping on slippery surfaces like ice, snow, mud, loose gravel, or manure.* If brakes are applied and the tractor skids, release the brakes and apply them again with a slight pumping action—on and off—until the tractor stops (Fig. 22).

5. *Be careful when towing and stopping heavy loads.* Remember that a combination of factors determines your stopping distance:

Fig. 22—Pump Both Brake Pedals To Stop In Slippery Conditions

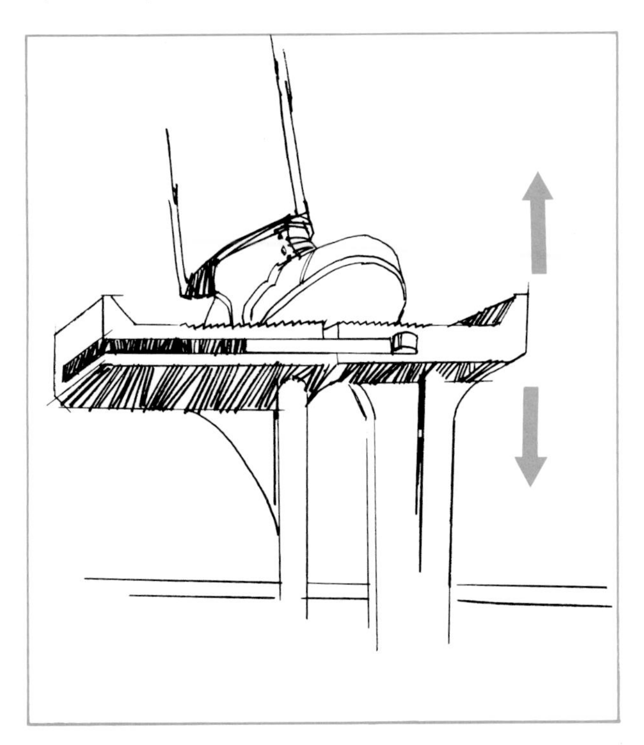

- **Your reaction time**
- **Travel speed**
- **Weight of the tractor and load being towed**
- **Efficiency and condition of the brakes**
- **Traction of the tires that have brakes**
- **Surface conditions**

These factors become important when towing heavy loads. To maintain your stopping ability and to avoid jackknifing towed equipment, observe these precautions carefully:

- If heavy, rolling loads are not equipped with their own brakes, keep travel speed slow and avoid hills and slopes (Fig. 23).
- Lock the brake pedals together for equal braking on each wheel.
- Tow heavy loads slowly — at less than road-gear speeds. Use safety chains.

Fig. 23—Individual Equipment Brakes Are Desirable When Pulling Heavy Rolling Loads

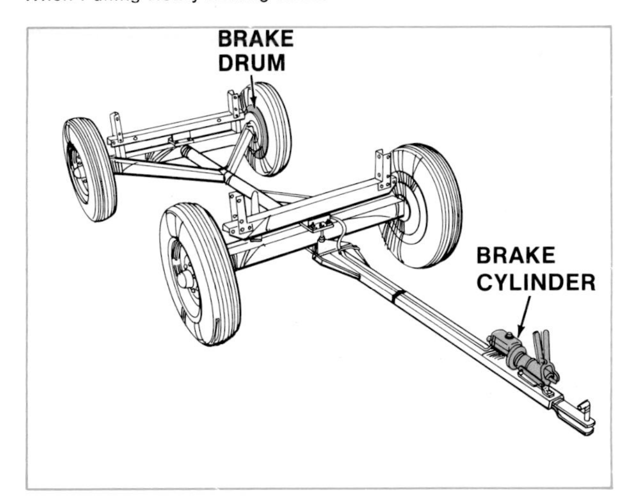

Fig. 24—To Avoid This, Set The Parking Brake And Shift Into Park

- Check brakes on towed equipment.
- Give yourself plenty of time and distance to stop.
- Slow down by reducing engine speed before applying the brakes.
- Travel more slowly in slippery conditions — on ice, mud, gravel, and manure.

6. *Shift to park or set the parking brake (Fig. 24).* If your tractor has a parking brake, use it. This is most important on hillsides and grades. Do not depend on leaving the transmission in one of the driving gears to keep your tractor from rolling. Many tractors are not equipped with parking brakes. Instead, one of the transmission shift levers is shifted to the park position. It locks the transmission with positive action, keeping the tractor stationary. Make it a habit to secure your tractor every time you leave it. Set the parking brake, or shift to park.
7. *Lower all equipment when leaving your tractor.* Don't rely on the tractor's hydraulic system to support a raised implement or machine temporarily, or for service or storage. Hydraulic lines may rupture, and seals and valves may fail. Children or others could move the controls when someone is under the machine. To eliminate any chance of an implement or machine falling unexpectedly, follow one or more of these procedures:
   a) Lower the equipment to the ground or onto sturdy blocks.
   b) Use mechanical safety stops or transport links, if provided.
   c) Use support stands if provided for the equipment.
8. *Turn off all electrical switches and remove the key.* Take the key with you to prevent unauthorized people from starting the tractor. See that all lights and electrical components are turned off. This will prevent a drain on the battery and starting problems the next time the tractor is used.

OPERATING THE TRACTOR

The safe operator is alert to the hazards of tractor operation. He uses safe operating practices *as a matter of routine.* He knows what types of accidents occur, and he takes every precaution to prevent them.

Types of tractor accidents people have:

- *overturn their tractors*
- *fall from tractors*
- *crushed (other than run over)*
- *run over*
- *collide with other motor vehicles*
- *entangled in power takeoff shafts*

Rush, fatigue, and preoccupation with other problems are some of the reasons that these people died. In other cases, the victims did not understand the hazards of tractor operation and became involved in accidents that, to them, were unexpected. Some accidents may become fatal if safety equipment is modified (Fig. 25). never modify safety equipment such as the ROPS.

Never modify structural members of the roll-over protective structure (ROPS) body by welding, bending, drilling or cutting as this might weaken the structure. If any structural member is damaged, replace the entire structure. Do not attempt repairs.

Fig. 25 — Damaged Safety Equipment Like A ROPS Must Be Replaced, Not Reused.

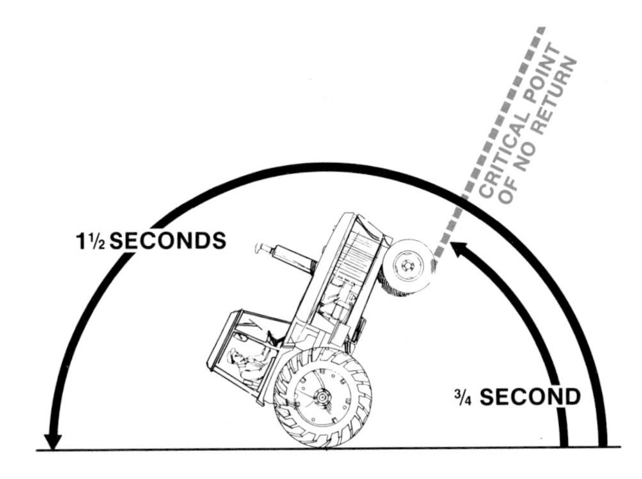

Fig. 26—Reaction Time In A Backwards Tip. You Can't Always Stop It In Time.

The safe operator understands the causes of accidents and takes *every* precaution to avoid them. Let's turn our attention now to the most common types of tractor accidents and see how they can be prevented:

- **Tractor upsets**
- **Falls from tractors**
- **Crushing and pinching accidents during hitching operations**
- **Collisions with other motor vehicles**
- **Getting entangled in PTO shafts**

TRACTOR UPSETS

Many fatalities result from tractor overturns.

The chances of surviving an upset without injury are not good, unless the tractor is equipped with rollover protection and seat belts. In a backwards tip, the tractor hood can hit the ground in less than 1 1/2 seconds after the front wheels begin rising. After the wheels begin to rise, the operator has less than 3/4 second to realize what is happening and to take preventive action. Usually, the tractor is past the *point of no return* before the driven can to do anything to keep it from falling on him (Fig. 26). You must prevent turnovers by knowing the causes and avoiding them.

Causes Of Upsets

- To understand the causes of upsets, learn how these forces act on a tractor:
- *Gravity*
- *Centrifugal force*
- *Rear-axle torque*
- *Leverage of the drawbar or hitch*

When a tractor is being operated, all these forces act at the same time. *Each one* is capable of upsetting the tractor by itself, but most upsets are caused by a combination of these forces.

GRAVITY

The earth exerts a pull. This pull is called the *force of gravity*. Gravity is measured in units of weight, such as ounces, pounds, and tons.

A tractor's *center of gravity* is the point where all parts balance one another. The center of gravity of a block of wood is easy to visualize, being in the center. If suspended by a string attached to the center, the block will hang straight (Fig. 27). But if the string's hitch point is at another location, the block will move until the hitch point and the center of gravity form a straight line. The block will then hang in a tipped position (Fig. 27).

A tractor's center of gravity depends on the location of its components. On conventional two-wheel-drive tractors, the rear wheels, differential, final drive, transmission, and cab weigh more than the rest of the tractor. The center of gravity is behind the midpoint of the tractor's length. And since most of a tractor's weight is carried above axle height, its center of gravity is above this level.

Fig. 27—The Block Tips Until The Line Of Suspension Passes Through The Center Of Gravity And The Attaching Point In A Straight Line

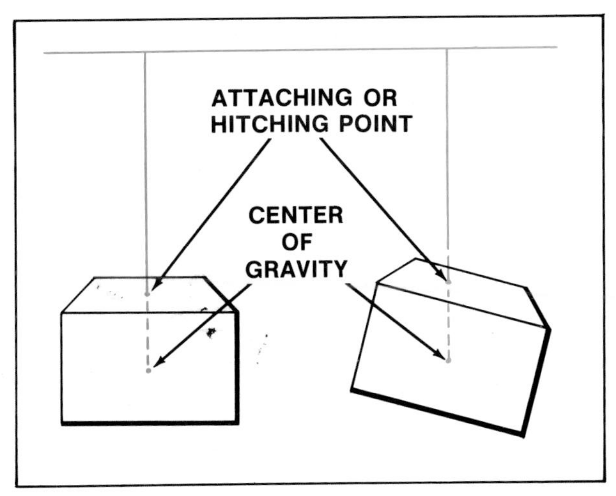

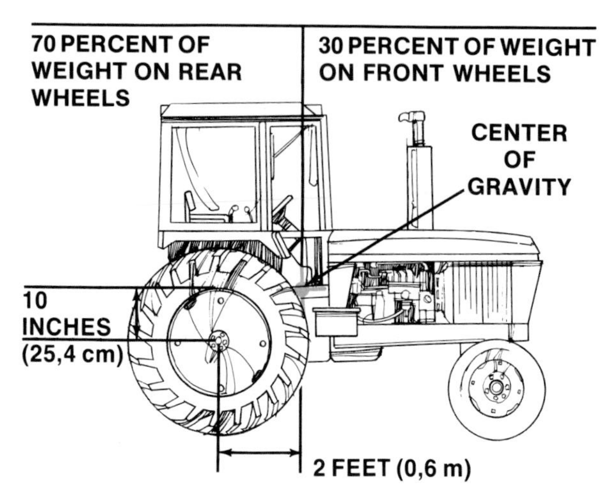

Fig. 28—The Center Of Gravity On A Two-Wheel-Drive Tractor

For a typical row crop tractor with two-wheel drive, the center of gravity is about 2 feet (0,6 m) in front of, and 10 inches (25,4 cm) above, the rear axle. This means that about 70 percent of the tractor's weight is on the rear wheels, and 30 percent on the front wheels (Fig. 28). And when rear-mounted weights or duals are added for increased traction, the center of gravity is moved even farther to the rear if weight is not also added to the front.

To help understand the role that the center of gravity plays in causing upsets, let's try to tip the block of wood we mentioned earlier. Before we do, let's attach a string and plumb bob at its center of gravity and place the block on a table edge with the plumb bob hanging below. Notice in Fig. 29A that the plumb bob hangs directly below the center of gravity, and that the entire bottom surface of the block is supported by the table.

If we tip the block a little, the plumb bob swings toward the edge supporting the block (Fig. 29B). If the block of wood is released, it will fall back to its original position. But if we tip the block far enough the block will tip over on its side (Fig. 29C and D). Tipping occurs when the block's center of gravity moves outside the limits of its *base of stability*.

Tractor upsets happen the same way. If the tractor's center of gravity moves outside its base of stability, the tractor will overturn.

A tractor's base of stability is determined by the location (width) of its rear wheels and the type of chassis support on the front wheels. Because the chassis of tractors with wide-front axles are supported on pivot points on the front axle, the base of stability is triangular, as it is for tricycle-type tractors, but the pivot point is higher on wide-front axles and thus provides more stability (Fig. 30).

To prevent upsets caused by the force of gravity, avoid all situations that might put your tractor in an unstable position. These situations are described in the section on preventing tractor upsets.

CENTRIFUGAL FORCE

An object in motion will travel in a straight line unless there is some force exerted upon it. The force that resists change in direction is called *centrifugal force.*

Fig. 29—Location Of The Center Of Gravity Determines If The Block Will Tip

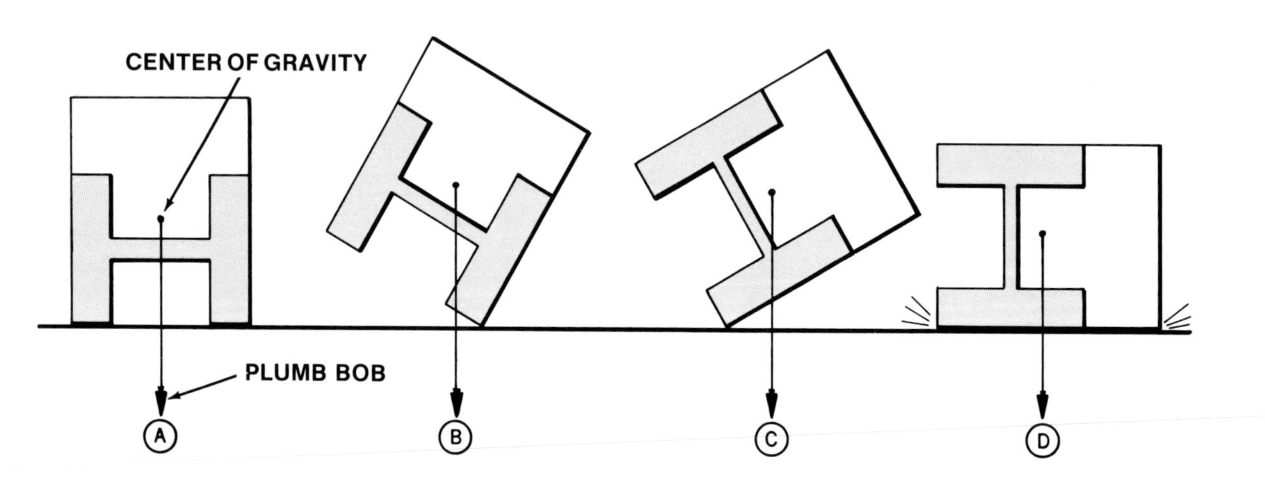

There are many illustrations of centrifugal force in everyday life. For example, whirl a weight around on the end of a string. The string is kept tight because it pulls in on the weight, keeping it in a circular path. But if the string is cut, the weight will travel outward in a straight line. In this example, centrifugal force straightens the weight's circular direction of travel into a straight line.

Let's look at the reverse situation and change the direction of travel of an object moving in a straight line to move it around a curve. To change its direction of travel, we must overcome the object's centrifugal force by placing a lateral force on it.

Consider a locomotive on a curve. Its forward motion makes the flanges on the wheels press out against the inside edge of the rails. The rail, in turn, pushes in against the flanges, forcing the locomotive to follow the curve and stay on the track.

Tractors are made to change direction the same way. When the operator turns the steering wheel on a conventional two-wheel drive tractor, its forward motion places an outward lateral force between the front tire treads and the ground's surface. If the tires do not skid, the ground resists this outward force forcing the tractor to turn. If the front tires do skid, of course, the tractor continues its travel.

Centrifugal force is a major factor in tractor upsets. If strong enough, centrifugal force can tip a tractor sideways using the outside wheels as pivot points (Fig. 31).

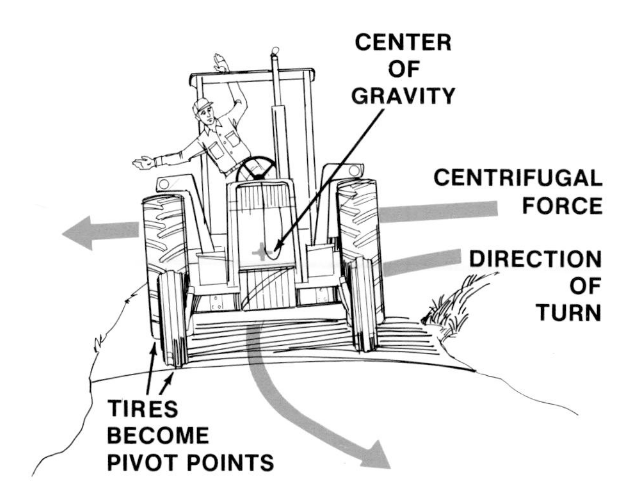

Fig. 31—Centrifugal Force Tries To Pivot The Tractor On Its Outside Wheels

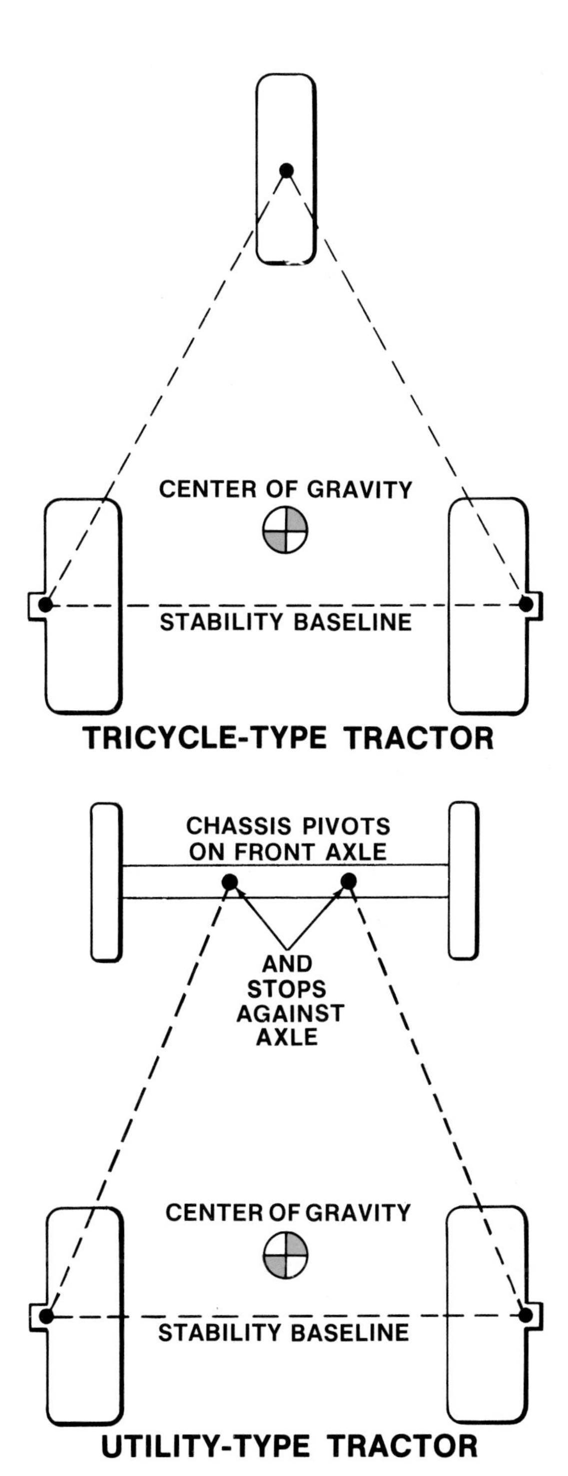

Fig. 30—Base Of Stability For Tricycle And Utility Tractors

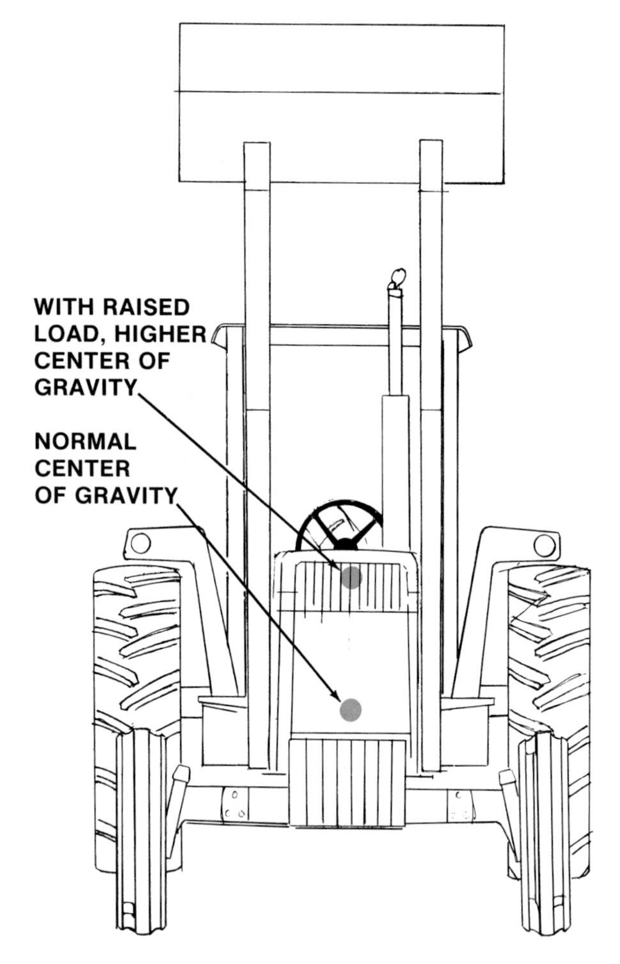

Fig. 32—Using Some Mounted Equipment Raises The Center Of Gravity

Centrifugal force varies in proportion to the square of the tractor's speed. This means, for example, that doubling tractor speed increases the strength of centrifugal force four times ($2^2 = 2 \times 2 = 4$). Tripling speed, say from 5 to 15 mph, increases the strength of centrifugal force nine times ($3^2 = 3 \times 3 = 9$).

Centrifugal force also varies in reverse proportion to the length of the turning radius. For example, if you make a short turn using half the turning radius of a wider turn, centrifugal force doubles.

Centrifugal force becomes more dangerous in tractor operation as the tractor's center of gravity is moved higher. This happens, for example, when a front-end loader is raised, or when spray tanks are mounted high on a tractor chassis (Fig. 32). The center of gravity moves forward as well as up when the scoop is raised (Fig. 32).

Remember these two points: short, quick, high-speed turns upset tractors, and tractor-mounted attachments that raise the tractor's center of gravity increase the risk of tractor upsets caused by centrifugal force.

REAR-AXLE TORQUE

Engaging the clutch applies torque (twisting force) to the rear axle of the tractor. Normally, the axle rotates and the tractor moves ahead. But if axle rotation is restrained in some way, for example when starting a heavy load or when the drive wheels are frozen to the ground, the twisting force of the axle tries to lift the front wheels off the ground and rotate the tractor backward *around the rear axle.* If it's easier for engine power to lift the front of the tractor than it is to move the tractor ahead or spin the wheels, the tractor will rear over backwards (Fig. 33A).

Because rear-axle torque reacts against the tractor chassis, the following practices *increase* the chances

Fig. 33—Hitching Mistakes That Can Cause Tractor Upset

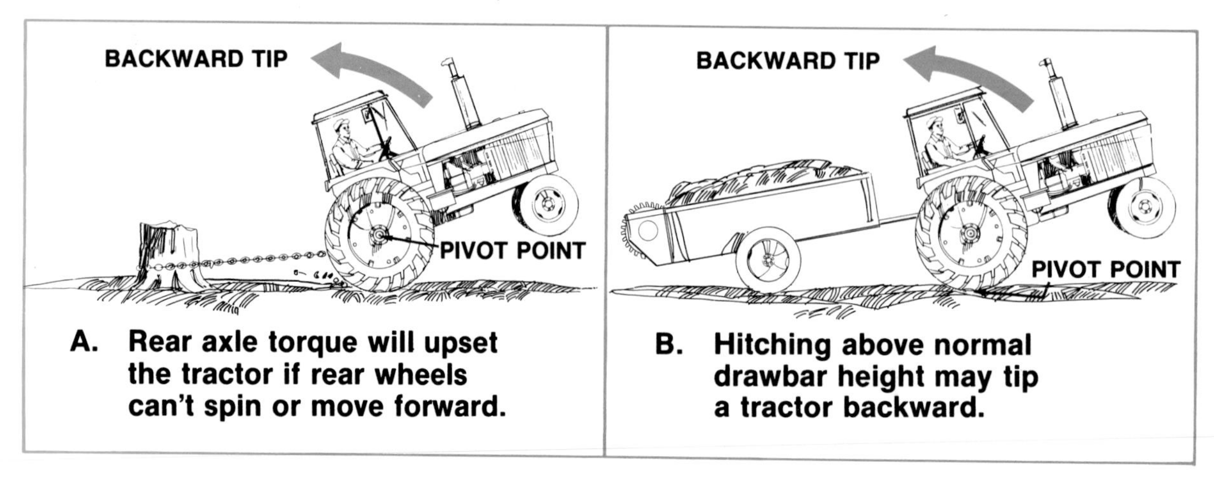

of tipping a tractor over backward when driving ahead:

- *Starting off in low gear with high engine speed*
- *Engaging the engine clutch quickly*
- *Accelerating quickly, especially when traveling uphill or pulling a heavy load*
- *Engaging power to drive wheels when unable to move forward or spin wheels.*

LEVERAGE OF THE DRAWBAR OR HITCH

When pulling a heavy load, the rear tires push against the ground with considerable force. At the same time, the load attached to the tractor hitch point pulls back against the forward movement of the tractor. When this happens, the point of contact between the rear tires and the ground serves as a pivot point, and the pulled load attempts to pivot the tractor, raising the front wheels off the ground (Fig. 33B).

The *draft* (amount of pull) and the height of the hitch point determine the force of the pivoting action. Increasing the pull and raising the hitch point increase the danger of upsetting the tractor.

An *excessive load* attached to a drawbar set at the *recommended height and with recommended ballast* will usually slip the rear wheels or stall the engine before overturning occurs, but even a *light load attached too high* can quickly tip a tractor backward. Drawbar heights are selected by tractor engineers to provide efficient and safe hitching points for towed loads. If the drawbar height on your tractor is adjustable, check your operator's manual for proper adjustment and for proper ballast.

Preventing Tractor Upsets

Put your knowledge of the causes of upsets to work to keep yourself alive and prevent tractor damage. Follow operating practices that *utilize the force of gravity to keep your tractor upright and minimize the forces that work to tip it over.*

About three-quarters of tractor overturn fatalities result from overturns to the side. Let's review the recommendations for preventing these, and then go on to those for preventing rearward upsets.

ROLLOVERS TO THE SIDE

1. *Set the wheel tread at the widest setting suitable for the job you must do.* This provides the widest possible base of stability for your tractor (Fig. 34).

2. *Lock brake pedals together before driving at transport speeds (Fig. 35).* Applying uneven brake pressure to the rear wheels during high-speed operation

Fig. 34—Wide Wheel Spacing Increases Lateral Stability

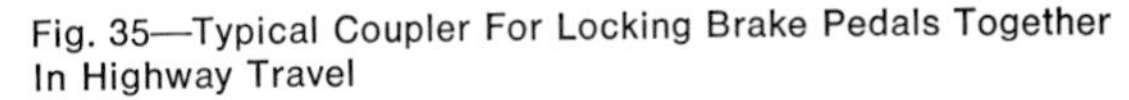

Fig. 35—Typical Coupler For Locking Brake Pedals Together In Highway Travel

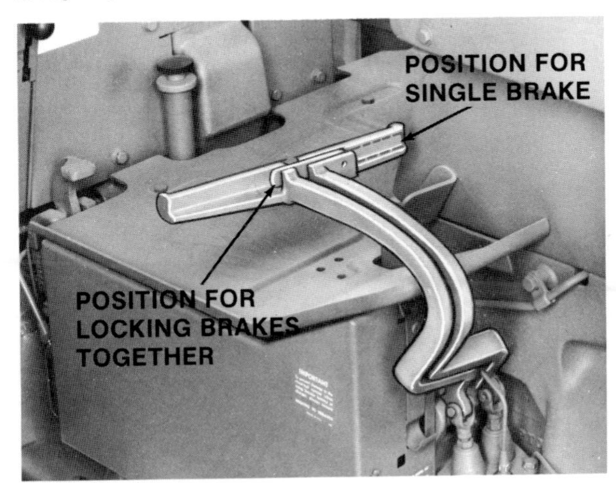

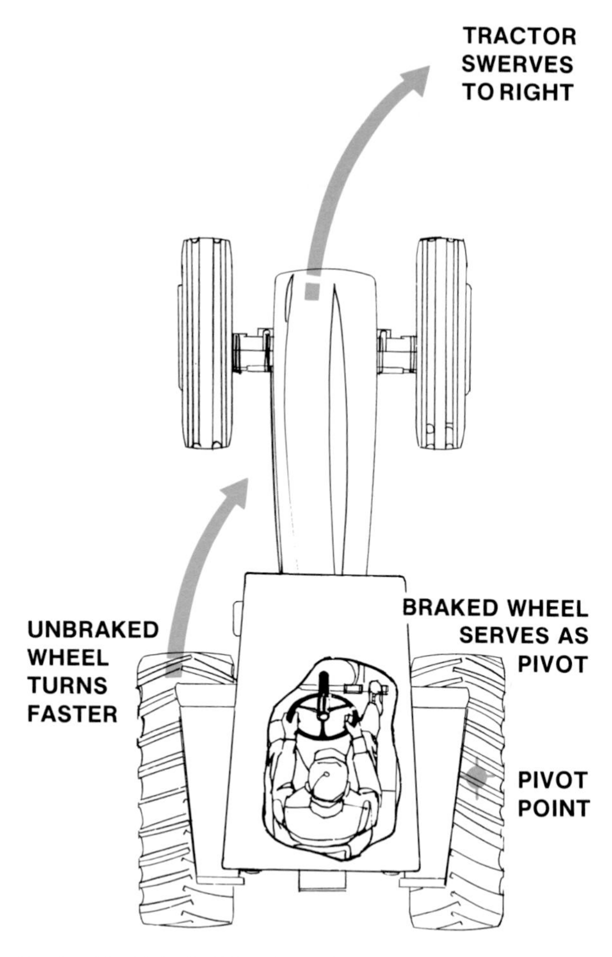

Fig. 36—The Unbraked Wheel Can Swerve A Tractor Out Of Control

can literally force a tractor to roll over. When only one wheel is braked, the other is driven at a correspondingly higher speed by the differential. Braking one pedal severely can force the tractor to swerve abruptly to the right or left into a rollover situation (Fig. 36).

3. *Restrict your speed according to operating conditions.* To avoid an upset, you need complete control over your tractor, and a travel speed slow enough for you *to see* and *to react* to hazards in your path. Always watch ahead for bumps, rocks, stumps, and other obstructions that could tip one side of your tractor *up* past its tipping point (Fig. 37). Also, watch for holes, ruts, and depressions that could drop one side *down,* putting your tractor in an unstable position.

4. *Don't let your tractor bounce.* This causes loss of steering control, and you do not have full control of the tractor.

5. *Drive slowly in slippery conditions.* If your tractor starts to skid sideways to your direction of travel, you could easily tip over in a ditch or crash into an obstruction. And beware of this hazard—if your tractor slides sideways and the tires hit an obstruction or a surface that brakes the sliding of the tires, your tractor may roll over.

6. *Pull heavy loads and equipment at safe speeds.* Reduce travel speed if towed equipment fishtails (weaves back and forth behind the tractor). The whipping action of the equipment is transferred to the front wheels of the tractor. If it becomes severe, or if you hit a slippery spot, the tractor could be thrown out of control into a sideways tip or into a ditch. You can get the same results when you try to stop a heavy load rolling behind the tractor. When the brakes are applied, the load pushes forward, forcing the tractor to skid. If it skids sideways, the tractor could overturn (Fig. 38).

7. *Turn safely.* Slow down before turning. Remember that centrifugal force affects the stability of your tractor (Fig. 31, page 149). The chances of tipping are greatest when crossing steep slopes while turning uphill, but tractors can also tip on level ground traveling at moderate speeds. Do not apply individual brakes for turning when driving faster than normal field-working speeds. Remember that quick, short, brake-assisted turns at high speeds overturn tractors. Here's the safest procedure: slow down by reducing engine speed *before* turning. Apply *both* brakes if braking action is required. Then turn as widely as you can with engine power pulling the load.

Fig. 37—Watch For Both Obstructions And Depressions That Might Upset The Tractor

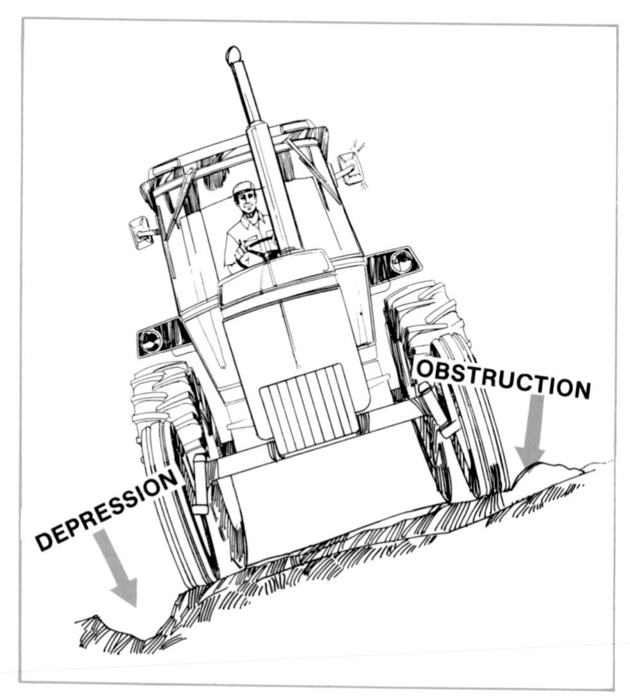

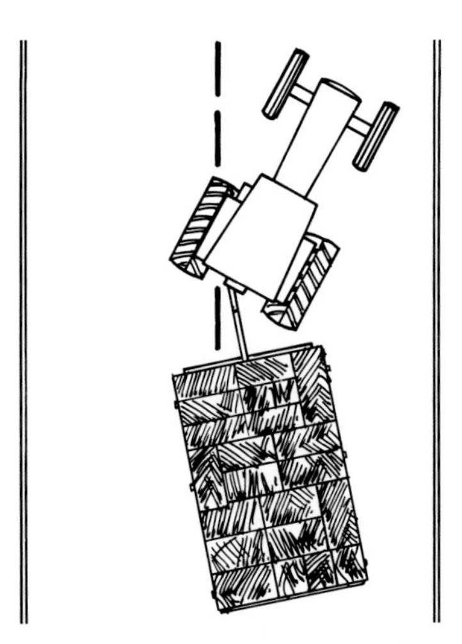

Fig. 38—Fishtailing Or Severe Braking At High Speed Can Cause Jackknifing And Rollover

Fig. 39—Shift To A Lower Gear On Level Ground, Before Going Downhill

8. *Use engine braking when going downhill.* Runaway tractors often tip over. Shift to a lower gear and reduce speed for downhill travel (Fig. 39). If in doubt as to what gear to use, select the lowest-speed gear. Shift *before* you start downhill. Don't disengage the clutch and try to shift after you have started. The brakes may not hold and *you may not be able to shift back into gear.* If the engine brakes too much, increase speed slightly with the throttle. If it doesn't brake enough, apply the tractor brakes. A number of accidents happen because the tractor does not have sufficient brakes to handle a towed load.

Some tractors "freewheel" and provide no engine braking in certain speed ranges. If your tractor is one of these, travel downhill using only those shift positions that provide engine braking action. Know what these shift positions are before you first drive the tractor. Check your operator's manual.

9. *Avoid crossing steep slopes if possible.* Tractor stability is reduced on steep slopes. If you must cross a steep slope, take every precaution to avoid rolling over. Drive slowly, avoid quick uphill turns, and watch for holes and depressions on the downhill side and for bumps on the uphill side. Keep sidemounted equipment on uphill side of tractor if possible (Fig. 40). Turn downhill if stability becomes uncertain. Space the rear wheels as far apart as you can to increase stability.

Fig. 40—Keep Side-Mounted Equipment On Uphill Side If Possible

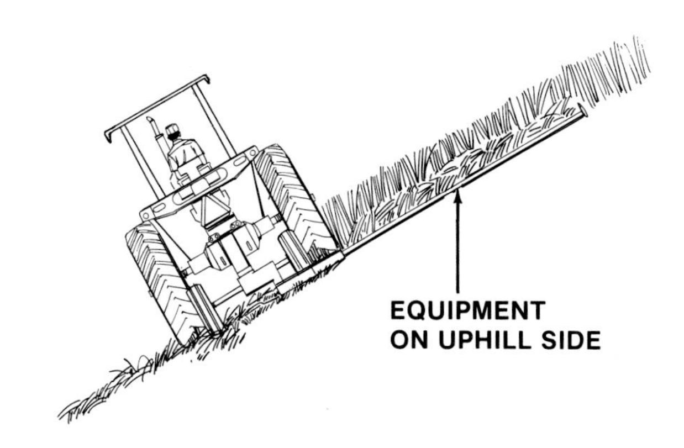

10. *Stay away from ditches and riverbanks.* Stay at least as far away from the bank as the ditch is deep. The weight of the tractor could cave the bank in if you approach or cross the *shear line* (Fig. 41). As you approach ditches and banks, look ahead rather than at your equipment. Give yourself plenty of room for turning and watch for holes, gullies, and washouts that could place your tractor in an unstable position.

11. *Operate front-end loaders cautiously.* Mounting a front-end loader on a tractor raises its center of gravity. In addition, loader-tractors often operate in confined areas that make short turns unavoidable. Both of these factors make loader-equipped tractors susceptible to rollovers caused by centrifugal force. To avoid an upset:

a) Keep the bucket as low as possible when turning and transporting (Fig. 42).
b) Watch carefully for obstructions and depressions.
c) Handle the rig smoothly, avoiding quick starts, stops, and turns.

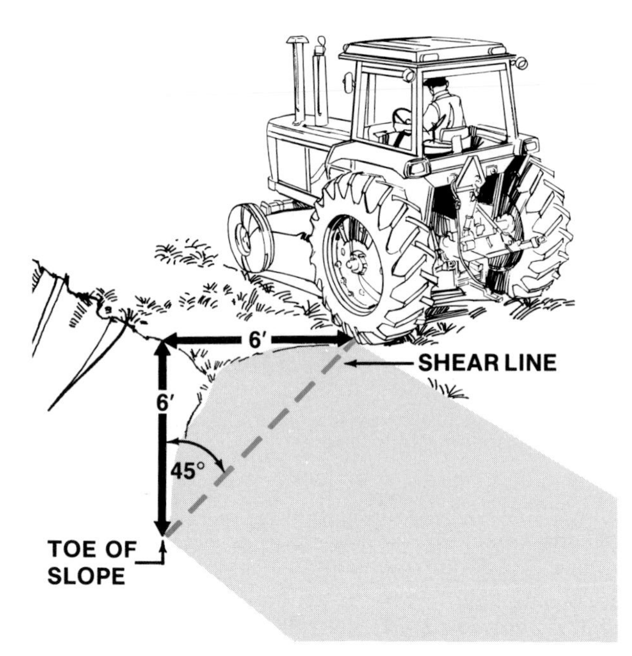

Fig. 41—When Operating Near Ditches And Banks, Always Keep Your Tractor Behind The Shear Line

Fig. 42—Always Lower The Bucket When Transporting

NOTE: Mounted loaders should be removed before the tractor is used for field or transport work. If this is not practical for a short job, keep the bucket positioned low, watch for obstructions, and drive slowly.

ROLLOVERS TO THE REAR

1. *Hitch towed loads only to the drawbar.* The regular drawbar or the drawbar attachment available for the three-point hitch is the only safe hitching point on a tractor for pulling towed loads. Never attach a towed load to the axle, to one of the lower links, or to the top link of the three-point hitch. Doing so can pivot the tractor into a rearward tip.

When using a drawbar attachment between the draft arms (lower links) of a three-point hitch, keep the hitch low, at the height recommended in your operator's manual. Install stay braces, if available, and lock the hitch control lever in position to keep the hitch from raising. Raising of a three-point hitch can be caused by hydraulics, or by the towed load pushing against the hitch when braking or traveling downhill.

NOTE: Holes are provided on the frames of some tractors for attaching mounted implements and machines. Do not use these holes to attach a clevis, chain, or other device to pull a towed load.

2. *Limit the height of three-point-hitch drawbars.* Don't pull with three-point-hitch drawbars positioned higher than 13 to 17 inches above the ground (See operator's manual). To avoid raising the hitch accidentally, install stay braces on the hitch and lock the hitch control lever in the desired position (Fig. 43).

3. *Use weights to increase stability.* Add front-end chassis weights to counterbalance rear-mounted implements and heavy vertical drawbar loads like two-wheeled trailers, manure spreaders, and rear-mounted loaders (Fig. 44). Do the same for pulling heavy loads uphill to help offset the loss of tractor stability. Add rear-wheel weights or tire ballast to counterbalance front-end attachments like spray tanks or front-end loaders. Refer to your operator's manual for recommendations.

4. *Start forward motion slowly.* Gunning the engine and jerking your foot off the clutch to start a heavy load is one of the surest ways to tip a tractor over backward. To start safely, use only enough speed to prevent engine lugging, and engage the clutch slowly. You can increase speed once you are underway.

5. *Change speed gradually.* Move the throttle lever slowly when changing engine speed. Accelerating a heavy load quickly, especially on uphill grades, can lift the front wheels off the ground, causing loss of steering control. And it could be the start of a rearward upset.

Fig. 43—Stay Braces Will Prevent Raising The Hitch Accidentally

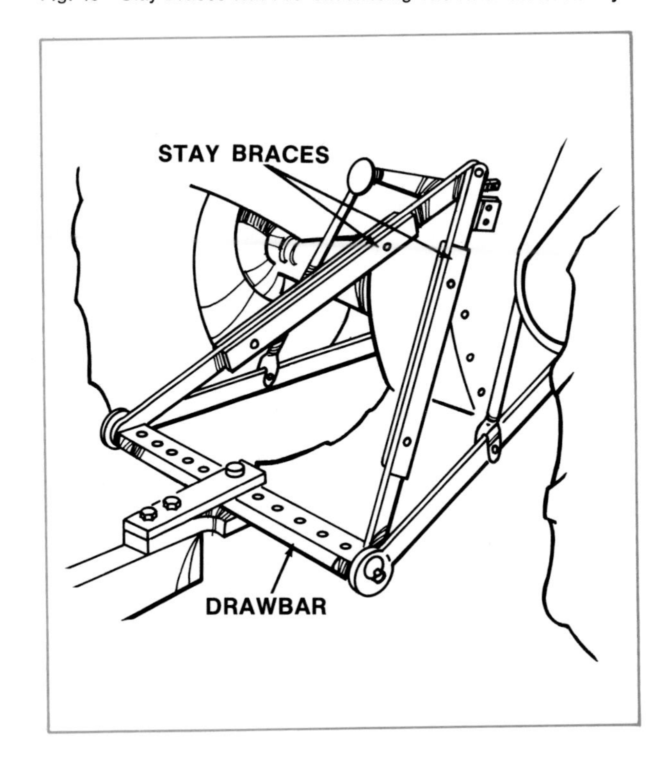

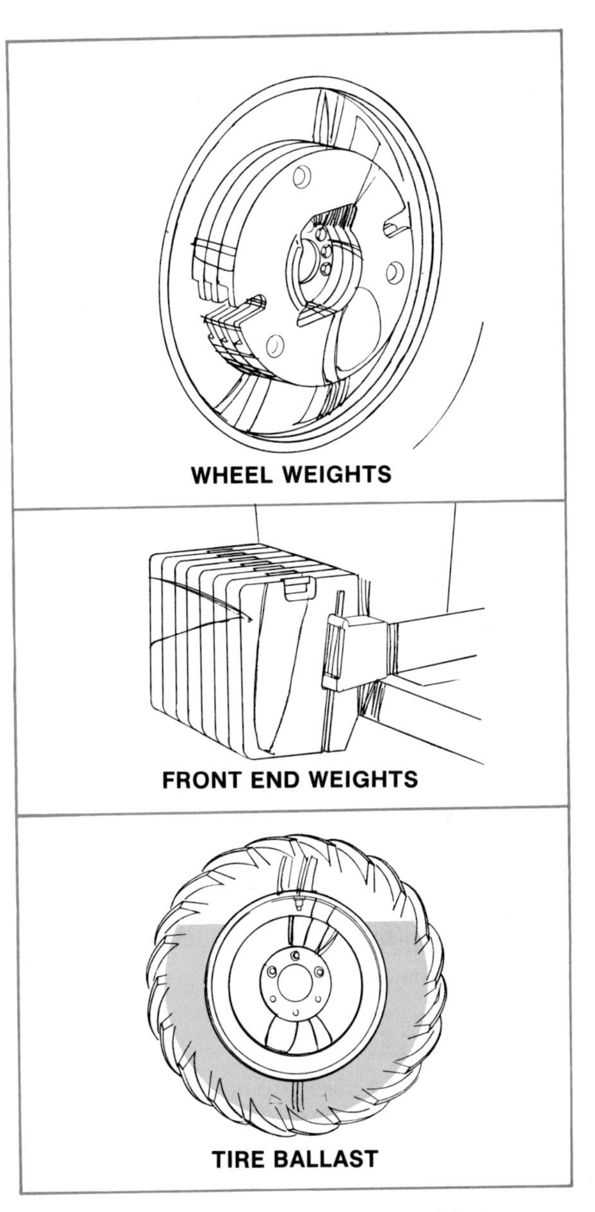

Fig. 44—Three Ways To Add Weight: As Wheel Weights, Front-End Weights, And Tire Ballast

Fig. 45—Direction Of Travel Up And Down Steep Slopes

6. *Brake cautiously when backing down a grade.* Braking hard will turn the tractor over backwards. If you must go up a steep grade, back the tractor up in *reverse,* and come down by driving forward (Fig. 45). If you must back down a slope, do it slowly, keeping the tractor in a low gear, so braking will not be required. If brakes are applied backing downhill, the tractor could rotate around the rear axle, tipping it backward. The faster the speed and the steeper the grade, the easier it is for the tractor to tip.

NOTE: If your tractor should ever roll backward down a steep grade with the clutch disengaged, think twice about engaging it. Engaging the clutch is nearly the same as applying the brakes, and could result in a backwards tip. If you find yourself in this situation, riding the tractor to the bottom of the slope without engaging the clutch or applying the brakes is usually the safest procedure.

7. *Plan for safe uphill pulls.* Pulling heavy loads uphill calls for a number of precautions, since the slope and drawbar leverage both work to tip the tractor backward. The safest procedure is to add front-end weights, set the drawbar in its lowest and longest position, engage the clutch slowly, and accelerate gradually.

8. *Drive around ditches.* Don't cross them unless it's absolutely necessary. If you must cross, recognize the hazards involved. Situations vary so much that it's impossible to outline all recommendations, but keep these points in mind:

a) Drive forward when going downhill, using low gear. If the slope is steep or slippery enough for the tractor to slide, don't go down.

b) Back the tractor up steep grades. If you can't, shift to a low gear and drive forward very slowly. Maintain uniform engine speed and try to avoid moving the throttle. Rapid acceleration could throw your tractor over backward. Keep in mind that the governor may accelerate your tractor automatically even though you don't move the throttle lever.

9. *Back your tractor out if it gets mired down in mud (Fig. 46).* If necessary, do any or all of the following to enable you to back out:

a) Dig mud from behind the rear wheels.

b) Unhitch any towed implement or machine and pull it away with a chain to a dry location.

c) Place boards *behind* the wheels to provide a solid base and try to back out slowly.

d) If necessary, get another tractor to pull you out. caution the other driver to engage his tractor clutch slowly and use his engine power cautiously to prevent tipping *his* tractor rearward.

When backing out is impossible, dig mud away from the front of all wheels, unhitch any towed load, if necessary, and drive *slowly* ahead. If you need a pull from another tractor, use a long chain (not a nylon rope) and caution your helper to engage his tractor clutch slowly and to use his tractor power cautiously.

Fig. 46—Follow Recommendations To Get Out Of This Situation Safely

your helper to engage his tractor clutch slowly and to use his tractor power cautiously.

CAUTION: Never put boards or logs in front of the drive wheels and attempt to drive ahead. If the drive wheels catch on them and cannot turn, the tractor may tip over backward.

10. *Back your tractor out if the drive wheels get lodged in a ditch.* Tractor stability is reduced by the tipped angle of the tractor, and the drive wheels are not as easily turned by the axle as they would be on level ground. Trying to drive the tractor ahead may rear it over backward.

Rollover Protective Structures

The development of *rollover protective structures* (ROPS) received a lot of attention from tractor manufacturers during the 1960s. Two major goals were to be met:

- *Limit most upsets to 90 degrees*
- *Protect the operator in upsets that tip the tractor beyond 90 degrees (Fig. 47)*

To meet these goals, two types of ROPS were developed: protective *frames* and protective *enclosures.*

Fig. 47—ROPS Are Designed To Limit Most Upsets To 90 Degrees, And To Protect The Operator In Upsets Beyond 90 Degrees

ROLLOVER PROTECTIVE FRAMES

Protective frames are generally two-post or four-post structures attached to the tractor chassis (Fig. 48). A seat belt is provided for the operator. In case of an upset, the seat belt holds the operator within the protective frame, to help keep him from being crushed by the tractor. Frames may be equipped with weather canopies to protect the operator from rain and sun. These canopies are designed for weather protection only—not to increase the overturn protection provided by the protective frame.

Fig. 48—Two-Post And Four-Post Rollover Protective Structures (Weather Shield On The Right)

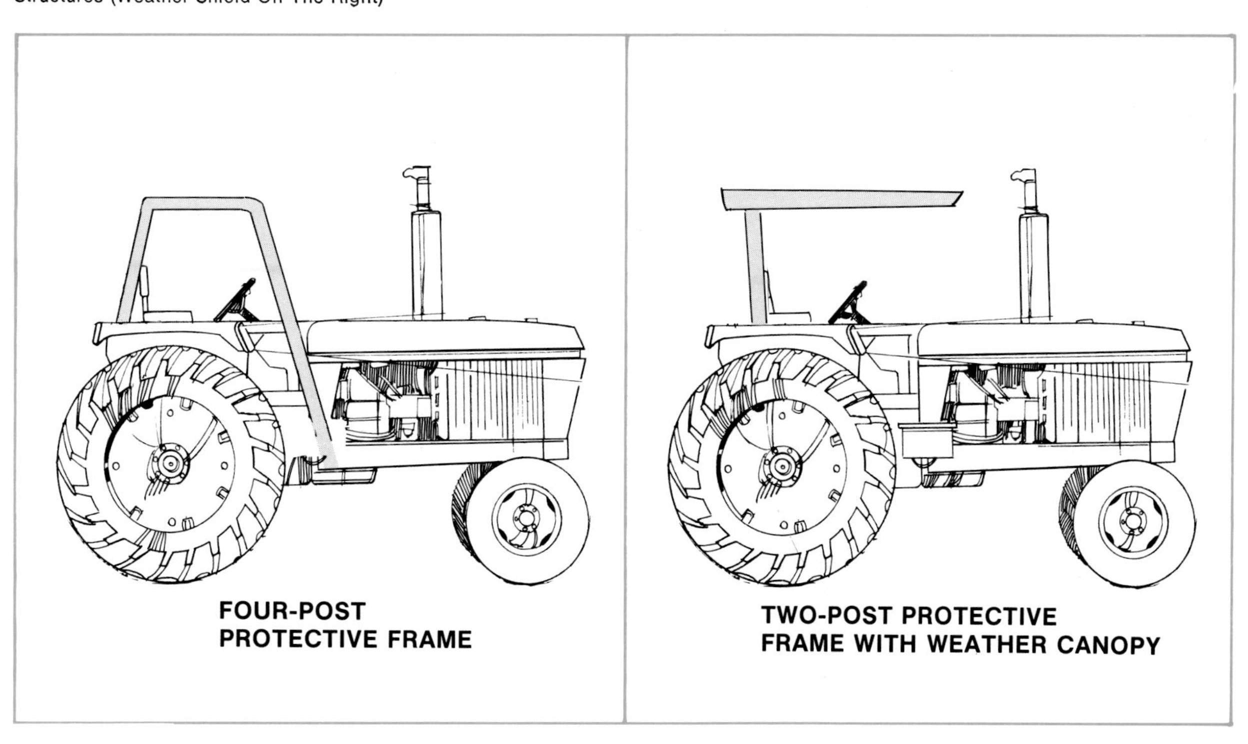

ROLLOVER PROTECTIVE ENCLOSURES

Rollover protective enclosures are protective enclosures built around a protective frame, or a strong metal frame is incorporated into their design (Fig. 49). In addition to protecting the operator during upsets, these enclosures provide other safety and health benefits. Since they're enclosed by windows and doors and have fans to bring in filtered air, enclosures are slightly pressurized to help keep dust from entering the compartment. Many enclosures reduce engine noise and vibration. They help prevent falls from the tractor. And they can be equipped with heaters, air conditioners, tape players, and radios to provide additional comfort and entertainment for the operator.

It is important to realize that some enclosures being used and sold are designed primarily to protect the operator from the weather. They are *not* designed to provide overturn protection. Consequently, these enclosures may be crushed if the tractor tips (Fig. 50). Read the technical manual to find out.

Enclosures and frames designed to provide overturn protection are built to meet the standards and regulations of various agencies. If an enclosure or frame has not been certified as meeting these standards, you have reason to question whether it will provide adequate protection in an upset. To check any enclosure or frame for overturn protection, either ask your dealer, check manufacturers literature, or look for a label on the frame or enclosure for evidence that it meets the standards of one or more of these organizations or governmental units:

1. The American Society of Agricultural Engineers (ASAE). 2. The Society of Automotive Engineers (SAE). 3. The State of California (or other state certification). 4. The federal government (such as established by the Occupational Safety and Health Act of 1970—OSHA).

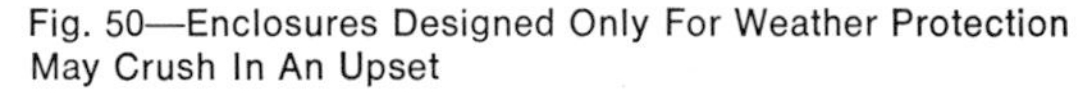

Fig. 50—Enclosures Designed Only For Weather Protection May Crush In An Upset

Fig. 49—The Heavy Metal Frame Of This Enclosure Provides Rollover Protection

Don't try to build your own frame or enclosure. You will not be able to test your design or the strength of construction.

Operating ROPS-Equipped Tractors

1. *Use the seat belt.* The belt is intended to hold you within the safety zone of your ROPS frame or ROPS enclosure if an upset occurs (Fig. 51). But do *not* wear a belt if your tractor is not equipped with a protective structure. If you do, you lose your chance of being thrown clear of the tractor in case of an upset.

2. *Replace all damaged members of the protective structure after an upset (Fig. 52).* A rollover places severe stress on the structural members of a frame or enclosure. Reuse of structural members that have been bent, buckled, or stretched is not recommended. If your frame or enclosure is damaged, consult your dealer and follow manufacturer's instructions for making repairs or replacement.

Fig. 51—Buckle Up *If* Your Tractor Is Equipped With ROPS

Fig. 52—This ROPS Enclosure Is Beyond Repair

Keep fasteners tight. Never modify structure members by drilling, bending, or cutting. It will weaken the structure.

3. *Stay alert.* Outside warning sounds from machinery, other people, livestock, or traffic may be muffled by the enclosure's sound-absorbing insulation. Be aware that an enclosure may isolate you from these warnings, and that you need to stay alert at all times.

4. *Beware of the surface conditions outside.* The warmth and shelter of a heated enclosure can make you forget how rough or slippery it really is. Exercise all precautions, even though you are warm and comfortable.

NOTE: ROPS Roll-Over Protective Enclosures can be retrofitted by the tractor dealer on many older tractors. ROPS must be fitted by the dealer to the tractor. Do Not attempt to modify a ROPS. If any structural member is damaged, replace the entire structure. Do not attempt to repair it.

Fig. 53—Use Steps And Handholds When Mounting And Dismounting

FALLS FROM TRACTORS

Falling from tractors is a common cause of serious accidents.

Falls are needless and preventable accidents. Falls occur from moving and parked tractors. Let's explore some methods for preventing these accidents.

Mounting And Dismounting

1. *Keep the steps and platform clean and dry.* Take time to clean off mud, ice, snow, grease, and other debris that accumulate on the platform and steps (Fig. 7, page 139). Don't carry tools or log chains on the platform. Equip your tractor with a tool box or use a trailer to carry tools for doing maintenance work.
2. *Do not jump from the tractor.* There is always the danger of catching your clothing on pedals, levers, or other protruding parts. And you could land on an uneven surface and injure your ankles, legs, or back.

3. *Use handrails, handholds, and steps to pull yourself up to the operator's platform (Fig. 53).* Try to keep three points on the machine at all times — either two hands and one foot, or two feet and one hand. Wear heavy-treaded boots.

   Before mounting power-steered, four-wheel-drive tractors that hinge in the middle of the tractor chassis, make sure that all people are clear of the tractor. When mounting, avoid moving or pulling the steering wheel. These precautions can prevent a bystander from being crushed or struck by the tires, chassis, or mounted implements which may swing when the steering wheel is turned.

Operator Falls

Operators sometimes fall from tractors and are crushed under the wheels or mangled by trailing equipment. This won't happen if you observe the following safety practices:

1. *Operate the tractor from the operator's platform only.* Fasten the seat belt if your tractor is equipped with a protective frame or enclosure. If you get tired of sitting, stop and take a break.

   Never operate the tractor while riding on the drawbar, sitting on the fender, standing on the steps, or sitting on the backrest of the operator's seat.

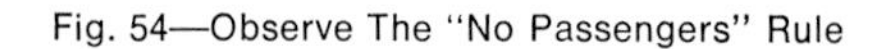

Fig. 54—Observe The "No Passengers" Rule

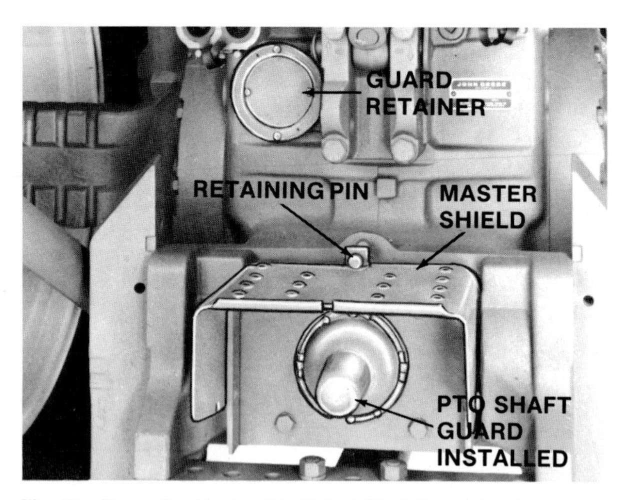

Fig. 55—Keep The Master Shield And Shaft Guard In Place.

2. *Maintain safe operating speeds.* Never drive so fast that the front wheels of the tractor bounce. Watch ahead for obstructions and avoid them. Slow down before making turns.

3. *Rest when you are tired.* Stop the tractor and lie down for a short nap if you feel tired or sleepy. Stop for 10 or 15 minutes every 2 to $2^{1}/_{2}$ hours. Don't drive when you feel like dozing. In this condition you are not alert enough to operate a tractor.

Carrying Passengers

Tractors are designed to carry one person, the operator (Fig. 54)! The proper place to ride is sitting in the operator's seat.

You will often be tempted to use your tractor to transport a helper. Children often plead for rides. *Do not give in.* An unexpected bump or turn could toss an extra rider right out the cab door and off the tractor. Your passenger cannot anticipate every tractor movement and brace against it. Furthermore, all the vibration and bumps are transmitted to the passenger. He can easily loose his grip and fall off.

Passengers can cause others problems, too. They can interfere with your operation of controls, they can accidentally move controls themselves, and they frequently distract the operator's attention.

Plan your work so that the need for carrying passengers will not arise. It's far safer to use another form of transportation. Provide another tractor for your passenger to drive if he's a qualified operator. Or use a pickup so your passenger can ride inside.

FOUR WHEEL DRIVE TRACTORS

Four wheel drive tractors with articulated steering pivot in the middle. NEVER stand between the tires when the engine is running. And, NEVER let anyone stand between the tires when the engine is running.

The steering is extremely sensitive. If you accidently move the steering wheel a bit while someone is standing between the tires, THEY WILL BE CRUSHED.

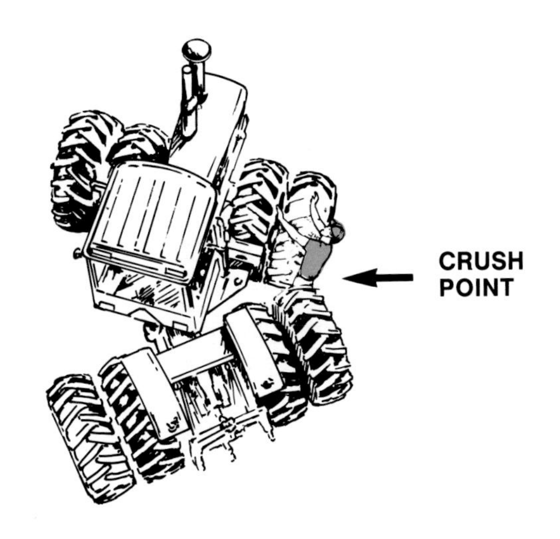

Be sure everyone is clear of the tractor at start up. Some movement can occur as the engine starts.

POWER TAKEOFF ACCIDENTS

Severe injury or death can result from accidents involving the power takeoff. There are two types. The most common is getting caught by the rotating shaft. And, on occasion, an operator is struck by a broken or disconnected shaft as it swings violently behind a tractor. Avoid these accidents by following proper installation, maintenance, and operating procedures.

Shielding

An unguarded PTO shaft is dangerous. It can catch on your clothing before you realize it. If you're lucky, your clothes will tear, freeing you without serious injury. If you're not so lucky, you could be strangled or mutilated by the high-speed shaft.

FOR THE TRACTOR

Master shields prevent accidental contact with the tractor stub shaft and the front universal joint of the

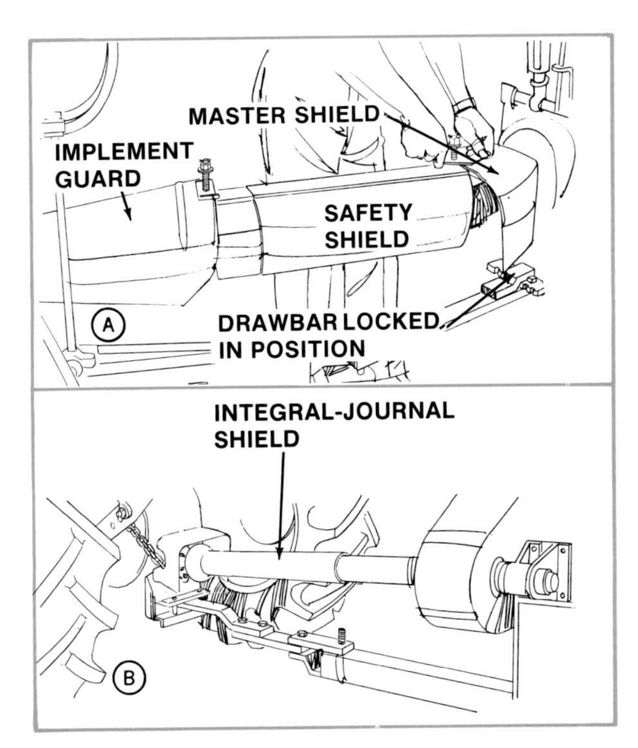

Fig. 56 — Integral Shield For PTO-Driven Machines DO NOT Rotate With The PTO Drive Shaft.

attached machine's driveline (Fig. 55). On some tractors, this shield also provides a point of attachment for older-style tunnel shields.

Tubular *shaft guards* are provided for many tractors to completely enclose the tractor stub shaft when the PTO is *not* being used.

CAUTION: Shiftable, multi-speed PTO's present a hazard to implements because they can be run too fast. Follow recommended PTO speed specifications in the technical manual.

FOR THE PTO-OPERATED MACHINES

Integral-journal shields completely enclose the PTO shaft (Fig. 56). These shields are metal or plastic tubes supported on bearings so the shields rotate independently of the shaft. When the PTO shaft is turning, they rotate with it. But if a person contacts the shield, rotation shield stops harmlessly while the shaft inside continues to spin. Keep these shields in place and in good condition, and they will protect you from the grabbing action of revolving shafts and universal joints.

The *fully shielded driveline* was introduced in 1972. It was designed to solve two problems. First, some operators do not keep the tractor master shield in place when the PTO is being used. This leaves part of the tractor stub shaft, the coupler, and universal joint attached to it exposed. Second, even when the master shield and the integral or tunnel shields are in place, the tractor-attached universal joint is often partially exposed.

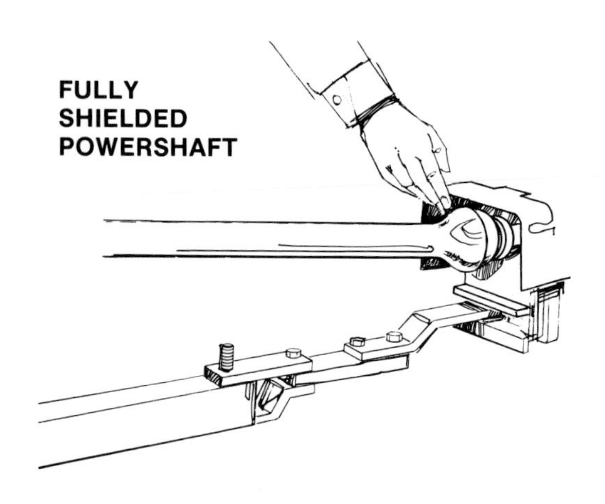

Fig. 57—The Fully Shielded Driveline Completely Encloses The Tractor-Connected Universal Joint And Coupler

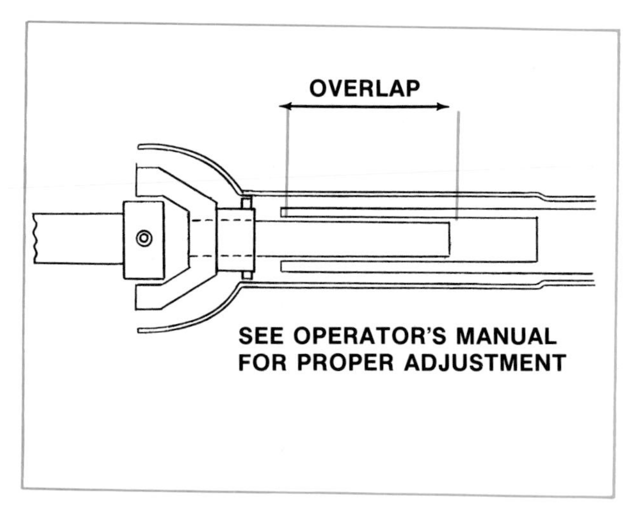

Fig. 59—To Avoid Separation And Compression, The PTO Driveline Parts Overlap

The fully shielded unit uses the rotating-type shield in combination with a complete shield provided for the tractor-attached universal joint and coupler (Fig. 57). When in place driven, rotating parts are not exposed to catch fingers, gloves, or clothing. If a person accidentally contacts the shield or coupler, rotation stops, giving complete protection.

Entanglement With The PTO

To prevent injury from entanglement with the power takeoff shaft, follow these safety practices:

1. *Always disengage the PTO, shut off the engine and take the key before getting off the tractor (Fig. 58).* This gives you three-way protection: from shaft rotation; from moving machine parts; and it prevents the unexpected engagement of power by another person when you are cleaning, lubricating, adjusting, or making repairs.
2. *Keep the master shield in place.* The master shield is only removed when required for special equipment with equivalent shielding.
3. *Check to see that integral shields are in good condition.* When the powershaft has stopped, you should be able to rotate the shield freely by hand. If the shield or bearings are damaged, make the necessary repairs immediately.
4. *Keep tunnel shields in proper alignment over the powershaft.* make sure they're securely latched in place and attached to the tractor master shield.

Fig. 58—Always Disengage The PTO, Shut Off The Engine, And Take The Key Before Dismounting From The Tractor

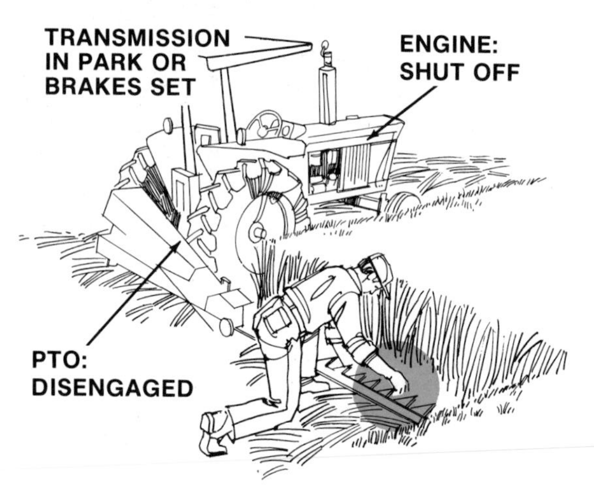

Fig. 60—Correct And Incorrect Assembly Of The PTO Shaft Assembly

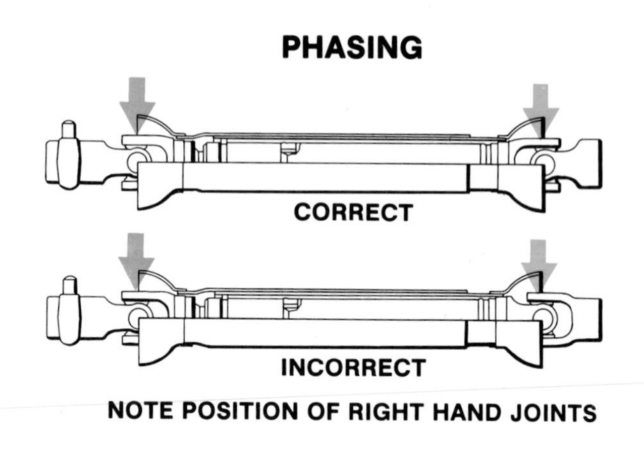

5. *Never step across a rotating powershaft.* When PTO-driven machines are running, always walk *around* the revolving shaft. Safety devices are usually reliable, but could malfunction.

Broken Or Separated Shafts

A PTO shaft may break or separate during operation if improperly used or adjusted (Fig. 59). If it does, the tractor-driven end swings violently behind the tractor, often causing severe tractor damage or injury to the operator. These accidents are most likely to occur when powershaft assemblies are improperly installed, when the tractor drawbar is incorrectly positioned, or when the PTO shaft is abused during operation. To prevent these accidents:

1. *Keep universal joints in phase.* Universal joints are properly phased or aligned when the end yokes are positioned in the same plane. Fig. 60 shows correct assembly. Some shafts can be separated; others can't. Separable shafts must be reassembled in phase to avoid driving the machine at irregular speeds and placing a severe strain on the shaft. Many shafts are designed so they cannot be reassembled improperly.

2. *Always use the driveline recommended for the machine.* Drivelines are designed for each implement. Drivelines overlap (Fig. 59 and 61). If they do not, they may separate when the rear tractor tires pass over a ridge on the ground. Or they may squeeze together so much that they bottom out when the tractor turns, or when the rear tractor tires drive through a depression or ditch. If this happens, the shaft and bearing supports are strained severely, and may be damaged or bent. Drivelines are also made with different cross-sectional diameters for low- and high-horsepower applications. For these reasons, use only the driveline recommended for each of your PTO-driven machines, and keep the shaft of each machine positioned properly as outlined in the operator's manual.

3. *Position the drawbar correctly.* You may have to position the tractor drawbar differently for each machine used in your farming operations. To do this:

a) Find these measurements in the tractor operator's manual (Fig. 62):

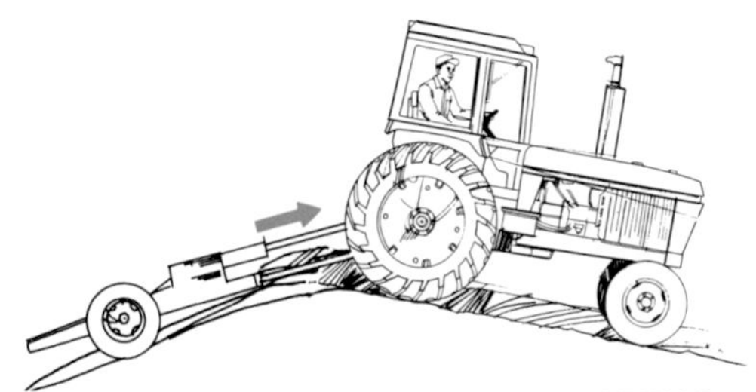

DRIVELINES LENGTHEN WHEN DRIVING OVER THE CREST OF A HILL

DRIVELINES SHORTEN WHEN DRIVING THROUGH DEPRESSIONS.

Fig. 61—Powershafts Must Lengthen Without Separating And Shorten Without Bottoming Out

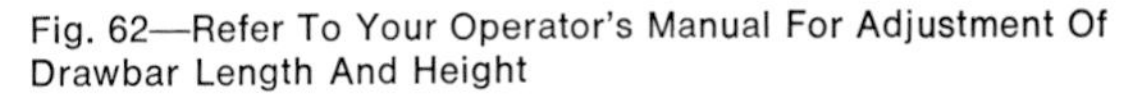

Fig. 62—Refer To Your Operator's Manual For Adjustment Of Drawbar Length And Height

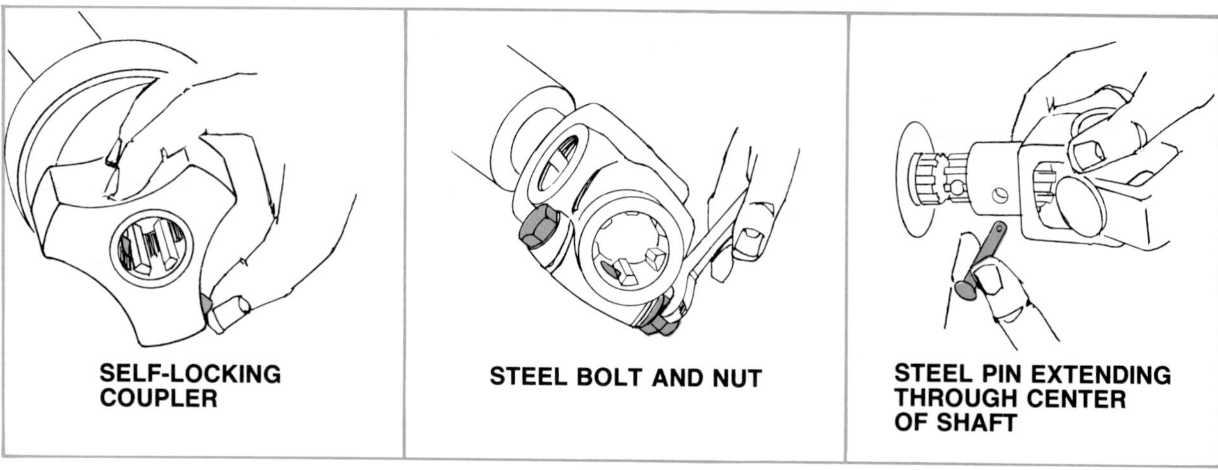

Fig. 63—Methods Of Locking Universal Joints To The Tractor PTO Shaft

• *The recommended tractor drawbar height for the PTO-driven machine you are using*

• *The recommended horizontal distance between the end of the tractor PTO shaft and the center of the drawbar hitch hole*

b) Using these measurements, position the drawbar parallel to and directly below the tractor shaft.

c) Lock the drawbar in this position.

4. *Make sure all connections are tight.* Fasten universal joints securely to the tractor shaft. After connecting the driveline to the tractor, check by pulling firmly on the driveline assembly to make sure it's securely fastened. Several locking devices are used (Fig. 63). Check all other universal joints on the shaft regularly.

Fig. 64—Align The Hitch Holes From The Tractor Seat If Possible

5. *Do not abuse the PTO shaft.* Specifically:

a) Avoid sharp turns when the driveline is rotating. Disconnect the driveline if the corner is too tight.

b) Avoid driving through ditches or depressions if telescoping will be excessive.

c) Do not overtighten safety clutches on PTO-driven machines.

d) Apply power to the driven machine gradually.

e) Do not jerk the PTO shaft by applying power suddenly to clear machines that are plugged.

HITCHING ACCIDENTS

Farm equipment is big and heavy, so be careful when hitching, unhitching, mounting, or removing it. Three risks are involved: crushing your hands and feet, being crushed between the tractor and the equipment, and being crushed by an implement or machine that falls unexpectedly.

Hitching accidents take many lives every year. To avoid them, follow these safe hitching and mounting practices:

Hitching

1. *Position your tractor for hitching from the operator's seat (Fig. 64).* Never operate the tractor from the ground. The risk of it jerking and pinning you is too great.

2. *Turn the engine off and shift to park or set the parking brake before dismounting to hook up.* This precaution is especially important on sloping ground. Pulling or pushing the tractor or equipment to align it could bring it rolling toward you with crushing force.

   If you have problems aligning the hitch, you may find this procedure helpful: back the tractor beyond the hitch point with the implement and tractor drawbars in line. Place the top of implement clevis on top of the tractor drawbar, and insert the drawpin in the implement hitch hole. Then, inch forward. When the holes align, the pin can drop into place.

3. *Shift to park and set the parking brake before permitting a helper to go behind the tractor to complete the hitch.* If you can't align the hitch by yourself, back the tractor past the hitch alignment point, shift to a forward gear, and then complete the hitch by driving forward (Fig. 65). This eliminates the risk of crushing your helper behind the tractor if your foot slips off the clutch. Capable operators take pride in being able to hitch and unhitch by themselves.

4. *Don't try to lift heavy equipment by hand.* Put a jack under the hitch or equipment frame to position the hitch for attachment to the tractor. Many implements and machines are equipped with a jack or stand to keep the hitch at the proper height. Use these if they are provided. For some implements, the remote hydraulic cylinder can be used to lift the hitch into position.

   When hitching, keep your fingers and hands away from pinch points. Never put your fingers in the holes in the tractor drawbar or in the holes in the equipment hitch when they are being aligned.

5. *If you are helping hitch: Never stand behind a tractor when it is backing up.* Stand to the side. Let the driver align the hitch and put the tractor in park or a forward gear. Then step in and make the hitch. The driver may inch forward to make the final alignment. But only on the assistant's signal!

6. *Avoid being pinched or crushed by a load-sensing hitch.* The three-point hitch draft arms of many tractors can be placed under three different modes of operation: floating, depth control, and draft or load control. When floating, the draft arms are raised and lowered, without interference of the hydraulic system, by the attached equipment as it follows the contour of the ground. When operating in depth control, the draft arms remain positioned at the height selected by the tractor operator. In draft or load control, the hitch is responsive to changes in the amount of draft (backward pull) of the implement. Gauge wheels are not necessary. When the draft decreases, as when plowing through light soil, the draft arms lower the implement. When draft increases, as when plowing through a tough spot, the arms raise the implement to reduce the draft of the implement on the tractor.

   If your tractor is equipped with a draft- or load-sensing hitch, move the hydraulic control lever to a position that maintains a fixed draft arm height before attaching equipment to a three-point hitch. If you don't, the draft arms may raise unexpectedly when the equipment's weight is placed on the hitch, pinching or crushing your hands, arms, or legs.

   NOTE: The names associated with the automatic control of the three-point hitch vary according to the manufacturer. Refer to your operator's manual.

Fig. 65—Drive Forward To Align The Hitch

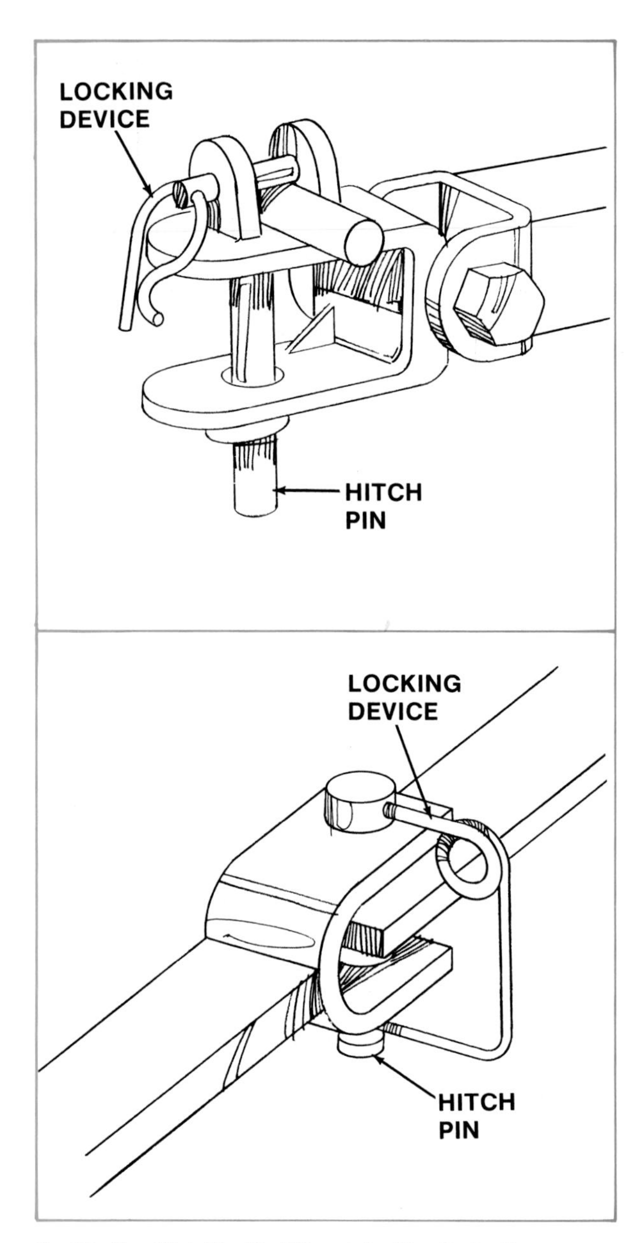

Fig. 66—Use Hitch Pins That Have A Locking Device To Keep Them In Place

7. *Secure the hitch with a locking device that can't jump out and release the trailed equipment.* Several types are available. Replace damaged pins (Fig. 66).

CAUTION: Trash and ground contact can unhook hitch pins.

8. *Attach a safety chain between the tractor and the equipment when transporting materials on public roads.* The safety chain will help guarantee the security of the hitch. Safety chains are available from dealers. The chain and its attaching points should be strong enough to pull the gross weight of the towed load. And it should have a safety hook to prevent accidental unlatching.

Attaching Equipment

1. *Use the proper hardware.* Be sure you have the right brackets and the recommended sizes and types of bolts for attaching equipment to your tractor. Then you'll be able to complete the job quickly and in a safer manner, and you will be sure that the equipment is mounted securely. Don't use undersized bolts, bolts of the wrong grade of hardness, or improvised brackets. Use the hardware recommended for the equipment.

2. *Use adequate jacks, hoists, or blocks to position the equipment (Fig. 67).* This will eliminate the need to physically lift, pull, or push heavy equipment into alignment for mounting. It will also reduce the risk of getting your hands and fingers pinched, or having the equipment fall or tip on you. Many implements have their own stands and jacks. If these are provided, use them to make mounting safer and easier.

3. *Stay out from under equipment if it's not adequately supported.* It should be fully attached to the tractor and resting on the ground, supported on solid blocks, or supported on the stands provided for the equipment. Do not rely on jacks. And don't rely on the tractor's hydraulic system to support raised equipment while you are under it. Hydraulic lines, valves, or seals may fail, or another person might accidentally lower the equipment by moving the hydraulic control lever. Remember that you may be pinched or crushed if the equipment tips, rolls, slides, or falls.

4. *Protect your arms and legs.* Watch for all possibilities of pinching or crushing your fingers, hands, arms, feet, and legs. Don't get them between the equipment and the tractor, between assemblies to

Fig. 67—Use Stands Or Blocks Provided To Support Equipment Adequately And For Ease Of Mounting

be fastened together, or between linkages that move in a scissors-like manner. Be careful when prying with aligning punches and prybars. These tools are brittle, and will snap if used with excessive force.

5. *Make sure safety locks and catches are fastened.* Many types of equipment and quick-attach frames are equipped with safety locks to secure the equipment in place. Be sure these are always in the locked position. If they aren't, the equipment may fall or tip when you get underway.

Unhitching Equipment

1. *Select a good location.* A firm, level, well-drained area away from livestock and traffic is best. In such an area, your equipment will not roll when uncoupled. It will be protected from damage by livestock and moving vehicles. And stands and jacks, or the equipment itself, will not sink as they would into mud or soft soil. If you must uncouple wheel-carried equipment on sloping ground, block the wheels adequately to keep them from rolling. And if you are in soft or muddy conditions, put boards under the support stands and the equipment, if necessary, to prevent it from sinking.
2. *Lower support stands and jacks, or block equipment.* Many implements and machines are not balanced when uncoupled from the tractor. They may tip to the front, rear, or sideways. To prevent tipping, lower all stands provided by the manufacturer before you uncouple the equipment from the tractor. If stands or other devices are not provided, support the equipment on blocks. The direction of tip of load-carrying machines like grain drills, fertilizer spreaders, and sprayers depends on machine design, weight of the load, and frequently, as with manure spreaders, the location of the load within the machine. If you unhitch a machine that's not completely empty, be sure it is adequately supported in the direction it may tip. Your operator's manual will give you more information.
3. *Use transport links or lower equipment to the ground (Fig. 68).* Many implements and machines are equipped with transport links or with safety locks to keep them supported in a raised position. If you store equipment in the raised position, engage these links and locks if they're provided, or use solid blocking. Do not rely on the pressure in the hydraulic system.
4. **CAUTION:** *A shift in the center of gravity can make the tongue raise or lower violently.*

ACCIDENTS ON PUBLIC ROADS

About 17 percent of all tractor fatalities occur on public roads and highways. Surprisingly, most accidents happen when operating conditions seem safest (Fig. 69). You can't afford to take chances, even when conditions seem ideal.

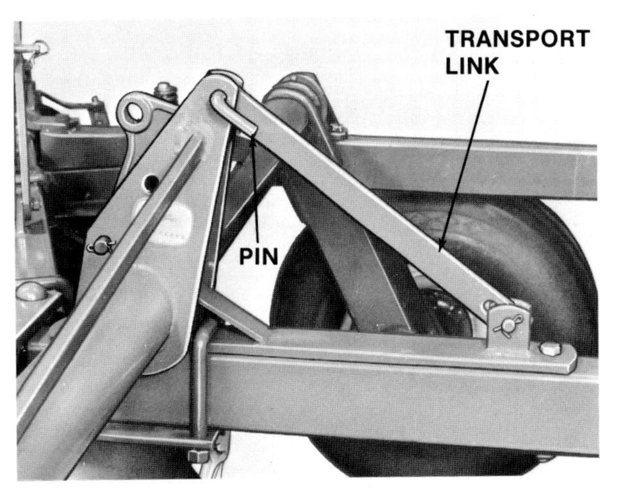

Fig. 68—Transport Links May Be Used To Keep Equipment In A Raised Position

Collisions with motor vehicles, driving off the road, and upsets are the most common types of tractor accidents on public roads. To prevent them, the operator must follow safety precautions *when getting his equipment ready* for highway travel and *when driving.*

Highway Tractor Safety Equipment

SMV EMBLEM

Most state vehicle codes require that slow-moving-vehicle (SMV) emblems be displayed on tractors and farm equipment traveling on public roads. If farm employees are driving, federal law (the Occupational Safety and Health Act of 1970) requires display of the emblem.

Fig. 69—Not As Safe As It Looks

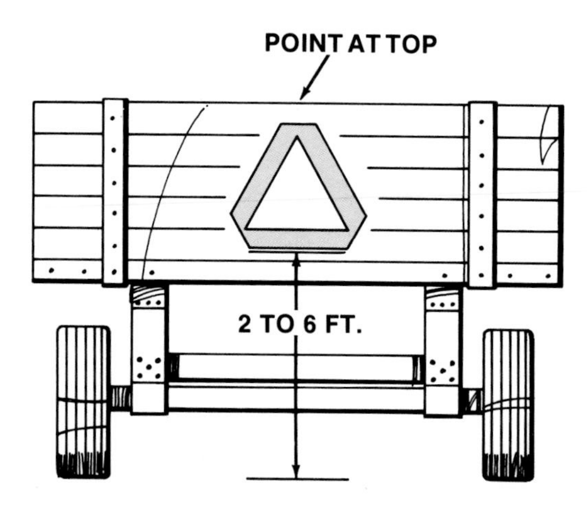

Fig. 70—The SMV Emblem Correctly Displayed

Many new tractors and self-propelled machines come factory-equipped with SMV emblems (Fig. 70). If you need to equip a tractor, self-propelled machine, or trailing equipment you presently own with an emblem, mount it near the center line, from 2 to 6 feet (0,6 to 1,8 m) from the ground, with the point at the top, and within 10 degrees of being perpendicular to the direction of travel. This position is necessary so the face of the emblem can be seen by drivers behind you, and so the reflective border will reflect back toward the headlights of other vehicles.

Fig. 71—Recommended Lighting And Marking For Tractors And Self-Propelled Machines (Check State And Local Regulations)

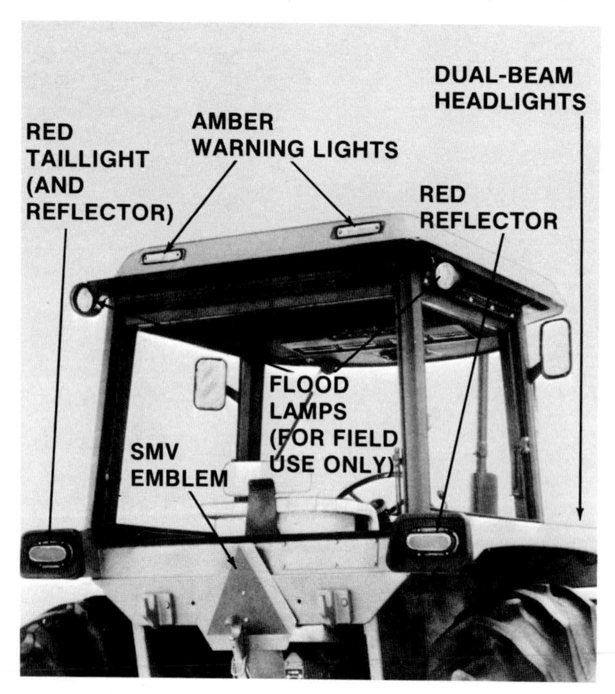

If you don't want to equip all your tractors and implements with SMV emblems, buy a mounting bracket for each, and transfer one emblem around as needed. The SMV emblem is *not* a substitute for warning lights and reflectors. Warning lights and reflectors make your equipment visible. The SMV emblem identifies your equipment as slow moving.

Replace SMV emblems when they lose their reflectiveness. Keep all SMV emblems in good condition, mounted securely in proper position, and clean. Emblems that are dirty, damaged, or improperly displayed do little good.

WARNING LIGHTS

Laws governing lighting requirements for slow-moving vehicles vary from state to state. Some states require steady-burning amber warning lights, but most require or allow flashing amber lights. Use your warning lights day or night. Generally, state laws conform to the standards for lighting and marking farm equipment recommended jointly by the American Society of Agricultural Engineers (ASAE) and the Society of Automotive Engineers (SAE). These standards recommend the following for tractors and self-propelled machines (Fig. 71):

- *Two white headlights*
- *At least one red taillight, mounted on the left-hand side*
- *At least two amber warning lights, visible from front and rear, mounted at least 42 inches high and at the same level*
- *At least two red reflectors, visible from the rear, and mounted on both sides*
- *Turning signal lights, visible from both front and rear, are being introduced*

For your equipment (Fig. 72):

- *If it extends more than 4 feet to the left of the center of your tractor, at least one amber reflector, visible to the front, mounted on the left-hand side*
- *If it extends more than 4 feet to the rear of the hitch point or more than 4 feet to either side of the tractor center line or if it obscures any rear light of your tractor, at least two red reflectors mounted on the left- and right-hand sides*

HEADLIGHTS

Forward visibility is necessary to keep from hitting obstructions or running off the road. You should have two headlights on your tractor. Keep these lights in working order and properly aimed to light the path ahead, but not blinding the eyes of approaching drivers. Check your operator's manual for instructions. Never run on a highway with the field work light on.

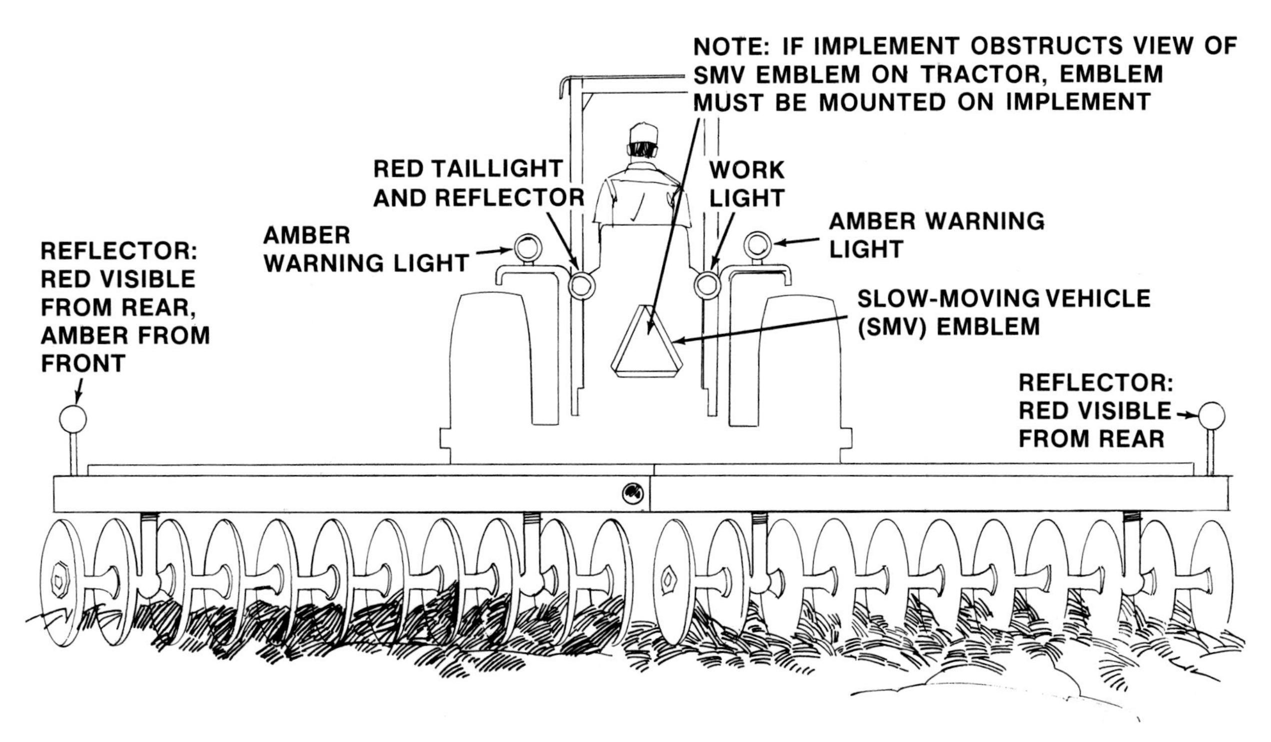

Fig. 72—Recommended Lighting And Marking For Towed Equipment (Check State And Local Regulations)

MIRRORS

Mirrors are effective safety devices (Fig. 73). They let you see all around you all the time.

Mirrors designed specifically for use on tractors without operator enclosures are not always readily available from farm equipment dealers. However, some mirrors designed for trucks or tractor cabs can be used by building a mounting bracket. Follow these suggestions:

1. Select a large mirror that has easily replaced glass.
2. Build a bracket to extend the mirror outward to obtain maximum rearward visibility.
3. Mount the bracket to keep it as free of vibration as possible, and to position the mirror where it is not likely to be hit by an obstacle, hit the operator if the tractor turns over, or interfere with mounting and dismounting.

Fig. 73—Mirrors Enable The Operator To See All Around All The Time

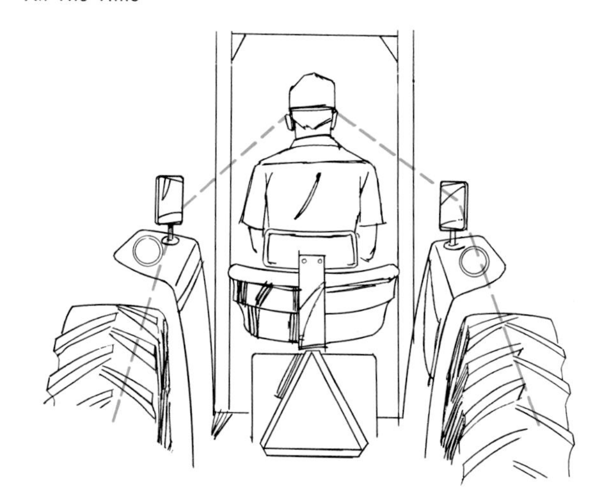

Fig. 74—Keep Your Tractor In Tip-Top Condition For Highway Travel

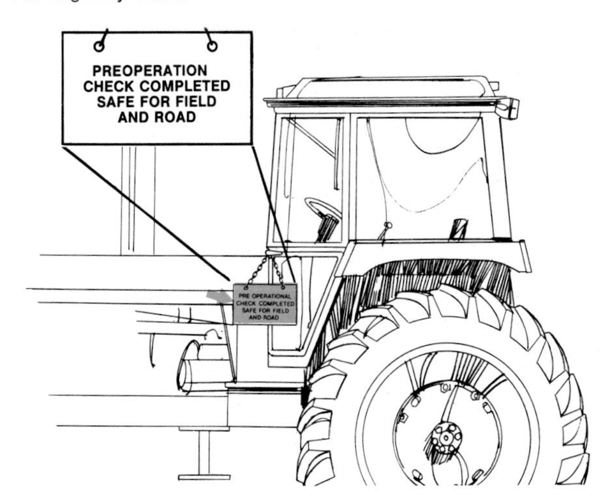

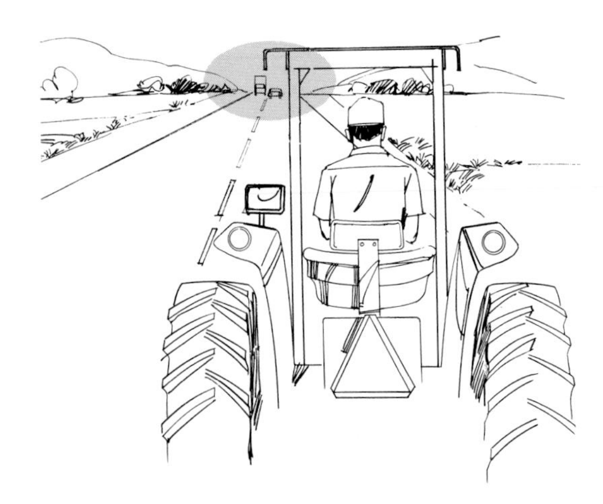

Fig. 75—Keep Your Eyes Moving To Watch For Hazards On The Road And Shoulders Ahead

TRACTOR CONDITION

Keep your tractor in top condition for highway travel (Fig. 74):

1. Brake pedals must apply equal braking force to each rear wheel without excessive foot pressure. Check pedal adjustment. Pedals get out of adjustment when used in the field where one brake is used more than the other.
2. Brake pedals should be locked together for highway travel.
3. Steering must be responsive, with no excess play in the steering linkage.
4. Tires must be in good condition, without cuts, cracks, or bulges that could cause a blowout.
5. All lights, reflectors, and SMV emblems must be in good condition: at least one red tail light on the left, two amber flashing lights visible from front and back, two red reflectors.

Fig. 76—This Tractor Operator Should Have Waited For The Road To Clear

DANGER!
WAIT FOR
ROAD TO CLEAR

Driving Practices For Highways

1. *Maintain full control of your tractor and equipment on the highway.* The safety practices discussed in this chapter that deal with safe travel speed, turning, braking, hitching, and preventing upsets are very important for safe highway travel.
2. *Stay alert.* Keep your eyes moving and watch for hazards in the road and on the shoulders ahead (Fig. 75). Use the mirror or look back frequently to check for traffic coming from the rear, but avoid pulling the steering wheel in the direction you turn your head. Keep in mind that motorists often misjudge the speed of farm equipment, and, to avoid hitting you from behind, they may be forced into panic braking or passing you on either side. Evaluate everything you see, forward and rearward, in terms of possible accident causes. Any time the situation looks potentially hazardous, take the necessary precautions.
3. *Wait for traffic to clear before entering a public road.* Wait your turn (Fig. 76). Don't take a chance and pull out in front of moving traffic. The tractor operator is usually injured severely or killed when involved in an accident with another vehicle. A car traveling at 60 mph on dry pavement needs approximately 600 feet to stop, and approximately 700 feet at 70 mph—*after* the other driver realizes you're in his way. Remember that it's not easy to estimate the speed of approaching vehicles, and that the other driver may not see you. Misjudgment on your part or lack of attention by the other driver could result in a very serious accident.
4. *Beware of blind intersections.* Slow down or stop before entering blind intersections where traffic from the right or left cannot be seen (Fig. 77). Intersections of country roads without stop signs

Fig. 78—Pull Over And Let Traffic By

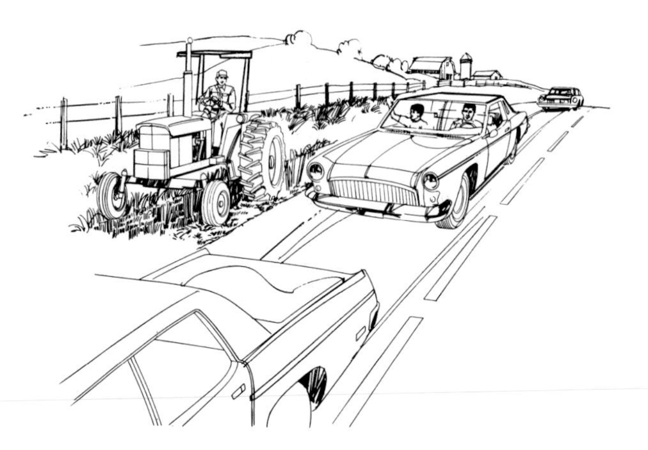

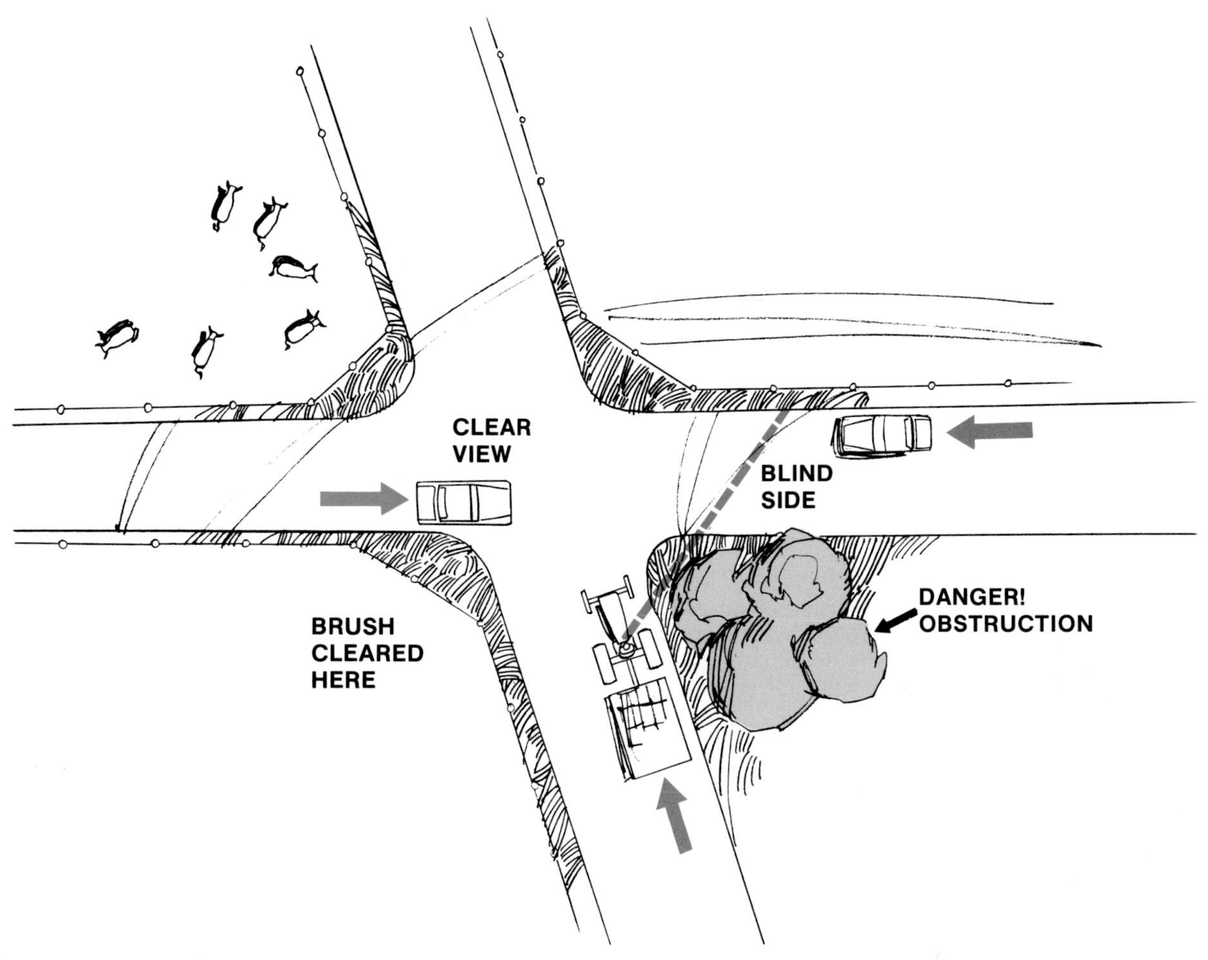

Fig. 77—Clear Blind Intersections Of Brush

are especially dangerous, since motor vehicles often speed through these intersections. Always be prepared to yield the right-of-way to other vehicles.

5. *Pull off the road and let traffic go by (Fig. 78).* Be sure to watch for holes in the shoulder to avoid upsets. Before turning into a traffic lane, maneuver the tractor into a position where you make sure no vehicles are approaching.
6. *Keep the approaching traffic lane clear.* If your equipment is wider than your traffic lane, keep it over on the shoulder to enable motor vehicles to pass in either direction (Fig. 79).

Fig. 79—Don't Obstruct The Approaching Traffic Lane

Fig. 80—Pull Off The Road And Stop. Don't Wave Them Past.

7. *Don't encourage motorists to take chances.* Pull off the road and stop to allow motorists behind you to pass (Fig. 80). Signaling them to pass is risky. They can't see well and the traffic situation could change so quickly that they might not be able to pass safely.
8. *Use arm signals or turn signal lights if available.* Amber warning lamps are used as turn indicators. When signaling, the indicator light blinks, and other lamp burns steady. Keep your driving behavior predictable by giving advance warning of your intentions to turn or stop (Fig. 81). This lets other drivers govern their driving behavior accordingly. Except when making left-hand turns, don't stop in a traffic lane. Pull over to the shoulder if a sound one is available. If not, continue traveling in the traffic lane until you find a safe place to turn off.
9. *Obey all traffic signs.* They warn drivers of hazardous conditions like intersections, hills, curves, narrow bridges, and railway crossings (Fig. 82). Heed these warnings. It's also wise to plan your escape from a possible accident in case other drivers fail to obey these signs.

TOWING EQUIPMENT

Extra care is necessary when towing equipment because you are controlling both the tractor and the implement behind it. The towed equipment must not weigh more than the tractor.

Most of the precautions for towing equipment have already been discussed in this chapter. Let's look at these again, along with a few others. First, we'll turn our attention to the precautions necessary for towing machines, implements, and wagons. Then we'll look at those that apply to towing tractors and self-propelled machines.

Towing Machines, Implements or Wagons

1. *Hitch only to the drawbar.* Prevent tractor upsets by hitching pull-type equipment only to the drawbar or to the three-point-hitch drawbar attachment (Fig. 83). Never pull from any other point on the tractor. This aligns the tractor's center of pull with the load, helps to keep the front wheels of the tractor moving straight ahead, and provides extra steering control over the towed equipment.

2. *Adjust drawbar height as recommended in your operator's manual.* Do not position a three-point-hitch drawbar attachment higher than the position recommended for the regular drawbar. Install stay braces, if available, to prevent accidental raising of the draft arms. Remember, the higher the hitch point, the more susceptible your tractor is to upsets.

3. *A 7-pin connector is installed on tractors so towed equipment tail light can be hooked up.*

Fig. 81—Standard Arm Signals Warn Others Of Your Intentions

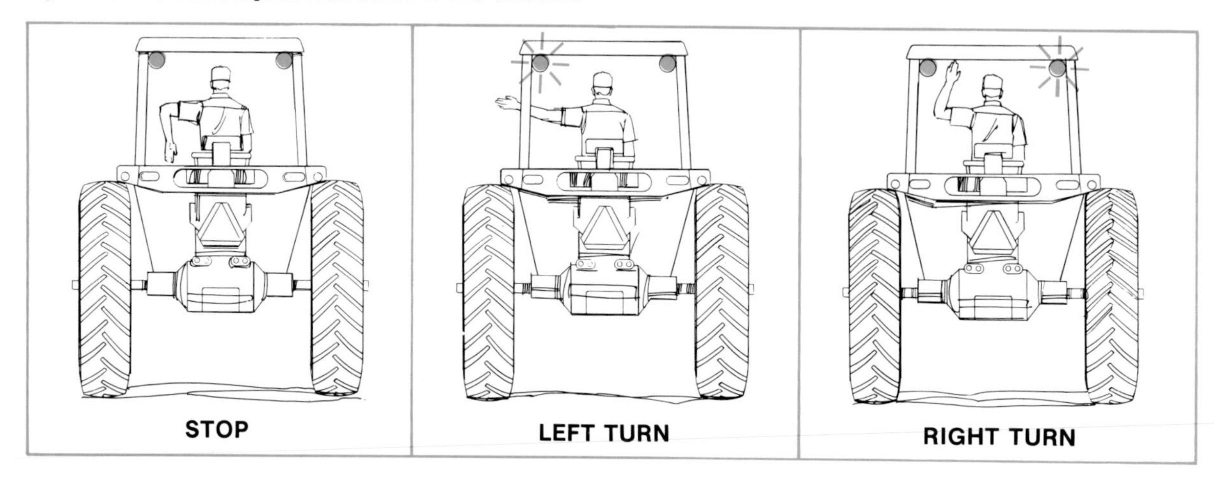

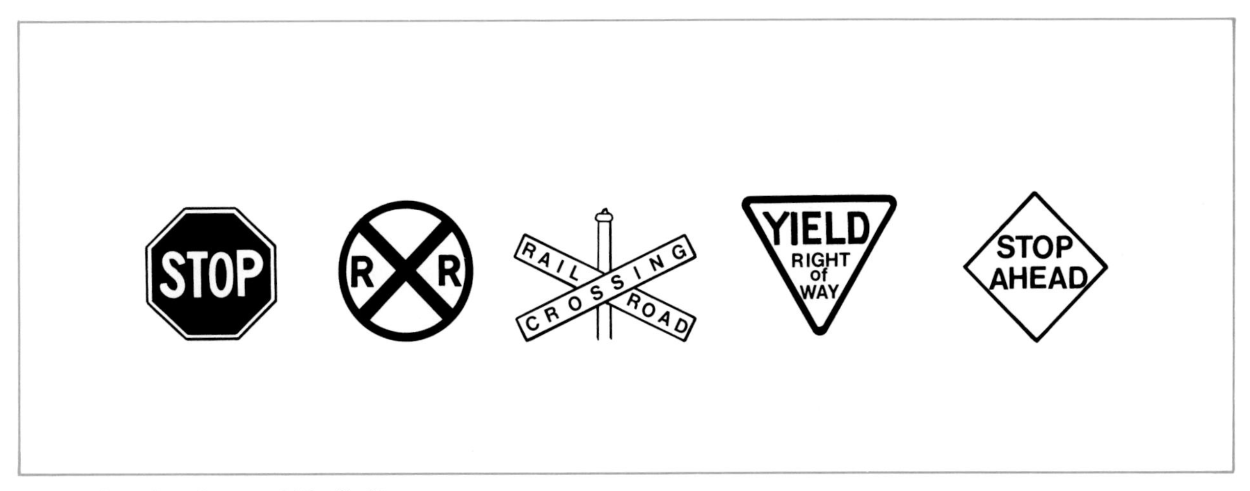

Fig. 82—Heed The Warnings Of Traffic Signs

Fig. 83—Prevent Upsets—Tow Only From The Drawbar Or Three-Point-Hitch Drawbar Attachment, Set at Recommended Drawbar Height

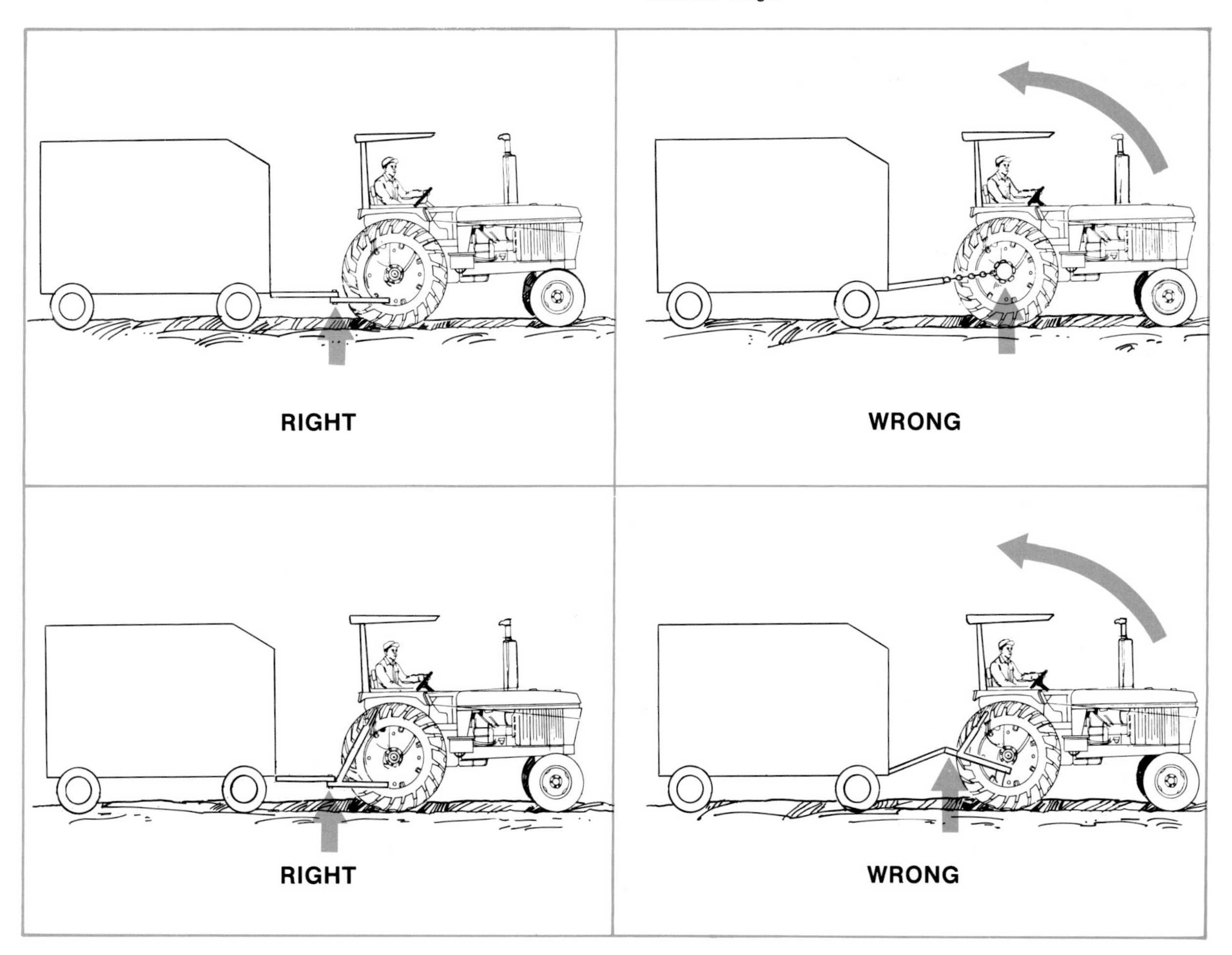

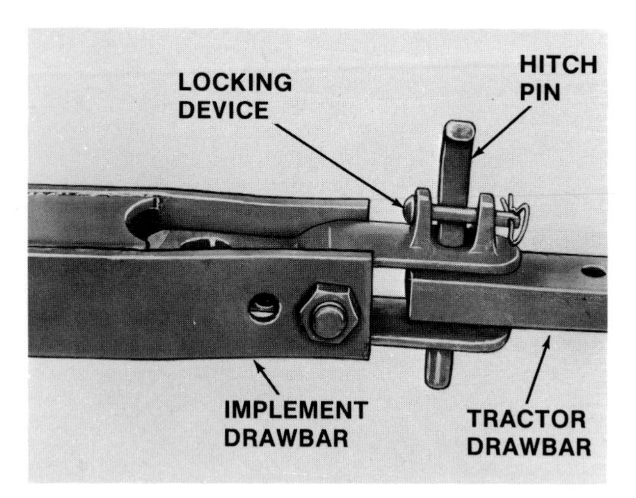

Fig. 84—Hitch Pins And Locking Devices Are Frequently Designed As An Integral Part Of The Hitch

4. *Use safety hitch pins.* Prevent the possibility of any drawbar-connected load coming unhitched from the tractor. Pins with different types of locking devices are available. Some machines and implements come equipped with hitch pins and locking devices incorporated into the design of the hitch (Fig. 84). Keep these pins with their machines and always use the locking devices provided. If you find yourself without a safety hitch pin, make your own, using a strong bolt, washers, and nuts.

NOTE: Locking devices are also essential to keep implements connected to the three-point hitch in place. If an implement pin in one of the lower draft links pulls out, one side of the implement will drop to the ground. If the implement drags sideways behind the tractor, it could be caught by the tractor rear wheel and carried up to the operator's platform. To avoid this, make sure that each lower draft link implement pin is locked in place with a positive locking device.

5. *Use safety chains on the public roads (Fig. 85).* They prevent wagons or other trailed equipment from rolling into ditches or into the path of oncoming traffic if the hitch uncouples. Safety chains or cables are available from your dealer. Follow the instructions provided for their installation and use.

If you make a safety chain, follow this procedure:

a) Select attaching points for the chain on the towed and towing units:

- *Strong enough to support the gross weight of the towed unit*
- *Within 6 inches vertically and laterally of the primary attaching point, where the drawpin couples the units together*
- *Close enough to the primary attaching point so that no more than 23 inches of chain, measured from the primary attaching point, will be required for fastening*

b) Select a suitable chain:

- *Strong enough to support the gross weight of the towed unit*
- *Just long enough to permit unrestricted turns and to fasten both ends to the towed and towing units*

c) Fasten the ends of the chains to the towed and towing units, using safety hooks or other devices that can't be opened accidentally.

d) Provide an intermediate support for the chain if the point of chain attachment is 6 inches in front of the primary attaching point on the towing unit, or 6 inches behind the primary attaching point on the towed unit. On the towed unit, the intermediate support should be at least half as strong in any direction as the points you select to attach the chain.

Fig. 85—Install A Safety Chain For Highway Travel

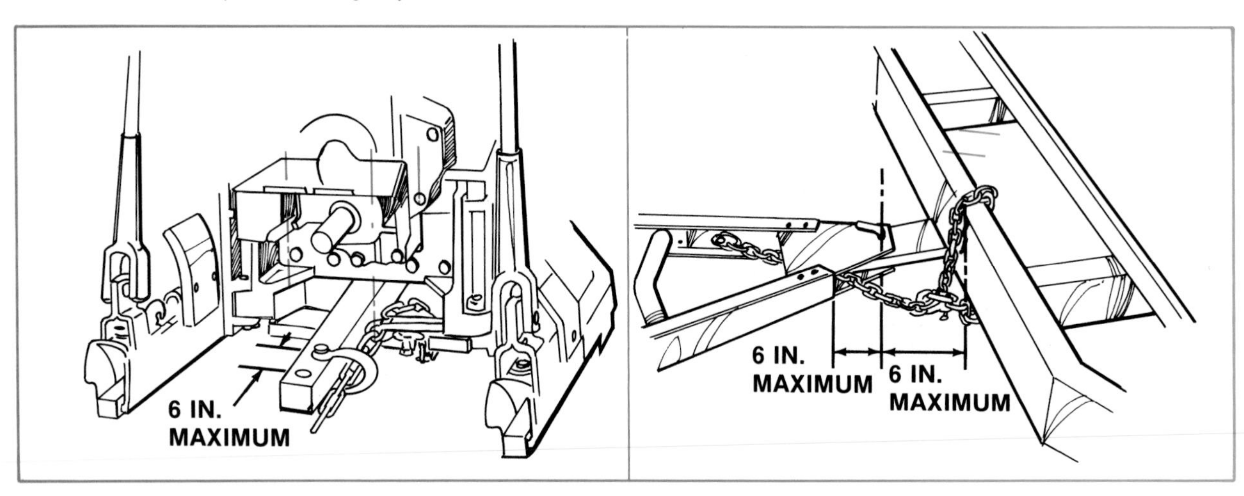

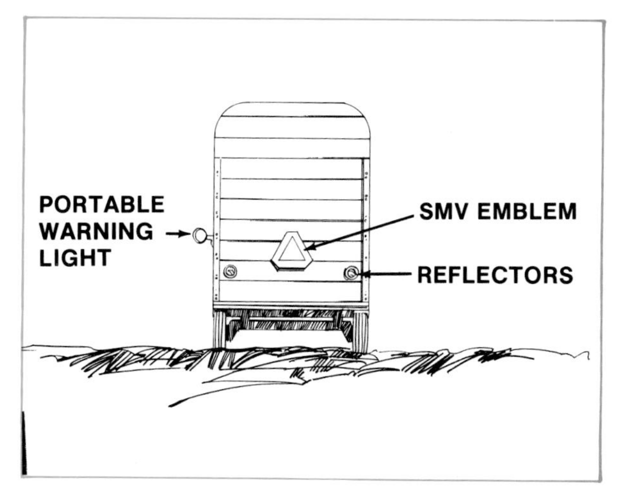

Fig. 86—Portable Warning Lamps Can Be Mounted On Equipment

Fig. 87—Implements Are Not Safe To Ride On, Especially On Public Roads

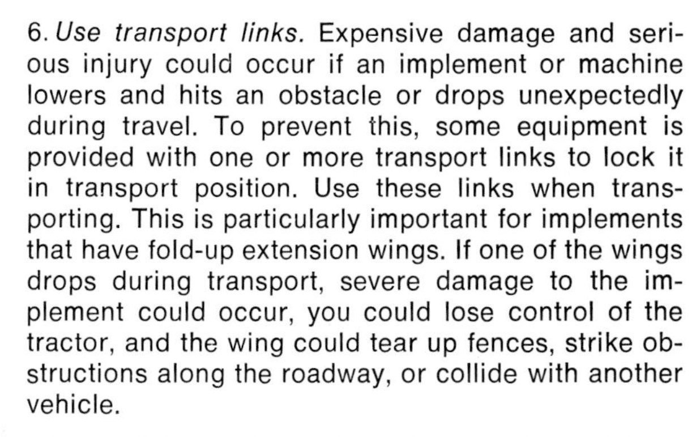

6. *Use transport links.* Expensive damage and serious injury could occur if an implement or machine lowers and hits an obstacle or drops unexpectedly during travel. To prevent this, some equipment is provided with one or more transport links to lock it in transport position. Use these links when transporting. This is particularly important for implements that have fold-up extension wings. If one of the wings drops during transport, severe damage to the implement could occur, you could lose control of the tractor, and the wing could tear up fences, strike obstructions along the roadway, or collide with another vehicle.

7. *Use safety warning equipment on public roads (Fig. 86).* Don't drive on public roads unless safety warning equipment is in place and functioning properly (see Highway Tractor Safety Equipment, page 167).

Mount at least two flashing amber warning lamps as far back on a towed vehicle as possible. Mount the lights at least 42 inches (107 cm) high. The lights should be visible from front and rear.

8. *Don't carry passengers on towed equipment.* This is almost as hazardous as carrying them on tractors (Fig. 87). Accident data show that farm wagons, for example, rank second to tractors as causes of personal injury, and that falls account for more than half of these accidents. Avoid carrying passengers on wagons if possible. If you must, insist that they remain seated. Start and stop slowly. Maintain a slow travel speed, watch for bumps and obstructions, and avoid quick turns. Use hand or electric turn signals for the benefit of passengers so unexpected movements won't catch them off guard.

9. *Maintain safe travel speeds.* The main dangers of traveling too fast are tractor upsets, collisions, and running off the road or into obstacles on the shoulder. To prevent these accidents, maintain a safe travel speed to keep you in control at all times. If your equipment fishtails, slow down to gain control (Fig. 88). Unloaded trailers fishtail more than loaded trailers.

Fig. 88—Slow Down If Equipment Fishtails

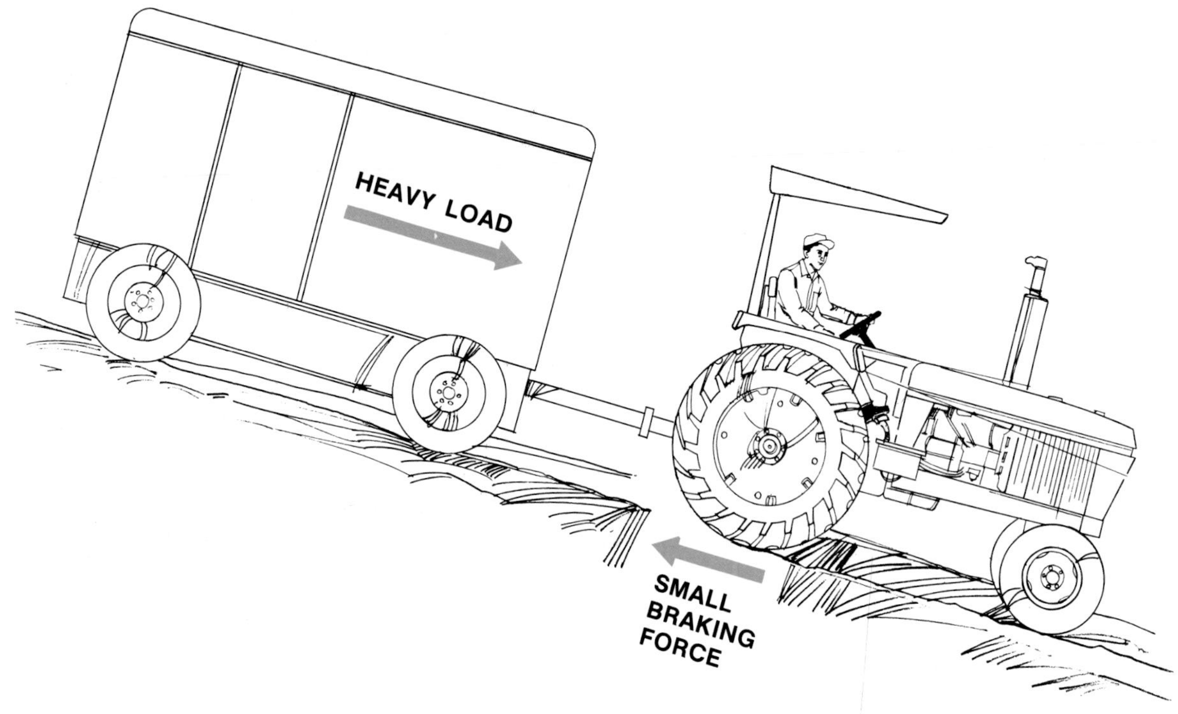

Fig. 89—Towed Loads Make It More Difficult To Stop, Especially On Slopes

10. *Don't rely too much on your brakes (Fig. 89).* Stopping distance increases with increasing speed, as the weight of the towed load increases, and on hills and slopes. When pulling a heavy load, tractors must be large enough to have enough braking power to handle the pulled load.

11. *Put wide machines in transport position.* Many machines have hitches that can be adjusted so the equipment is towed more in line behind the tractor (Fig. 90). Placed in transport position, this type of hitch reduces the width of the machine for safer towing. Do not drive half on and half off the road. If traffic backs up, pull off and stop to let traffic by.

12. *Watch out for overhead obstructions.* Remember that towed equipment has *height* as well as *width.* Always watch for overhead obstructions like tree limbs and power lines.

Towing Tractors And Self-Propelled Machines

Driving tractors and self-propelled machines or hauling them on a truck is usually more desirable than towing them. Towing *doesn't* save travel time, since tractors and self-propelled machines should never be towed faster than they could move under their own power. There are other problems, too. Finding strong and conveniently located hitch points is not always

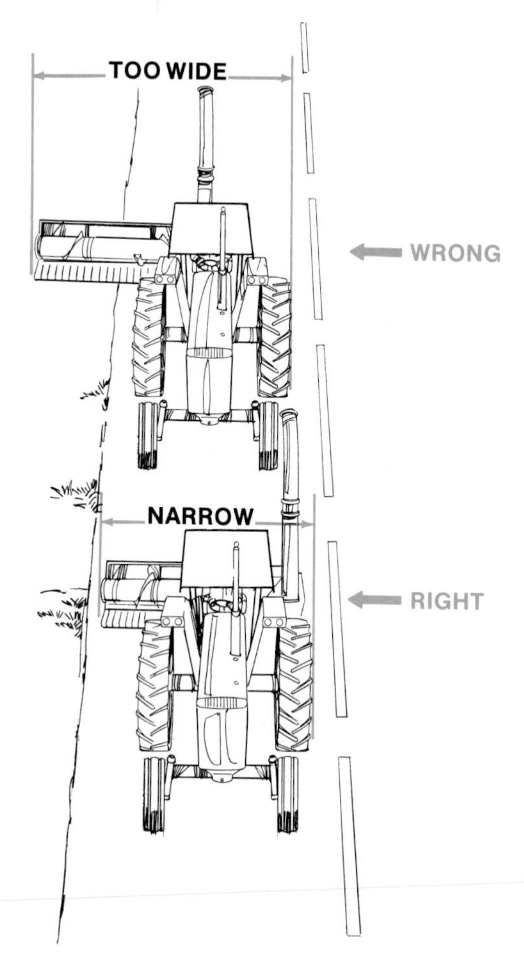

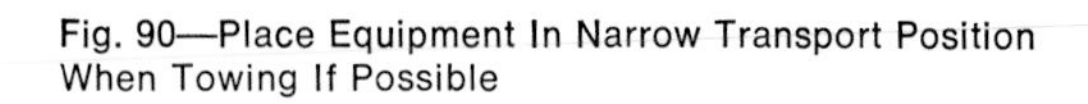

Fig. 90—Place Equipment In Narrow Transport Position When Towing If Possible

easy. Damage to the transmission, differential, or final drive of the towed unit can occur if the recommendations outlined in the operator's manual are not followed. And careful coordination of movement between the towing vehicle and the towed unit is necessary to prevent a collision between them.

If it is necessary to tow a tractor or a self-propelled machine, follow these practices:

1. *Prepare the unit for towing.* Follow instructions given in the operator's manual (Fig. 91). This is the only way to make sure that lack of lubrication will not damage bearings and gears in the transmission, differential, or final drive, and to prevent other damage.

a) Place all transmission shift levers in neutral or tow position and disengage the PTO drive.

b) Disengage the rear wheels from the transmission drive line, if recommended in the operator's manual. This may involve the simple shifting of a lever, or removing the drive shafts from some self-propelled machines.

c) Run the engine to obtain hydraulic pressure for the power steering and brakes and to lubricate the transmission (on some tractors).

2. *Hitch safely.* Find a hitch point strong enough for towing and located as close as possible to the center line. Avoid hitching to any linkage of the steering mechanism, or to the front wheel knuckles of utility-type tractors. Don't pull from the front pedestals of tricycle-type tractors. These pedestals are not designed for towing and may break if subjected to shock loading. When attaching chains to self-propelled machines, avoid bending sheet metal. Use a rigid tow bar, if possible, to control speed and steering of the towed unit. If you use a chain, select one strong enough and long enough to give the driver of the towed unit adequate stopping time and distance to avoid a rear-end collision with the towing vehicle (Fig. 92). Never use a nylon rope. Nylon stretches and produces a sling shot effect which jerks the towed vehicle suddenly. Nylon also stretches tight before it parts and snaps violently.

Always attach tow chains with hooks up. Then if the hook spreads under load, if flies down, not up.

Fig. 91—Follow Instructions In The Operator's Manual To Prepare The Unit For Towing

Fig. 92—When Towing, Use A Chain Long Enough To Prevent A Rear-End Collision

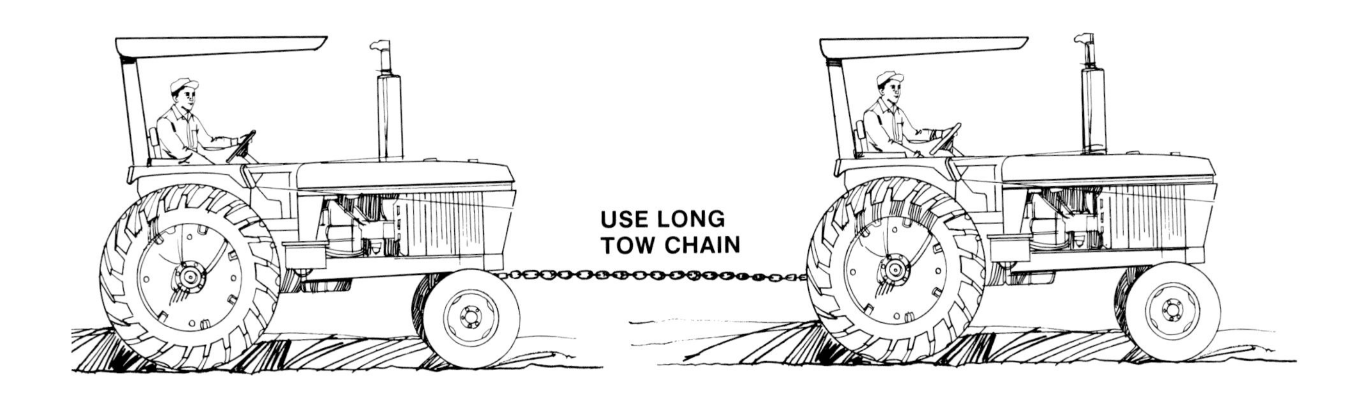

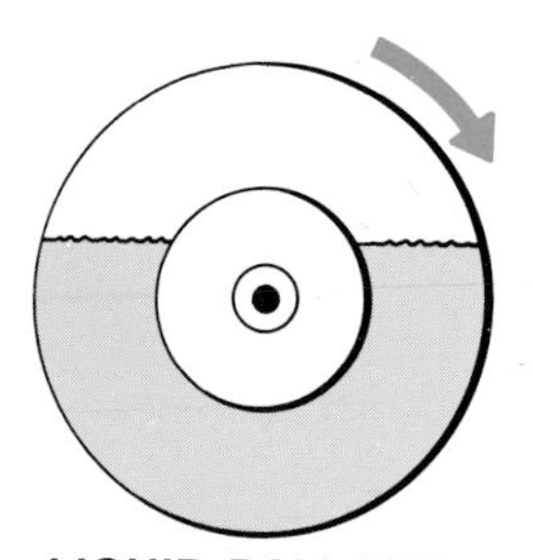

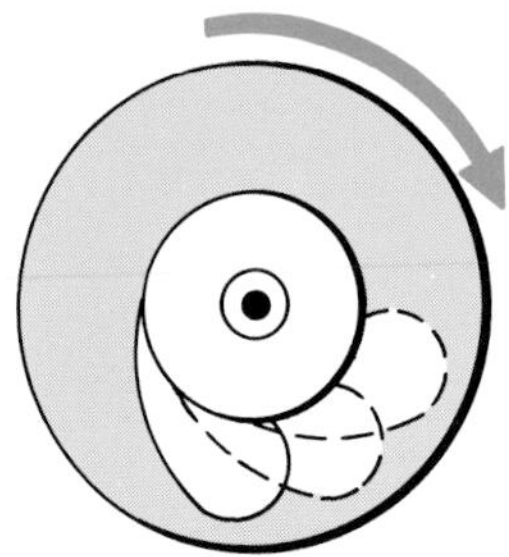

Fig. 93—Liquid Ballast Rotates With The Tire At Higher Speeds, Making Control Of The Tractor Difficult

- *Loss of control of the towing vehicle due to the speed and weight of the towed unit—especially when going downhill*
- *Loss of control of the towed unit caused by liquid ballast rotating and surging in the tires (Fig. 93)*
- *Heating and seizing of the brakes on the towed unit that could make it swerve off the road*
- *Damage to drive line gears and bearings because of lack of lubrication at high rotating speeds (Fig. 94)*

CAUTION: Machines which have power brakes or steering should not be towed except with a rigid tow bar unless someone is in the operator's seat and the engine is running to provide hydraulic power. Check the operator's manual for details.

IMPORTANT: Some tractors should not be towed faster than 5 mph. Be sure to consult your operator's manual.

Fig. 94—Upper Gears And Bearings In Some Transmissions Are Not Lubricated When The Tractor Is Towed

4. *Follow other safety practices.* Review the safe operating practices outlined in the preceding sections on highway safety and towing equipment. They also apply to towing tractors and self-propelled machines.

SAFETY FEATURE CHECKLIST FOR TRACTORS

Tractor manufacturers have made impressive and effective improvements to make tractors easier and safer to operate (Fig. 95). Most of these improvements fall within four classifications:

- **Visibility and recognition**
- **Operator comfort and ease of operation**
- **Protection from operational hazards**
- **Increased stability**

Here is a list of these safety features within each classification.

Visibility And Recognition

- Headlights
- Taillights
- Rear-mounted worklights
- Flashing amber warning lights
- Turn signals
- Reflectors
- SMV emblems
- Windshield wipers and tinted glass
- Rearview mirrors
- Horns

Operator Comfort And Ease Of Operation

- Adjustable, shock-absorbing seats
- Steps and handholds for mounting
- Adjustable steering wheels
- Conveniently located controls
- Hydraulic power and coupling devices to control heavy implements
- Fans, filters, air conditioners, and heaters (in cabs)
- Weather shields
- Labeled controls and instruments using universal symbols
- Noise reduction (in some cabs)

Protection From Operational Hazards

- Rollover protective structures (ROPS)
- Seat belts (used with ROPS)
- Fenders
- Steps and handholds for mounting

TYPICAL SAFETY FEATURES AVAILABLE ON MOST NORTH AMERICAN MANUFACTURED TRACTORS

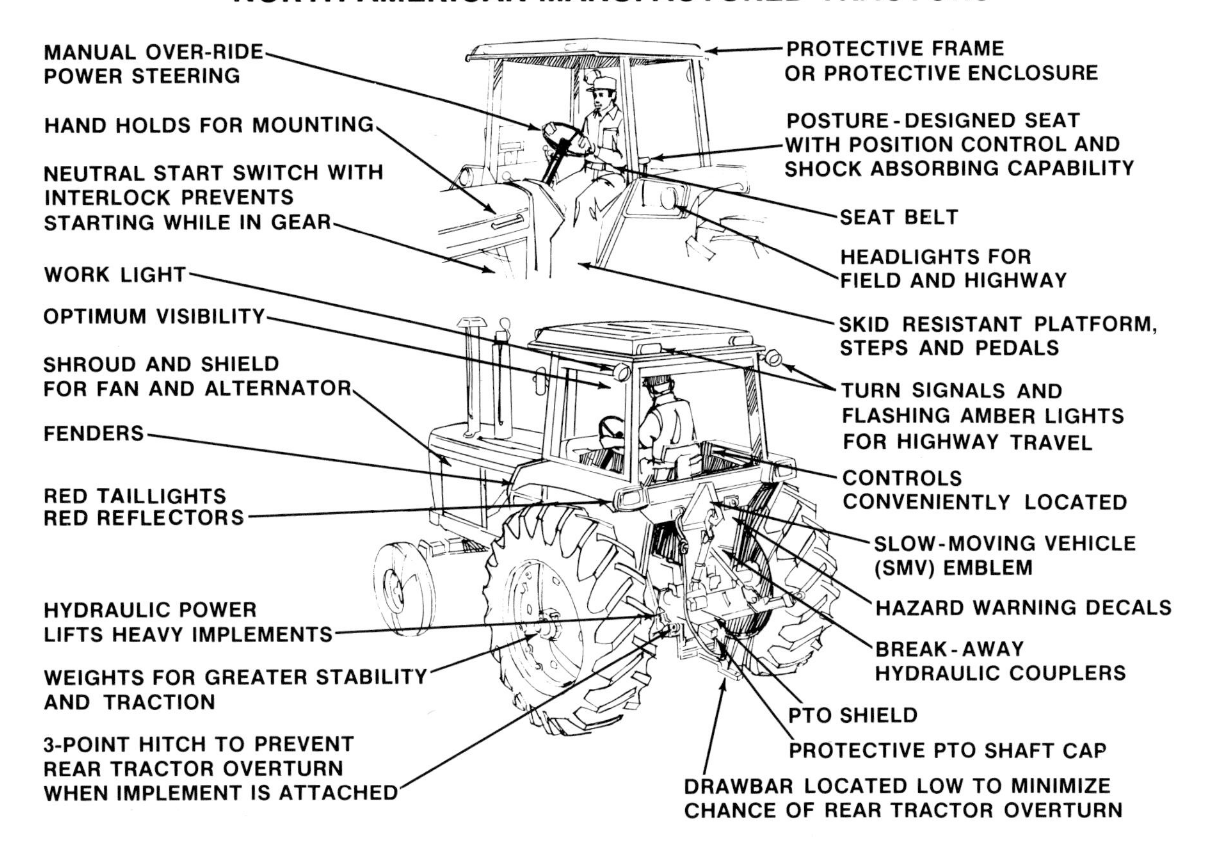

Fig. 95—Safety Features: Keep Them Working And Use Them!

- Skid-resistant platforms and steps
- Shields for rotating parts
- Powered, self-adjusting brakes
- Key starting switch
- Transmission park position or parking brake
- Power steering
- Neutral-start switches
- Tachometers and speedometers
- Standardized controls
- Hazard warning decals

Stability

- Low-mounted drawbars
- Front-end weights
- Rear-wheel weights
- Tire ballast
- Easily adjusted wheel spacing
- Interconnected brake pedals

These design features, and others, are intended to make tractor operation easier and safer. To benefit from them:

1. Know what they are.
2. Know how they function.
3. Keep them in good condition.
4. Use them, and make sure others do also.

Your tractor may not have all of these features. If not, and if you feel that some are essential for safety in your particular operation, ask your farm equipment dealer if they can be obtained for and installed on your tractor.

SELF-PROPELLED MACHINES

Self-propelled machines are mobile power units that incorporate harvesting, planting, spraying, and other working units into their design. For this reason, they involve two types of operating hazards—those aris-

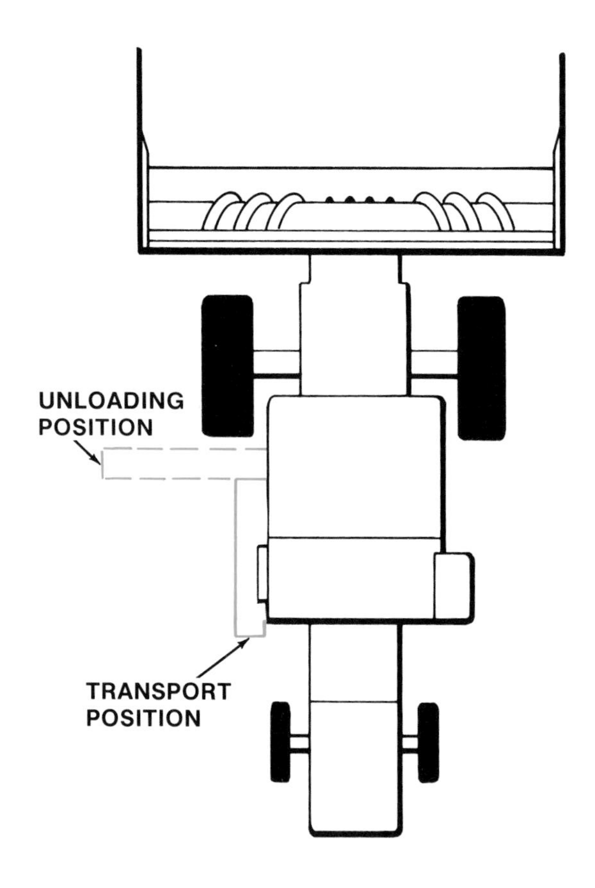

Fig. 96—Place The Unloading Augers In Transport Position Before Transporting A Combine

Fig. 97—Headers Carried Higher Than Necessary Shift Machine Weight Off The Rear Wheels, Decreasing Steering Control

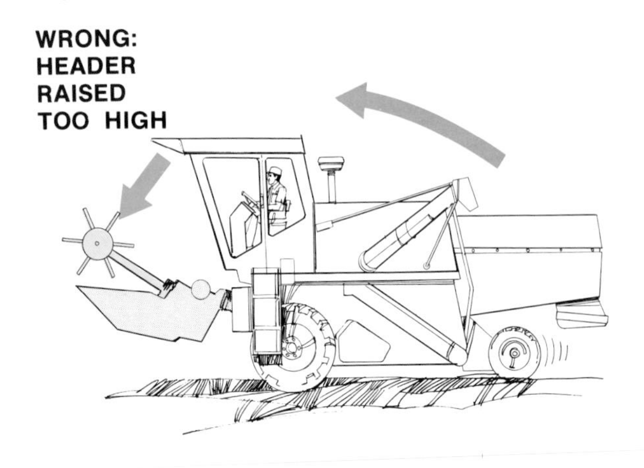

ing when the machine *is being driven as a mobile unit,* and those arising from *the operation of the working unit.*

To avoid accidents caused by the working unit, review Chapter 3 and follow all of the recommendations to protect you from the hazards of moving and sharp-edged machine components. Read Chapters 8, 9 and 10 for information on the safe operation of harvesting units.

Let's look now at important recommendations for driving self-propelled machines safely as mobile power units:

1. *Follow the recommendations for safe tractor operation.* They also apply to the operation of self-propelled machines. These are especially important:

- **Make preoperational checks**
- **Refuel safely to prevent fires and explosions**
- **Start and stop cautiously**
- **Maintain safe travel speeds**
- **Prevent upsets, collisions, falls, and hitching accidents**
- **Drive safely on public roads**
- **Perform maintenance according to operator's manual**

2. *Prepare the machine for transport.*

a) Empty grain tanks and hoppers. This lowers the center of gravity of most machines and makes them more stable. It also relieves the machine of the strain of carrying the weight of the harvested product.

b) Place all unloading augers and elevators in the transport position. This decreases the width of the machine and reduces the chances of hitting obstructions (Fig. 96).

c) For good visibility, lower the header to a height that will clear obstructions on the ground or roadway. Headers are heavy, and the higher they are carried, the more they shift the center of gravity forward, making the rear wheels lighter (Fig. 97). You need as much weight as possible on the rear wheels for effective steering.

Instructions for carrying the header in transport position will vary with the kind and make of machine. Check your operator's manual for instructions.

NOTE: It is often desirable to remove the header from the machine and transport it by truck or implement carrier. If possible, follow this procedure if the machine is to be transported more than a short distance, if the header is much wider than a normal traffic lane, if traffic might be hazardous, or if the shoulders of the road are narrow and in poor condition.

d) Lock both brake pedals together for highway transport, see Fig. 35, if an interlock is provided, to maintain equal braking on the drive wheels. Self-propelled machines can swerve and go out of control just like tractors.

e) Inspect the safety warning devices before driving on public roads. Check to see that all warning lights are in place and functioning. Display an SMV emblem, making sure that it is clean, and that reflectors are, too.

3. *Look before moving the machine.* Don't run the risk of running over someone. You have good long-range vision from the operator's platform, but, on many self-propelled machines, the area immediately surrounding the machine is not within the operator's line of sight (Fig. 98). After an individual has been run over by a self-propelled machine, the operator responsible for the accident often explains that he "didn't see him." Don't get yourself into this situation. *Before you start off, make sure you know exactly where all other persons are.*

4. *Maintain safe travel speeds.* Self-propelled machines are heavy, big, wide, high, and bulky. Don't drive them too fast. Travel slowly enough to maintain steering control, to prevent bumping and jarring of the machine on rough ground, and to give you time to stop.

To be safe when travelling:

a) Slow down when passing obstructions on either side. Have a helper guide you if you're not absolutely sure that you can pass safely.

b) Slow down gradually. Reduce engine speed and use the variable speed drive (if your machine is so equipped) before applying the brakes. Remember that slowing down shifts machine weight forward, decreases the weight on the rear wheels, and makes steering control less effective.

c) Avoid panic stops by watching your driving environment in all directions. There are two good reasons for this: you might not be able to stop in time to prevent striking some obstacle, and, during panic stops, you may lose all steering control. A self-propelled machine *will not turn* if the drive wheels are braked solidly, or if the rear wheels are lifted off the ground.

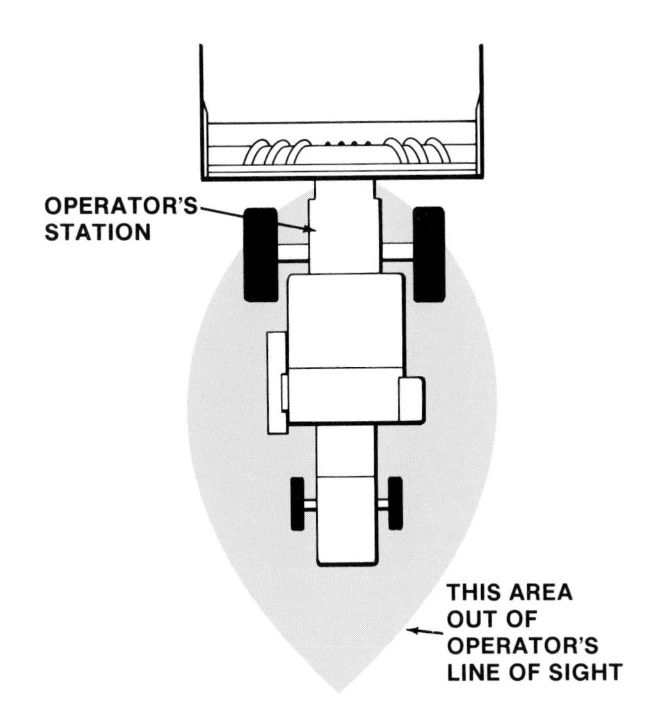

Fig. 98—You Can's See The Ground Near Some Self-Propelled Machines

5. *Avoid collisions on the highway.* The extra width and limited rearward visibility on some self-propelled machines can lead to collisions with other vehicles.

Fig. 99—Turn Off The Road So Others Can Pass

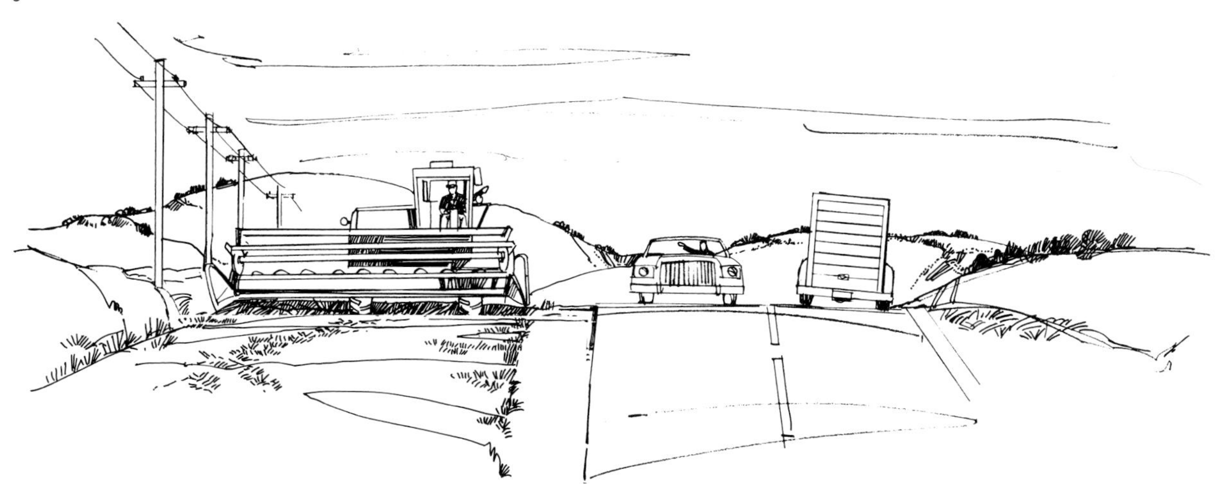

Machines with wide platforms or pickup attachments, and with grain tanks that obstruct rearward vision, should be driven with special care. If possible, drive on public roads when traffic is the lightest. Drive within your own traffic lane to keep the passing lane clear for traffic to pass. If motor vehicles can't pass from behind, turn off the road at your first opportunity and let them go by (Fig. 99).

FOUR WHEEL DRIVE TRACTORS

Four wheel drive tractors with articulated steering pivot in the middle. NEVER stand between the tires when the engine is running. And, NEVER let anyone stand between the tires when the engine is running.

The steering is extremely sensitive. If you accidentally move the steering wheel a bit while someone is standing between the tires, THEY WILL BE CRUSHED.

CHAPTER QUIZ

1. True or false? If a safety defect shows up in a preoperational check, you should get the necessary repairs during your next trip to your farm equipment dealer.
2. The safest way to start the engine is:
   a) With the transmission in neutral.
   b) With your foot disengaging the clutch.
   c) With the transmission in neutral, the clutch disengaged, and the brakes applied.
3. True or false? Fuel mixtures are only explosive when there is a high concentration of fuel vapor in the air.
4. What colors would you paint new safety cans to be used for gasoline and diesel fuel?
5. True or false? Carbon monoxide has an easily recognized odor.
6. True or false? Tractor drivers should use the same arm signals as the motor vehicle drivers in their own state or province.
7. Which is the best answer?
   Using a lower gear when going downhill:
   a) Saves wear and tear on the brakes.
   b) Prevents overspeeding the engine.
   c) Lets you use both engine and brakes for slowing down.
8. When should towed loads be equipped with their own individual brakes?
9. Name three of the most frequent types of tractor accidents.
10. Explain why widening tractor wheel spacing reduces the chances of a sideways tip.
11. Name two practices that could overturn a tractor to the rear.
12. What accident can happen if a towed load is connected higher than the recommended drawbar height?
13. If only one tractor brake is applied, what happens to the speed of the other tractor wheel?

    Which of the following statements is false?
14. a) You should condition your driving to driving conditions.
    b) Locking brake pedals together helps prevent upsets.
    c) If tractor stability becomes uncertain on a slope, you should turn downhill.
    d) Fast travel speed is often a contributing cause of sideways upsets.
    e) When stuck in the mud, it's best to try to drive out forward first.
15. How can you tell the difference between a cab that provides rollover protection and one that does not?
16. Name the locations on a tractor where passengers can ride safely.
17. What type of shield has most recently been developed for PTO assemblies?
18. True or false? You should always disengage the power and stop the engine before leaving the operator's platform to clean, adjust, or service a PTO-driven machine.
19. When hitching a towed load, how can you protect helpers from injury in case your foot slips off the clutch?
20. Give two reasons why hydraulically lifted equipment might fall unexpectedly.
21. Which of the following statements are true?
    a) Displaying the SMV emblem while driving on public roads is required by federal law.
    b) The SMV emblem is a substitute for other safety warning devices.
    c) Amber warning lights should be visible from both front and rear, except where prohibited by state or local regulations.
    d) Tractor drivers should always exercise their legal right to stay on the road.

e) Signaling drivers behind you to pass is both safe and courteous.

22. True or false? The high position of the operator's cab on self-propelled machines provides operator visibility all around.

23. True or false? Accident data shows that farm wagons rank second to tractors in personal injury accidents.

24. List two safety warning devices that should be visible to drivers approaching your tractor from the rear in the daytime.

25. At what two locations are weights placed to increase tractor stability?

26. Why is it desirable to empty grain tanks or hoppers before transporting self-propelled machines?

6
Target: Tillage and Planting Safety

Fig. 1—Tillage And Planting Operations Seem Safe—But Any Job Done Carelessly Can Result In An Accident

INTRODUCTION

Some tillage and planting operations can be hazardous if done carelessly (Fig. 1).

This chapter covers specific safe operating practices for tillage and planting equipment. More information on the operation and adjustment of this equipment can be found in the appropriate John Deere FMO manuals and in manufacturers' operator's manuals.

SIZING FOR SAFE OPERATION

MATCHING EQUIPMENT TO JOB

Attempting a job that's too big for your equipment often results in a feeling of being hurried. You might run undersized equipment too fast and work long hours to try to get the job done. As a result, you could get tired and operate your machinery under unsafe circumstances or take chances. Use equipment that's appropriate for the job you're doing (Fig. 2).

MATCHING EQUIPMENT TO TRACTOR

Serious mismatching of equipment and tractor is rare, because it is obvious to the operator. Dangerous mismatching usually occurs when a tractor is undersized for an implement. Like using a four-bottom plow on a tractor intended to pull three bottoms.

Fig. 2—Match Size Of Equipment To Working Conditions

Trying to work with an underpowered tractor puts excessive strain on the tractor. The tractor's brakes aren't capable of handling the extra load. The tractor is unstable because the extra weight shifts the center of gravity. You are asking for an accident!

MATCHING HYDRAULIC CONNECTIONS

It is possible to connect some hydraulic couplings that aren't intended to mate. A common error is to interchange hose ends on the auxiliary cylinder so the control valve operates in reverse. Instead of the implement being raised when the control is pulled for raising, it goes down. This can be hazardous because the implement motion is opposite to that expected. This problem is discussed in more detail in Chapter 2.

After you attach hydraulic hoses between a tractor and implement, cautiously test to see if you get the proper result. If not, correct it at once.

HITCHING

Basic principles for hitching implements to tractors are covered in Chapter 5. Certain safety considerations will be discussed briefly here:

- **Position of the implement**
- **Backing up**
- **People between the tractor and implement**
- **Safety hitch pins**

POSITION OF THE IMPLEMENT

For the easiest hitching, an implement should be parked on a firm, flat surface free of other objects (Fig. 4). Hitching is more difficult if the ground is uneven or if jacks or blocks tilt the implement at an improper hitching height. Poor implement position can lead to frustration and operator injury when hitching. Block mounted implements securely before lowering the hydraulic lift.

A little forethought in unhitching can save fingers, toes, backs, tempers, and a lot of time when rehitching.

When disconnecting plows or planter units from a tractor:

1. *Use hydraulic lift arms to support integral hitches for multiple planting units when hitching or unhitching.*

2. *Don't try putting hitch pins in place from the operator's seat while the tractor is in gear.*

3. *Set brakes or put the transmission in park before getting off the tractor.*

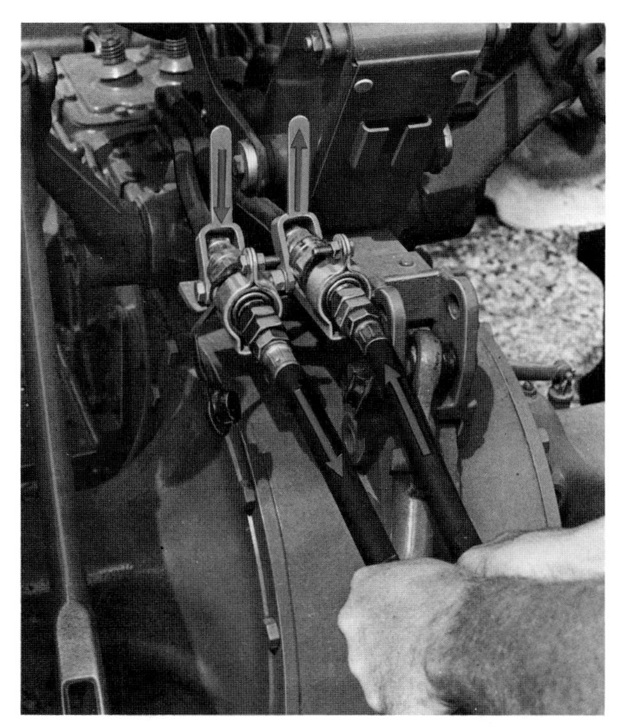

Fig. 3—Color-Code Painting Of Mating Parts Is One Way To Prevent Mismatching

4. *Be sure you have enough headroom when backing up inside buildings.*

5. *Check the operator's manual for specific suggestions on hitching and parking implements.*

BACKING UP

Position the tractor directly in front of the implement before starting to back toward it for hitching. Back-

Fig. 4—Implement On A Flat Surface For Easy Hitching

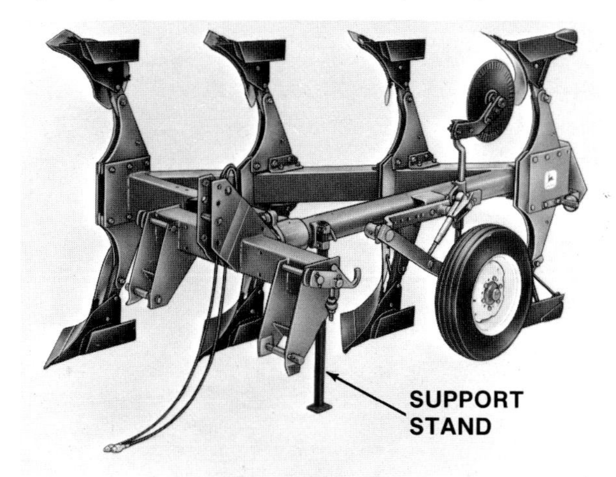

ing up to an implement at an angle makes hitching more difficult, especially with three-point hitch linkages (Fig. 5).

Fig. 5—Back Tractor Straight Up To Machines With Integral Hitches

An exception to this rule is in one-pin hitching of an implement with a steerable tongue, like a one-way plow or wagon. The side-to-side movement of the tongue can allow for some misalignment of the tractor.

Hitching is much easier if you have a dependable assistant. Both of you should understand and use hand signals, described in Chapter 1. To avoid injuring the assistant, plan the hook-up so he is between the tractor and implement only when you move the tractor slightly forward to make the final lineup. Many operators prefer to hitch alone, removing the possibility of pinching their assistant.

If no assistant is available, a snap-close hitch, self-centering hitch, or telescoping-tongue can ease hitching.

PEOPLE BETWEEN THE TRACTOR AND EQUIPMENT

Keep people out of the area between the implement and tractor, especially when backing up to hitch (Fig. 6). There is always the possibility of your foot slipping off the clutch. Or the implement may move, catching someone between the tractor and implement and crushing them. There is also a pinch point between hitches that can catch a finger or hand.

The experienced assistant steps in to make the final pin connection only when the tractor is in park and the drawbar and hitch are in final position. He must be alert to avoid unexpected tractor or implement movement.

CAUTION: When unhitching, the tongue of some implements will whip and can injure you.

Fig. 6—Have Helpers Stand To One Side When Hitching Equipment

LOCKING HITCH PINS

Pins used to connect an implement to a tractor should be the proper size and secured with a clip or pin. Don't use makeshift pins such as a long bolt. They can break, bend, or jump out under load. Use safety hitch pins that are easy to remove but have springs or clips to keep them in place (Fig. 7).

Three-point hitch links should also be secured with restraining clips or pins. Wire, spikes, or bolts are not suitable. A three-point hitch link that comes loose under load or in transport can cause serious injury.

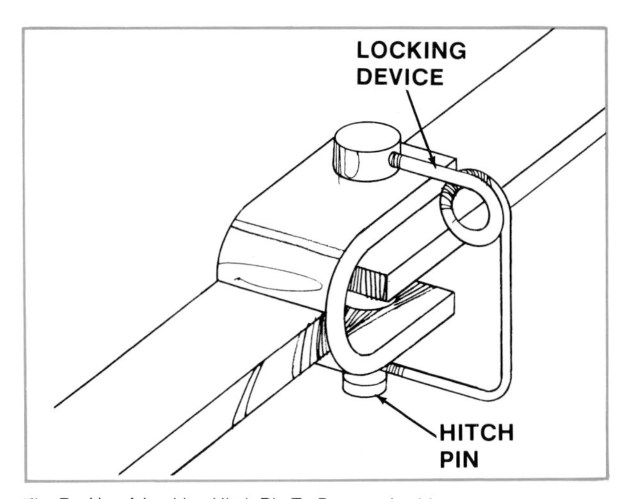

Fig. 7—Use A Locking Hitch Pin To Prevent Accidents

MACHINE OPERATION

Good tillage and planting jobs begin with adequately sized, properly maintained equipment and safe operating practices. Tillage equipment that is in bad shape or improperly adjusted pulls harder and wears faster, resulting in early wearout, failure, frequent plugging, and other problems that can lead to accidents.

Size, shape, and slope of fields affect the pattern of tillage and planting movement. Avoid dangerous working situations like extremely tight turns where the tractor tire may catch the implement tongue or frame, causing damage or forcing equipment up onto the operator's station (Fig. 8). Working on steep slopes where a sideways or rearward upset is likely or getting too close to fences and other obstructions is also risky.

MACHINE ADJUSTMENT

Safe and efficient machine operation demands that the machine be properly adjusted. Improperly hitched plows cause poor steering control and penetration. Improperly set coulters can cause a plow to plug. And trip beams that fail to trip may result in a bent beam or broken bottom. Such difficulties may cause you to become frustrated or angry, and this could lead to accidents. See Chapter 2 for more information on how emotions contribute to accidents.

Always lower implements to the ground or use jacks, blocks, transport links or lock pins when it's not in use or when you're working on the machine (Fig. 9).

Check outrigger locking mechanisms on fold up tillage equipment. The mechanisms can fail and let the equipment fall.

Fig. 9—Lower The Implements To Ground When Not In Use, And When Servicing Or Adjusting, To Keep It From Falling On A Worker

Fig. 8—Avoid Tight Turns During Tillage Operations

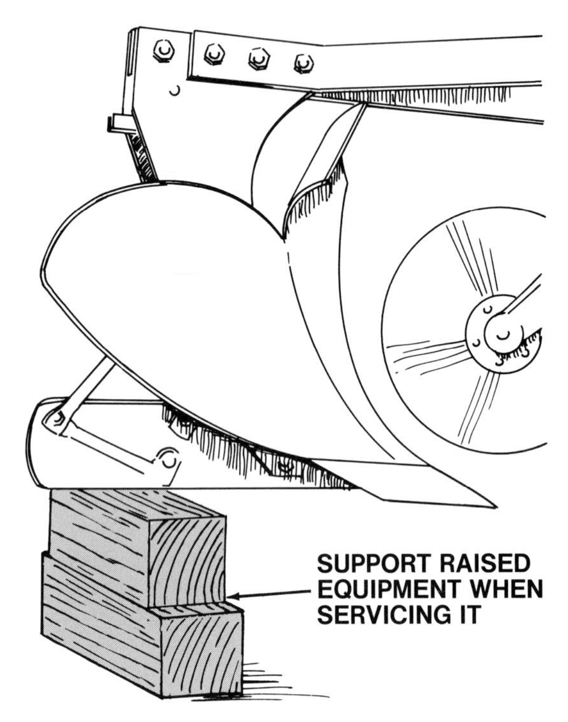

Fig. 10—If A Machine Must Be Serviced In The Raised Position, Use Transport Locks and Block It Up

Fig. 11—Use Lights, Reflectors, And An SMV Emblem To Warn Other Motorists

This will prevent the possibility of someone tripping the machine unexpectedly and getting caught. Check your operator's manual for details. If you must service machinery in the raised position, use blocks, jack stands, or other support on firm ground (Fig. 10). Do not depend on the hydraulic system to hold the implement up.

TRANSPORTING TILLAGE AND PLANTING EQUIPMENT

Towing wide, slow-moving agricultural machinery on public roads is potentially hazardous because the travel speed is much less than cars and trucks on the road. Chapter 5 contains guidelines to follow. Here are a few important ones:

1. *Equip your machinery with adequate lights, reflectors, and a bright, clean slow-moving-vehicle emblem (Fig. 11).* If the towed implement hides the flashing amber warning lights on the tractor, mount amber lights on the towed equipment. Use the 7-pin connector on the rear of the tractor.

2. *Put the machine in as narrow a configuration as possible.* Many wide machines have a special transport position (Fig. 12). Some states require a special permit to transport equipment that's more than 8 feet wide. Check your local and state regulations.

3. *If an integral hitch is used for a double planter or harrow, the implement should be transported in tandem.*

4. *Use transport links to take the load off the hydraulic cylinder when equipment is moved more than a mile or two.*

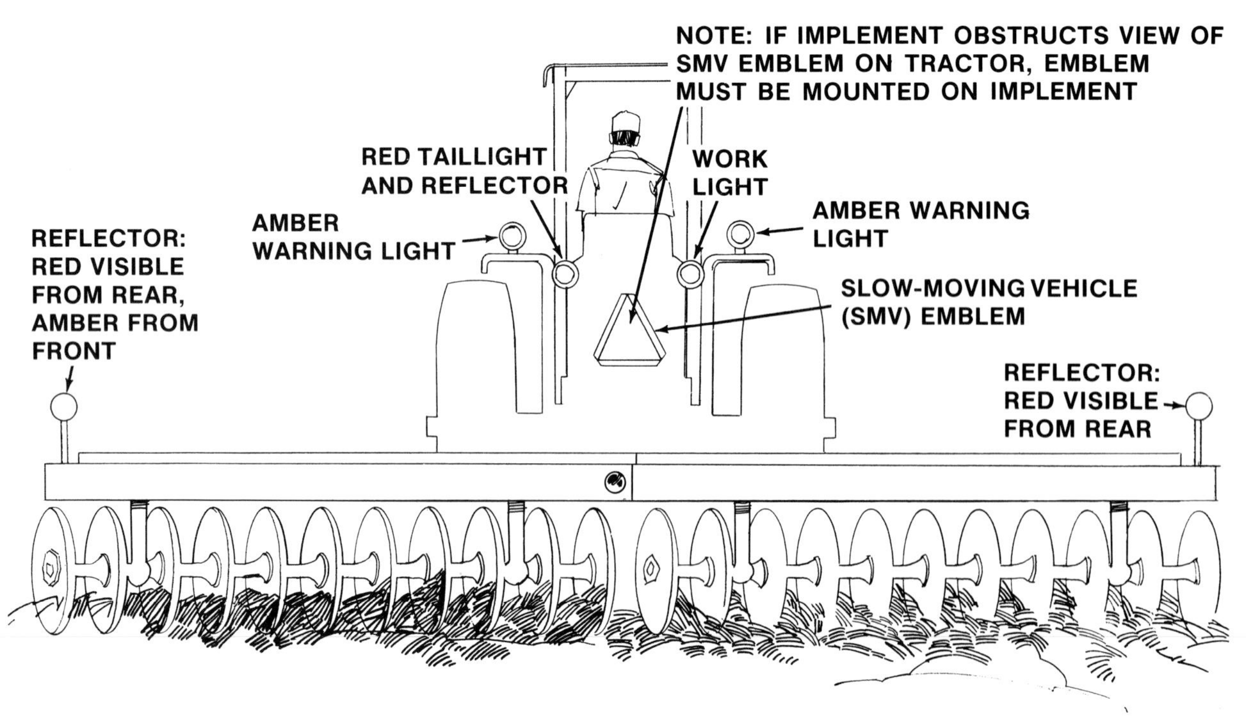

Fig. 12—Put Machines In Narrow Transport Position Before Traveling On Public Roads

5. *Use flashing warning lights, unless prohibited by local regulations.*

6. *Keep equipment to the right of the center line.* It is courteous to pull over when possible to allow cars to pass if oncoming traffic prevents them from using the passing lane.

7. *Watch power lines, especially when transporting large fold up equipment.*

TRANSPORTING BY TRUCK OR TRAILER

When equipment must be moved some distance, loading it onto a trailer or truck is often the best thing to do. Follow these safe practices:

1. *Use adequate loading ramp or a loading dock if possible (Fig. 13).* Trying to drive onto a trailer from a ditch bank is risky. It invites tractor upsets. Many operators have been injured or killed and pieces of equipment have been damaged or tipped over as a result of this method of loading.

2. *Secure the load with chain binders.* Be sure they are tight. A 1- to 2-foot pipe extender may help pull the binder tight. Wire the handles to prevent loosening in transit. If chain binders are not available, use rope, wire, blocks, or a winch cable. Check the load after traveling four or five miles and every 50 to 100 miles thereafter to make sure it's not coming loose. Also, check it after rough road bumps.

3. *Check with the State Police or sheriff's office for height and width regulations.* Know and observe the laws in your state.

4. *Display the proper flags, lights, and reflectors to alert other motorists.* See Chapter 5 for details.

AGRICULTURAL CHEMICALS

Know the hazards of the fertilizers, insecticides, herbicides, and other chemicals you use during tillage and planting operations. Study the label for correct use and follow precautions. Avoid any contact with dust, spray mixes, and vapors, especially from

Fig. 13—Use A Loading Dock Or Ramp To Load Equipment—Not A Ditch Bank

deadly organic phosphates, which are fast-acting nerve poisons.

Use a respirator, rubber gloves, and protective clothing. Avoid skin contact or breathing dusts and vapors, even when using less toxic chemicals.

Anhydrous ammonia is dangerous, too. Eyesight can be destroyed in seconds by anhydrous ammonia burn, unless the eyes are flooded with water immediately and rinsed continuously for at least 15 minutes.

Use eye and face protection, gloves and an apron when working with anhydrous ammonia (Fig. 14). See Chapter 7 for more information on handling chemicals.

ENVIRONMENTAL HAZARDS

The environment can present problems in tillage and planting operations.

HILLS

The possibility of a tractor upset increases with the steepness of a slope. It's safest to back the tractor up slopes.

Sideways upsets are common on steep slopes. If you have to cross a slope, drive slowly and look for uneven areas or holes and avoid them. Rollover protective structures — frames or enclosures — protect the operator in case of an overturn, and are especially important on sloping fields. This does not excuse an operator from using his skill, knowledge, and understanding to maintain maximum stability for given conditions. Space the wheels as far apart as you can for stability.

See Chapter 5 for more information on tractor upsets.

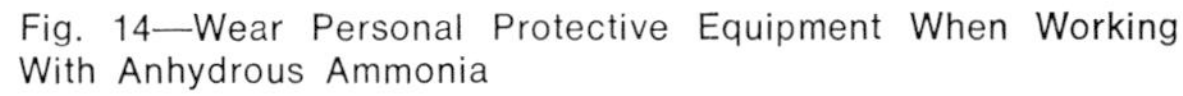

Fig. 14—Wear Personal Protective Equipment When Working With Anhydrous Ammonia

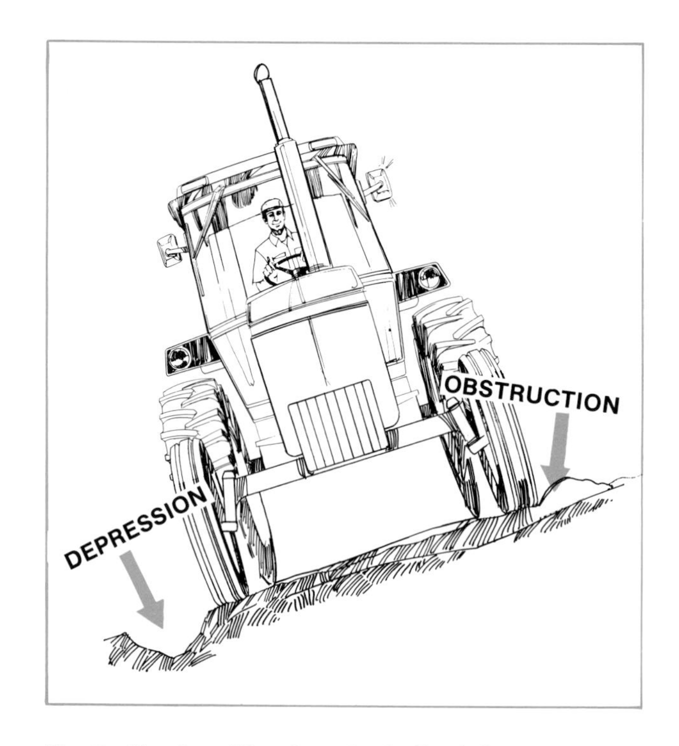

Fig. 15—Slow Down When Operating On Rough Or Sloping Terrain

ROUGH GROUND

High speed and rough ground lead to fatigue, accidents, and damaged equipment (Fig. 15). Occasionally, the operator loses control of his machine and may be bounced off due to a combination of speed and rough ground if he isn't wearing a seat belt.

Fig. 16—Make Sure Your Tractor And Equipment Have Adequate Lights For Field Operation

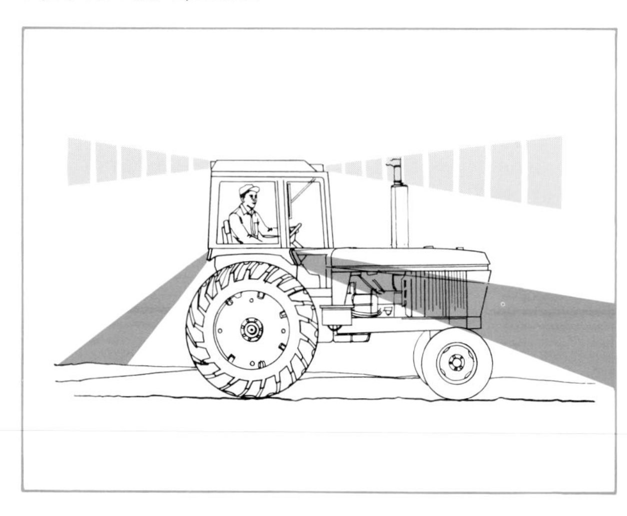

Look ahead for rough ground. Anticipate bumps and slow down. Mark sharp breaks, gullies, and ditches ahead of tilling and planting, and stay away from them. You can't run harvest equipment near these areas without risking an upset. Leave a grass-covered margin around these gullies and ditches for safety and soil conservation.

STONES AND OTHER OBSTACLES

Large stones, buried stumps, or logs can damage equipment. Plow points can be broken if you hit large stones, buried roots, or other obstacles. Beams and bottoms may be bent, and disk blades and bearings are frequently broken. Similar damage can happen to planters, although they are usually able to roll up and over the obstacles, minimizing damage. Anything that keeps an operator from following his normal work plan is inefficient, expensive, and can result in errors and accidents.

TEMPERATURE

Tillage and planting operations are usually done at times of the year when the temperature is relatively mild. The usual temperatures encountered have little direct effect on the operation.

Occasionally, tillage and planting must be done when the temperature, although not extreme, is far from the comfort zone. Above 80°F and below 40°F, human performance efficiency decreases. When the temperature is cold, a person will usually be so uncomfortable that he will voluntarily stop work to get warmed up or put on warmer clothes before any bodily damage is done. Heat stress, however, can occur without a person recognizing it.

More information about temperature and its effects on human performance can be found in Chapter 2.

DARKNESS

The danger in working at night is in not being able to see or be seen clearly. The major nighttime hazard in operating tillage or planting equipment in a field is limited sight of obstructions or the equipment itself. Good lights on the tractor, both in front and to the rear, will help (Fig. 16). Without adequate working lights, a rock, fence, stump, or ditch can be hit before it's seen. Equipment may malfunction, and you won't realize what has happened until it's too late.

DRY FIELDS

Dust is a problem in dry fields. Pulling a disk harrow or drag for several hours in a cloud of dust can be annoying and unhealthy.

Sometimes it's possible to work a field so that the wind blows the eye- and lung-irritating dust away from you. If this isn't possible, a filter respirator may help. In hot weather filters can be almost as annoying as the dust, but it's more healthy.

The ideal solution is an airconditioned tractor cab that filters and cools the air and lets you work in comfort.

WET FIELDS AND MIRED TRACTORS

Plowing and disking in wet fields can be tricky. Your tractor can become stuck almost before you realize it. Once forward motion stops, the tractor is stuck. Spinning the wheels only gets the tractor in deeper, increasing the difficulty and danger of getting out.

Here are some tips for freeing a mired tractor. For more information, see Chapter 5.

1. *Try to back the tractor out of the mud.*

2. *Dig away the earth to provide clearance for the tractor chassis and a sloping ramp for the tractor to move up on.*

3. *Use a second tractor, pulling straight forward or rearward from solid ground.*

4. *Fasten the chain or cable securely to the front frame or axle of the mired tractor.*

5. *Never chain a log or a plank across the wheels, thinking that the tractor will climb out on it. It can cause an upset.*

6. *If the ground is too muddy to be dug away, and no other help is available, leave the tractor alone until the ground dries out enough for it to be freed.*

CHAPTER QUIZ

1. True or false? Most mismatching of equipment and tractor involves tractors that are *slightly* oversized or undersized for the equipment.

2. Why is it hazardous to interchange the hose ends on the auxiliary cylinder?

3. Which of the following are *not* safe practices to use when unhitching a plow or planter from a tractor?

a) Set brakes or put the transmission in park before getting off the tractor.

b) Put links in place from the tractor seat, making sure the tractor is in gear.

c) Fill fertilizer tanks to improve stability.

d) Check to see that you have adequate headroom before backing up inside buildings.

4. True or false? A bolt and nut makes a good safety hitch pin.

5. Which of the following are good safety guidelines for transporting tillage and planting equipment on public roads?

a) Equip the machine with adequate warning lights and an SMV emblem.

b) Put the machine in transport position.

c) Use transport links to take the load off the hydraulic system.

d) All of the above.

6. True or false? Loading equipment onto a trailer from a ditch bank is an acceptable alternative if you don't have a loading ramp.

7. List five potential environmental hazards that may affect tillage and planting operations.

8. When trying to free a mired tractor, you should:

a) Try to drive forward slowly out of the mud.

b) Try to back the tractor out of the mud.

c) Chain a plank to the wheels so the tractor can climb out on it.

7
Target: Chemical Equipment

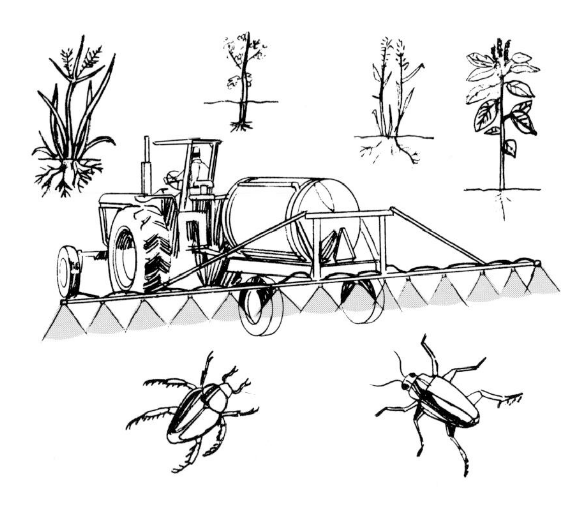

Fig. 1—Modern Agriculture Uses Many Chemicals

INTRODUCTION

Agriculture is chemistry. We use chemicals to feed crops, control weeds, control fruit set on trees, and control insects. Chemicals contribute greatly to the high productivity of farming.

Chemicals can however be harmful if used improperly. They can injure or kill people, domestic animals, wildlife, and desirable plants, and contaminate water and air.

Fig. 2—Clothing Protects Against Fertilizer Burn

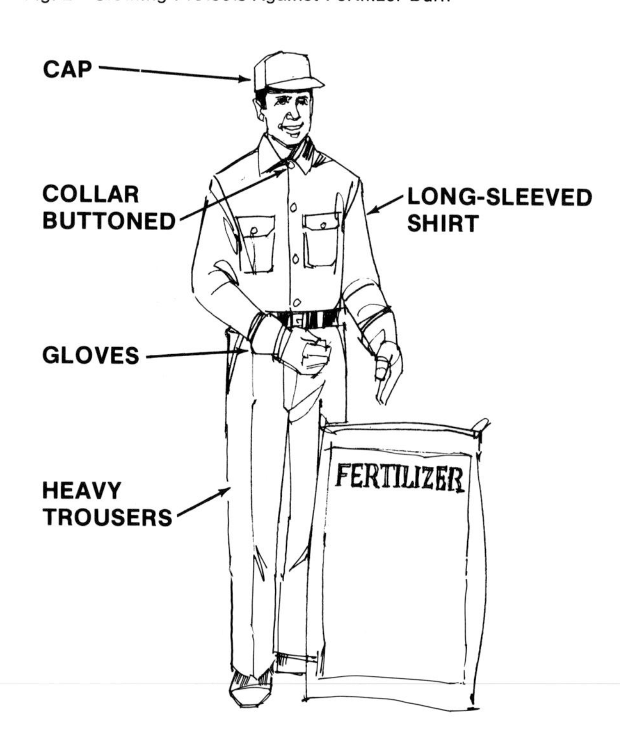

Who has accidents with chemicals? There are two groups:

- **People who are not aware of the hazards**
- **People who are aware of the hazards, but because of carelessness, hurry, or for some other reason, don't use safe practices**

In this chapter, you will learn to recognize many of the hazards involved in working with chemicals, and how you can protect yourself and others from the dangers involved in chemical accidents and exposure.

CHEMICAL HAZARDS

Chemicals used in agriculture are handled both by hand and by machine. Like other machinery, chemical application equipment can be hazardous if it's not operated properly. Hazards involved in chemical application machinery operation fall into three groups:

- **Hazards common to all machines**
- **Hazards specific to chemical equipment**
- **Hazards involved in the use of some agricultural chemicals**

Hazards common to all machines are discussed in other chapters of this manual. The other hazards are covered in this chapter.

FERTILIZERS

Fertilizers are materials used to supply plants with nutrients needed for growth. There are two basic types — chemical fertilizers and animal manures. Chemical fertilizers are available in solid (dry), liquid, and gaseous forms.

DRY CHEMICAL FERTILIZER

Dry fertilizer is hygroscopic—it attracts moisture. It can draw water out of skin, and leave it red and sore. These skin burns are usually a minor discomfort, but they can be painful to people with sensitive skin. Dry fertilizer can also get into and irritate the mouth, nose, and eyes. The scalp is another area that's often very sensitive to the effects of dry fertilizer.

Prevent fertilizer burns by keeping the dust off your skin. Wear a long-sleeved shirt buttoned at the collar and a cap or hat to keep dust out of your hair (Fig. 2). Change clothes daily—more often if they pick up a lot of dust. Wash your face, hands, arms, and other exposed skin areas several times a day with soap and water—don't let fertilizer stay in contact with your skin.

Let the wind blow dust away from you. Stand upwind when filling hoppers (Fig. 3). Drive crosswind in the field, if possible, so dust is blown off to one side. If you can't stay out of the dust, wear goggles to protect your eyes, and use a filtered respirator to keep dust out of your lungs. You should be able to get a respirator from a local bulk fertilizer dealer or from a well-equipped drugstore.

Fertilizer Spreaders

Centrifugal broadcast spreaders throw fertilizer particles at high speeds. These particles can be painful if they hit you, and can get into your eyes, ears, or mouth. Stay away from the back of the machine when the spreader is operating (Fig. 4).

If the spreader plugs, it's tempting to jump off and go back to see what's wrong. Don't do it! Besides getting hit by flying particles, you could get tangled in the spinner mechanism. *Always* stop the machine and shut off the engine before doing any inspection or maintenance work.

Use a safe speed when pulling loaded spreaders. They are heavy. If you drive too fast, the unit can veer out of control while going downhill or around corners. These spreaders are not normally equipped with brakes, so the tractor brakes alone must do the stopping.

Fertilizer equipment is often cleaned with diesel fuel. The fumes can ignite and burn. Always work outdoors or in a well-ventilated area, and stay away from flames. Do not allow smoking in the work area.

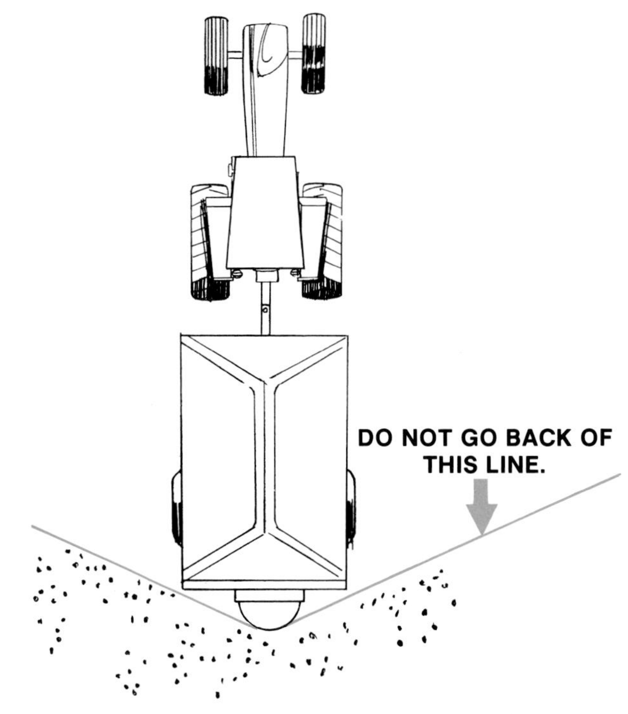

Fig. 4—Stay Away From Spinners On Fertilizer Spreaders

ANHYDROUS AMMONIA

Anhydrous means "without water." Anhydrous ammonia is dry or pure undiluted ammonia that is often

Fig. 3—Wind Can Keep Dust Away From You

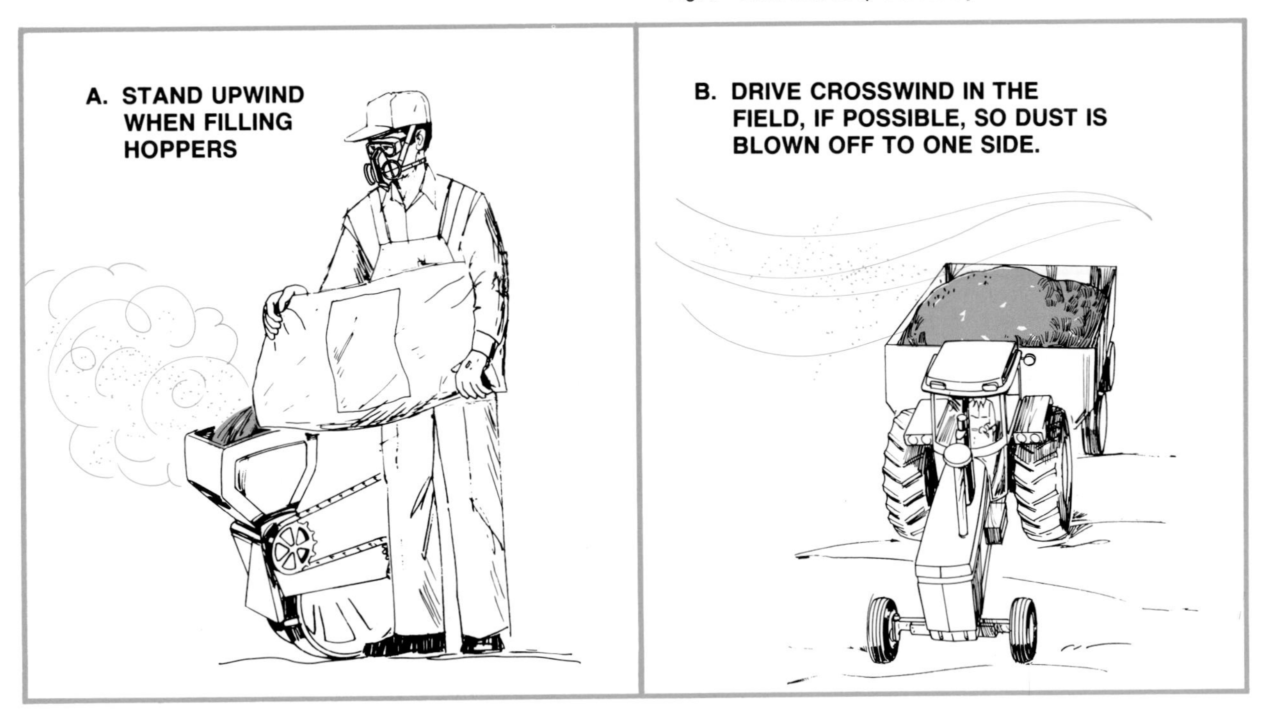

Fig. 5—Rinse Off Ammonia With Clean, Fresh Water

Fig. 6—Use Specially Designed Anhydrous Ammonia Equipment

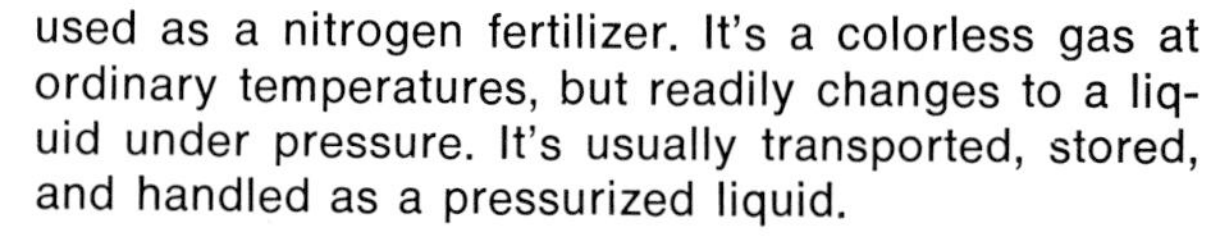

used as a nitrogen fertilizer. It's a colorless gas at ordinary temperatures, but readily changes to a liquid under pressure. It's usually transported, stored, and handled as a pressurized liquid.

Anhydrous ammonia is a strong alkali, and can cause painful skin burns. Even mild exposure can cause irritation to tissues of the eyes, nose, and lungs. Permanent damage to eyes can result. Prolonged inhalation can cause suffocation leading to death. And because of the low boiling point, anhydrous ammonia can cause burns by freezing as well as by caustic action.

One good thing about ammonia is that its sharp, pungent odor is an excellent warning signal. You won't voluntarily stay in an area filled with its fumes.

If ammonia gets on your skin or in your eyes, rinse immediately with clean water. Continue for at least 15 minutes to be sure that all the ammonia is washed away (Fig. 5). See a doctor.

If clothing is frozen, run water over it until it is thawed before you remove it.

If a person is overcome by ammonia fumes and stops breathing, get him to fresh air and administer artificial respiration. Get medical help as soon as possible.

CAUTION: Never use LPG tanks to store anhydrous ammonia.

Handling Anhydrous Ammonia

Anhydrous ammonia is stored under pressure. There are three important things to remember in order to handle this material safely:

1. Use proper equipment.

2. Take care of the equipment.

3. Follow safe practices.

How do you do this?

1. *Use equipment that is specifically designed for use with anhydrous ammonia (Fig. 6).* Be sure there are no copper or brass parts, since ammonia is corrosive to copper alloys. A tube or fitting weakened by corrosion can burst unexpectedly and spray you with ammonia.

2. *Take care of your equipment.* Check it before use to make sure all joints are tight, valves are working, and fittings are in good shape. Repair or replace weak hoses or other parts that don't work or are weakened due to corrosion or damage. Don't take chances with weak parts. Failure could mean serious injury.

3. *Replace safety relief valves that don't work.* These valves are designed to open if pressure exceeds a preset level. However, if a valve does not close properly the entire contents of the tank may escape. Replace a defective valve; do not attempt to repair it. If you want to test a valve to make sure it's working properly, consult your machinery dealer.

4. *Keep tanks coated with light-colored paint.* A good coat of paint prevents rust that can clog valves. The light color reflects sunlight and helps keep down the temperature and pressure in the tank. The best paint to use is a high-reflectance white paint made especially for ammonia tanks.

5. *Take care of hoses.* They can burst unexpectedly. Check hoses and connections at least once a year. Store hoses in a cool, dry place when not in use. Hang them with the open ends down. Don't bend them sharply over a nail or some other projection. Hang them over an old wheel rim or similar curved object. Replace any hoses that are worn or cracked.

6. *Keep heat and fire away.* Anhydrous ammonia is nonflammable under most conditions, but with high concentrations, the vapors can ignite. Treat it like a flammable gas. Flame-cutting or welding on anhydrous tanks should be done only by experts. Don't

do it yourself or have anyone else do it unless they've had special training in working with ammonia tanks. Don't smoke or allow any smoking around ammonia equipment.

7. *Wear rubber gloves and tight-fitting unvented goggles when transferring anhydrous ammonia (Fig. 7).* Most accidents occur during transfer operations, and eyes are most often affected. Goggles designed for use with grinders or other tools are not satisfactory because the air vents may allow ammonia to get behind the goggles.

8. *Remove clothes that have ammonia on them and wash the skin beneath thoroughly.* If clothing has frozen to your skin, thaw it loose by pouring water on it before removing it.

9. *Close hand wheel valves by hand.* Don't use a wrench, or you could break the stem or damage the seat. Either could let ammonia escape.

10. *Know how to operate equipment correctly.* About 85 percent of anhydrous ammonia accidents result from using improper procedures. They happen because of a lack of training or failure to follow prescribed practices. And a lot of anhydrous ammonia equipment is rented. If you rent yours, have the owner show you how to operate it safely and supply you with an instruction manual. If you own your own equipment, study the operator's manual carefully. Follow the work practices recommended by the manufacturer.

11. *Keep a supply of fresh water handy for washing (Fig. 8).* Federal laws require that at least five gallons of water be carried on nurse tanks transporting anhydrous ammonia. A lot of people carry 5 gallons of water in the tractor cab too. Make this a habit. If your equipmient doesn't have a water container, build a place to mount one. The container should have a label on it reading, "Water. Change daily. Do not drink. For washing only." Keep the water fresh and clean. Empty the container completely and refill it every day. Anhydrous ammonia is readily absorbed into water when the machine is surrounded by atmosphere containing some ammonia. If the water is not changed regularly, it can become contaminated with ammonia and be useless or even hazardous to use for washing in case of an accident.

12. *Carry a plastic squeeze bottle of fresh water to flush eyes in case of accident (Fig. 9).* There isn't enough water in the bottle to flush eyes thoroughly for 15 minutes, but it will help while you're on your way to the water in the tractor or on the nurse tank. Change the water daily, since it absorbs ammonia and becomes contaminated.

13. *Do not exceed 25 mph when transporting anhydrous ammonia on public roads.*

Fig. 7—Wear Goggles Or A Full Face Shield, Rubber Gloves And A Rubber Apron When Transferring Ammonia

Fig. 8—Keep Fresh Water Handy On Ammonia Equipment

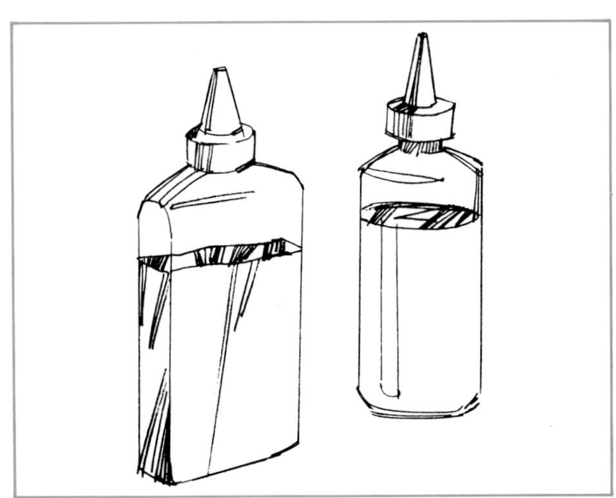

Fig. 9—Keep A Container Of Fresh Water In Your Pocket For Flushing Eyes

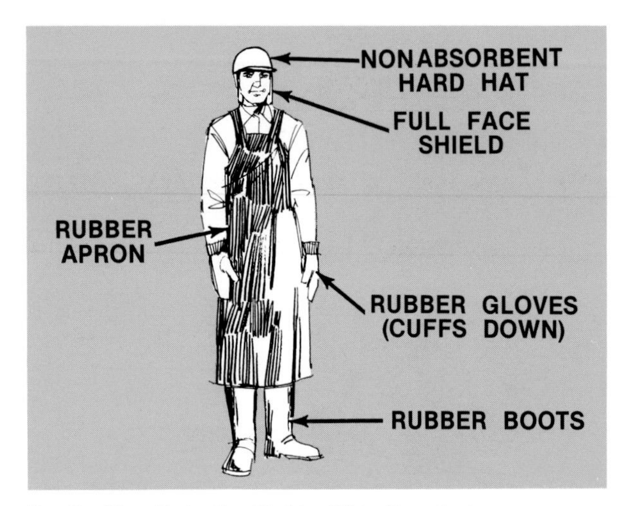

Fig. 10—Wear Protective Clothing When Transferring Aqueous Ammonia

NOTE: Get a copy of the anhydrous ammonia rules and follow those that apply to your situation. Contact OSHA headquarters in Washington, D.C. or an anhydrous ammonia dealer.

AQUEOUS AMMONIA

Aqueous (aqua) ammonia is a popular liquid fertilizer. A solution of anhydrous ammonia in water, it's similar to household ammonia, but more concentrated. Aqua ammonia is not usually stored under pressure. But don't consider it strictly a nonpressure fluid. On hot days it vaporizes and builds up pressure in the supply tank and lines.

Hazards of handling aqua ammonia are similar to those for anhydrous ammonia. It has the same powerful choking odor, and strong solutions can cause severe burns.

As with anhydrous ammonia, getting aqua ammonia into your eyes can be very dangerous. Vision is nearly always impaired, and blindness can result.

Many accidents occur when aqua ammonia is transferred from the nurse tank to the field applicator tank. Protect yourself during transfer operations by wearing rubber gloves and tight-fitting goggles or a full face shield (Fig. 10). Your ammonia dealer can tell you where you can buy this safety equipment. Also, keep plenty of clean water in the work area for washing in case of an accident.

Get medical attention if you're exposed to ammonia gas. An exposure that seems unimportant at the time can develop into a problem weeks later.

Handling Aqueous Ammonia

Aqua ammonia users often fabricate their own equipment, using components originally intended for other uses. If you do this, be sure to use equipment that is compatible with ammonia. Don't use any copper or brass parts. Ammonia is corrosive to copper and brass. Select components made of high-strength iron or steel, and keep your equipment in good condition.

Don't overfill containers. They should not be over three-quarters full. Vaporization during a hot summer day can cause the pressure to rise enough to pop off the lid or burst the tank. A coat of the same light-colored paint used on anhydrous ammonia tanks will

Fig. 11—Let Light-Colored Paint And The Shade Help Keep Ammonia Tanks Cool

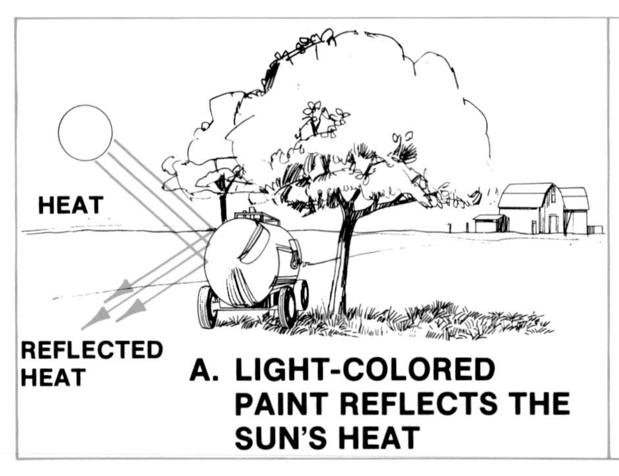

reflect much of the heat and help prevent pressure buildup. The tank should be parked in the shade whenever possible to keep it cool (Fig. 11).

Stay away from fumes when working with aqueous ammonia equipment. Work on the upwind side when transferring material between tanks. Drive crosswind, when possible, to stay away from the fumes.

ANIMAL MANURE

Gases are often given off when animal manure is stored. The most common gases given off from liquid manure are: ammonia, carbon dioxide, methane, and hydrogen sulfide.

A man can suffocate in a manure pit or building because of gases given off by manure. The gases displace air until there is no oxygen to breath.

Carbon dioxide is an odorless gas that is a normal part of the air we breathe. However, it exists in the air at a very low concentration (about .03 percent). When it's present in higher concentrations, it displaces the air so that less oxygen is available. Concentrations of three to six percent can cause heavy, labored breathing, drowsiness, and headaches. A 30-percent concentration can cause death by suffocation.

Methane gas is nontoxic. Concentrations as high as 50 percent only cause headaches. However, methane is highly flammable. It ignites readily, and methane-air mixtures can explode.

Hydrogen sulfide is released rapidly, and is most dangerous when liquid manure is first agitated. It has a foul odor, similar to rotten eggs. It causes headaches, dizziness, and nausea, in concentrations as low as .5 percent. Exposure to a 1 percent concentration can result in unconsciousness or death. Although you can smell very low levels of hydrogen sulfide, continued exposure dulls your sense of smell and you may not know that you're in danger. Several deaths have been attributed to this gas.

Follow these rules to be safe from manure gases:

1. *Know the effects of each of the gases described above.* Any time you detect one or more of the symptoms, get to fresh air immediately. A delay could be fatal.

2. *Provide maximum ventilation to keep gases away from people and animals whenever a tank is agitated.* If a power failure has occurred and lasted for several hours, open all windows and doors and get all people and livestock out of the building before they suffocate (Fig. 12). It is best to have an emergency generator.

Fig. 12—Turn On The Ventilation System For Maximum Ventilation When Agitating Liquid Manure Tanks Under Buildings

3. *Don't allow any smoking or other fire source in or around the liquid manure tank.* Methane-air mixtures can be very explosive.

4. *Never go into a liquid manure storage unless it has been thoroughly ventilated, a rope and harness are attached to you, and enough people are standing by to pull you out.* If you have to go into the tank, wear a self-contained breathing apparatus (Fig. 13). Better yet, hire a professional who is thoroughly trained in the breathing equipment.

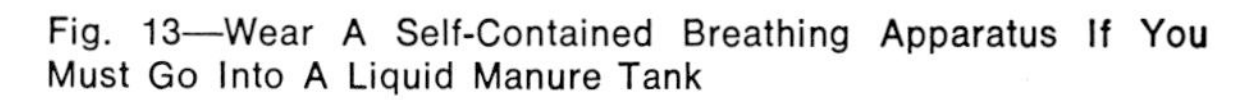

Fig. 13—Wear A Self-Contained Breathing Apparatus If You Must Go Into A Liquid Manure Tank

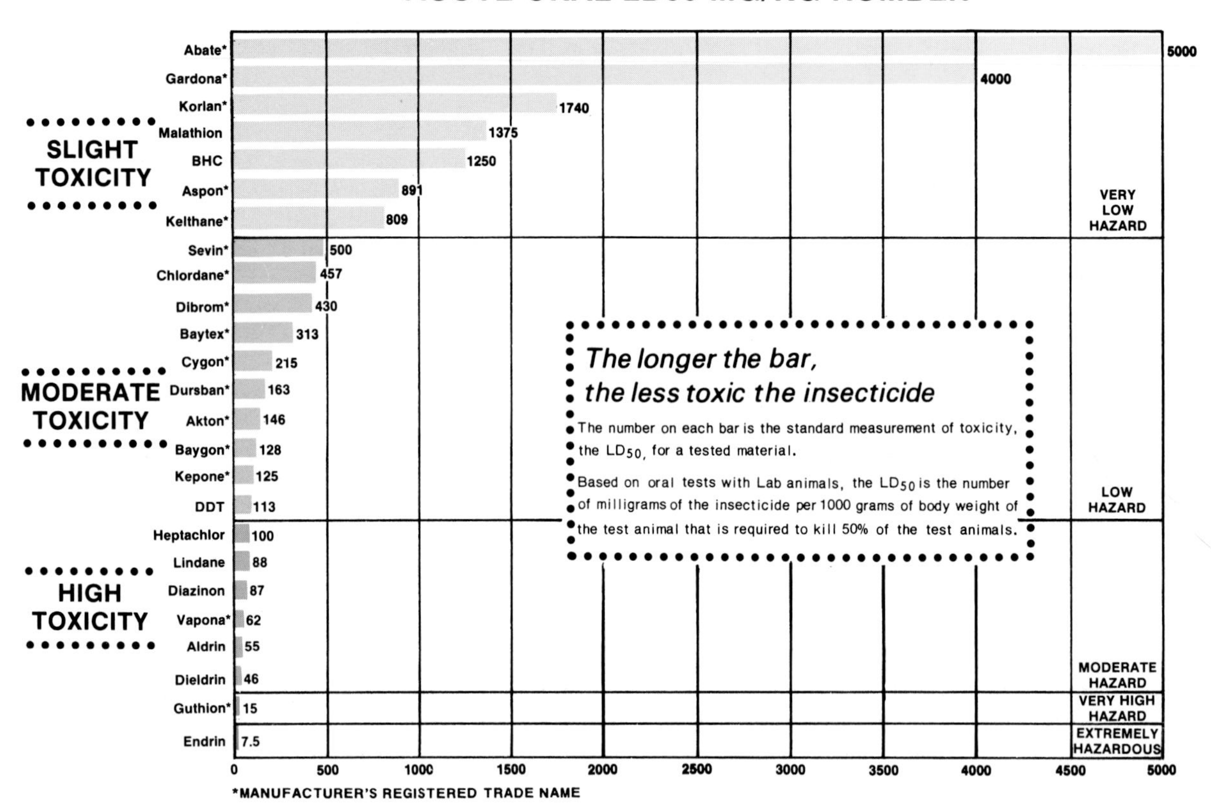

Fig. 14—Relative Oral Toxicity Of Insecticides

PESTICIDES

Pesticide is a broad term for a large number of chemicals used to kill pests. Some specific pesticides and the pests they control are: insecticides (insects), fungicides (fungi), herbicides (weeds), miticides (mites), rodenticides (rodents), avicides (birds), and bactericides (bacteria).

Pesticides control pests that:

- **Compete with man for food or feed**
- **Cause injury to man, his animals, or plants**
- **Carry diseases to man, his animals, or plants**
- **Cause annoyance to man, his animals, or plants**

Pesticides are poisonous. They have to be in order to kill undesirable plants, animals, or insects. So they are also toxic (poisonous) to desirable organisms, including man. Exposure to a sufficient amount of almost any pesticide can make a person ill. And many pesticides are so toxic that very small quantities can kill a person. If you are working with pesticides, you should be aware of the dangers.

Experience has shown that the greatest likelihood of accidents occurs at the following times:

- *At the beginning of the season when inexperienced personnel are handling the chemicals*
- *When a new pesticide is being introduced*
- *During the first prolonged hot spell of the season (Men get tired of using uncomfortable protective clothing and respirators during hot weather, and often take them off, running the risk of overexposure)*
- *At the end of a long, hard day (especially at the end of the season, when men become careless and may have already suffered some mild exposure)*
- *Late in the season, especially if the work load has been prolonged and heavy. People who have experienced repeated small exposures to pesticides, perhaps enough to cause some cholinesterase depression but not severe enough to cause obvious symptoms, may acquire additional exposures, the effects of which are cumulative with the previous one. Also, equipment and protective devices that*

functioned well earlier in the season may have become faulty.

ACUTE AND CHRONIC EXPOSURE

If someone is doused with a pesticide or swallows some pesticide and immediately loses consciousness, there is no doubt about the cause of the illness, and proper medical treatment can be administered. "One-time" cases like this are examples of acute exposure.

Chronic exposure is a low-level exposure over a longer period of time. The effect is a mild, slow poisoning. The symptoms are less severe than in acute cases, and they're often difficult to isolate as having been caused by pesticides.

Chronic exposures can be every bit as dangerous as acute exposures. The greatest risk when working with a pesticide is a combination of the two types. If you've been receiving daily exposure from a contaminated hat or contaminated shoes, for instance, your body will be less able to deal with an acute exposure, such as spilling pesticide on yourself.

KINDS OF EXPOSURE

Before a pesticide can harm you, it must be taken into your body. Toxic materials can be taken into the body in three ways:

- **Oral—through the mouth and digestive system**
- **Dermal—through the skin**
- **Inhalation—through the nose and respiratory system**

ORAL EXPOSURE

If the toxic material is taken into the mouth and swallowed, it is called oral exposure. This kind of poisoning may occur because of an accident, but it is more likely to be due to carelessness. If you blow out a plugged nozzle with your mouth, smoke or eat without washing contaminated hands, or eat fresh fruit that has recently been sprayed with a pesticide, you receive oral exposure. Oral exposure can be either acute or chronic. The seriousness of a particular exposure depends on the oral toxicity of the material and the amount swallowed.

Acute oral toxicity is expressed by an LD50 value (Fig. 14). This is the number of milligrams of pesticide per kilogram of body weight that kills 50 percent of a group of test animals (usually white rats or rabbits). For example, 113 milligrams of DDT per kilogram of body weight can kill one-half of a test animal population. The smaller the LD50, the more toxic (dangerous) the chemical.

There is some uncertainty about applying LD50 values from tests on rabbits or rats to predict a material's toxicity to humans. However, most experts agree that the *relative* toxicities of the materials are about the same for humans as they are for rats.

What does the LD50 value mean in terms of danger to man? If you know the LD50 of a material you're working with, you can find out how much of it would be likely to cause death by using this formula:

$$\text{LD50} \times .0016 \times \frac{\text{body weight (lbs)}}{100} = \text{Weight (ounces)}$$

of pesticide likely to cause death.

Let's look at DDT for an example. The LD50 of 113 milligrams per kilograms multiplied by 0.0016 equals 0.18 ounces per hundred pounds of body weight. So if you weigh 100 pounds and swallow 0.18 ounce of DDT, you could expect to have about a 50-percent chance of dying from the exposure to the material. A person who weighs twice as much (200 pounds) might tolerate twice as much material (about 0.36 ounce) and have the same chance of dying.

Pesticides can also be rated on a relative hazard scale. The ratings range from very low hazard to extremely hazardous, depending on the LD50 rating of the material. The hazard ratings and the amount of material that would probably kill the people are shown in Fig. 15. The hazard ratings of some common insecticides are shown in Fig. 14.

Fig. 15—The "Probably Lethal" Dose Of A Pesticide Depends On Toxicity Of The Chemical And The Weight Of The Victim

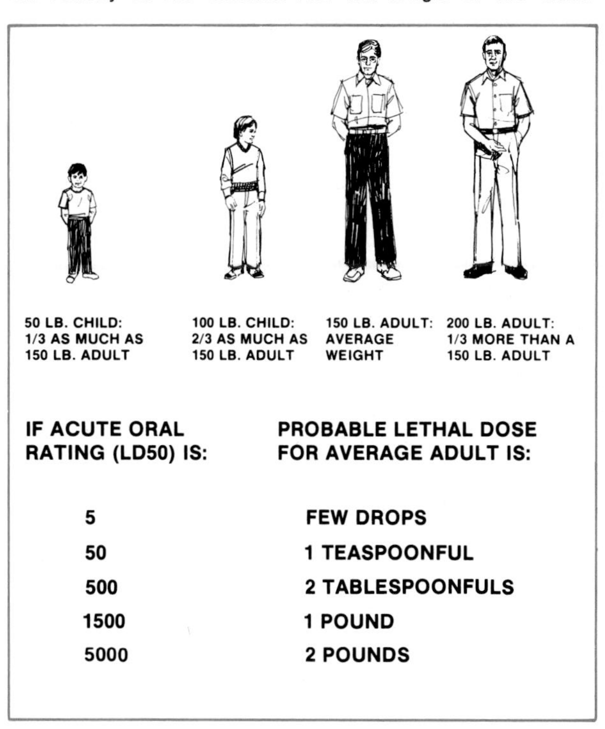

IF ACUTE ORAL RATING (LD50) IS:	PROBABLE LETHAL DOSE FOR AVERAGE ADULT IS:
5	FEW DROPS
50	1 TEASPOONFUL
500	2 TABLESPOONFULS
1500	1 POUND
5000	2 POUNDS

There is also chronic oral exposure. There is no good measurement of its seriousness, but following the suggestions in the next section will help you avoid chronic as well as acute exposure. Remember, even a brief contact between any contaminated material and your mouth can be dangerous.

Avoiding Oral Exposure

Most oral exposure is a result of accident or carelessness. Proper precautions can eliminate nearly all cases:

1. *Always keep pesticides in their original, labeled containers and never reuse pesticide container for anything.*

2. *Never store pesticides in containers that originally held food, such as soda pop bottles.* Small children will drink almost anything that's in a pop bottle.

3. *Keep drinking water containers away from the pesticide work area.*

Fig. 17—Wash Carefully After Working With Pesticides

Fig. 16—Post Warning Signs In Fields That Have Recently Been Sprayed—Especially As Required By Some Regulations For Certain Chemicals

4. *Do not drink or fill water containers from the water hose used to fill the sprayer tank.* The end of the hose could be contaminated.

5. *Do not eat anything touched by spray.* Some pesticides have a safety interval that must be observed between last application and the time when the crop may be sold and eaten. If you eat something that has been sprayed before this safety interval has passed, you will be eating the pesticide, too.

To prevent dangerous pesticide residue from remaining on food or feed, apply the pesticide well before harvest. Observe the time limit printed on the label.

6. *Post a sign where you spray (Fig. 16).* The sign should be conspicuous. Anyone tempted to eat crops out of that field will be aware it has been sprayed.

7. *Wash thoroughly with soap and water before putting your hands to your mouth or eating anything after working with pesticides (Fig. 17).* Residue on your hands can be transferred to food and then eaten. Use the following procedure:

(1) Wet hands with water.

(2) Lather hands with a detergent that won't irritate your skin. (Soap may not be effective in removing oil-based chemicals.) Wash under fingernails (and toenails, if appropriate).

(3) Rinse off detergent.

(4) Repeat.

(5) Air dry your hands or use disposable towels that have not been exposed to the chemical.

8. *Do not smoke around pesticides.* Smoking materials absorb pesticides and transfer them to your lips. Food and snacks like candy and gum should not be in the work area.

9. *Clean sprayer nozzles with a low pressure air hose* (under 30 psi) (Fig. 18). Direct the spray away from yourself and others. Protect your eyes by wearing chemical splash goggles. Some operators try to clean nozzles by holding them to their mouths and blowing through them. This is dangerous, since residue on the nozzle can be taken directly into the mouth and swallowed or absorbed through the lips. Don't let chemicals touch your mouth.

Fig. 18—Clean Clogged Nozzles With An Air Hose—Don't Blow Through Them

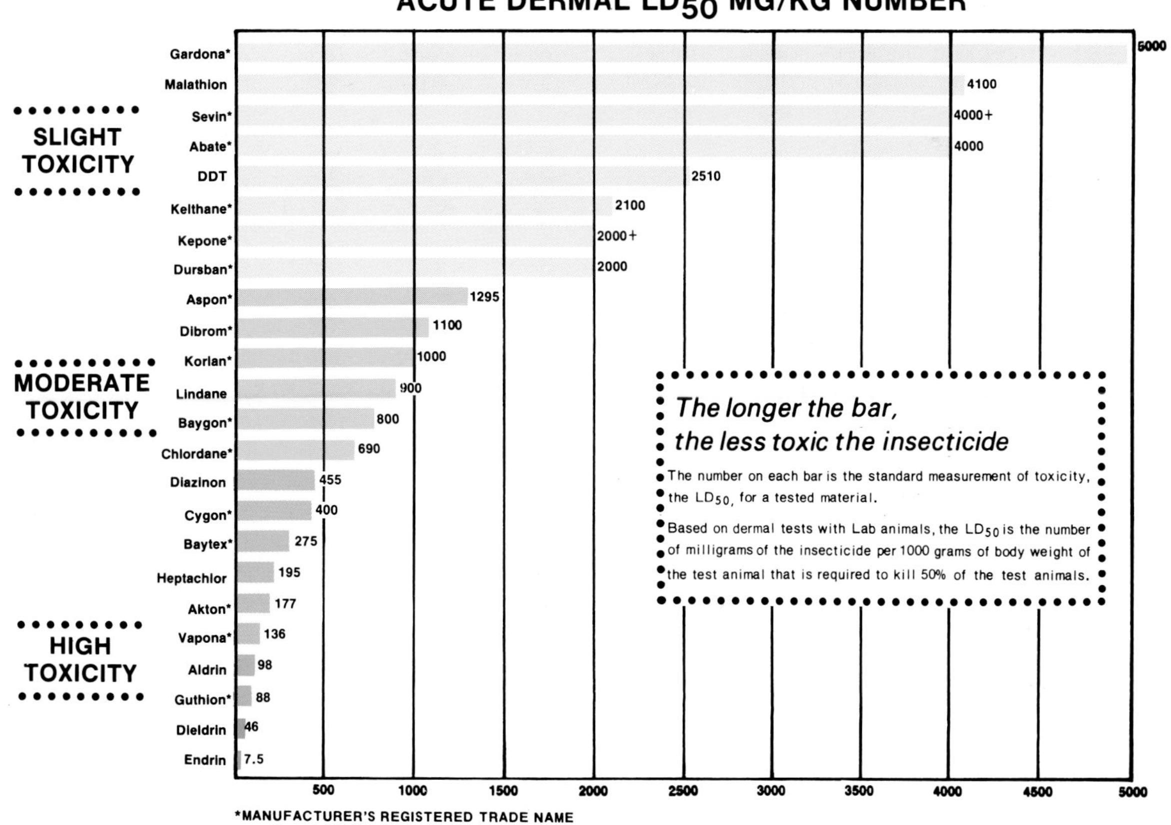

Fig. 19—Relative Toxicity Of Insecticides By Acute Dermal LD50 MG/KG

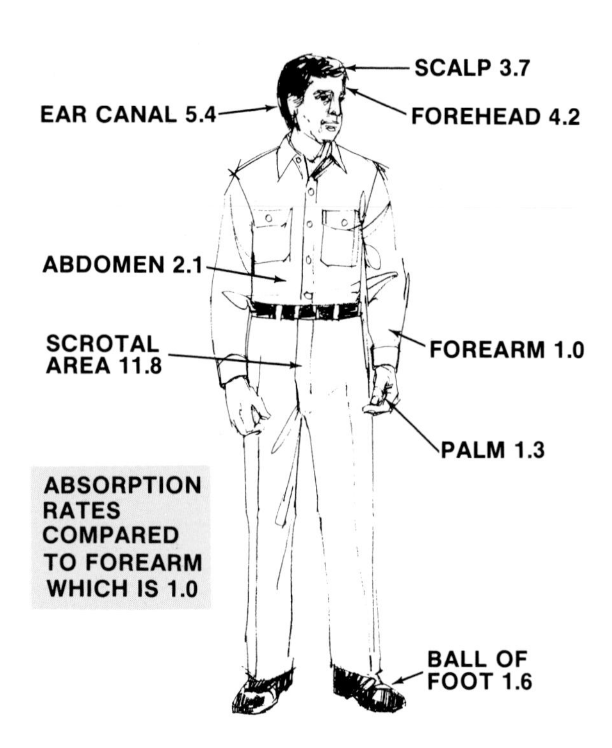

Fig. 20—Dermal Absorption Rates Are Higher For Some Parts Of The Body Than For Others

DERMAL EXPOSURE

Dermal exposure is skin contamination. It can occur any time a pesticide is mixed, applied, or handled, and it's often undetected. Its seriousness depends on:

- **Dermal toxicity of the material**
- **Rate of absorption through skin**
- **Amount of skin area contaminated**
- **Amount of time the material is in contact with the skin**

Like oral exposure, dermal exposure can be acute or chronic. Acute exposure can be expressed as an LD50 value (Fig. 19).

Rates of absorption through the skin are different for different parts of the body (Fig. 20). For example, absorption is over 11 times faster in the lower groin than on the forearm. Absorption through skin in the scrotal area is rapid enough to approximate the effect of injecting the pesticide directly into the bloodstream.

At this rate, the absorption of pesticide through the skin into the bloodstream is more dangerous than swallowing it. You should also be aware that it is not just the pesticide that's dangerous. Many pesticides are carried in oil-based materials. If this oil gets into the bloodstream, it can be fatal. The oil crosses the skin barrier as rapidly as the pesticide in highly sensitive parts of the body. Wear rubber aprons or rubber trousers to protect the lower part of your body.

Absorption continues to take place through all of the affected skin area as long as the pesticide is in contact with the skin. Seriousness of the exposure is increased if the contaminated area is large or if the material is left on the skin for a long time.

Avoiding Dermal Exposure

Avoid dermal exposure by keeping pesticides away from your skin. Use protective clothing and equipment. Take special care to avoid chronic exposure, because pesticides can build up in your body and the effects may not be immediately apparent.

Some ways that dermal exposure can occur are:

- **Splashing**
- **Mixing**
- **Spills**
- **Drift**
- **Contaminated clothing**
- **Reentry**

SPLASHING

Liquid pesticides often splash when the sprayer tank is being filled, and often land on the upper part of the body. Wear protective clothing when filling tanks—a waterproof rubber coat and a wide-brimmed hat. Protect your eyes with goggles.

Wear a hard hat, not a soft cloth hat that absorbs chemicals. As the farmer perspires, the chemicals in the hat are absorbed through the scalp.

MIXING

Spray material can be mixed in the sprayer tank or in a transfer container. If mixed in the sprayer tank,

Fig. 21—Two Steps To Safe Pesticide Mixing

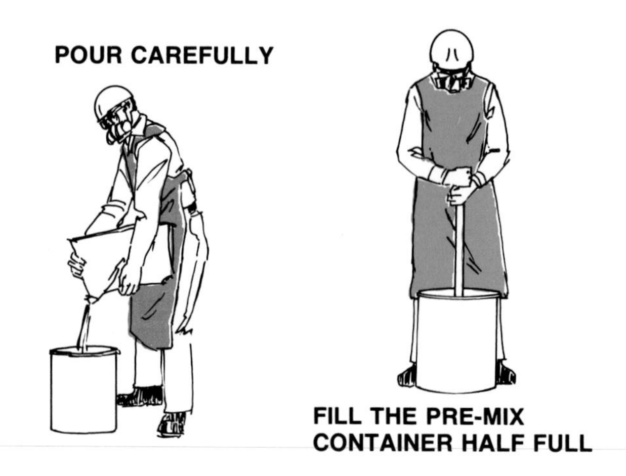

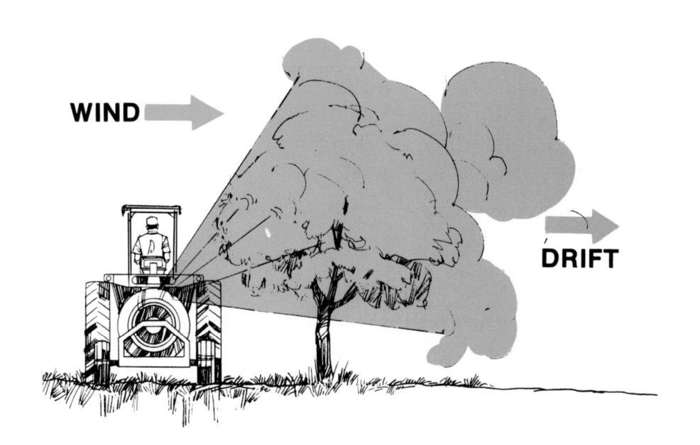

Fig. 22—Small Pesticide Particles May Drift With The Wind—A Definite Hazard

fill the tank about half full of water, then add the chemical. Mix until it is thoroughly blended, then add the remaining water. If the material is being premixed in a transfer container, fill the container about half full of water, then add the chemical (Fig. 21). Stir this mixture, then add it to the water in the sprayer tank.

Some chemicals can form *invert emulsions* if they're not mixed properly. An invert emulsion is a thick, mayonnaise-like mixture that is very difficult to clean out of a sprayer. Trying to clean the sprayer and get rid of the emulsion can expose the operator to a high risk of accidental contamination. Keep invert emulsions from forming by following the mixing directions on the container label carefully.

SPILLS

Clean up and dispose of all spilled chemicals promptly. If you walk through spilled material, your footwear could be contaminated. This is a common cause of chronic exposure, so wear rubber boots to protect your feet. Procedures to use when cleaning up spills are discussed later in this chapter.

DRIFT

Every sprayer produces a range of particle sizes. Some are small enough to drift away with the wind for long distances (Fig. 22). They may land on your skin, causing dermal exposure. To avoid dermal exposure from drift, stay away from the flying particles and wear goggles, a filtered respirator with the correct filter, and protective clothing. Protective clothing is especially important when the spray material has a low dermal LD50 (less than 250 mg/kg).

Even though some tractor cabs are spray particle proof, don't depend on all tractor cabs to provide protection against all drift. Small spray particles and vapors can pass through a damaged air-filtering system or openings on some kinds of cabs. Wear protective gear even if you have a cab around you. Clean or replace tractor cab filters after spraying.

To minimize the hazards of spray drift:

1. *Use nozzle tips that have the largest openings that will give an acceptable application of material.* Larger drops will be formed, which do not drift as far as small drops.

2. *Operate your sprayer at pressures as low as possible (below 50 psi for field sprayers).* Low pressures also tend to cause larger droplets to be formed.

3. *Wait until wind speed is below 3½ miles per hour.* The droplets will not drift as far in a lower wind. The best way to determine wind speed is by using an instrument called an anemometer. Your spray dealer will probably be able to tell you where you can purchase an inexpensive model.

4. *Keep others away from the area where the spray may drift.*

CONTAMINATED CLOTHING

Clothing can be contaminated by splashing, drift, and spill. Do not wear contaminated clothing. Wash it thoroughly first. Put on clean, fresh clothing every day. And change during the day if you think your clothing has absorbed pesticide.

Tractor cushions also get contaminated. Wash all cushion covers frequently. The rate of absorption is very high on the buttock area.

The sweat band of a hat can also hold contaminants and expose you.

REENTRY

Dermal exposure can occur if you walk through an area that has been sprayed before the reentry interval has expired. Don't let anyone go into an area that has been sprayed with a material that has a specified reentry interval until that interval has expired. Check chemical labels and spraying regulations (local, state, OSHA, etc.). Post signs all around a sprayed area. If you cannot avoid going into a sprayed area, wear rubberized protective clothing on all parts of your body likely to come into contact with the vegetation that has been sprayed.

INHALATION EXPOSURE

Inhalation exposure results from breathing in pesticide vapors, dust, or spray particles. Like oral and dermal exposure, it is more serious with some pesticides than others.

Fig. 23—Stay Clear Of Smoke From Burning Pesticide Containers

The safest procedure is to avoid breathing in any pesticide.

Poisoning through the lungs is more common in confined areas like greenhouses than it is outdoors, because the pesticide is concentrated. It can occur outdoors if the concentration of material is high or if a highly volatile material like TEPP is used.

Some of the ways that inhalation exposure can occur are:

- **Smoking**
- **Breathing smoke**
- **Drift**

Fig. 24—A Chemical Cartridge Respirator Has Dust Filters And Gas-Absorbing Canisters

Smoking

Smoking contaminated tobacco is extremely dangerous. The pesticide is vaporized by the heat of the burning tobacco and drawn directly into the lungs with the smoke. Cigarettes can be contaminated by spills, splashes, drift, and other means. Don't carry smoking materials when you are working with pesticides. And wash your face and hands thoroughly before smoking.

Breathing Smoke

Whenever anything that has been contaminated by pesticides is burned, stay away from the smoke (Fig. 23). It contains vapors from pesticide material and is dangerous to breathe. Stand upwind while burning containers, and stay as far as possible from the fire. Keep people and animals out of the smoke. (See precautionary note in the section on disposal of empty containers.)

Some pesticides contain material that can explode when burned. Check the label.

Drift

Small drifting particles can be inhaled directly into the lungs. This is especially dangerous when low-volume spraying is used, since this system uses concentrated spray material that's more dangerous than a dilute mixture. Also, for a more uniform application, the nozzles and system pressure are selected to produce a high percentage of small particles, so the probability of drift is high.

Some tractor cabs do not provide adequate protection against drift, especially if they are damaged or designed improperly. Small spray particles can pass through the air filtering system and get into the cab. Wear protective gear even if you have a cab around you.

Use a chemical cartridge respirator for protection from low concentrations of pesticides (Fig. 24). Use one that has been approved for the material you are using by the U.S. Department of Agriculture, the Bureau of Mines, or NIOSH. Maintain your respirator according to instructions. Use cartridges that are effective for the material you are spraying, and change them at specified intervals.

DRY PESTICIDES

GRANULAR PESTICIDES

Hazards associated with dry granular pesticides are a lot like those for granular fertilizers, except that pesticides are more toxic.

Because of their form, granular pesticides *seem* safer than liquid pesticides, and this sometimes leads users to relax their use of safety precautions. Don't do it! Unless proper safeguards are followed, the possibility of chronic exposure from dry pesticides is great. Keep the dust away from your skin and hair. To do this, follow the procedures outlined in the section on dry fertilizers.

Mixing Powdered Pesticides

Whenever you mix or pour a powdered material, stay out of the dust. Try not to create dust. Keep the container low when you are pouring from it to avoid dust clouds.

WETTABLE POWDERS

Wettable powders are dustier than granular pesticides. Use a chemical cartridge respirator when working with them. And always wear rubber gloves when handling these toxic powders.

Always open packages of dust or powder from the top. When pouring, hold the container low, close to the receptacle. Pour slowly to avoid billowing dust.

Once a mixture of a wettable powder has been prepared, continue spraying until the sprayer is empty. Stopping for lunch or some other reason gives the powder time to settle out. Remixing cannot be relied upon to put all of the powder back in suspension, and variation in strength of the application can result. This can be hazardous to the operator, since he may be exposed to material that's more concentrated than he expects. Also, the material may be more difficult to clean from the sprayer. This also exposes the operator to a high risk of accidental contamination.

SYNERGISM

Synergism occurs when the combined action of two compounds produces an effect greater than the individual additive effects of the compound. For example, say pyrethrin alone would kill 10 insects, while piperonyl butoxide alone would kill 5 insects. If these two chemicals were combined, you might expect them to kill 15 insects, but, due to the synergistic effect, 50 insects could be killed.

This increased toxicity affects all three types of exposure. Use particular caution whenever you use combination sprays. Never make up your own mixtures. Use only those combinations recommended by your state cooperative extension service or some other reputable agency.

Fig. 25—Read And Follow Label Instructions on Pesticide Containers

LABELS

A pesticide label is your most readily available and accurate source of information. Read it *twice* before handling a chemical for the first time, and review it *every time* you use the material (Fig. 25).

The label includes instructions for use, warranty statements, and other valuable information. This information is included on the label for your protection and to help you use the material in the package correctly. These labels are important. The information required for the registration of a label generally costs the pesticide manufacturer $8 to $10 million. *Read and follow all instructions given on the pesticide label.*

Labels don't generally give the toxicity of a pesticide as an LD50 value. And the toxicities for oral, dermal, and inhalation exposure are not itemized. However, you can tell how potentially dangerous a material is by looking for certain signal words that indicate what "danger group" it's in.

KEEP OUT OF REACH OF CHILDREN
POISONOUS IF SWALLOWED
OR ABSORBED THROUGH SKIN
FLAMMABLE • VAPOR HARMFUL
Do Not Get in Eyes on Skin or on Clothing
Do Not Breathe Vapors or Spray Mist
Do Not Take Internally
Keep Away from Heat and Open Flame
Keep Container Closed

Fig. 26—High Toxicity Warning

WARNING

IRRITATING TO SKIN AND EYES
CAUSES BURNS ON PROLONGED CONTACT
HARMFUL IF SWALLOWED

Avoid Breathing Spray Mists
Keep Out of the Reach of Children
Do Not Get in Eyes on Skin or on Clothing or Shoes

Fig. 27—Moderate Toxicity Warning

Group 1: The word *danger* is required on all labels for *highly toxic* materials. The label also carries the words *poison* and *warning* and instructions for handling. The antidote is given, along with instructions to call a physician immediately in case of exposure (Fig. 26).

Group 2: These products are *less toxic* than those in Group 1. The label carries the word *warning* and gives handling instructions (Fig. 27).

Group 3: These products are of *low to moderate toxicity.* The label carries the word *caution,* plus instructions for avoiding hazards (Fig. 28).

Group 4: Some products are of low enough toxicity that they don't carry any of these signal words. However, not even this group is always completely safe.

Fig. 28—Low Toxicity Warning

CAUTION

KEEP OUT OF REACH OF CHILDREN
ABSORBED THROUGH SKIN
CAUSES EYE IRRITATION
HARMFUL IF SWALLOWED
FLAMMABLE LIQUID

PESTICIDE APPLICATION EQUIPMENT

This section tells how to avoid hazards involved in using liquid pesticide application equipment (sprayers).

CONNECTIONS

Loose connections can leak and contaminate the outside of a sprayer. They can also come apart, letting the pesticide splash the operator.

Inspect all connections each season before the sprayer is used. They should be leakproof. When using the sprayer check it at least once a day for loose connections, leaking valves, and other problems (Fig. 29). Wear rubber gloves to tighten leaking joints and replace leaking valves or gauges. Clean up and wash the work area, gloves, and tools when the job is done. Use procedures outlined in the disposal section of this chapter.

CALIBRATION

Calibrate your sprayer frequently to be sure it's working properly. Before calibrating, pump plenty of clean water through the machine to rinse out as much residue as possible. Then use clean water to calibrate it. Don't stand where spray could splash or drift onto you. There is always some residue on sprayer parts, so wear rubber gloves to protect your hands.

REPAIRS

Be careful to avoid exposure from spray residue on sprayer parts whenever the equipment is dismantled for repair or modification. Use rubber gloves when disconnecting lines and valves. And always relieve the

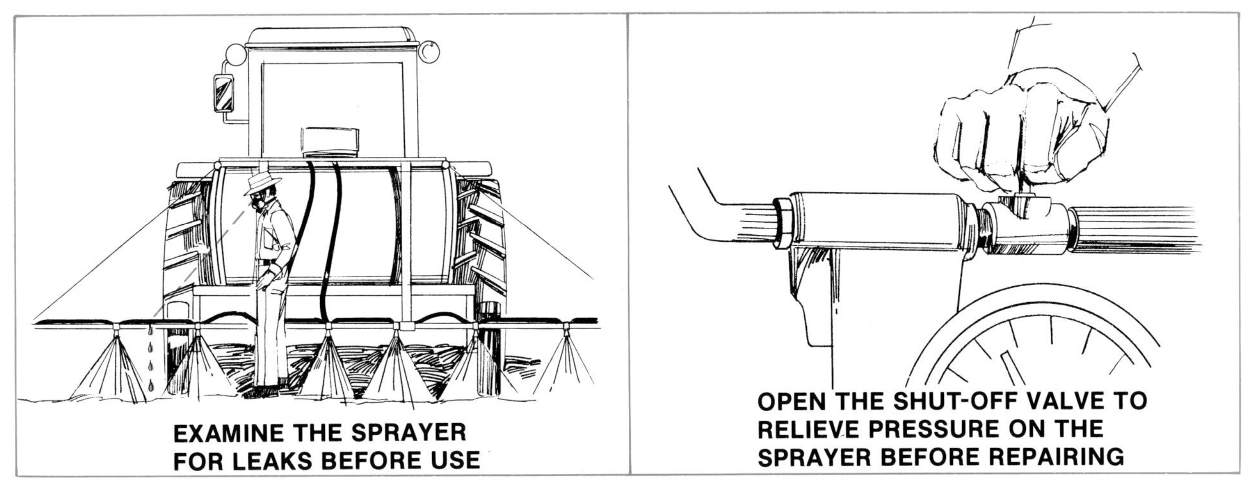

Fig. 29—Examine The Sprayer For Leaks Before Use. Open The Shutoff Valve To Relieve Pressure On The Sprayer Before Adjusting

system pressure, or the spray material might squirt out unexpectedly. Escaping fluid under high pressure can have enough force to penetrate your skin, which could cause serious injury and contamination of the bloodstream. Even if the fluid is not under high enough pressure to penetrate your skin, it could still get into your eyes, ears, nose, or mouth. Wear a face shield. Stay clear of dripping parts. Rinse the sprayer thoroughly with fresh water before disassembling it.

PLUGGED NOZZLES

In case of clogging or other nozzle malfunction, shut off the sprayer and release the pressure. Don't attempt to unplug the nozzle in the field. Keep spare tips available for replacement. Later, clean clogged nozzles with an air hose as described earlier in this chapter.

DECONTAMINATION OF MACHINERY

A sprayer should be carefully cleaned after each different pesticide has been used, when the season is over, and when any repairs have to be made. If spray material is spilled on the machine while mixing or loading, the outside of the machine should be decontaminated immediately (Fig. 30).

The most important step in machinery decontamination is a thorough washing with soap (or a mild detergent) and water, followed by a complete rinse with plenty of water. A steam cleaner can be used if available. Compacted chemical deposits can usually be removed by wetting and scrubbing with a stiff-bristled brush. Pay particular attention to tires and hoses when cleaning sprayers. Dry material can be removed conveniently with a vacuum cleaner.

Take special care while cleaning to avoid splashing your eyes, hands, or clothing. Wear protective clothing, goggles, or better yet, a full face shield.

Fig. 30—Decontaminating Chemical Application Machinery

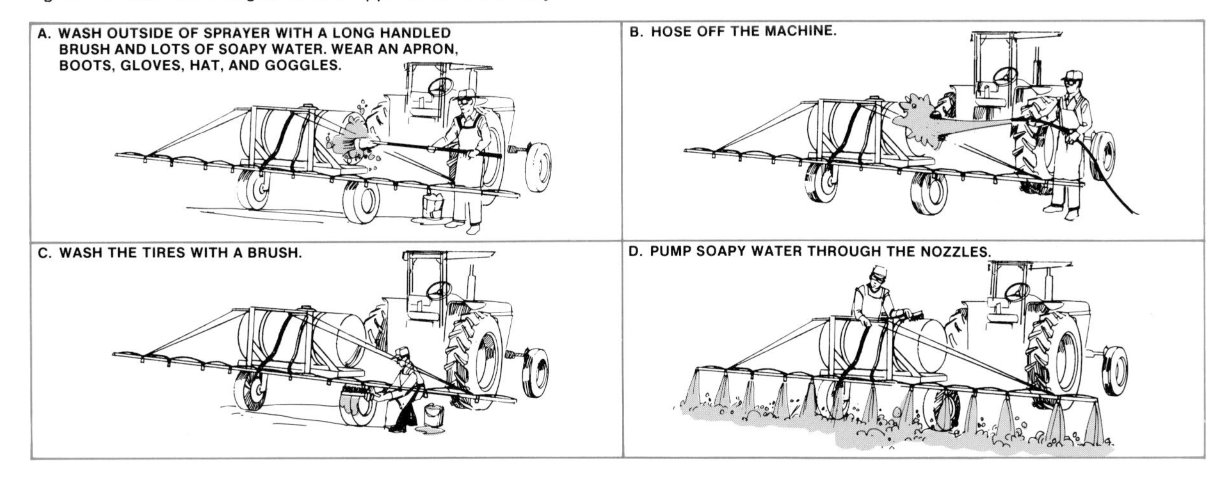

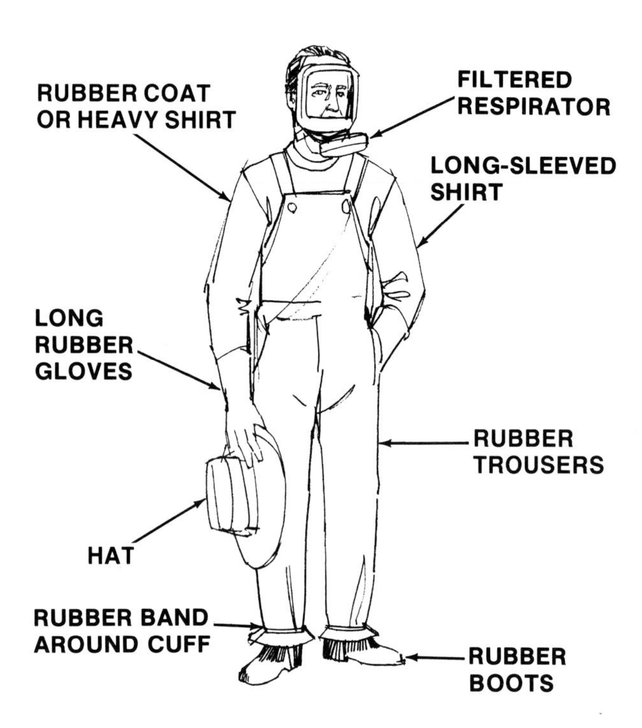

Fig. 31—Wear Water-Repellent Protectve Clothing When Working With Pesticides

Use a strong soap-and-water solution to clean the inside of a sprayer if you've been using carbamates or organophosphates. If organochlorines have been used, the wash water should be mixed with acetic acid (vinegar). Do not use soap with the organochlorines.

Sprayer parts should be washed in the same mixture as recommended for washing the inside of the sprayer.

PROTECTIVE CLOTHING

Protective clothing should always be worn when handling, mixing, or applying pesticides (Fig. 31). Protective clothing covers the body except for the face. Water repellent clothes give the best protection. The next best protection is given by coveralls thick enough to rpevent penetration by pesticides. Full length sleeves are necessary. Don't roll up sleeves!

Leather absorbs pesticides and is extremely hard to decontaminate. Use rubber gloves and boots! Wear a wide brimmed, water repellent hat to protect your head and neck, but be sure it isn't made of an absorbent material like felt. The common billed cap does not give adequate protection.

Wash gloves, boots, hat, and waterproof clothing at the end of each day. Coveralls should be changed regularly, at least once a day, and more often if contaminated. All other clothing should be changed daily. Shower at the end of each working day.

GLOVES

Pesticides must not be handled without gloves. The best protection for the hands is unlined, rubber gauntlet gloves. Unlined gloves are easier to clean, and the gauntlet prevents wrist exposure. If you are applying organophosphates the gloves should be of natural rubber. Organophosphates like parathion can penetrate some synthetic rubber gloves. Keep the glove gauntlets folded to keep liquids from running down your arm.

APRONS

Always wear a rubber or plastic apron when mixing or loading. It should be long enough to cover your boot tops. Wear a natural rubber apron when working with organophosphates.

BOOTS

Always wear rubber boots when handling pesticides or walking through recently sprayed fields. Never wear leather or canvas boots. Keep trouser legs over your boots so that runoff will go outside the boot.

GOGGLES

Unless a full face respirator is being worn, there is always danger to the eyes. Be sure to wear goggles when mixing or loading pesticides. The goggles should be the unvented, chemical splash, nonfogging type.

RESPIRATORS

Wear a respirator whenever there is a risk of dust or vapor being inhaled. In addition, any person applying organophosphates or carbamates should wear a mask or respirator approved for the particular pesticide. Never wear a simple gauze dust respirator. It will only filter out dust or liquid particles, not gases or vapors.

The respirator should at least cover the mouth and nose, and it should fit well and comfortable. There are two main types — chemical cartridge respirators and gas masks. Cartridge respirators are light, and give protection from dust and gases. They have a filter to collect dust or mist, and an absorbent cartridge to remove gases or vapors. The filters and cartridges have a limited capacity, and must be replaced if you smell the chemical or at the specified number of hours of use.

The full-face gas mask or respirator has a large cartridge and filter canister. It has built-in goggles, and covers the entire face. Use one when highly toxic materials are being used. It is also needed when premixing is done in an enclosed area. After removing cartridges and filters, clean the face plate of the respirator with soap and water, then rinse and dry it with a soft cloth. When not in use, respirator cartridges and filters should be stored in plastic bags in a clean, dry place.

In situations where there is a lack of oxygen, use a supplied-air respirator or a self-contained breathing apparatus.

See Chapter 3 for more information on respirators.

PESTICIDE STORAGE

Pesticides are usually purchased before they are needed, and must be stored on the farm between purchase and use. Store them properly to prevent possible damage or hazard to:

- **People, especially children**
- **The environment**
- **Pets and domestic animals**
- **Materials and containers**

Store pesticides in a separate building, if possible, away from other farm structures. It should be located on high ground with good drainage to avoid water running in, and be readily accessible to all types of traffic (Fig. 32). Make provisions for water runoff during fire fighting.

Lock the storage area securely to keep children, pets, and irresponsible adults out. Label it with warning signs like those in Fig. 33. Make your own sign or ask the pesticide dealer for one.

Fig. 33—Chemical Storage Should Be Locked And Clearly Marked

Fig. 32—Store Pesticides In A Separate Building On High Ground

Fig. 34—Store Pesticides In Original Containers

Fig. 35—Keep Dusts And Wettable Powders Up Off The Floor

Follow these tips for pesticide storage:

1. *Always store pesticides tightly closed in their original containers so there is no doubt about the contents (Fig. 34).* Never store pesticides in a container originally used for food products. This is a common cause of accidental poisoning, particularly of children. They think that the contents of the container are the same as the original contents, not a pesticide.

2. *Stack containers only when the bottom one is strong enough to support the stack without splitting open.* A broken container lets material spill out, making cleanup and decontamination necessary.

3. *Read the label for storage instructions.* Keep pesticides in a warm area if they're in glass containers or if freezing could alter their chemical makeup. Store granular materials and wettable powders off the floor to minimize caking, which results when moisture from the floor gets into the container (Fig. 35). Caked material can cause splashing when the powder is added to the mixing water.

4. *Store herbicides away from insecticides and other materials.* Never store pesticides with food, feed, seed, or other farm supplies if contaminated material would present a hazard.

5. *Use material in a package only if you are sure about its contents.* Stored material should be identified only by the container label. Don't try to identify chemicals by smelling them or by looking closely into containers. This could be very hazardous because of inhalation exposure.

6. *Don't store materials unnecessarily.* Purchase only the amount you will use in one year. Dispose of:

- **Materials that are no longer registered** (Pesticides must be registered with the federal government in order to be legal for sale. If this registration is not in effect, the material is not legal and should not be used. Check the registration status at your county extension agent's office.)
- **Empty containers**
- **Spilled materials**

Most pesticides have a limited storage life. It is not generally advisable to keep powdered pesticides longer than one year, because the chemical makeup may change. An unopened package, however, can be

Fig. 36—Wear Protective Clothing When Working In The Pesticide Storage Area

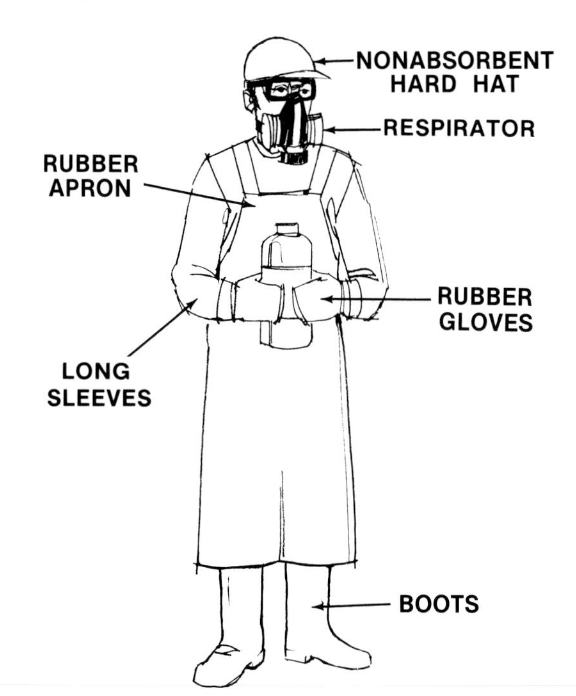

expected to have a somewhat longer storage life. If you doubt the strength of a pesticide, call a pesticide disposal company to take it away.

7. *Keep fire away from the storage area.* Some pesticide formulations contain flammable solvents. This is another place where smoking must be prohibited.

8. *Wear protective clothing and equipment whenever you work in the storage area (Fig. 36).* Keep your hands away from your eyes and mouth, even if the container is tightly closed. There is always the chance of some pesticide residue being on the outside of the container.

PESTICIDE DISPOSAL

Use extreme care when disposing of pesticides. They are hazardous wastes. Do not contaminate ground water, surface water, or air.

Plan for the disposal of these kinds of materials:

- **Leftover chemicals**
- **Spilled materials**
- **Empty containers**

Your dealer can help you with disposal plans. Most dealers have contact with disposal companies.

LEFTOVER CHEMICALS

Try not to get stuck with leftover spray mixture. Estimate on the short side when mixing. Then you will be able to make an accurate judgment of the amount of mixture needed to finish the job.

Leftover spray mixture should not be kept on the farm. Storing it in odd containers can be dangerous, especially to children and animals. Get rid of it in an approved method. See your dealer for the approved method in your location.

SPILLED MATERIALS

If you spill liquid pesticide on a concrete floor or hard surface, keep the liquid from spreading. Surround it with a ring of absorbent material like dried clay (Fig. 37). Then clean it up and dispose of it in an approved method.

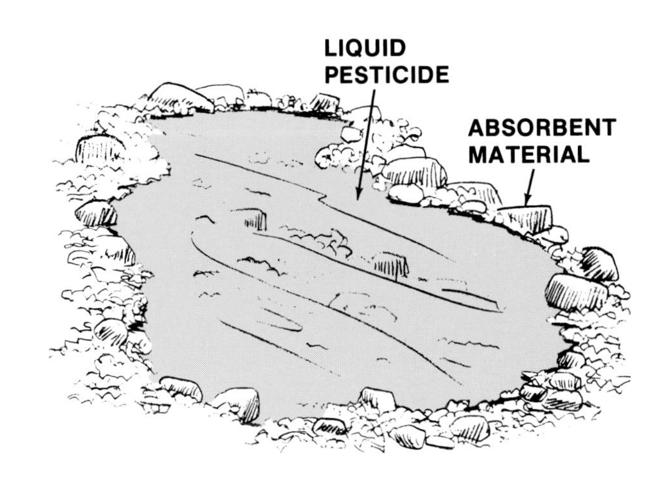

Fig. 37—Surround Liquid Pesticide Spills With A Ring Of Absorbent Material

Fig. 38—Scoop Material Into A Container

Add more clay to soak up the liquid, and scoop the material into a container (Fig. 38). Solid spilled material can often be picked up and used. If you store this material, use the original labeled container.

After all solid material has been collected, clean the area thoroughly. Scrub with water and detergent if a chlorinated hydrocarbon was spilled (Fig. 39). Use lye and water to clean up after organic phosphate spills. After scrubbing, hose down the area (Fig. 40). Repeat this procedure several times. Clean up all water (Fig. 41). Dispose of the rinse water and the spilled chemicals in an approved way. Your dealer can help you with disposal problems or put you in contact with a disposal company.

Use this method to clean up spills on wooden floors, too. The higher absorbency of wood may require more washing and hosing. A spill on soil can be treated by digging up the soil and disposing of it. Contact your dealer for approved disposal methods and companies.

Pesticide Safety Teams

If you have a large spill or if a spilled material is highly toxic, it may be wise to call in outside assistance. Pesticide safety teams have been organized to assist in the prompt cleanup and decontamination of large chemical emergencies, like warehouse fires or transport accidents. They can be called upon for assistance and advice in any kind of emergency connected with pesticides. The pesticide safety teams maintain a

Fig. 39—After Collecting Spilled Material, Scrub With Water And Detergent

Fig. 40—Hose Down Spill

24-hour answering service. Their telephone number is **800-424-9300** CHEMTREC. In some states the Department of Agriculture provides this service. Keep this number handy. Paint it on a wall. Put it on a sign posted inside the pesticide storage area and near your household telephone.

The pesticide safety teams have been set up to handle emergency situations. However, if you have any kind of a question you feel is important, and you cannot get an answer, call this number. They can refer you to someone who will be able to take care of your problem. The number can be called at any time of the night or day from anywhere in the United States. It is *not* a toll-free number.

Fig. 41—Clean Up Rinse Water

EMPTY CONTAINERS

Chemical containers are never really empty. They always contain a residue of hazardous material. Keep them in a locked storage area and dispose of them at least once a year in a way that poses no hazard to humans, animals, or the environment.

Use a rinse-and-drain procedure to clean containers (Fig. 42). Ultimate disposal is accomplished by a chemical disposal company. Unless the label cautions against it or local regulations forbid it, burn paper packages and cartons. Choose a day when the fumes will be blown in a safe direction, and stay away from the smoke. The fumes can be very poisonous. Bury the ashes at least 18 inches deep in an area where the water table is at least 6 feet below the surface.

IMPORTANT: Burning may not be permitted in your area. Check with your local air quality control board and the state air quality control board before burning any pesticides or pesticide containers. If you are not satisfied with the answer you get, check with the Environmental Protection Agency in Washington, D. C.

Containers are usually returned to the supplier or removed by a licensed hazardous waste disposal company. Do not cut up the drum with a cutting torch — it could cause a fire or explosion. And don't convert empty pesticide drums into feed troughs, water storage tanks, or raft floats. They cause livestock poisoning and water contamination.

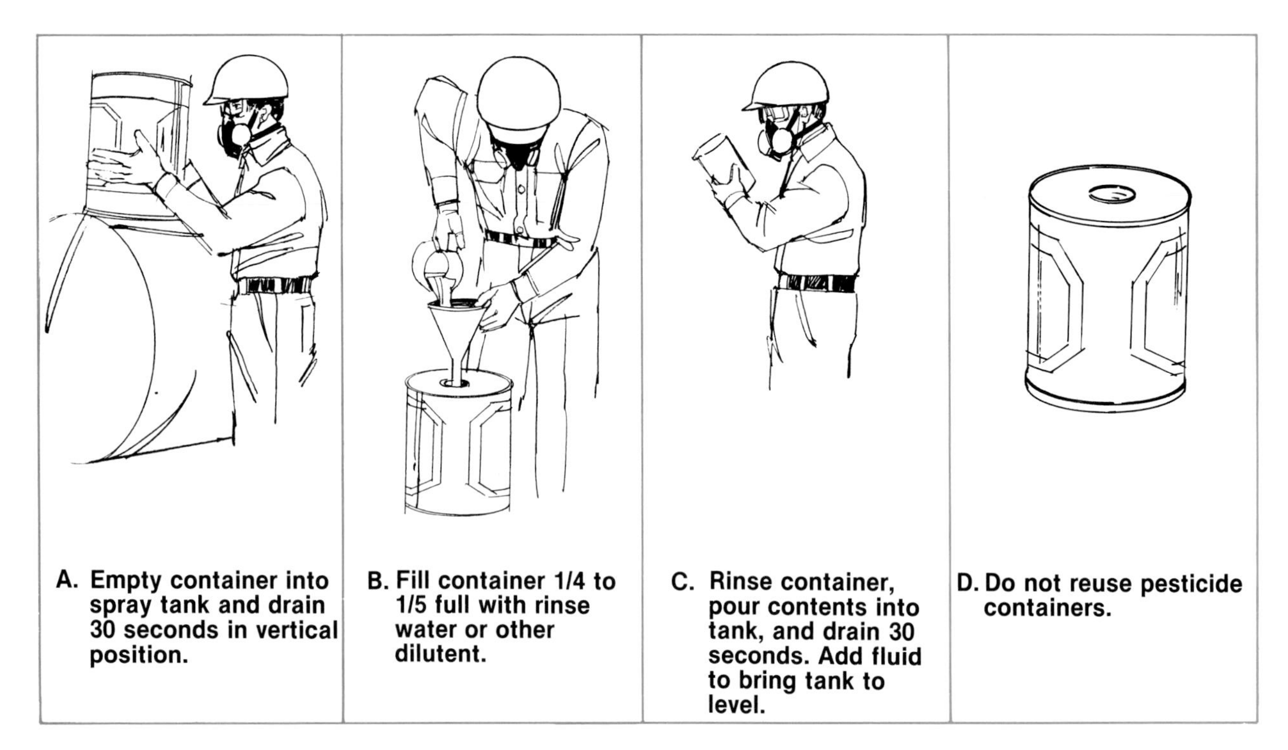

Fig. 42—Rinse And Drain Pesticide Containers Before Disposal

PESTICIDE POISONING

If you or anyone using pesticides show any of the signs of poisoning, get to a doctor immediately.

Poisoning symptoms depend on the particular chemical. The label on the pesticide containers for highly toxic materials lists poisoning symptoms. Read the label and know the symptoms to watch for. The following symptoms are associated with some of the more common pesticides.

CHLORINATED HYDROCARBONS

(aldrin, chlordane, dieldrin, DDT, etc.)

Chlorinated hydrocarbons act as stimulators to the central nervous system. They produce irritability, convulsions, or coma. Nausea and vomiting usually occur, but may not if the dose is large.

DINITRO GROUP

(dinitrocresol, dinitrophenol, Elgetol, etc.)

Early symptoms may include an extreme feeling of euphoria, night sweating, insomnia, restlessness, thirst, and fatigue. In more severe cases, profuse sweating, intense thirst, yellow skin, rapid pulse and respiration, fever, apathy, convulsions, or coma can occur.

ORGANOPHOSPHATES

(diazinon, Guthion, malathion, parathion, TEPP, etc.)

Symptoms depend on the strength of the formulation, specific material, intensity, duration, and rate of exposure.

Initial symptoms usually include headache, weakness, blurred vision, perspiration, and nausea. Abdominal cramps, vomiting, and excessive salivation may be noticed. Diarrhea may occur. Breathing is difficult, and the throat and chest may feel constricted. In severe cases, muscular twitching may be present, along with pupil dilation. The illness can progress to loss of consciousness or convulsive seizures.

People who handle organophosphate insecticides should have regular checks for cholinesterase activity. Organophosphate compounds, which are related to nerve gases, are toxic to insects, animals and man because they interfere with the cholinesterase enzymes that control nerve impulse transmission. Have your doctor give you a baseline cholinesterase level blood test before you begin working with the pesticide (Fig. 43). Then have cholinesterase tests at weekly or biweekly intervals during the season. In this way, evidence of over-exposure can be detected, and steps can be taken to correct the situation before it becomes dangerous. Follow recommendations your doctor gives including eliminating contact with the pesticide if the cholinesterase level has shown evidence of dropping.

Self-diagnosis and self-treatment are extremely dangerous. Don't take drugs, even drugs as common as tranquilizers or aspirin, without first discussing it with your doctor. The interaction between the pesticide and the drug could be very dangerous.

There is no drug that you can take to prevent the effects of the organophosphate pesticide. Some operators take atropine tablets to prevent organophosphate poisoning. This is dangerous. Atropine has a place in treating proven clinical cases of poisoning, but it can be dangerous if taken otherwise. Atropine is a potent drug in its own right, and if taken in the absence of organophosphate or carbamate poisoning, it may lead to serious illness. Furthermore, atropine taken in this way can mask early symptoms of poisoning, so that further exposure is permitted. The consequences of this could be very serious.

For people who often operate away from home base, use of the "medic alert" system is recommended. Everyone working with the pesticide should wear a tag attached to his clothes giving this information:

- **Name of the pesticide being used**
- **Name and phone number of the physician to contact in case of an accident**

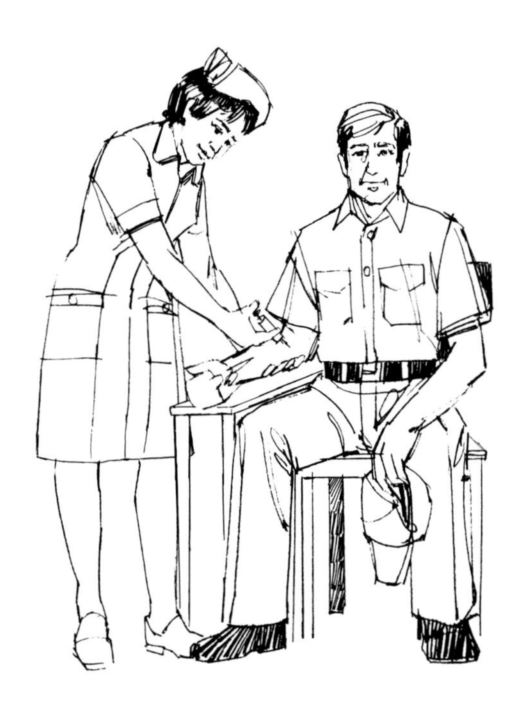

Fig. 43—Have A Blood Test For Baseline Cholinesterase Before Working With Organophosphate Pesticides

POISON EMERGENCIES

In spite of all the precautions used in handling pesticides, poisoning cases still occur. You should be ready to handle any kind of foreseeable emergency. Know the measures to follow in an emergency.

No matter what pesticide is involved, follow these steps:

1. *Call a doctor at once.* If you can't reach a doctor, call the police, local hospital, or a poison control center. Keep their phone numbers posted on the wall of the pesticide storage and near your home phone. Supply the following information:

- **Pesticide involved**
- **Type of work the victim was doing**
- **How poisoning took place (swallowing, skin contact, breathing fumes)**

If you can't reach emergency help and are doubtful about first aid, get the victim to the emergency room of the nearest hospital. Have someone else call to alert the staff and give them the necessary information. Take the container with you.

2. *Give the victim first aid as advised by the doctor or hospital.* Instructions will also be given on the container label. If possible, send the label along with the victim when he's taken to the hospital.

- **In case of poisoning by swallowing, instructions frequently are:**

If the swallowed material is caustic, or if it has a hydrocarbon solvent, don't induce vomiting unless specifically advised to on the label. Vomiting in these cases can cause even more serious problems. Induce vomiting only if advised to on the label. To cause vomiting, have the victim drink an emetic, such as two tablespoons of salt in a glass of warm water. Warm soapy water will also do the job.

Fig. 44—Keep A List Of Emergency Phone Numbers Posted Near Your Phone

• In case of poisoning by skin contact, instructions generally are:

Remove clothing immediately and wash the affected area thoroughly with soap and warm water. Flush the skin with plenty of water.

• In case of inhaled poisons, instructions usually are:

Remove the victim to fresh air. Apply artificial respiration if necessary. Maintain body temperature, but don't let the victim get overheated.

If a poison inhalation victim is in an enclosed area, do not go in to get him unless you are wearing a respirator. The gas-absorbing canister type or the self-contained (air tank) type would be best. If you don't wear this safety equipment, you'll probably end up as a victim, too.

• In case of chemical in the eyes, instructions often say:

Wash the eye with plenty of clean water for at least 15 minutes. Take the victim to a doctor or hospital emergency room.

CHAPTER QUIZ

1. Why should copper or brass parts never be used in ammonia application equipment?

2. What is the first thing you should do if you get ammonia on your skin or in your eyes?

3. True or false? You are less likely to cause drifting when you use low system pressure.

4. Water carried on ammonia equipment should be changed:

a) Daily

b) Every other day

c) Weekly

d) Not at all

5. True or false? The danger from manure gases is greatest when liquid manure is agitated.

6. The three types of exposure to pesticides are ____________, ____________, and ____________.

7. The combined effect of a mixture of two pesticides is sometimes greater than the individual effects added together. This is called:

a) Invert emulsion

b) Toxicity

c) Synergism

d) Calibration

8. If a sprayer nozzle plugs in the field, the safest thing to do is:

a) Shut off the sprayer, release the pressure, and replace the nozzle with a spare.

b) Shut off the sprayer and unplug the nozzle by carefully blowing through it.

c) Continue spraying, and unplug the nozzle after you've finished the job.

9. Who would you contact (and how) if you accidentally broke open a drum of dieldrin?

8
Target: Hay and Forage Equipment

Fig. 1—Plant Maturity Stage and Weather Conditions Usually Rush Operators To Harvest Quickly—Increasing Chances For Injury

INTRODUCTION

Hay and forage crops comprise more acres of cropland in the United States than any other crop. There are three factors involved in hay and forage operations that are important to your safety:

- **Timing of operations and crop conditions**
- **Hay and forage equipment**
- **Field terrain**

TIMING OF OPERATIONS

Maturity and moisture level are more important with hay than with forage coming from row crops.

Weather is also an important factor in haying. Because of delays caused by rain or extremely fast drying conditions, part of the crop may be lost.

These factors sometimes rush operators into disregarding safety. Faster hay and forage harvesting is possible. But the operator must remember safe operation of the machine is more important than hurried operation. An operator in a hurry, disregarding safe operating procedures for machinery, can make hay and forage harvesting hazardous.

HAY AND FORAGE EQUIPMENT COMPONENTS

Most hay and forage harvesting machines have several moving parts and one or more forms of cutting action (Fig. 2). Like other types of PTO and hydraulically powered machinery they are potentially hazardous.

1. Study your operator's manual, especially safety instructions.

2. Lubricate and maintain machinery as specified in the manual.

3. Wear close-fitting clothes.

4. Follow safe operating procedures.

5. Follow local regulations and use SMV emblems.

A third potential accident factor, along with a "hurried attitude" and the potentially dangerous nature of hay and forage equipment is the terrain.

FIELD TERRAIN

Many times, hay and forage crops are grown on ground too rough for row-crops (Fig. 3).

1. *Remove stumps and stones* or mark them clearly to prevent upsets, breakdowns, and dangerous driving conditions.

2. *Inspect ditches* for undercutting.

3. *Plan harvesting so equipment travels downhill on steep slopes to avoid overturning.*

4. *Space tractor and equipment wheels as far apart as possible* for stability.

Because hay and forage harvesting is strenuous, and frequently performed during hot weather, "field preparation" of the operator is also important. See Chapter 2 for guidelines for reducing the dangers of heat stress.

Fig. 2—Moving Components Are Potentially Dangerous To The Careless Operator

Fig. 3—Know The Terrain For Safe Operation of Equipment

Fig. 4—Always Disengage Power And Stop The Engine Before Working On A Mower Or Conditioner

TYPES OF HAY AND FORAGE EQUIPMENT

Just as there are different methods and systems of storing hay and forage crops, there are a wide variety of machines to get the job done:

- **Mowers and Conditioners**
- **Windrowers**
- **Balers and Bale Ejectors**
- **Bale Handling Systems**
- **Hay Stackers**
- **Stack Movers**
- **Big Balers**
- **Forage Harvesters**
- **Hay Cubers**
- **Forage Wagons**
- **Blowers**

Each type of machine will be discussed.

MOWERS AND CONDITIONERS

There are four danger spots.

- **Power take-off**
- **Cutterbar**
- **Crimping or crushing rolls**
- **Gathering reels**

Fig. 5—Mower Cutterbars Are Hinged In Two Ways

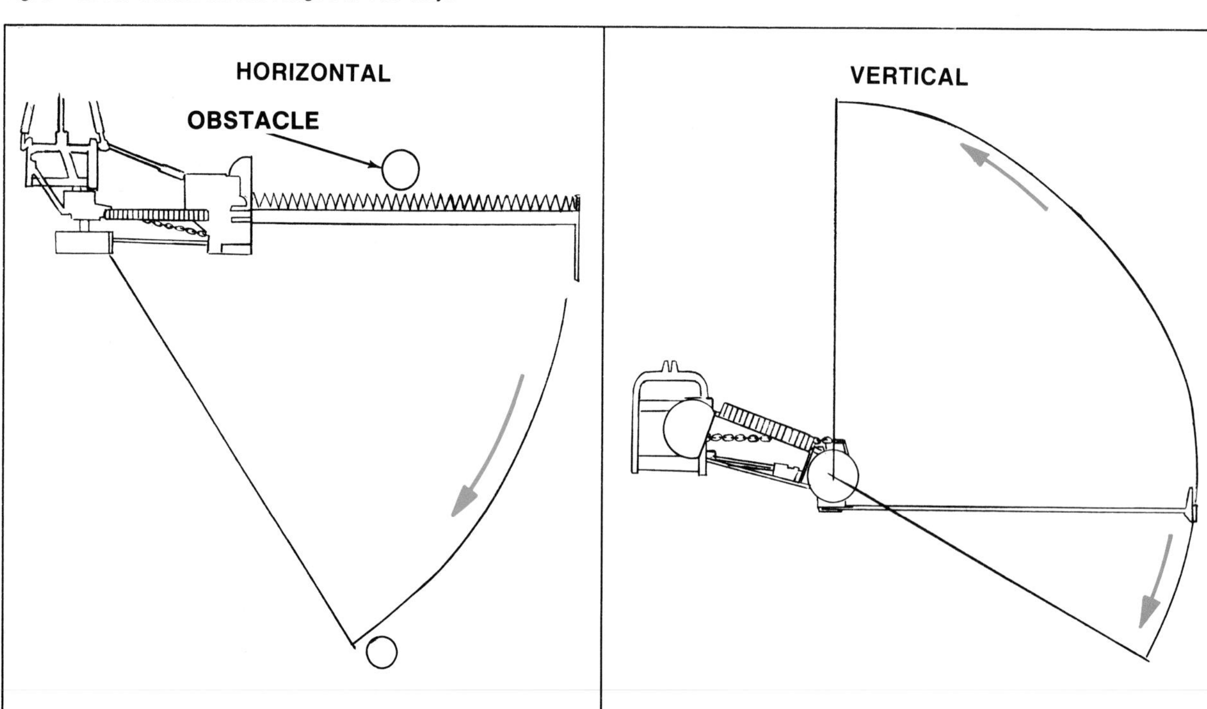

POWER TAKEOFF

Always make sure the machine is hooked up correctly. Never attach a 540 rpm mower or conditioner to a 1000 rpm PTO tractor. Operating a mower or conditioner at the wrong speed can cause machine failure and possibly injure you.

Be extremely careful when working near PTO shafts. Make certain all safety shields are in place before you operate. Rotating PTO shafts that are not properly shielded are potential killers.

Even proper shielding does not provide complete protection. If mud or metal burrs are on the shaft, it may be enough to entangle your clothing. For this reason, never approach a PTO shaft that is still in motion. Always disengage the PTO, shut off the engine, take the key, and wait for all moving parts to stop before working near the PTO shaft or any other part of the machine (Fig. 4).

CUTTERBAR

Even though precautions are taken to insure correct operation of mowers and mower-conditioners, *plugging* or *trash build-up* on cutterbars may still occur.

To safely unplug the cutterbar or remove trash, follow these steps:

1. Stop and disengage the PTO.
2. Raise the cutterbar and back up.
3. Shut off engine and engage the parking brake or shift the transmission into park (or neutral).
4. Pull hay away from cutterbar.
5. Check the cutterbar for broken guards or knife sections.
6. Start engine and engage PTO at low speed.
7. Ease mower into standing hay and resume operation.

Mower Cutterbars

The cutterbar on mowers is hinged in two directions (Fig. 5).

- *Vertical Arc—for transport and vertical travel during operation*
- *Backward—for avoiding damage to mower when encountering obstacles*

Fig. 6—Handle Cutterbars Carefully From The Back To Avoid Cutting Fingers

FIELD OBSTRUCTIONS

When field obstructions are encountered, a safety spring release allows the cutterbar to swing back to avoid damage (Fig. 5). To relatch the cutterbar, disengage the PTO, raise the cutterbar and back the tractor. With some mowers it may be necessary to turn the front wheels to the left to provide more momentum to relatch the cutterbar. Make sure no one is standing near during this operation.

TRANSPORT POSITION

Transport position requires shutting off the tractor and manually lifting the cutterbar by the grass board or backside to a vertical position. Lock the cutterbar into place with a pin or another device provided (Fig. 6). Always handle the cutterbar from the back, keeping hands away from the knife section. A small knife movement can easily cut your fingers.

Fig. 7—Stay Clear of Reels Unless the PTO And Engine Are Off To Avoid Entanglement

REELS

Most of the safety precautions applying to mowers and cutterbars apply to mower-conditioners. However, in addition, mower-conditioners have reels located over the cutterbar to deliver the crop to the crimping rolls. The reel is an extra reason, in addition to the cutterbar, to shut off the engine and disengage the PTO before going near it (Fig. 7).

Reels also have tines to grab the crop and direct it into the mower-conditioner. The tines can easily hook or grab your hand or clothing, and drag you into the cutterbar. Always make sure the engine is off and the PTO disengaged before unplugging or working on the cutterbar or reel.

Fig. 8—Never Allow Anyone In Back Of Conditioner Rolls While Operating—An Object Could Be Thrown And Cause A Serious Injury

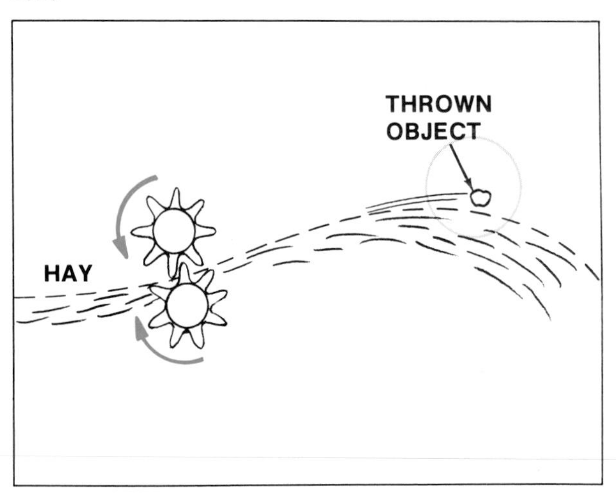

CRIMPER ROLLS

Conditioners and mower-conditioners use crimping or crushing rolls to condition hay so it may dry faster. The conditioning rolls are PTO-powered. They pull the hay between them, and throw the hay out of the back of the machine.

For this reason, never allow anyone to stand near the back of a conditioner, or mower-conditioner during operation. The crimping rolls may pick up a stone or other object and throw it out the back injuring someone (Fig. 8).

Never work on or around crimping rolls unless the engine is off and the PTO disengaged. The rolls can easily grab your hand or clothing if you leave the machine operating. Don't take a chance.

WINDROWERS

Windrowers save valuable time in haying operations. They can mow, condition and windrow hay in one trip over the field in a similar way to mower-conditioners. But windrowers present two additional safety considerations:

1. Windrowers use an auger or canvas draper to carry hay to the center of the platform.

2. Windrowers may be self-propelled as well as drawn.

AUGER PLATFORMS

The auger platform poses potential safety hazards not only because it is an additional moving part, but because it may get plugged up by crops (Fig. 9). Operators often get careless when unplugging machines.

The potential hazard of the auger lies in its shearing and pinching action. The same precautions in unplugging, servicing, repairing and inspecting mower-conditioners should be followed with windrowers. Never work on the platform, including cutterbar, reel or auger, without first disengaging the PTO and shutting off the engine.

SELF-PROPELLED WINDROWERS

As with any other piece of mechanized farm equipment, self-propelled windrowers require a responsible and mature operator. Become thoroughly familiar with the machine and controls before beginning field operations (Fig. 10). The aspects of safe operation to be considered are:

- **Pre-Operation**
- **Field Operation**
- **Dismounting**

Each phase of operation deserves special attention as to the safest methods and procedures. For general safety procedures on self-propelled equipment, refer to Chapter 5.

Pre-Operation

Take these precautions before starting the windrower to avoid accidents:

1. Be sure no one is near the windrower.
2. Place the variable ground speed control lever in neutral.
3. Place steering levers in neutral.
4. Disengage platform drive clutch.

By following these procedures, the operator can prevent sudden movement of the machine or parts of the machine during starting and avoid accidents.

Field Operation

Alertness to field and crop conditions is one key to operating windrowers safely. Also, a machine in good operating condition gets the job done faster and safer because of fewer operating problems or repairs. Proper operation includes:

- **Safe windrower speed for conditions**
- **Correct operation of steering mechanisms**
- **Correct platform height**

WINDROWER SPEED

Windrower speed is determined by the terrain and the density of the crop. When operating over rough terrain or on hillsides, take care to avoid holes or obstacles that can tip a windrower or throw you from the machine.

Crop density also affects the speed at which you operate the windrower (Fig. 11). In heavy crops, high operating speeds cause frequent clogging. The more often you must unplug the machine, the higher the chances for an accident. Also, remember that by taking the crop in at the proper speed you'll save time. Frequent stops to clear the machine almost always consume more time than harvesting at a slower speed.

STEERING MECHANISMS

Steering mechanisms on windrowers consist of steering levers and a steering wheel (Fig. 10). The steering levers are for sharp turns at the end of the field. Adjust speed according to the type of turn to be made. Use the steering wheel to obtain trim and for transport.

Make turns on hillsides with care. Reduce speed and make turns with caution to avoid overturning.

Fig. 9—Auger Platforms on Windrowers Present Extra Safety Hazards Due To The Pinching Action Of Augers

PLATFORM HEIGHT

The height at which you operate the platform will vary from crop to crop. For instance, if you windrow grain for combine harvest, the platform will usually be operated much higher than for forage crops. When harvesting hay crops, the platform may be operated very near the ground. Keep a close watch for rocks or other obstacles that could be hit or picked up by the platform. Irregularities in the ground also present safety hazards.

Manufacturers install skid plates and include other design characteristics on platforms to keep the platform from scalping or digging into the ground. However, if the windrower is operated on steep slopes over rough terrain or goes through a ground depres-

Fig. 10—Be Familiar With Windrower Controls Before Operating To Avoid Losing Control Of The Machine

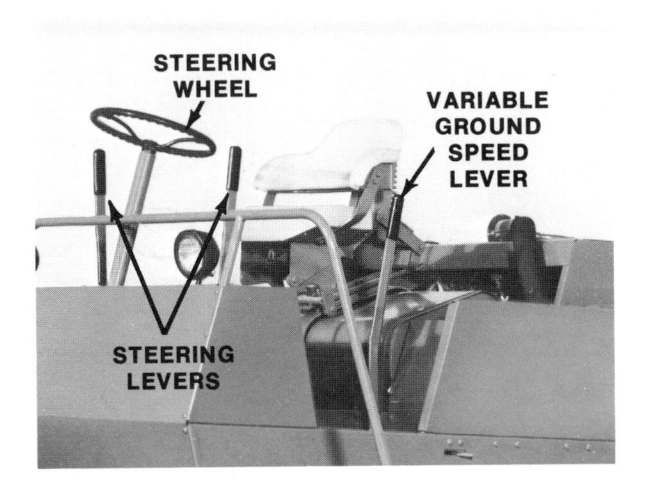

Fig. 11—Adjust Speed To Crop And Field Conditions To Avoid Plugging Of The Machine And For Safer Operation

sion, the platform may dig into the ground. This may cause the windrower to veer or the machine may stop suddenly. If this occurs you may be thrown against the controls or completely off the operator's station. Keep these potential hazards in mind. Slow down and maintain maximum control.

The careful operator keeps a watchful eye on the cutterbar and platform as well as watching where he's going.

Don't allow passengers to ride on the windrower. Sharp, fast turns and rough terrain can easily throw them off.

Dismounting Machine

Before dismounting the machine for any reason:

1. Move ground speed control lever to neutral.
2. Disengage platform drive clutch.
3. Reduce throttle to idle and turn off ignition.
4. Engage the parking brake.

If you plan to leave the machine for any period of time, remove the ignition key and lower the platform to the ground. This will keep an unauthorized person from starting up the machine and from accidentally lowering the platform, perhaps pinning someone under it or damaging the windrower.

BALERS AND BALE HANDLING SYSTEMS

Hay balers revolutionized hay making in the United States in the 1930's. With labor becoming even more scarce and costly in the 70's, producers once relying on manual labor to handle hay are turning more to labor savers, such as bale ejectors, bale accumulators, automatic bale stackers, and the large package hay systems.

As with any machine doing the job better and faster, the operator must be familiar with the mechanisms involved and with safety precautions. The machines we will be looking at from the safety standpoint in this section are:

- **Balers and bale ejectors**
- **Bale handling systems**

Though these machines greatly reduce the time and labor needed for haying, the operator must be more conscious of safety hazards. Baling usually puts stress on the operator to get the job done quickly and efficiently (Fig. 12). Weather conditions can change rapidly and the value of the crop may be cut due to conditions that are too wet, too cool, too hot or too dry.

But, no crop, no matter how large or how good, is worth injury or death. Careless operation that saves time is foolish. Slow down and work safely.

Fig. 12—Baling Places Stress On The Operator To Hurry To Beat The Weather

Fig. 13—Moving Parts Which Must Handle Tough Crops Can Injure Careless People

BALERS AND BALE EJECTORS

The two common types of balers making the traditional or small bales are:

- *Rectangular-type balers*
- *Round-type balers*

Rectangular-Type Balers

Balers producing rectangular bales use a pickup head and auger to feed the hay to the bale chamber for compression and bale formation (Fig. 13). As the hay enters the bale chamber, two knives slice it and distribute it uniformly in the bale chamber.

These and other components can injure you if you do not take all safety precautions during operation and servicing of balers. Make sure all such parts are completely stopped before you get near them. A baler in good mechanical condition adds to your chances for safe and efficient operation.

Proper operation of the baler reduces the number of times it needs repair, and therefore, the number of potentially dangerous field repairs you must make. Repairs also take time away from field operation and increase the stress factor on you to get the job done.

Proper *timing* of all moving parts is essential for successful baler operation and for avoiding serious damage or breakage of moving parts. Timing is usually accomplished by mean of chains or sprockets. Consult your operator's manual for specific timing instructions and other adjustments.

Fig. 14—Bale Ejectors Throw Bales With Considerable Force

Correct *ground speed* is important to safe baler operation. If you travel too fast, the baler can become overloaded. If operating over rough terrain, the knotter may miss tying bales if the baler is bouncing too hard.

If the need should arise to adjust, inspect, or repair the baler, follow these procedures for your safety:

- *Disengage all power*
- *Shut off the engine*
- *Wait for the flywheel and all other moving parts to stop*

If you need to test the bale knotter, the proper procedure is to disengage the PTO, shut off the tractor engine, remove the key, and then turn the flywheel by hand after tripping the knotter drive. With the flywheel turning slowly, you can observe the knotter going through a tying cycle in slow motion. But, be sure you just watch while the parts are moving. Keep hands away!

Other recommendations for safe baler operation are:

1. *Never allow anyone to turn the flywheel while someone is working on the machine knives.* Moving parts can easily injure someone.

2. *Be sure bale twine or wire is properly spliced and threaded in the machine to avoid knotter problems.*

3. *Never pull anything out of the knotter while it's in operation.* It can easily entangle you and cause a serious injury.

4. *Keep children and animals out of the field during operation.*

5. *Replace shields and guards*

Small Round Balers

Round, or rolled bales, are approximately 36 inches long by 16 to 30 inches in diameter (914 mm by 406 to 762 mm). They are formed by rolling hay between belts. Bale density is controlled by belt tension and feeding rate. The bales are wrapped with twine.

Practice the same safety precautions used for rectangular baling. Always be sure the PTO is disengaged, the engine shut off, and the key removed before dismounting to service or adjust the baler.

Take extra care when stacking and transporting round bales. They roll easily.

Bale Ejectors

Bale ejectors, or throwers, were first introduced on rectangular balers during the early 1950's. Three mechanisms are used.

- *Hydraulic ejectors*
- *High-speed belts*
- *High-speed rollers*

Fig. 15—Bale Elevators Make Work Easier—But Treat Them With Respect For Your Safety

The bales are ejected into wagons with sides. There is no need for a man to stack them on the wagon (Fig. 14). The force of an ejected bale could injure a person. Do not allow anyone to ride on the bale wagon. Never allow anyone to stand behind or work on the ejector while the PTO and engine are operating or while a bale is in the ejector. See page 235 for big round baler safety.

MANUAL BALE LOADING

Manual bale loading can be done safely if it is done carefully (Fig. 15). The nature of wagons and bale handling means the operator and bale handlers alike should take extra care. Be sure everyone is aware of these potential hazards:

1. Starts and stops can cause bale handlers to fall off of the wagon or truck.

2. Accidently stepping off the wagon or truck while loading bales.

3. Falls from the wagon or truck could mean getting run over as well as fractures, sprains and concussions.

4. Tossing bales can easily knock someone off balance.

It is just as important to practice safety while transporting bales. Travel at a safe speed and do not allow riders on top of the hay or on back of the wagon or truck.

Fig. 16—Bale Accumulator

BALE HANDLING SYSTEMS

Several bale handling systems have been developed to speed handling of rectangular bales. These systems also eliminate some physical handling of bales in varying degrees of gathering, transporting and storing baled hay. Other basic systems besides bale ejectors are:

- *Bale accumulating and handling systems*
- *Automatic bale handling systems*

Bale Accumulating and Handling Systems

Bale accumulating and handling systems generally consist of two pieces of equipment. The accumulator is attached to or trails the baler or may be pulled independently (Fig. 16). It assembles the bales in groups of 8 or 10. When the accumulator is filled, the cluster of bales is deposited on the ground without stopping the forward progress of the baler.

The second phase of the system is usually stacking or loading the groups of bales with a special attachment for tractor loaders. Hydraulic powered steel tines are thrust into the bales to hold them together as a unit for stacking and loading.

This same tractor-mounted unit can usually be used to unload the bales and store them in ground-level storage, either in a building or in open stacks.

Special considerations for safe operation of these mechanisms include:

- *Hydraulic operation*
- *Avoiding accumulator surface*

HYDRAULIC OPERATION

Hydraulic operation safety, as discussed in Chapters 3, 4 and 5, applies as most bale accumulators are hydraulically operated. Long hydraulic hoses connect hydraulic power from tractor to accumulators which trail behind the baler. Carefully attach hoses when hooking behind balers. Inspect them often to detect leaks.

ACCUMULATOR SURFACES

Accumulator surfaces get slick (Fig. 17). Bales sliding over the surface polish the metal. Dew makes it stick.

Automatic Bale Handling Systems

Both categories of systems handle bales automatically, but must be used properly to avoid injury.

Automatic bale handling systems mechanically pick up bales from the field and stack the bales (Fig. 18). Automatic bale stackers may be drawn or self-propelled.

They transport the load to storage. The bales can either be stacked at ground level or with some machines. They are unloaded mechanically one at a time.

Keep others clear when loading and unloading bales to avoid injury. Because automatic bale handling systems rely heavily on hydraulics and chain conveyors, observe the following safety precautions:

Fig. 17—Be Careful If You Must Step Onto Accumulator Surfaces —Polishing Action of Hay Makes Them Slippery

1. *Keep hydraulic fluid clean and hydraulic system in good repair.* Many safety hazards can be created by dirty or contaminated hydraulic fluid or damaged hydraulic system components.

2. *Periodically inspect and check operation of hydraulic control valves, mechanically activated switches, hoses, and flow dividers.* Repair or replace any faulty equipment immediately. See page 73 and 74 for an examination procedure for pinhole leaks.

3. *Be cautious when operating bale accumulating and handling equipment on steep slopes.* Avoid dumping clusters of bales on steep slopes because using a tractor loader to load bales where it is too steep could easily cause the tractor to tip over.

4. *Be sure tractor brakes or self-propelled bale handler brakes are uniformly adjusted* to avoid uneven braking and a possible upset.

5. *Maintain the desired tension on all chains and conveying belts* as recommended in the operator's manual.

6. *Keep all shields in place to guard against injury.*

BIG PACKAGE HAY SYSTEMS

Hay production was revolutionized in the early 1970's by the growing use of large-package haying systems. The biggest impact on safety with these new haying machines is that only one person is required to handle and transport the hay.

The two common big package haying systems are:

- **Stacking systems**
- **Big-bale systems**

This section will cover basic safety considerations. Always consult your operator's manual for specific details on safe operation of the big package hay systems.

STACKING SYSTEMS

Stackers operate by picking up hay or stover either with windrow pickup or flail pickup heads. The hay is blown or conveyed into a cage or container. Different techniques are used to compress hay and increase stack density. Two common methods to increase stack density are:

- **Hydraulic compression**
- **Blowing hay to compress the stack**

Safety procedures for operating stackers include:

1. *To avoid falls, do not allow riders on stackers.*

2. *Lower the pickup head when the stacker is not in use.* The head could accidently be lowered by someone and injure anyone under it.

3. *Place blocks under the pickup head when working under or around it.* This will prevent the pickup head from falling on or rapidly descending on anyone under it.

4. *To avoid entanglement and injury, do not lubricate, adjust, or clean the stacker while it is running.*

5. *Always block the wheels when working on, under or around the stacker* to prevent it from rolling on you or someone else.

6. *Reduce speed on steep hills or rough terrain.* A full stacker is top heavy and could turn over.

7. *Be sure your tractor has enough front-end ballast.* Some stackers with a full load may weigh approximately eight tons and could cause your tractor to tip backward if not properly ballasted.

8. *Before unloading, check to make sure no one is near the rear of the stacker.* The door mechanisms could hurt or pin someone under the stack (Fig. 19). Unload on level ground.

STACK MOVERS

Stack movers use chains or links revolving on beams to load and unload stacks (Fig. 20). Hydraulically operated, the conveyor chains are in the open. Do not

Fig. 18 — An Automatic Stacker Uses Many Chains And Hydraulic Hoses — Do Not Feel For Pinhole Leaks. Carefully Inspect The Hoses Regularly And Keep All Shields In Place.

Fig. 19—Be Sure Everyone Is Clear Before Opening The Rear Door Of Stack Wagon

Page 20—Keep Others Clear When Operating Stack Movers

allow anyone near the mover while it is in operation as they could become entangled in the chains.

Take the same safety precautions with stack movers as taken with stackers. In addition, place an SMV emblem and proper reflectors on the back of a stack being transported because the tractor's emblem is blocked from view.

BIG ROUND-BALE BALERS

Big bale machines form bales weighing from 500 to 2,000 pounds (227 to 908 kg) or more. Do not eject large bales on slopes where they may roll away (Fig. 21). Because the bale is heavy, once it starts moving it has a tremendous force that can damage fences, other machinery, or injure people.

CAUTION: Tractor-mounted pick up devices can cause instability. Also bales have rolled down on the operators.

1. Stop the engine and remove the key before dismounting to service the baler.

2. Do not eject big bales where they might start rolling.

3. Observe all safety precautions applying to PTO and hydraulically operated machinery.

4. Do not let anyone stand near the rear of the baler when a bale is being ejected.

5. Do not feed or unplug feed rolls with the engine running. The rolls will jerk your hand in before you can let go of the hay.

6. Block up rear gate so it can't fall on someone.

The large bales may be transported by tractor loaders, 3-point hitch carriers or by trailing bale movers. Always use a bale clamp to hold the bale. Proper tractor weighting is important to avoid tipping backwards. Keep everyone clear of the area while loading and transporting the bale. See Chapters 5 and 11 for details on loader safety and tractor weighting.

FORAGE HARVESTERS

Forage harvesters cut and chop hay or grain for feed.

Three types of forage harvesters are:

- **Cylinder cut**
- **Flywheel cut**
- **Flail cut**

Fig. 21—Do Not Allow Anyone To Stand Near The Rear Of The Baler When A Bale Is Being Ejected. The Bale May Roll.

Both cylinder and flywheel forage harvesters have self-sharpeners. The use of self-sharpeners requires extra precautions because sharpening is done while parts are moving. Be aware of these potential hazards. Except for flail harvesters with flail pickup an integral part of the machine, most forage harvesters are available with different types of pickup heads:

- **Direct-cut or mowerbar head**
- **Windrow pickup head**
- **Row-crop pickup head**
- **Corn head**

All pickup attachments and the flail chopper have moving parts which are designed to cut and handle crops. Never approach them while the engine is running. They can also cut people.

GENERAL SAFETY PRECAUTIONS

Safety for operation of tractor-pulled and self-propelled hay and forage equipment was discussed earlier in this chapter. Many of the same safety suggestions apply to self-propelled and tractor-drawn forage harvesters:

1. *Never stand behind or under the discharge spout while the harvester is operating (Fig. 22).* Hard objects become dangerous projectiles coming out of the spout.

2. *Be sure the harvester is completely stopped before hooking up wagons to avoid being hit by objects from the spout.*

3. *Never clean, oil, or adjust the harvester when it is running.* Make sure the engine has stopped.

Fig. 22—Never Stand Under Or Behind Discharge Spouts While They Are Operating

4. *Disengage power rear-wheel drive on self-propelled harvesters while driving on icy roads or highways to avoid losing control.*

5. *Remember, the auger, fan, and cutterhead may continue to rotate for several minutes after power is shut off.* To avoid injury, do not open doors or shields until these parts have stopped rotating.

6. *Keep doors and shields latched tightly during operation.* Objects are thrown from the cutter with great force. The shields must be in place to deflect these objects.

7. *Review Chapter 10 for safety suggestions for operation of forage harvesters with corn heads.*

Self-Sharpener Safety

Operating the built-in knife sharpener is an operation that must be performed while the harvester is running (Fig. 23). Some machines allow the doors over the knife reel and auger to be closed during the sharpening process. Others require the door to be open so the sharpener can be operated. If your machine's doors must be open during the sharpening process, use extreme care. Make sure your feet are on dry, solid ground. Keep others clear and do not be distracted from your work. A moment's carelessness can easily result in a lost hand or arm.

Make sure no objects can fall into the knife reel. Because the blades turn at very high speeds, the object could be thrown a long ways. Wear protective eye glasses to guard against particles of forage and steel thrown. Study the operator's manual for specific instructions and safety rules (Fig. 24).

Here are additional safety procedures for sharpening knives:

1. Park the tractor and harvester on level ground to prevent rolling. Set parking brake.

2. Shift the feed roll shift lever to neutral.

3. Make sure the self-sharpener is in correct position according to operator's manual.

4. Close and latch auger and cutterhead doors before starting the tractor if your machine is so designed.

5. After starting the tractor, engage the PTO, adjust to proper rpm for sharpening, and then dismount to operate the sharpener.

6. Stay away from any moving parts.

7. When sharpening is completed, disengage the PTO and turn off the tractor. Make sure the cutterhead has stopped rotating before inspecting the knives.

Knife Balance and Safety

If adding or removing knives, the primary concern is locating knives so the cutterhead or flail knives are in balance and properly adjusted. If not balanced and adjusted, excessive vibration and wear will result. Tighten cutterhead bolts with a torque wrench to manufacturer's specifications. A loose knife can be hurled through metal shields with deadly force and be thrown long distances.

Always block the cylinder cutterhead completely before adjusting, removing or replacing knives, to prevent dangerous knife movement (Fig. 25). The cutterhead turns easily, and since it is heavy it has high inertia. It tends to keep on rotating. It can easily cut a hand or fingers.

HAY CUBERS

Hay cubing is practiced in areas where hay can be cured to approximately 10 percent moisture content. Uniformly low moisture level is necessary for efficient and workable operation of the cuber.

Most hay cubers are self-propelled, but stationary type cubers are also used. General safety suggestions for self-propelled cubers are covered in Chapter 5 under "self-propelled machinery."

Special safety considerations for self-propelled cubers include using extra caution around ditches or hills. With a full water tank for wetting the hay, hay cubers tend to be top heavy.

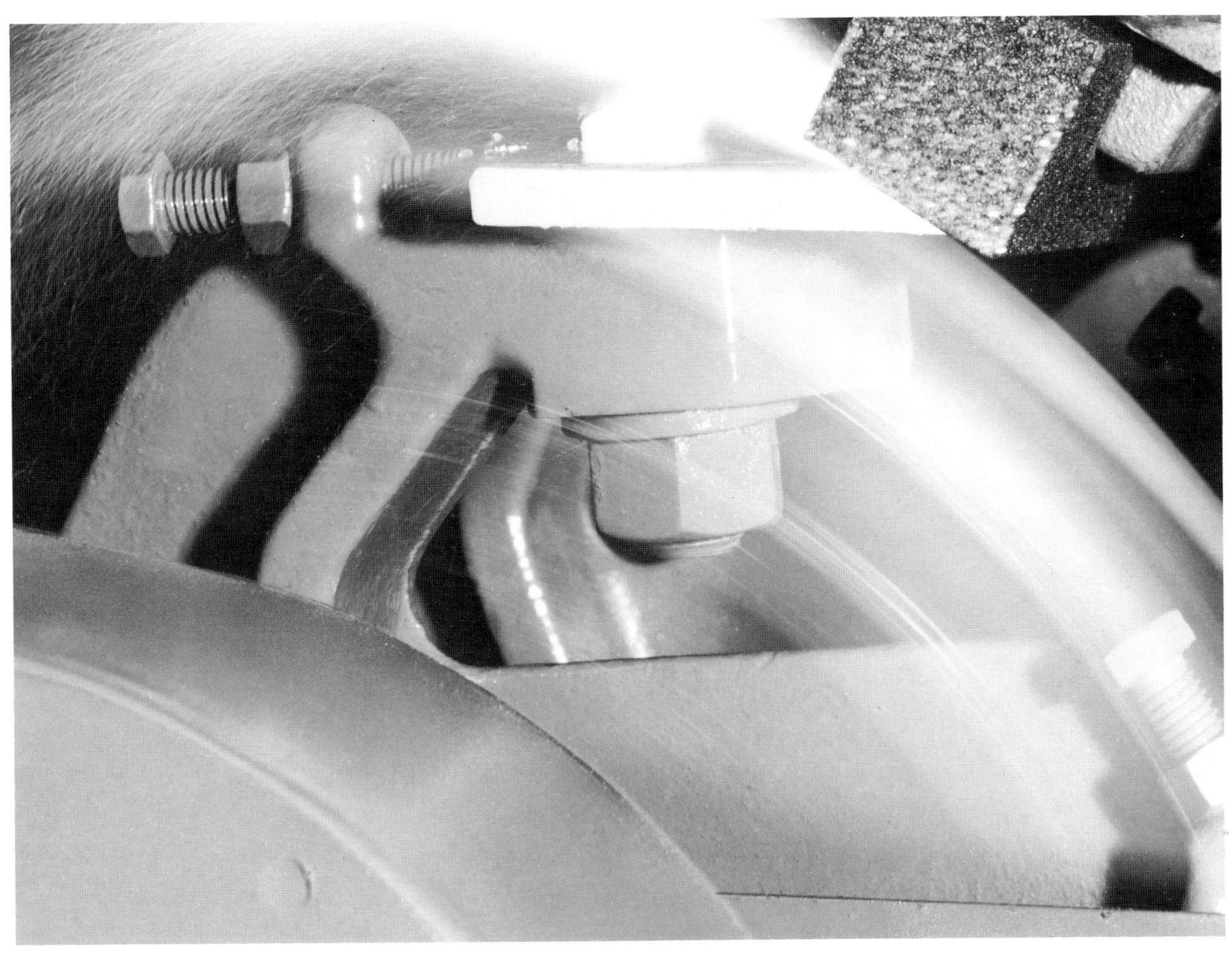

Fig. 23—High-Speed Cylinder Mounted Knives May Have Built-In Sharpeners

Fig. 24—Follow Your Operator's Manual For Safe Knife Sharpening Procedures

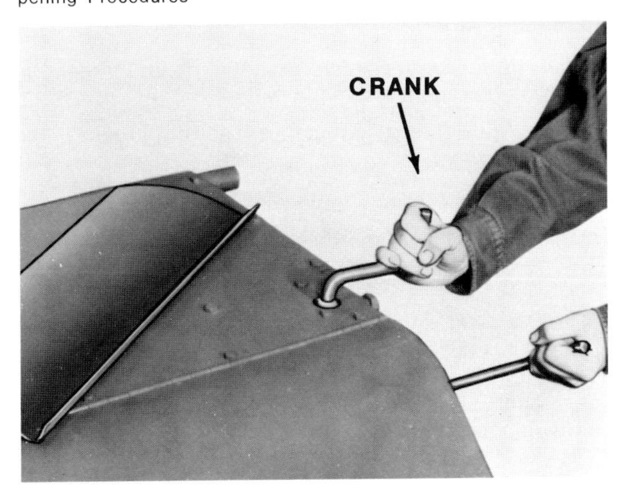

Fig. 25—Always Block Cutterhead Before Adjusting or Removing Knives

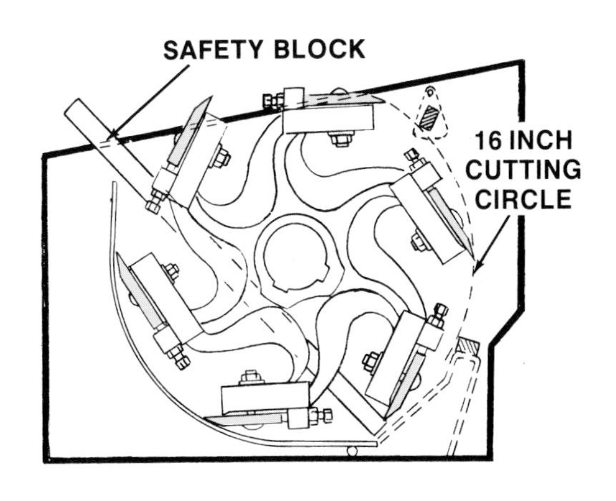

Fig. 26—Be Sure Others Stand Clear

Because high-dump wagons are often used with hay cubers for easy transfer of cubes into trucks, safety in unloading should also be considered. Always make sure no one is standing around or near the high-dump wagon when unloading (Fig. 26). Consult your operator's manuals for specific safety and operating instructions.

FORAGE WAGONS

Forage wagon safety is covered in Chapter 11. When hooking up wagons to balers or forage harvesters, always exercise great caution. If the wagon is equipped with a telescoping tongue, have the man hooking up the wagon stand clear until you have stopped backing up. He can then use the telescoping tongue to hook up the wagon.

BLOWERS

The most important safety precautions for working around forage blowers are (Fig. 27):

- *Stay clear of all moving parts*
- *Keep all shields in place*
- *Never climb into the hopper or use hands or feet to force forage into a blower.*

Failure to follow the above precautions could result in your entanglement in moving parts.

Before lubricating, adjusting, or unplugging the blower, always:

- *Disengage power*
- *Shut off the tractor engine and take the key*
- *Wait for the blower fan to stop*

If the blower is PTO-operated, fasten the blower hitch securely to the tractor drawbar. Vibration could cause the blower to move causing the PTO shaft to part and rotate dangerously; or it could bring the blower pipe down on you.

BLOWER PIPE

Another important aspect of blower safety is the assembly of the blower pipe, and safe procedures to solve plugging problems. Blower pipe should be assembled on the ground and then raised with a rope and pulley. Attach the rope to the mid-section of the pipe and make a half hitch near the top of the pipe.

Carefully raise the pipe into position, tie off the rope, and secure the pipe at top and bottom. Raise the deflector from inside the silo to eliminate the need for transferring the unit from the outside to the inside at the top. This reduces the chances of a fall. If using a telescoping pipe, always attach the smallest diameter section to the blower to reduce the chances of plugging.

SILO GAS

Silo gas is an important safety consideration. Chapter 11 covers the dangers of silo gas. When operating blowers in a shed next to a silo, be sure the shed is vented. Gases may accumulate inside the shed.

CHAPTER QUIZ

1. What are 3 factors in hay and forage operations affecting your safety?

2. (Fill in the blank.) Plan hay and forage operations downhill to avoid________________.

3. True or false? "Gathering reels pose no real hazard to the quick operator."

4. Multiple choice. Always handle cutterbars from the:

a) front

b) bottom

c) back

5. True or false? Windrower controls consist of only two steering levers.

Fig. 27—There Are Many Moving Parts in a Forage Handling System. Look For Hazards

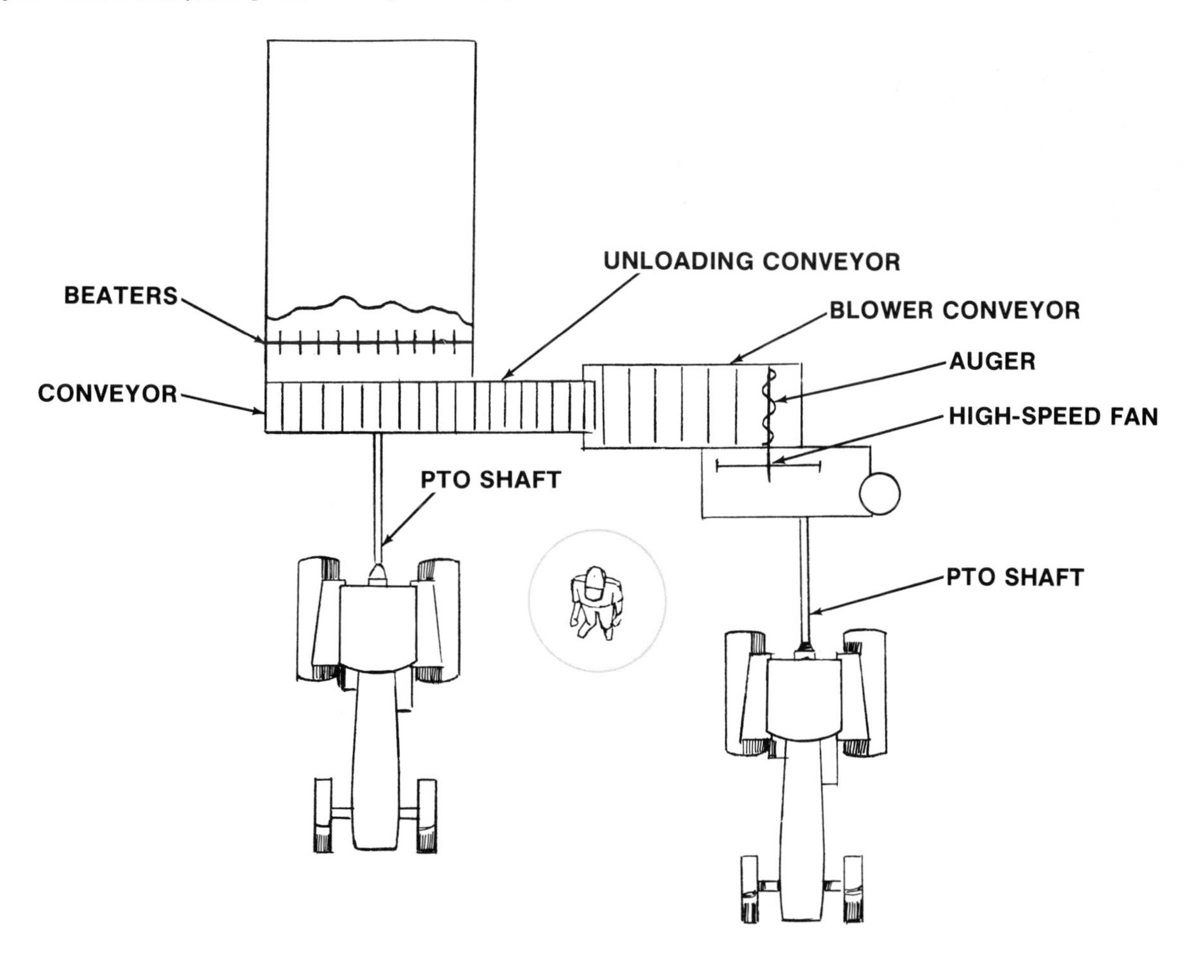

6. (Fill in the blanks.) Safe procedure for any machine before dismounting is to disengage the power and turn off the ______________________________ .

7. True or false? "Bale ejectors are safe to work on if there is no bale in the ejector."

8. Why are bale accumulators unsafe to step or ride on?

9. True or false? "The rear doors of stackers are not dangerous to anyone during unloading."

10. (Multiple choice.) During operation, fasten PTO-operated blowers to:

a) special stand

b) the silo

c) tractor drawbar

9 Target: Grain Harvesting Equipment

Fig. 1—Operators Make Or Break Profit When It Comes To Using Safe and Efficient Grain Harvesting Methods

INTRODUCTION

Good safety habits are a must for anyone who operates combines, corn pickers, and other grain harvesting machines. Engineers develop safer machines every year, but safe *use* of the machine is still up to you.

The major causes of accidents are factors the designer has no control over.

Interviews with victims who suffered amputations, skeletal damage, or partial loss of sight show that the most frequent causes of accidents were:

- **Conscious acceptance of risk by the operator**
- **Hurrying to meet deadlines**
- **Preoccupation**

Accidents can't all be avoided by simply observing standard safety precautions and installing shields. Good design, proper training, safe operation, alertness, and good crop conditions go hand in hand to prevent accidents.

A safe and efficient grain harvest is a key to profit for a year of work and expense (Fig. 1). Most cereal crops and beans are harvested with combine harvesters that cut, thresh, and separate the grain in one operation. Some crops, particularly cereal grains in the northern part of the United States, are cut and allowed to dry in a windrow before harvesting with a combine. Windrowers for grain are similar to those for hay. Chapter 8 covers safe practices for these machines.

This chapter covers safe practices for:

- **Machine preparation**
- **Field preparation**
- **Adjusting and servicing combines**
- **Driving the combine**
- **Attaching combine headers**

- **Combine field operation**
- **Moving combines on public roads**
- **Header attachments**
- **Corn heads**
- **Corn pickers**

MACHINE PREPARATION

Grain harvesting machines have many drives and other moving parts that need regular adjustment and maintenance. Wrong row spacing on a corn head increases operator fatigue, which can lead to accidents (Fig. 2).

Start getting these machines ready for harvest in the off season, or at least several weeks before you'll be using them. An early start is necessary because:

- *It takes time to get these machines into efficient and safe operating condition.*
- *Lead time may be involved when ordering replacement parts.*
- *Other field operations near harvest time compete for time needed for machine preparation.*
- *Major adjustments may be necessary to convert a machine to a particular crop.*
- *Special attachments may have to be mounted.*

Fig. 2—Adjust Corn Head Row Spacing Correctly To Avoid Operator Fatigue

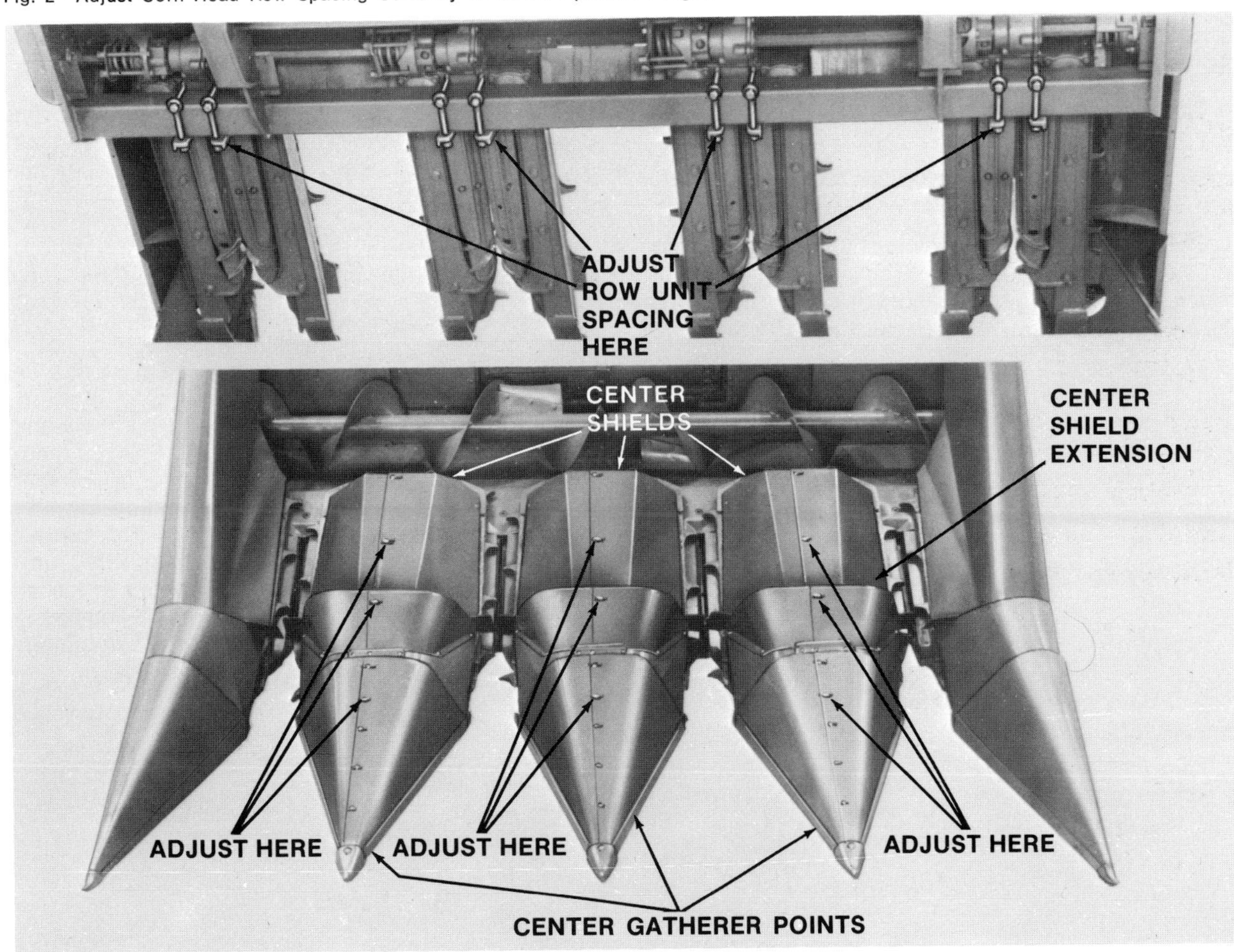

Make sure that all PTO covers, safety stands, and shields are installed on the machine before taking it to the field.

Refer to the FMO manual on Combine Harvesting and your operator's manuals for getting these machines into top condition for a safe harvest (Fig. 3).

Fig. 3—Read The Operator's Manual For Recommended Settings

FIELD PREPARATION

Preparing a field for harvest begins long before the crop is mature. It's easy to see obstacles in a field during tillage and planting operations, but they might not be visible at harvest time (Fig. 4). For example, a stub post in an old fence line may not cause any trouble during planting, but might cause severe damage or personal injury later if hit by a combine. It could also cause costly down time and put extra stress on *you,* the operator, perhaps leading to personal injury.

STONES

Stones in the field can be hazardous during harvesting operations, especially in:

- *Direct-cut crops that must be cut low (soybeans, for example)*
- *Corn, where the row-gathering units may strike or pick up stones*
- *Windrowed crops, where stones can be taken into the machine with the windrow*

Large stones or stumps can cause upsets, especially on slopes. Mark posts, stones, stumps, and other obstacles that can't be removed at planting time, with a tall pole or stake so their location can be seen clearly in the mature crop. You can then drive around the obstacles, avoiding damage to equipment and sudden stops or upsets that could cause personal injury.

Fig. 4—When Planting, Check Field For Obstacles That May Be Hard To See During Harvesting

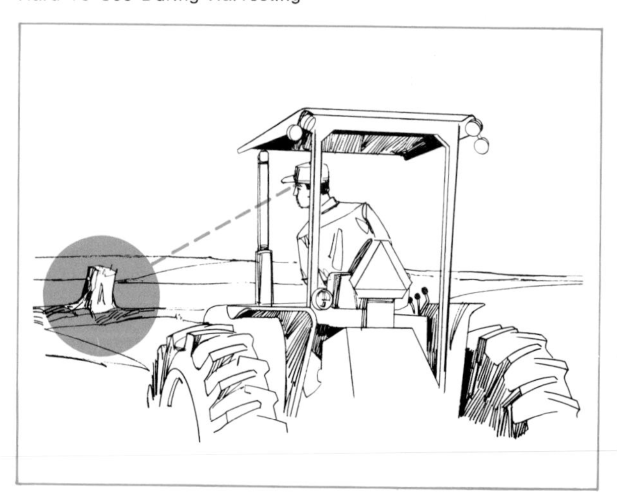

DITCHES

Ditches cause special problems. What was a safe distance from ditches at planting time may not be safe during harvesting operations. Heavy rains can undercut banks or form new ditches. An eight-row planter may have been used, which would have kept the heavy tractor away from the bank. But if the crop is harvested with a combine with a four-row corn head, the combine's weight will be much closer to the bank than the tractor's was (Fig. 5). This problem is even more serious when the ditch doesn't run at right angles to the rows because turning at the end of some of the rows may require the combine to travel too near the edge of the bank. In general,

Fig. 5—Combine Is Too Close To The Bank—A Dangerous Situation

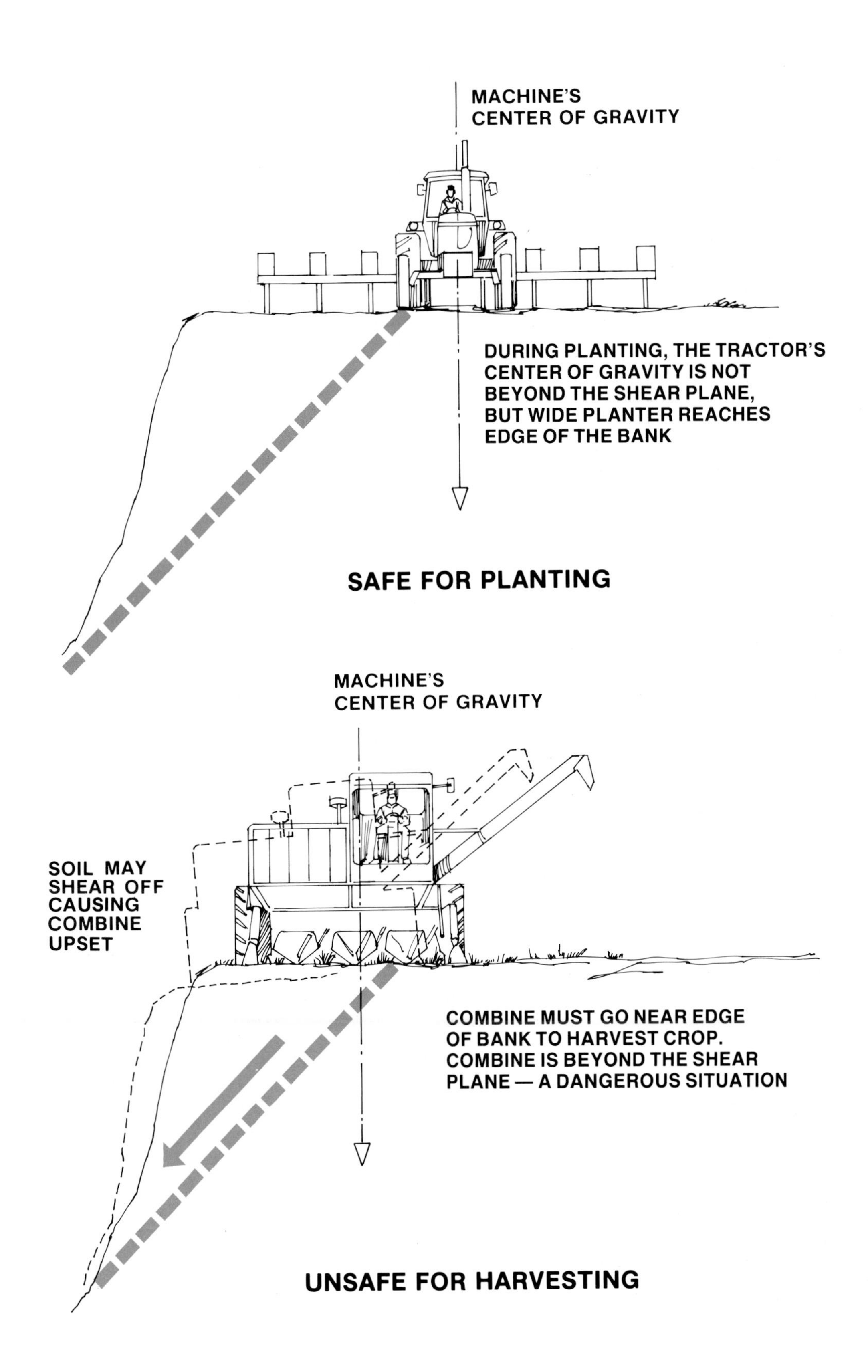
MACHINE'S
CENTER OF GRAVITY
DURING PLANTING, THE TRACTOR'S
CENTER OF GRAVITY IS NOT
BEYOND THE SHEAR PLANE,
BUT WIDE PLANTER REACHES
EDGE OF THE BANK
SAFE FOR PLANTING
MACHINE'S
CENTER OF GRAVITY
SOIL MAY
SHEAR OFF
CAUSING
COMBINE
UPSET
COMBINE MUST GO NEAR EDGE
OF BANK TO HARVEST CROP.
COMBINE IS BEYOND THE SHEAR
PLANE — A DANGEROUS SITUATION
UNSAFE FOR HARVESTING

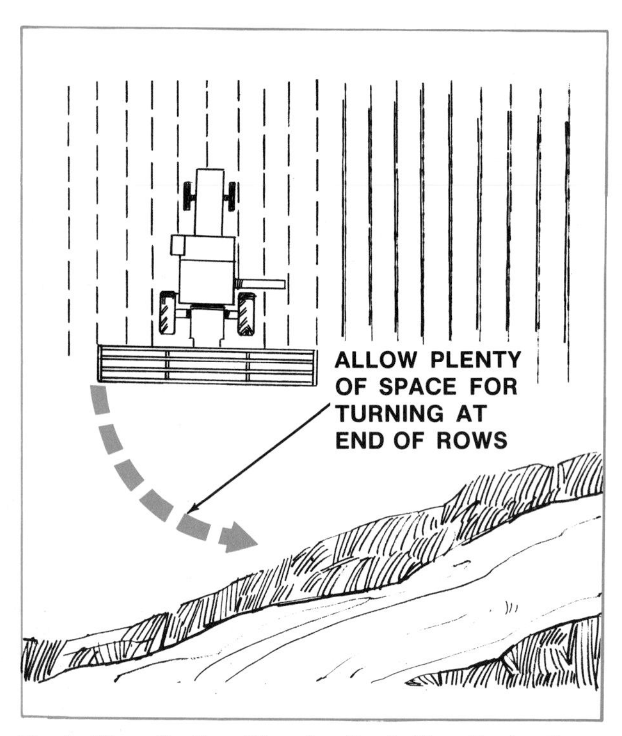

Fig. 6—When Planting, Allow One-Fourth More Turning Space Than That Required By Your Largest Piece Of Equipment

allow at least one-fourth more turning space than that required by the largest piece of equipment you'll be using (Fig. 6).

WEEDS

Weeds cause problems during many grain harvesting operations. They can wrap around rotating drives and plug machines, causing costly delays (Fig. 7). You may become tired and be more likely to take chances—like dismounting to unplug the machine without shutting off the power. A sound weed control program can give you a safer harvest and a higher yield.

PLAN FOR HARVEST

1. *Remove as many posts, stones, stumps, and other obstacles from the fields as possible.*
2. *Mark any remaining obstacles clearly.*
3. *Plant the crop to allow extra turning space near ditches.*
4. *Check for undercut embankments and new ditches before harvesting.*
5. *Prepare the edges of the fields for smooth turning and maneuvering of harvesting equipment.*

Fig. 7—Weeds Can Create Hazards For Operators. Use An Effective Weed Control Program

Fig. 8—Service Combine Before Going To The Field

6. *Plant crops with row spacing that matches row spacing of the gathering unit of the harvester.*

7. *Make sure the number of planter rows is a whole-number multiple of the number of rows on the harvester.*

8. *Use adequate weed control practices.*

ADJUSTING AND SERVICING COMBINES

Only a properly serviced machine is safe and efficient to operate. Faulty wiring can cause fires. A loose traction drive belt on a self-propelled machine can cause loss of control on a hill. Complete servicing and maintenance before starting work for the day, and as required during the day.

1. *Adjust and service the combine before going to the field (Fig. 8).*

2. *Lubricate and fuel the machine after it cools in the evening.* Rust may form on moving parts or moisture may condense in the fuel tank if a combine is left out overnight without servicing.

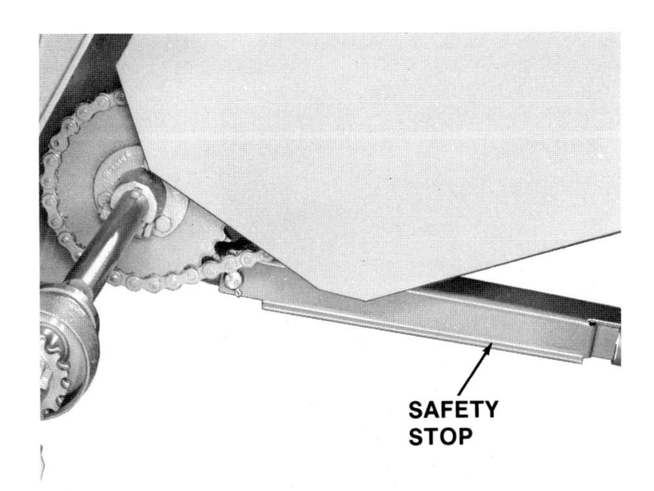

Fig. 9—Always Latch The Header Safety Stop Or Put Blocks Under The Header When Working Under It

3. *Block all parts before working under a header.* You may need to raise the header to work on the cylinder and other parts. Do not rely on the hydraulic system. **Be sure the header latch is fixed in place or that proper blocking is used** (Fig. 9).

4. *Shut down both the machine and the engine when working on harvesting machines.* This is especially important for self-propelled combines. They have one or more main belt drives that turn whenever the engine is running, even when other components are shut off, like the variable speed belt drive to the traction transmission shown in Fig. 10. Any moving parts

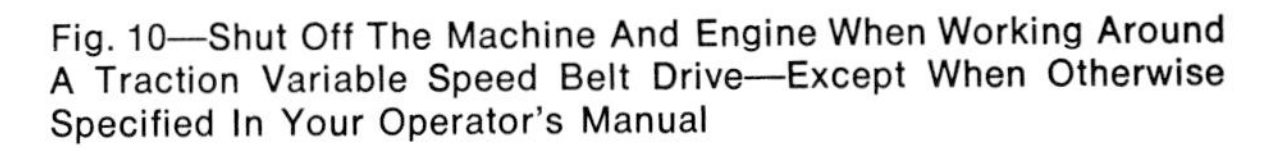

Fig. 10—Shut Off The Machine And Engine When Working Around A Traction Variable Speed Belt Drive—Except When Otherwise Specified In Your Operator's Manual

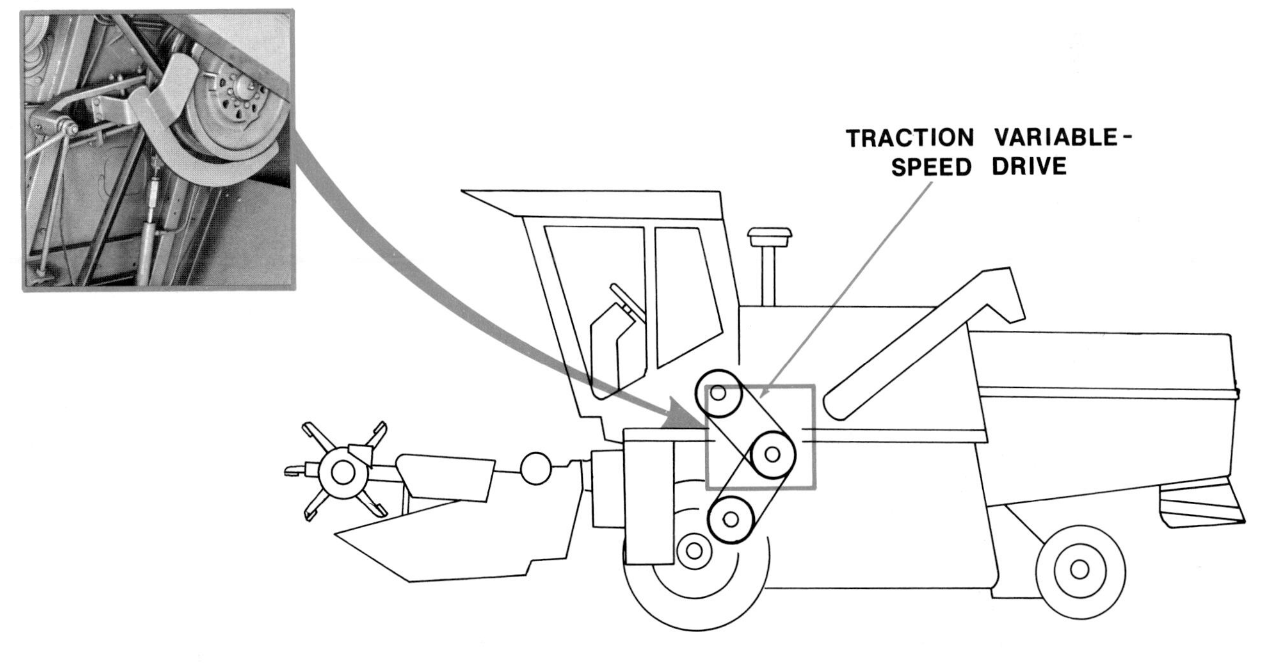

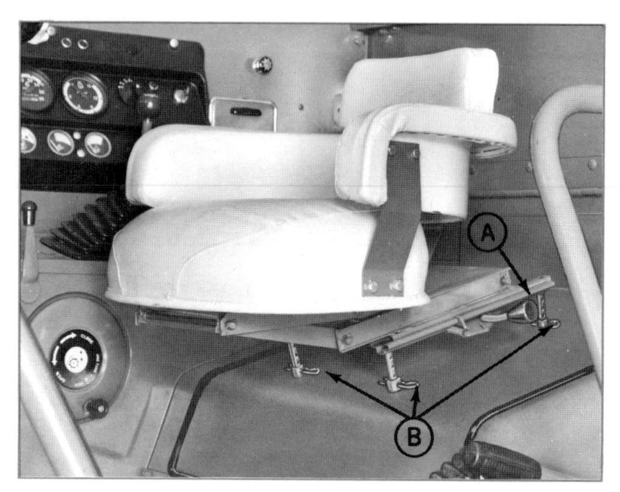

Fig. 11—Use Seat Adjustments For Comfort

can be hazardous for anyone servicing the machine. Follow the rule: Disengage all power, shut off the engine, and take the key.

5. *Don't be tempted to remove weeds from a rotating shaft or from the header while the separating mechanism is running.* However, changing the speed of the variable speed cylinder or fan must usually be done while they're running. Shielding is usually provided for safety. But don't make any other adjustments with the machine running, even if it seems convenient. Follow instructions in the operator's manual.

6. *If two or more people are needed to position, adjust, or service a machine, make sure everyone knows what's being done and how to communicate with others.* Have the person on the ground use standard hand signals when helping you position the combine (see Chapter 1). And be sure you know where everyone is before starting or moving the combine.

7. *When working on a harvesting machine, always keep in mind that someone could show up unexpectedly.* If you're working on the inside of the machine, remove the ignition key and post a sign saying that the machine is being serviced. Keep everyone away from the controls while you're working on the combine unless you need their help.

8. *Replace all guards.* Clean platform and steps. Use handrails when climbing into the cab.

DRIVING THE COMBINE

OPERATOR'S SEAT

The operator must be comfortable and have the controls within easy reach. Adjust the seat to your height and reach.

All combines have adjustable seats that can be raised and lowered and moved forward and rearward. Some can be folded up or moved out of the way so the operator can stand while driving. A typical seat mounting is shown in Fig. 11.

Fig. 12—Adjust The Steering Column To Fit Your Arm's Reach

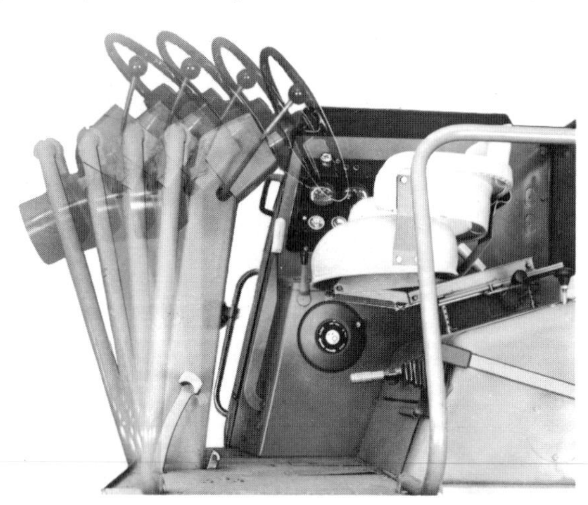

Fig. 13—Orientation For A Self-Propelled Machine

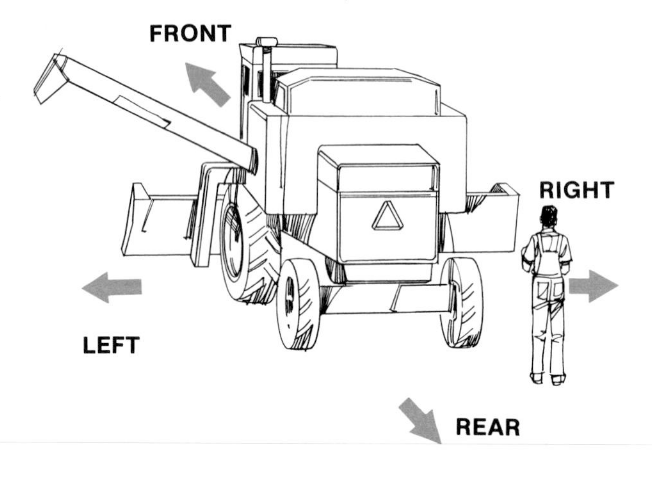

STEERING

Most combines also have an adjustable steering column for individual arm lengths or for standing position. Adjust the steering column and move the column (Fig. 12).

The wheels that steer many combines and other self-propelled machines are mounted on the rear of the machine. Right and left on the machine are designated as shown in Fig. 13. Be careful when making a turn near obstacles. The rear of the combine may swing around and strike something.

Most combines have power steering which makes guiding the combine down rows easy. However, the rear end swings around so far the back of the machine can hit fence posts even when the header has plenty of clearance. Leave at least one-fourth more room for turning at the ends of fields than necessary for normal turns (Fig. 14 and Fig. 15).

BRAKES

Two brake pedals brake individual drive wheels (Fig. 16). When used separately, these pedals assist in turning. When used together, they stop the combine in a straight line.

To stop the combine, press on both brake pedals. Uneven application of brakes will cause the combine to swerve to one side, especially at high speed on the road. This could result in an upset. Lock the brake pedals together for highway driving.

When using the brakes to make sharp turns, slow down to a safe speed. Start turning the steering wheel *before* applying the brake to assist turning. Otherwise, the rear wheels will skid sideways and the turn will be more difficult.

Fig. 16—Two Brake Pedals Are Provided That Control Individual Drive Wheels

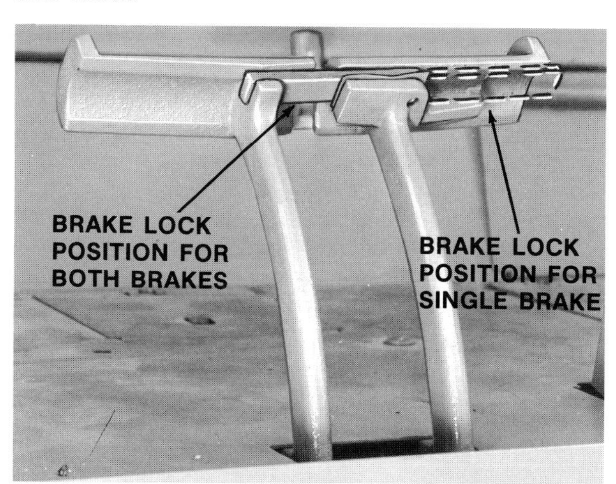

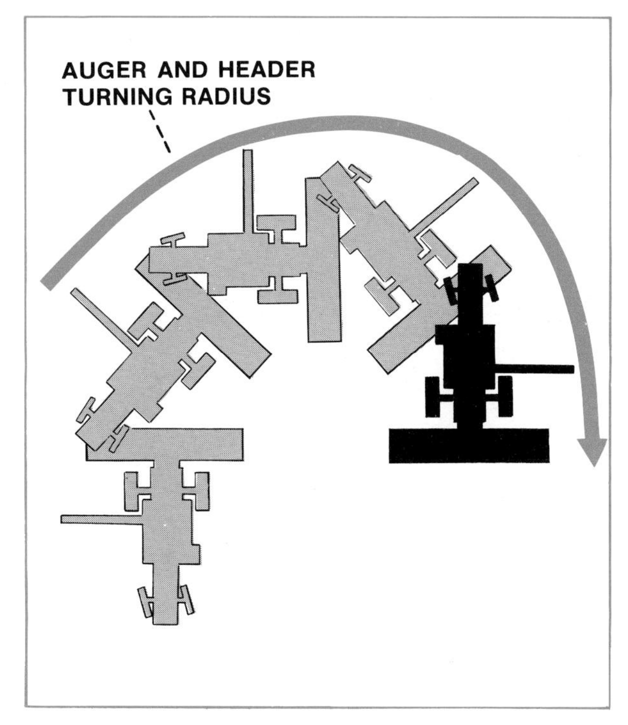

Fig. 14—Turning Machines With Rear Steering Requires Practice

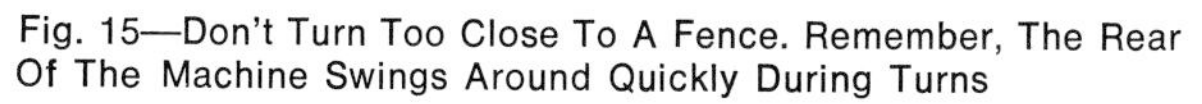

Fig. 15—Don't Turn Too Close To A Fence. Remember, The Rear Of The Machine Swings Around Quickly During Turns

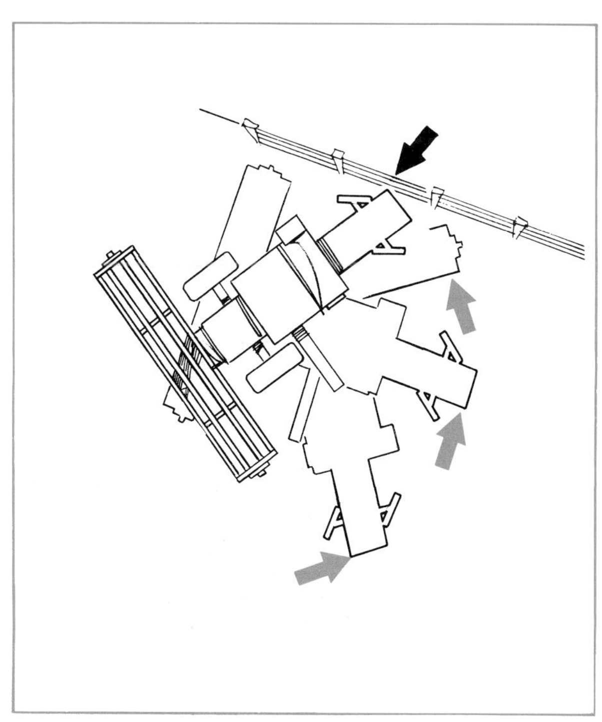

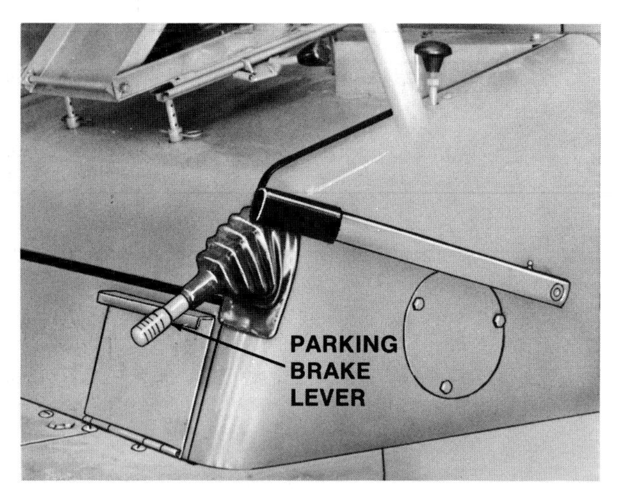

Fig. 17—Use The Parking Brake

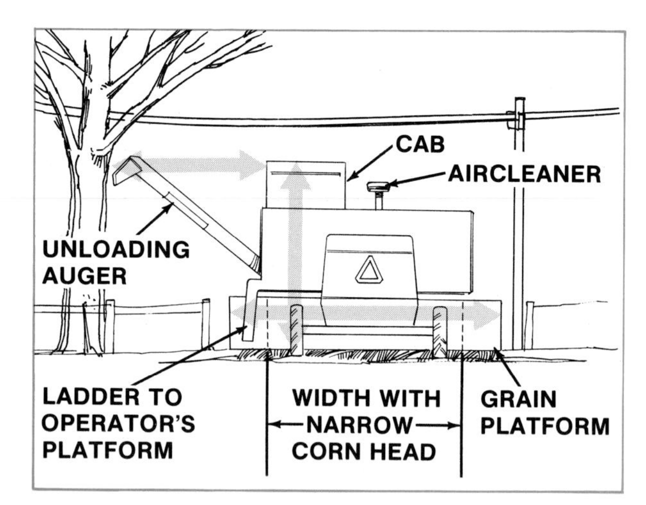

Fig. 18—How Big Is Your Combine?

Always reduce travel speed before applying brakes. Quick stops can result in the combine nosing forward.

Parking Brake

Some machines have a parking brake to lock the wheel brakes so the machine can't move when left unattended. Never attempt to move the machine with the parking brake engaged. Some combines have a warning light that tells when the parking brake is engaged.

To engage the parking brake (Fig. 17), pull the lever up. To disengage, push the lever down.

Fig. 19—A Large Machine Blocks The Operator's View Of Some Areas

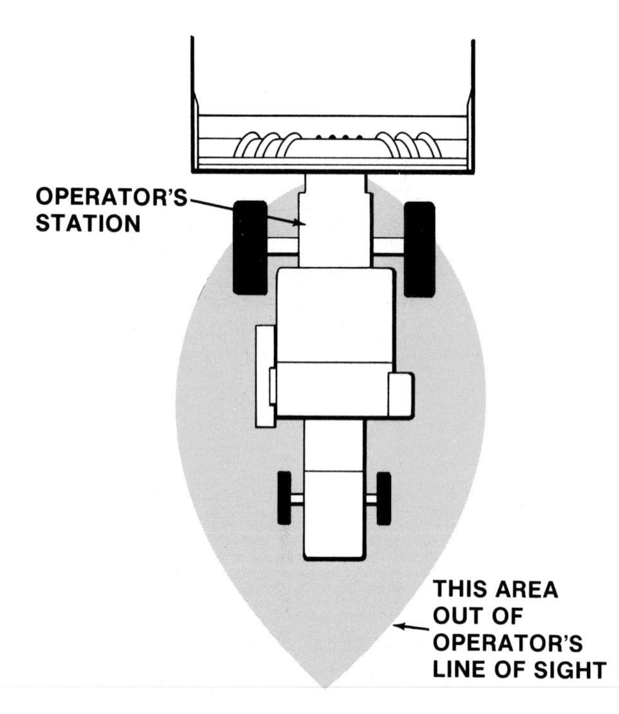

SHIELDING

Combines have several cover plates, drive shields, and guards to help make them as safe as possible to operate. Lock all of these devices into place before using the combine.

LADDERS AND PLATFORMS

Ladders and platforms provide access to the operator's station and service areas of the combine. Keep the steps and walking surfaces free of grease and dirt. Use handrails for safe mounting and dismounting. Be careful when climbing steps or walking on platforms on frosty mornings. Frost-covered sheet metal is slippery. And don't use the operator's platform as a place for storing loose tools or lunch boxes.

The ladder leading to the operator's platform on some self-propelled combines may be hinged for lifting during operation or transport. Fasten it in place when lifted so that it doesn't fall on someone approaching the combine. Lock the ladder down to prevent falls when using the ladder.

COMBINE SIZE

Know how big your combine actually is. The header is usually the widest and lowest part of the combine

(Fig. 18). Don't ram it into rocks and stumps. And don't get too close to fences and trees.

Look for parts on the sides of the combine that extend beyond the header, especially when you're using a narrow corn head. Know the position of the unloading auger and the ladder to the operator's platform.

The highest point on a combine is usually the cab or grain elevator, but could be the radiator screen, exhaust pipe, air cleaner or CB antenna. Check to see which part extends the highest on your machine. Make sure there's plenty of clearance under power lines and trees.

Because of a combine's large size, visibility to the rear is usually limited (Fig. 19). Make sure everyone is clear before backing or turning the combine. Blow the horn before starting the engine.

ATTACHING COMBINE HEADERS

The most important factor in the safe and easy attachment of headers is their position. Use blocks and jacks. Follow the operator's manual to position headers for the right height and angle of attachment. Here's how to attach a header:

1. *Align the combine and drive in slowly.*

2. *If you have to move the header to complete the connection, stop the combine and turn off the engine.*

3. *Use a long pry bar.*

4. *Locate pinch points and keep your hands away from them.*

5. *Insert all connecting pins and bolts and connect all drives and hydraulic lines before moving the combine.*

6. *Raise all jacks and stands to their highest positions.*

7. *Remove all blocking materials and store them to use when you remove the header.*

8. *Add extra weight to the rear wheels if needed for stable machine operation.* Check your operator's manual.

The best way to have the header in the right position for attachment is to plan for reattachment when you're removing it. Here's how:

1. *Situate the header so that both the header and the combine are level and on firm surface, preferably concrete (Fig. 20).*

2. *Allow enough room for the combine to be driven straight into the header.*

3. *Use blocks or stands to hold the header firmly in the correct position.*

4. *Release all pressure in hydraulic lines before disconnecting them.*

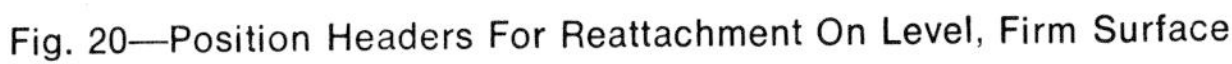
Fig. 20—Position Headers For Reattachment On Level, Firm Surface

Fig. 21—Strive For Uniform Feeding To Prevent Plugging Problems

5. *Clean out any trash or dirt between the header and the combine header mounting.*

COMBINE FIELD OPERATION

Proper machine operation is important for safe and efficient harvesting. This means reducing harvesting problems and dealing safely and effectively with those that do occur.

FEEDING

Uniform feeding of a crop in good condition provides a smooth flow through the combine and results in fewer breakdowns (Fig. 21). Non-uniform feeding can cause plugging.

For uniform feeding:

1. *Use good tillage and planting practices to help produce a uniform crop.*

2. *Cut the crop evenly.*

3. *Make sure the header is the right size for the combine.*

4. *Select the proper header speed and ground speed so the crop will flow freely across the header toward the threshing cylinder.*

The crop should be dry enough for good threshing without wrapping around the cylinder or other rotating parts. Early harvesting gives weeds less time to develop into a problem for the combine, but late harvest can sometimes be beneficial with weedy crops, especially where frost comes early enough to kill the weeds.

COMBINE CAPACITY

A combine's harvesting rate can be limited by separator capacity size, size of feeding and threshing mechanisms, available engine power, and ground conditions. If you try to increase the harvesting rate above any of these limits, you could plug the machine or cause a breakdown.

Do not speed, especially on rough ground. Make sure your combine is under control at all times.

PERFORMANCE CHECKS

Don't dip your hand into the bin when the machine is moving. There may be augers running below the surface of the grain. And never get into the bin while the combine is running.

To check tailings, use the inspection ports provided on most combines, where you can safely and conveniently observe the material being returned from the shoe to the threshing cylinder. Many self-propelled combines have inspection ports near the operator's station. Use the ports provided for that purpose. Never try to collect a tailings sample by opening the door at the bottom of the tailings elevator where fingers could get into the elevator.

Checking grain losses at the rear of the machine can be dangerous. An electronic grain loss sensor is safer and more accurate. But if you have to check grain loss by hand, use a container like the one in Fig. 22 to collect a small sample. Don't walk ahead of the rear wheel or too close to the side of the machine. Checks should be made only on smooth ground while the combine is moving straight forward. Turning of the machine or bumpy ground could cause the checker to fall and be injured. The operator and checker should be fully aware of the procedure and time of the checks.

Disconnect the straw spreader or chopper before making performance checks at the back of the combine (Fig. 22). The straw chopper has to be moved out of the way or removed to make the checks. It's dangerous to collect samples at the back with the chopper running, and it rethreshes the crop, which makes evaluation impossible.

Fig. 22—Disconnect Straw Spreader Before Checking Combine Performance To Avoid Being Struck By Flails

PLUGGING

Plugging the header, cylinder, or separator causes increased stress on the operator and subjects him to the hazards of unplugging the machine. Cylinder plugging is usually the result of weeds or overloading.

You can sense the load on the cylinder by checking the tachometer. Load on the engine is a good indication of cylinder load. Slow down the combine and avoid taking in slugs. Remember, the chance of plugging increases as cylinder speed decreases. After you've unplugged the cylinder a few times, you should be able to tell when it's about to overload.

Running the combine separator slower than recommended in the operator's manual will result in less effective harvesting, and ultimately, plugging of the separator and conveying mechanisms. Adjust and equip your combine properly for the particular crop being harvested to reduce plugging problems.

If the combine does plug, use the reverser to clear it. If it will not clear, stop the combine mechanism as quickly as possible. Shut the engine off and remove the key (Fig. 23). Then, to unplug the machine:

1. *Make sure the separator drive is not in gear.*

2. *Take off the inspection plates so you can locate the plugged material easily.*

3. *Remove the plugged material, using tools or hands protected by heavy gloves.*

4. *Get your hands and fingers out of the mechanism before anyone attempts to turn the drives.*

5. *Turn the main drive shaft or cylinder with tools provided for the purpose.*

6. *After the mechanism has been freed, remove all tools from the inside of the combine.*

7. *Remove tools used to turn the drives.*

8. *Replace all inspection plates.*

Fig. 23—Shut Off Engine And Remove The Key Before Unplugging The Combine

Fig. 24—Skip Wet Areas In Fields To Avoid Getting Stuck. Return Later When Ground Is Dryer And More Firm

ADVERSE OPERATING CONDITIONS

Down Crops

Use the right equipment on the combine for down crops to avoid damage to the combine and reduce operator fatigue. Special attachments are available for down crops, including pickup reels, special dividers, and gathering devices.

Be especially alert for ditches, washouts, and holes in the field. A down crop covers them.

When handling down crops, reduce speed so the long straw will be taken in slowly and the header will feed evenly. Slower speed will keep the straw from tangling and bunching up and reduce the chances of plugging. You'll finish faster and safer if you can avoid plugging.

Fig. 25—Remove Dust And Chaff From Engine Compartment Area To Maintain Cooling Efficiency And Reduce Fire Hazard

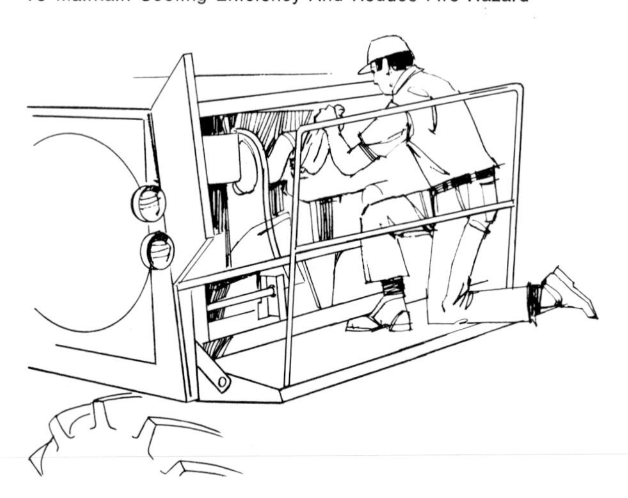

Muddy Fields Or Weedy Crops

When you harvest, skip the muddy and weedy areas in your fields. Return to them later when there's less pressure on you and the field is dryer and firmer (Fig. 24).

Special equipment, such as powered rear wheels, wide flotation tires, and half tracks helps harvest muddy fields. When installing these attachments, block the machine carefully and do the job on a hard surface.

Keep steps, platform, and controls clean when working in mud, snow, or ice. Reduce slipping hazards. Keep foot pedals and other controls clean, too.

Dry Crop Conditions And High Temperatures

Check for overheated bearings that could burn out and start a fire in dry chaff. Remove excess dirt and chaff from engine, particularly around the exhaust system (Fig. 25). Severely overloaded or slipping belts can cause fires. Watch speed indicators to catch slipping belts, and service all belt drives regularly.

Operators who have experienced combine fires know the value of a good fire extinguisher. Just one fire costs more than all the fire extinguishers you will ever need. Use a multi-purpose dry chemical extinguisher.

A low-hanging exhaust system or an engine backfire from a car or truck can start a fire in dry stubble fields. Take precaution if you drive a car or truck into a field. Catalytic converters start a lot of fires in wheat fields. Drive forward — never back up with a truck or car. The exhaust pipe can load up with straw, creating a fire hazard. And after a trip with heavy loads or at high speeds, let the vehicle cool for a few minutes before entering a dry field.

Fig. 26—Hillside Combines Automatically Level Themselves On Steep Slopes

Operating On Hillsides

Operating any machine on a hillside demands good equipment and good operator control. The hillside combine is designed for safe harvest on very steep slopes, such as those in the Pacific Northwest. These machines level themselves automatically (Fig. 26). However, level-land combines *may* be operated safely on many minor slopes used for crop production if these practices are followed:

1. *Operate the combine smoothly.* Avoid quick changes in speed or direction, and don't brake or turn sharply, even at low speeds. Remember that combines have a high center of gravity and can be upset.

2. *If space permits, make a large loop when turning to go up a hill (Fig. 27).* Turning and braking to go uphill after following along a hillside can make the drive wheels spin out and cause side slipping and possible loss of control of the machine.

3. *Don't attempt to shift gears when going up or down a hill because most transmissions cannot be shifted back into gear while the machine is moving.* You have much better continuous control when you change speed slowly with the variable speed traction drive than when you brake or shift gears.

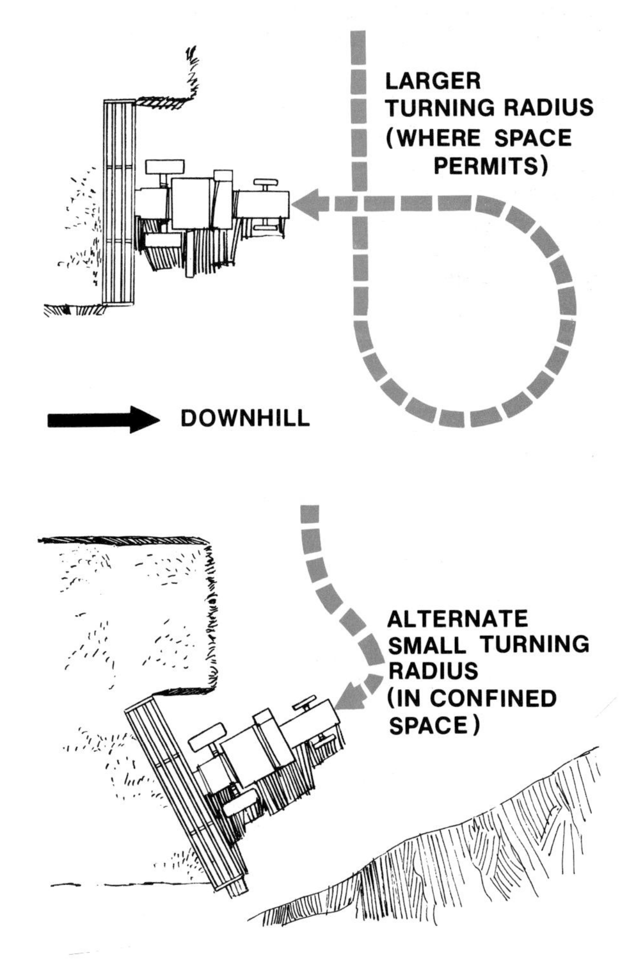

Fig. 27—Where Space Permits, Make A Large Loop When Turning On A Hill

Fig. 28—Watch Out For Pinch Points When Positioning The Unloading Auger

4. *Select a transmission gear to operate in mid-range when combining on steep slopes.* If the drive slips when the variable is at its lowest speed position, increase the variable speed slightly to maintain control. Adjust the traction drive belt periodically, and recheck it if there is any indication of slippage. Check its tightness throughout the speed range.

5. *Maintain a low center of gravity for machine stability on side hills.* The more grain in the grain tank, the higher the center of gravity on most combines, so don't overfill the tank. And don't use a tank extension above that recommended by the manufacturer. Such extensions can make the equipment top-heavy, and the added weight could put too much strain on the combine axle.

6. *Be alert for ditches and holes.* If the wheel on the downhill side drops, it could cause an upset.

These practices apply to hillside combines too. Hillside combines can be used safely on much steeper slopes than level-land combines. Follow these rules for using the leveling system on hillside combines:

1. *Change slope slowly.* Allow time for the hydraulic system to level the combine. Upsets can be caused by allowing the leveling system to react too quickly to rapid changes in slope.

2. *Don't operate on slopes steeper than those the leveling system was designed for.* Slopes that are too steep can cause upsets for hillside combines just as for level-land machines. Most combines have signals to indicate when the leveling limit has been reached.

3. *Use the manual control wisely.* Tilt the machine in the right direction for leveling. Prepare for changes in slope.

Unloading The Grain Tank

Position the auger for unloading before beginning to harvest. When moving the auger, be sure that both the swing path for the end of it and the connector joint are free of obstructions (Fig. 28). Keep your fingers away from pinch points in the connector during positioning. Keep the combine on level ground so the weight of the auger can be handled with minimum effort. This is especially important for the long unloading augers used with wide combine headers.

No one should be in the grain tank when the combine engine is running. If there's room for grain to get to the unloading auger, there's room to get a hand or foot caught in the auger.

Fig. 29—Use The Mirror When Unloading On-The-Go

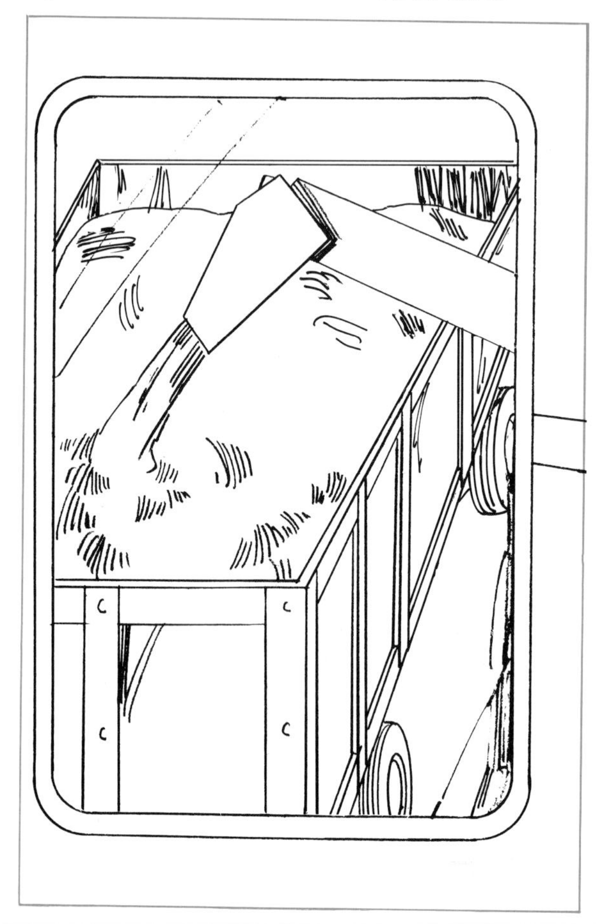

If grain bridges during unloading, stop the auger and shut off the engine before trying to free the grain. Use a small shovel or stick to break the bridging.

Never use your hands or feet to remove trash or to push the last bit of grain into the unloading auger. You will get caught in the auger and be pulled in before you can react. Stop the machine and the engine, take the key, and use a broom.

Unloading On-The-Go

If the combine grain tank is emptied on-the-go, cooperation between the combine operator and the hauler is essential to avoid accidents between the combine and the truck or wagon. The hauler is responsible for positioning the truck or wagon for unloading without getting too close to the combine. He must be prepared for unexpected stops of the combine and leave plenty of room for the combine to turn at the ends of the field. The combine operator must stop unloading in time for the hauler to turn corners and drive around obstacles safely.

Unloading on-the-go should only be used:

- *When the crop is standing and uniform*
- *When the combine is being operated at less than maximum capacity so that a constant ground speed can be maintained*
- *When the combine is traveling in a straight direction and the field is relatively flat*
- *When there is plenty of room for unloading before reaching the edge of the field, a fence, or other obstruction*

Have a rear-view mirror positioned so you can see the end of the unloading auger and the truck or wagon at the same time (Fig. 29).

Straw Choppers And Spreaders

Straw, stalks, and other flying material thrown from choppers can injure nearby people. Make sure everyone is away from the discharge area of the machine before and during operation.

Discharge from straw choppers can be dangerous, because the material is thrown at a high speed. Small stones and other heavy objects can be picked up and thrown by the machine. Corncobs can hurt, too.

Material from a chopper is usually thrown back before it spreads out, which means that walking along the side of a moving machine is less hazardous than standing in one place and letting it pass you (Fig. 30).

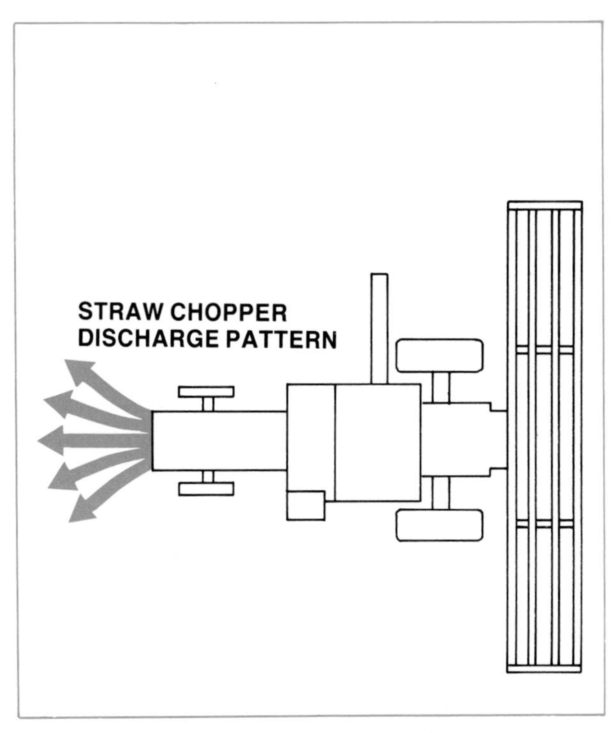

Fig. 30—Straw Chopper Discharge Pattern—Stay Clear!

Chopper rotors are heavy and may continue running after the rest of the machine has been shut off. Disengage all power, shut off the engine, and allow plenty of time for the rotor to stop before removing covers, inspecting, or cleaning. Keep your hands away from the discharge area of the chopper whenever the combine is running.

Check and service choppers regularly to keep them properly balanced. Imbalance in the rotor causes severe vibration and may lead to blade breakup, damage to the combine, or serious injury to nearby persons. Maintain balance by:

- *Removing excess dirt from the rotor*
- *Replacing defective blades*
- *Installing new blades in balanced pairs as directed in the operator's manual*

Most straw spreaders throw material out from both sides and the back of the machine. The speed of the flying materials is less than that for a chopper, but it

Fig. 31—Stay Away From The Straw Spreader Discharge When It's Operating

still can be dangerous, especially to eyes. Stay away from the straw spreader discharge when it's operating (Fig. 31). Disconnect or remove the chopper or spreader if you have to be near the back of a combine for performance checks. Eye protection is necessary for anyone working near a combine when the straw chopper or spreader is running.

Shutdown Procedure

When leaving a harvesting machine, even for a short time, make sure the header is down on the ground, supported by solid blocks, or locked in the up position with the safety latch. Shut off the engine, set the parking brake, and leave the drive in gear to keep the combine from moving. Remove the ignition key to keep children or unauthorized people from starting it up.

MOVING COMBINES ON PUBLIC ROADS

Moving a self-propelled or pull-type combine on roads and highways requires special care, especially for larger machines. Any time a combine is to be moved on roads, whether driven, towed, or hauled, it should be properly prepared (Fig. 32). To do this:

1. *Empty the grain tank to reduce weight and lower the center of gravity.*

2. *Move the unloading auger to the transport position.*

3. *When practical, remove the header if it's wider than the basic machine, and transport it on a truck or implement carrier.*

4. *Be sure SMV emblem, all reflectors, and lights are in proper working order and that they comply with state laws.* Check with the police or sheriff if you have any questions.

5. *Measure the height and width of the machine and write this information near the operator's platform with paint or a wax pencil for quick reference.*

Fig. 32—Preparing A Combine For Travel On Public Roads

Other preparation of the combine depends on road conditions, distance moved, and method of moving. Check the operator's manual for specific procedures. Consult local or state officials for regulations on moving large machines on public roads.

When driving a self-propelled machine on the road, you need good visibility both to the front and to the rear. Use rear-view mirrors.

Have a car or truck drive in front and behind the combine to warn traffic on public roads. Stay in your lane. Pull off, stop, and let traffic pass. Do not drive on the shoulder!

Fig. 33—As Combine Turns, Rear Of Machine Can Swing Into Traffic Lane. Watch All Turns—Wait For Traffic To Clear At Intersections

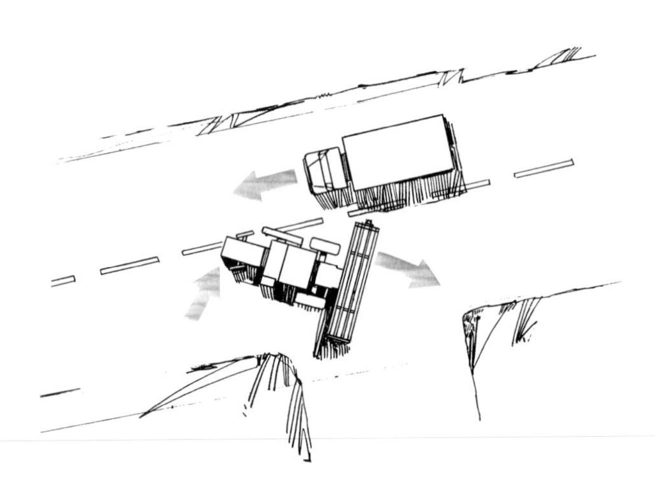

Because the wheels for steering are in the back, self-propelled machines often fishtail when turned quickly at transport speeds. Steering the combine to the right will whip the rear of the combine to the left, and vice versa. If you steer suddenly to the right when meeting oncoming traffic, the back of the combine will swing out into the path of oncoming traffic (Fig. 33).

If you slow or brake the combine too rapidly, you could lose some steering control (weight on rear wheels). This problem is most noticeable when driving a combine with a corn head or some other heavy header, or with the header raised high (Fig. 34). In this case, most of the combine's weight will be on the drive wheels. The following practices will reduce the problem:

1. *Install rear wheel weights as recommended in the operator's manual.*

2. *Keep the header as low as possible.*

3. *Use the variable speed drive or engine throttle to slow the machine.*

4. *Reduce speed before you need to apply brakes.* And always transport with the brake pedals locked together.

Towing pull-type combines at transport speeds can be hazardous because of side forces on the tractor when stopping too quickly. Side forces from slowing a combine quickly may cause a tractor to skid, especially on loose gravel (Fig. 35). Slowing down while turning a corner can cause jackknifing . So, slow down before the corner so the towed combine doesn't get out of control.

Self-propelled combines should be towed on roads only if it's absolutely necessary. And even though it may be easier to tow from the rear, towing from the front is preferred because there's better visibility and stronger structural members for fastening the tow chain. Follow instructions in your operator's manual.

When hauling a combine on a truck or trailer, fasten the combine at each drive wheel and at the rear. Lock the brake pedals together. Use wheel chocks to block each side of the drive wheels. Radiator screens, cab tops, or elevators may have to be temporarily removed to comply with maximum transport height requirements. Low truck trailers are available for hauling with minimum overall height. When hauling a machine, be sure you know the height, width, and weight of the load, and check local and state regulations. Watch for narrow culverts and bridges and low-clearance objects like power lines and bridges.

If the header is wide enough to cross the center line, a car should travel in front and behind the combine flashing its lights as a warning.

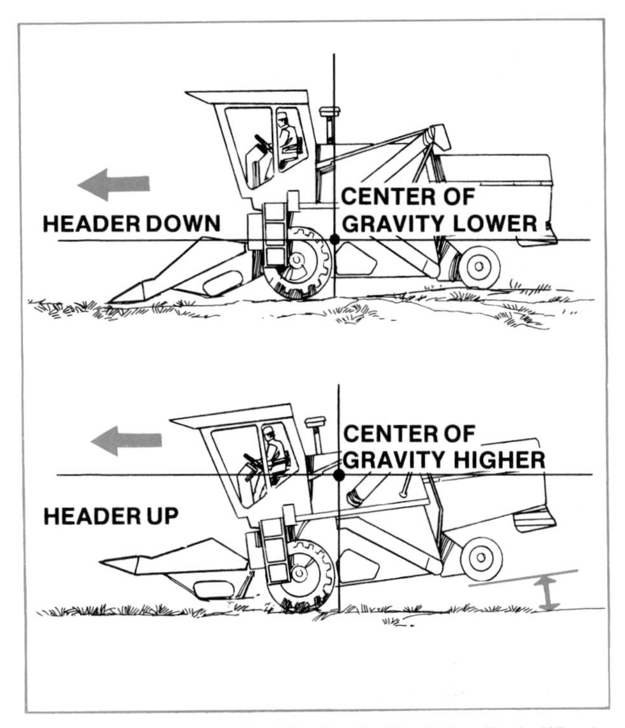

Fig. 34—Slowing Too Rapidly Tends To Raise Back Wheels When The Header Is High

Fig. 35—Side Forces From Slowing A Pull-Type Combine Can Make A Tractor Skid

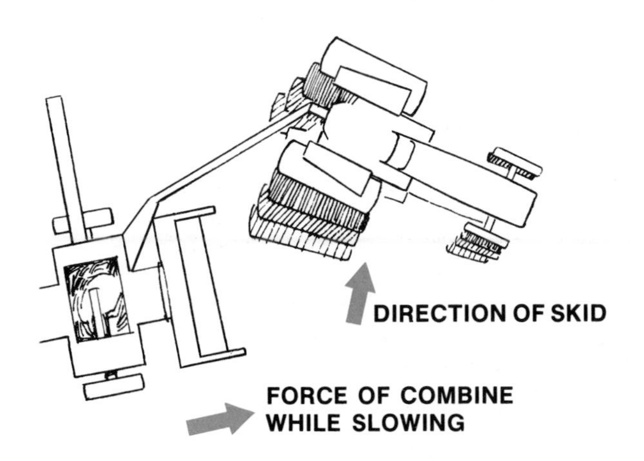

HEADER ATTACHMENTS

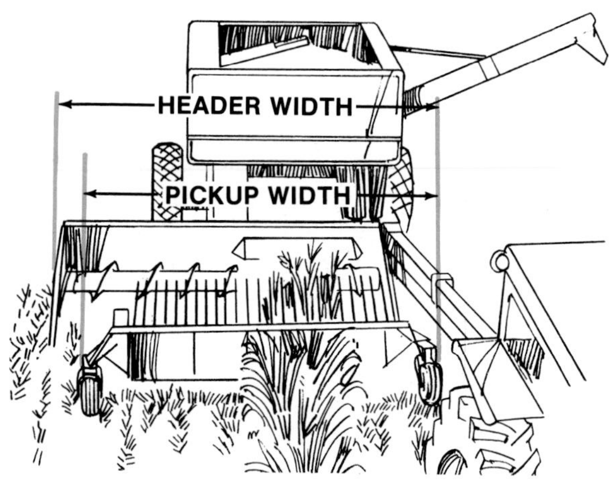

Fig. 36—Watch Both Machine Widths (Header And Pick-Up) When Using A Windrow Pickup

WINDROW PICKUPS

A problem with using a windrow pickup on a standard header is that it's usually quite a bit narrower than the header. This means you have two machine widths to watch—a narrow one for picking up the crop, and a wider one that may hit obstacles along the windrows (Fig. 36). Keep an eye on the header to avoid fences and other obstructions along the edge of the field while concentrating on picking up the windrow.

Before traveling at road transport speeds, check pickup support chains to make sure they're securely hooked and in good condition. Having the pickup drop while moving at a fairly high speed could damage the combine and injure you. Also, remember to add the amount of rear weight specified by the manufacturer before using or transporting the pickup.

Fig. 37—Heavy Header Attachments May Require The Addition Of Rear Weights For Stability

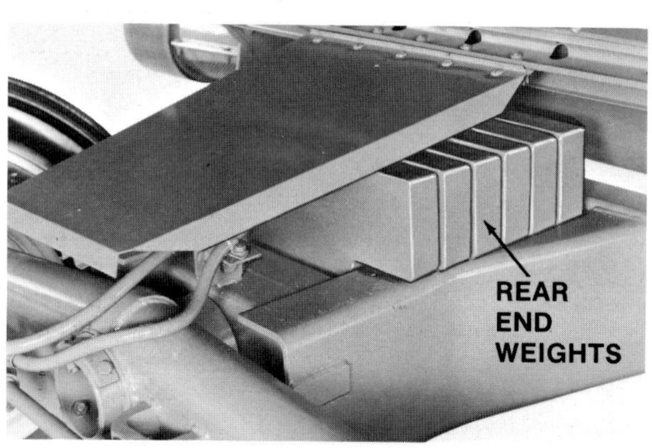

SPECIAL ATTACHMENTS

Several attachments can be used on combine headers to increase their effectiveness in gathering various crops. Many of these attachments require some modification of the grain header. In some cases, they're heavy and you'll need a hoist or some other lifting device to install them. For example, you should use a hoist to remove a bat reel and install a pickup reel. Using the right equipment saves time and reduces chances of injury.

Most attachments increase the weight on the front part of the header, so you may need additional weight at the rear of the combine for stability (Fig. 37).

Correct installation of header devices is also important for proper operation of the rest of the hydraulic system, including steering, traction drive speed control, and other critical functions.

CORN HEADS

Proper operation of snapping rolls, snapping bars, and gathering chains on corn heads is important for the smooth flow of materials into a combine. Improper functioning of these parts can cause uneven feeding, plugging, more down time, and increased risk for the operator. Trying to unclog snapping rolls without shutting off the machine is a leading cause of serious injuries during corn harvesting.

Follow the recommendations in the operator's manual to adjust the corn head properly:

1. *Tapered stalk rolls do not require adjustment to prevent plugging. To prevent plugging on corn heads with straight stalk rolls, the rolls must be set at a spacing which will aggressively feed the stalks*

through the rolls without breaking off the stalk.

2. *Deck plates or snapping bars must be spaced far enough apart that stalks will pass through freely and at the same time prevent ears of corn from getting into stalk rolls.*

3. *Gathering chains must be timed and tensioned according to manufacturer's instructions in order to effectively move the stalks into the corn head.*

Stalk roll speed must be comparable to ground speed. This is important for good stalk movement through the stalk rolls. If stalk roll speed is too fast, many stalks will be chewed off which can cause plugging. If the stalk roll speed is too slow, stalks will be pulled out of the ground, plugging the stalk rolls.

If the corn head plugs:

1. *Stop.*

2. *Back up the combine.*

3. *Disengage all power and shut off the engine before trying to unplug it.*

4. *Remove as much of the material as you can by hand.* If rolls are wrapped with vines or other green or tough material, cut them free with a heavy knife, cutting away from your body (Fig. 38).

5. *Start the machine to move the remaining material through the rolls.*

6. *If wrapping persists, shut off the engine again and adjust or replace the snapping rolls or vine cutter. Follow instructions in operator's manual.*

REPLACING SNAPPING ROLLS

Some rolls are made of cast iron, while others have replaceable steel bars. When cast iron rolls become worn, the entire roll has to be replaced. Replace

Fig. 38—Disengage Power And Shut Off Engine Before Removing Materials From Snapping Rolls

snapping rolls while the corn head is on the combine—not while it's being stored. The combine holds the corn head rigid and provides power for turning corn head drives for safe and easy installation of the rolls. Block the header before working under it. Use the safety latch to hold it in the raised position. And keep your hands and fingers away from the many pinch points in the rolls and gathering chains.

VISIBILITY IN THE CORNFIELD

Always be on the lookout for movement ahead of your harvester. It's hard to see people walking in a cornfield. Harvesting noise usually signals them to stay back, but children or animals may be attracted by the noise and run toward it without realizing how fast the combine is moving (Fig. 39). Keep children completely away from the fields when you're harvesting.

Harvesting noise drowns out speech, and you may have to use hand signals to communicate with others

Fig. 39—Watch Out For People Or Animals When Combining

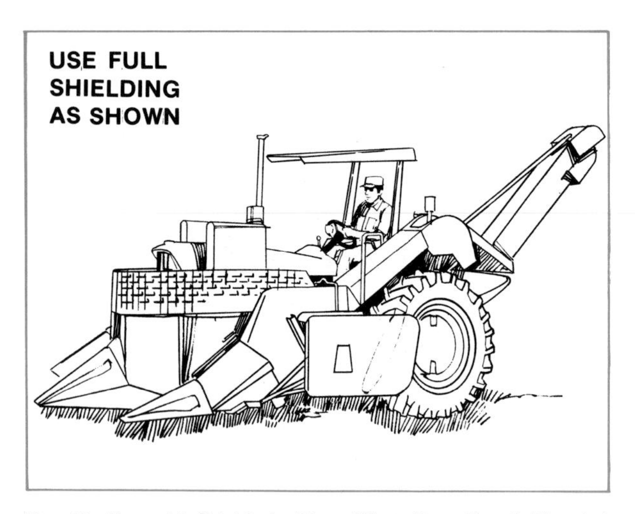

Fig. 40—Keep All Shields In Place When Operating A Mounted Corn Picker

(see Chapter 1). Make sure everyone knows your intentions and stands clear before the machine is moved.

CORN PICKERS

The safe use of corn pickers is similar to that of combine corn heads, discussed in the previous section. Snapping and husking rolls are dangerous if the operator attempts to clean them while they're running. Keeping shields in place is especially important on mounted corn pickers where the operator sits close to the gathering units and power drives (Fig. 40).

Fig. 41—Use A Hoist Or Get Help When Attaching An Ear-Corn Elevator

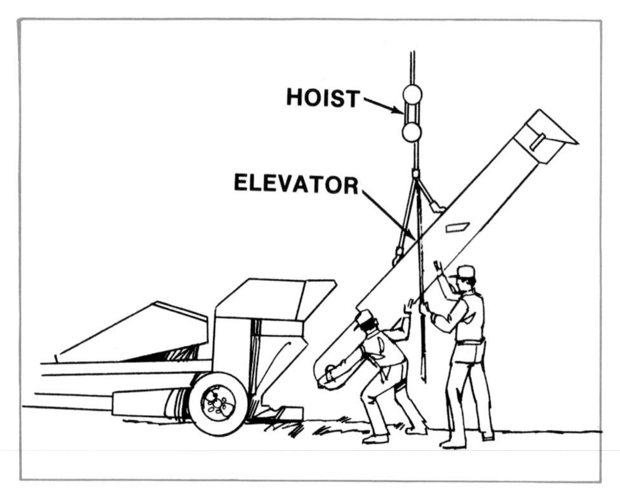

PREPARATION

It's not difficult to get a pull-type corn picker ready for harvest if it was in good shape when last used and was protected from the weather. If the gathering units or ear-corn elevators were removed for storage, they'll have to be reattached. With the header properly positioned and blocked, this should be relatively easy.

The elevator, although it doesn't weigh much, often requires a hoist or two people to install it (Fig. 41). Be sure to get help to avoid injury. Because elevators are long and must be raised high, good lifting practices are important. Use a hoist to support most of the elevator's weight and use manual force only to position it. Use the same procedure when attaching an elevator to a tractor-mounted picker.

Attaching tractor-mounted pickers may take as much knowledge and effort as setting up a new machine. Good setup practices should be followed. Check the operator's manual, and use hand and power tools and stands as recommended. Care and forethought in removing mounted equipment pays off when attaching the picker the next year.

Installation of all power transmission shields is very important, because they protect people against personal injury.

To prevent fires on mounted pickers:

1. *Enlarge or raise the radiator screen for effective cooling.*

2. *Shield leaves from the hot exhaust manifold and muffler (Fig. 42).*

3. *Clean leaves and trash from hot spots.*

4. *Mount a multi-purpose dry chemical fire extinguisher on the equipment to use in case of fire.*

OPERATION

For effective and safe picker operation:

1. *Maintain a ground speed for snapping ears midway along the rolls.*

2. *Adjust the snapping rolls and gathering chains for good stalk feeding without breaking the stalks.*

3. *Replace snapping rolls as required for good feeding action.*

Snapping Rolls

Maintaining the right relationship between gathering chain speed and ground speed is important. When ground speed is too slow, stalks are fed too far back on the rolls and ears may catch in the stalk ejector and cause plugging at the back. When ground speed is too fast, stalks won't move back far enough, causing snapping and plugging near the front of the rolls and gathering chains. When these mechanisms get plugged, operators take chances and get hurt. Keep them operating smoothly.

Because spiraled rolls aren't as aggressive as fluted rolls, they often have more of a plugging problem. If snapping rolls plug, *disengage the power and shut off the engine* before trying to get stalks or weeds out. If there's a lot of plugged material, separate the rolls. You can do this from the operator's station on many mounted units. Remove as much material from the rolls as you can before starting the machine. Cleaning fluted rolls by running the machine is more effective than it is with spiraled rolls.

One hazard of snapping rolls is that a weed or stalk can catch between the rolls and stop. Then if it's moved slightly, the weed or stalk can feed through the rolls very quickly—faster than someone holding onto it two or three feet (600 to 900 mm) from the rolls could possibly release it (Fig. 43). This is how some operators are caught when they think they're reasonably safe. *Never attempt to clean snapping rolls while the machine is running.* Adjust rolls properly and use correct ground speed to reduce plugging problems.

Turning

A tractor hitched to a pull-type picker and wagon needs a lot of space for turning. If equipment is turned too sharply, the wagon's steering mechanism can be damaged or the hitch broken free from the picker. Someone could be seriously injured by an

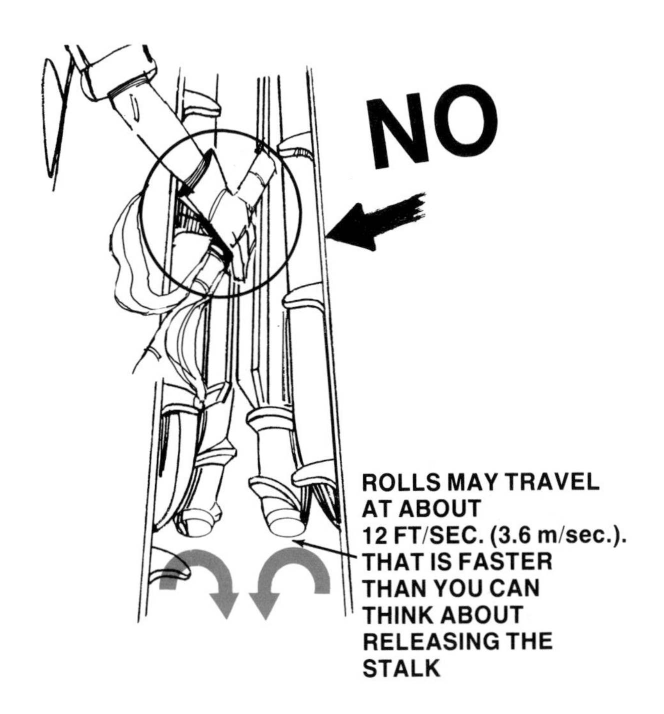

Fig. 43—Snapping Rolls Can Pull You In Faster Than You Can Let Go. Don't Try It!

Fig. 42—Make Sure Shields And Screens Are In Place To Prevent Fires

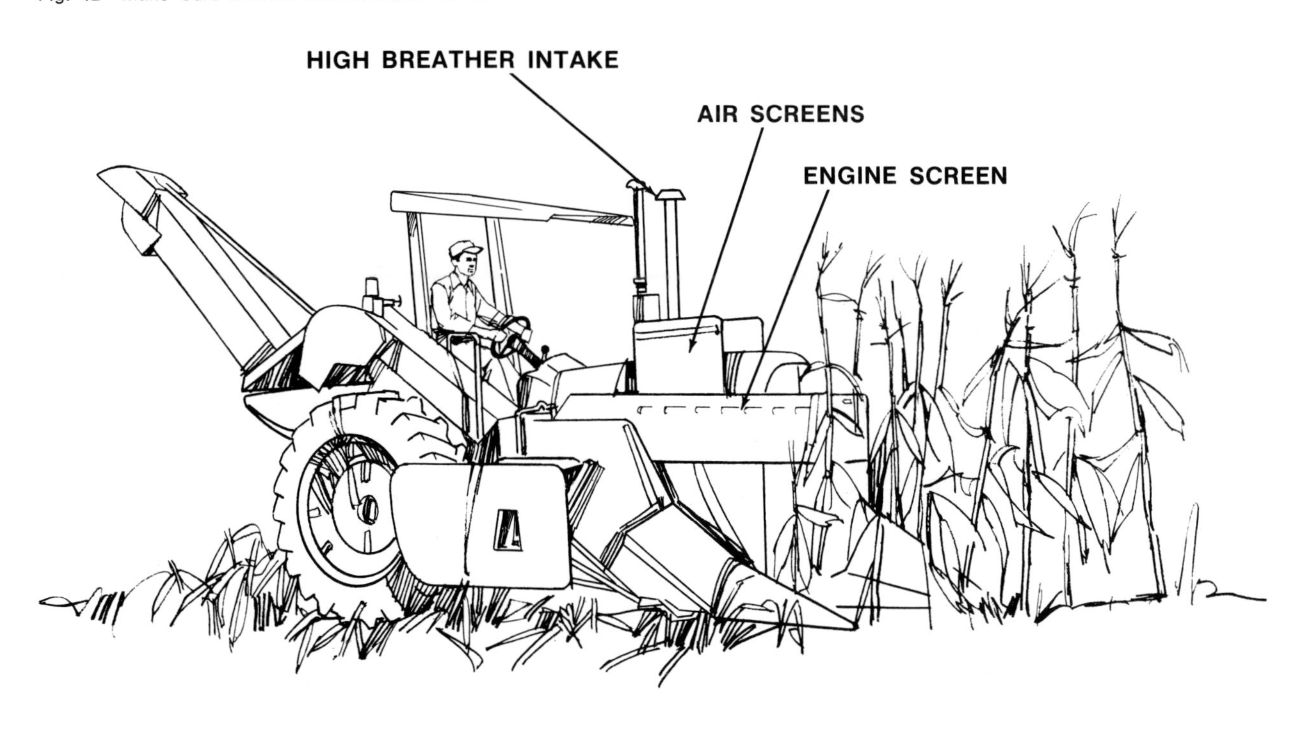

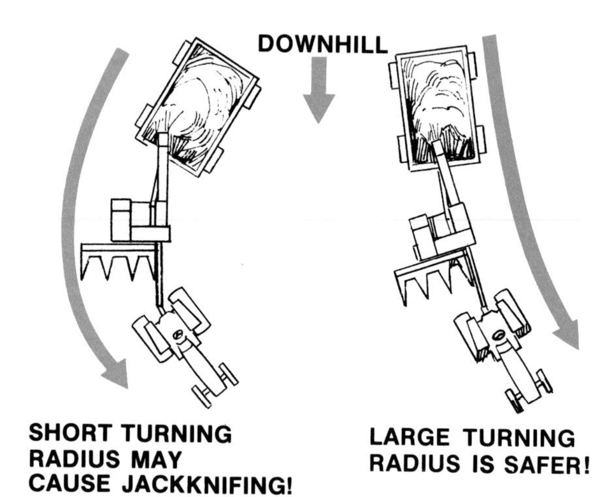

Fig. 44—Make Wide Turns On Downward Slopes To Avoid Jackknifing

Fig. 45—Use Protective Equipment When Operating A Corn Picker Under Moldy Conditions

uncontrolled wagonload of corn. Sharp turns cause the wagon to create large side forces on the picker, and these can cause skidding of the picker and loss of control, particularly when turning on a down hill or side hill with a fully loaded wagon.

Under some conditions, the offset loading of a pull-type picker adds to the problem. Take special care when moving down steep slopes to avoid jackknifing the picker and wagon (Fig. 44). Give yourself plenty of room for turning at the end of the field.

A tractor with a mounted picker has more weight on the tractor wheels, and this can improve the control of a trailed wagon. Sharper turns are possible, but avoid turning so sharply that you damage the wagon's steering mechanism or tear up a tractor tire by catching the side of a wagon.

Fig. 46—Riders In Wagons Behind Corn Pickers Can Be Hit By Ears Thrown By The Elevator And They Can Fall Off—Insist On No Riders

Personal Protective Equipment

When harvesting conditions are especially dirty, eye and respiratory protection may be needed (Fig. 45). This may be of greatest concern when operating machines which place the operator near the crop gathering devices. And noise from the many gears, chains, and other drives may also make ear protection desirable, especially when units are operated several hours a day.

Mounting

Climbing on or off a tractor equipped with a mounted picker can be hazardous if the steps and platforms are not in place. Keep all steps and handholds in place and in good repair. Keep steps clear to prevent slipping. Keep all shields in place at all times. Watch where you put your hands.

Riders

Keep riders out of wagons behind corn pickers. When ear corn is harvested, the ears can hit a rider (Fig. 46). Or the elevator that carries the corn to the wagon could injure a rider, especially when turning or harvesting on rolling ground. And wagon fall-off accidents are frequent. Do not stand while operating corn pickers. You cannot adequately control the machine if a sudden emergency occurs.

HUSKING MECHANISMS

Husking beds have pairs of rolls made of steel or rubber designed to grasp and pull the husks. These rolls can also catch gloves, fingers or hands. Ear retarders above the rolls may consist of pressure wheels with rubber paddles or a chain-and-slat mechanism. Most of this equipment is open, and can cause injury if you are careless. *Disengage the power and shut off engine before doing any cleaning or adjusting around the husking bed.*

Plugging isn't as much of a problem with husking rolls as it is with snapping rolls, but damp or green material sometimes gets wrapped around them. If this happens, disengage the power, shut off the engine, cut the material free with a strong knife, and pull it out by hand (Fig. 47). Wear well-fitting gloves, because corn husks and leaves have sharp edges that can cut. And weeds or trash wrapped around the rolls can injure bare hands.

SHELLING MECHANISMS

If inspection of the sheller is necessary, disengage the power, shut off engine, and see that the sheller has stopped completely before opening any covers or doors. If the cylinder plugs, use procedures similar to those recommended for unplugging combine cylinders. Using the right tools, protecting hands with heavy gloves, and avoiding pinch points are musts for unplugging a sheller safely.

TRAILED WAGONS

Size and type of picker and weight of the tractor are important in selecting wagon size. A large wagon can be pulled safely behind a tractor equipped with a mounted picker. But when a small tractor is used with a one-row, pull-type picker, a relatively small wagon should be used. More important than the size of the wagon is the weight of the load. Sizing the wagon for the maximum safe load behind the picker eliminates the temptation to overload it.

If your tractor is too light, you can't control the load effectively when going downhill. Jackknifing of the corn picker and wagon is a potential hazard, even when a large tractor is used to pull a lightweight picker if you don't maintain proper control. If the picker is turned on a hillside so the wagon is pushing from the side, the tractor has little control over either the picker or the wagon, and jackknifing can result.

To avoid equipment damage during picking, consider these three factors:

- **Length of wagon tongue**
- **Height of the front and sides of the box**
- **Radius of the wagon's turning circle**

Selecting the right tongue length and box height will eliminate interference between the wagon and the picker under most operating conditions. The shorter a wagon is able to turn, the less chance there is of damaging the wagon's steering mechanism while turning. Refer to the operator's manual for help in selecting wagon and tongue sizes.

Hitching wagons can be hazardous when two people are involved. The most important rule is never to allow anyone to stand between the picker and the wagon while the picker is being backed into position. Telescoping tongues that allow the operator to extend the wagon tongue to make the connection are helpful. This process involves backing the picker until the tongue latches. Hitches that connect auto-

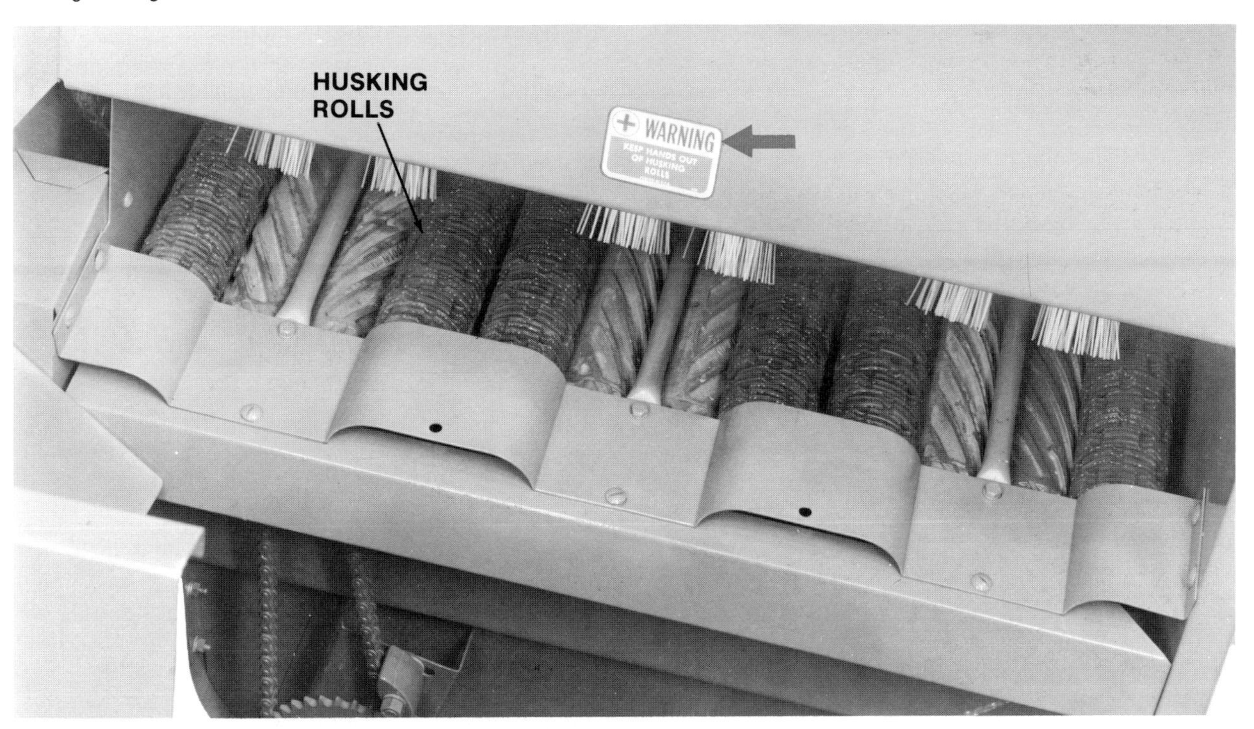

Fig. 47—Clean Husking Rolls Only After Disengaging Power And Shutting Off Engine

matically when the picker is backed into the wagon also provide a safe solution to the problem, if the hitch is well designed.

When a second person must assist, have him stand away from the hitch point and signal the operator (Fig. 48). Be sure you can see your helper at all times. Use standard hand signals to get the picker into position for hitching (see Chapter 1).

POOR HARVESTING CONDITIONS

Corn harvesting often takes place in late fall and even into the winter. Frozen ground, snow, ice, and mud are just a few of the conditions that can complicate harvesting.

When harvesting under poor conditions:

1. *Wear comfortable, well-fitting clothing appropriate for weather conditions.*

2. *Keep steps and platforms clear to prevent falls.*

3. *Disengage all power and shut off the machine before dismounting to avoid the danger of slipping into drives or other moving mechanisms.*

4. *On wet ground, use a larger tractor to pull the picker and wagon, or don't fill the wagons to full capacity.*

CHAPTER QUIZ

1. Fill in the blank. The best time to start getting a combine ready for harvest is ________ ________ before harvesting begins.

2. List five things to do to prepare a field for safe harvesting.

3. True or false? Traction drive belts on combines run only when the transmission is in gear.

4. Describe the hand signal used to indicate that you want the combine operator to lower the corn head.

5. True or false? The best way to prepare for attaching a combine header is to position it properly when it's removed.

6. You are driving a self-propelled combine on a two-lane road and making a right-hand turn into a field. The back of the combine will:

a) Move to the right away from oncoming traffic.

b) Move to the left into oncoming traffic.

c) Move straight forward.

7. Why is it important to have the combine on level ground when manually positioning an unloading auger?

8. Under what conditions is unloading on-the-go a suitable practice?

STAY CLEAR OF HITCH AND DIRECT OPERATOR BY USING HAND SIGNALS

Fig. 48—Helpers Should Stand To One Side Of The Hitch And In View Of The Tractor Operator

9. List the advantages of servicing a combine at the end of a day's operation.

10. True or false? When driving a combine on the road at transport speeds, you should raise the header as high as possible to avoid obstructions.

11. Fill in the blank. You may need to change ______ on the rear of the combine when you attach a different header or header attachment.

12. List three requirements for the safe and effective operation of ear-corn snapping units.

13. Why is material caught in spiraled picker snapping rolls a hazard?

14. True or false? Proper selection of wagon size for trailing behind a pull-type corn picker depends only upon the power of the tractor pulling the picker.

15. What are two of the most important practices to prevent personal injury when working around harvesting machines?

10 Target: Other Harvesting Equipment

INTRODUCTION

Several specialized machines are available for harvesting. We will cover three of these machines in this chapter:

- **Cotton harvesters**
- **Potato harvesters**
- **Sugar-beet harvesters**

Some safety practices apply to the operation of any of these machines:

1. *Put the mechanism or transmission in neutral or park position before starting the engine.* Engaging the starter with these systems in gear may result in a sudden, unexpected movement that could injure the operator or others.

2. *Avoid high-speed stops and turns, uneven terrain, and soft ground.* Operation under these conditions may result in an upset.

3. *Keep all safety shields in place when operating the harvester.* They keep your hands and clothing out of moving parts.

4. *Equip your harvester with a multi-purpose dry chemical fire extinguisher.*

5. *Wear close-fitting clothing and stay alert to keep from becoming entangled in moving parts.*

6. *Keep hands and feet away from moving parts.*

7. *Never carry extra passengers, especially children.*

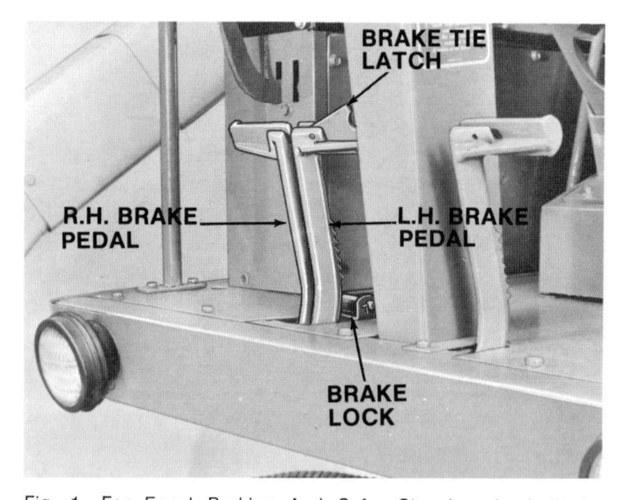

Fig. 1—For Equal Braking And Safer Stopping, Lock Brake Pedals Together Before Operating At Transport Speeds

Fig. 2—Self-Propelled Cotton Stripper

8. *Learn hand signals and use them to communicate with others during operation.*

9. *Plan rest breaks to reduce fatigue.*

10. *Lock brake pedals together before operating at transport speeds (Fig. 1).* Keep brakes adjusted evenly. This is important when moving any equipment at transport speeds to maintain equal braking pressure on both wheels.

11. *Use warning lights, reflectors, and SMV emblem when transporting the harvester.* Let other motorists know you're on the road. When possible, plan your operation to avoid nighttime travel.

12. *Always disengage the PTO and shut off the engine when leaving the tractor.*

13. *Disengage the power, shut off the engine, and wait for all parts to stop moving before lubricating, adjusting, or servicing the harvester.*

CAUTION: You can't shut off the engine on some older cotton pickers. The spindles must be running.

14. *When lubricating parts that must be in operation when serviced, wear close-fitting clothing, follow manufacturer's instructions, and concentrate on the job you're doing.*

15. *Always use safety stops or lift cylinders, supports, and blocks.* Keep the machine from falling on you.

COTTON HARVESTERS

The mechanical cotton harvester is a large machine with many fast-moving parts. Three common types are shown in Figs. 2, 3, and 4. Skillful operation is necessary for the safety of people on and around the machine and for effective machine operation. Some general recommendations are outlined in the first section of this chapter. This section covers potential hazards and safe practices for:

- **Field operation**
- **Chokedowns**
- **Service**
- **Fire hazards**

Fig. 3—Pull-Type Cotton Stripper

Fig. 4—Tractor-Mounted Cotton Stripper

FIELD OPERATION

Many cotton harvesters are equipped with large overhead baskets to store cotton. Cotton is dumped from the baskets into trailers. Follow these safety practices:

1. *Lower the basket before driving away.* When the basket is raised, the harvester becomes more top-heavy and may cause an upset if the machine is moved across a field (Fig. 5).

2. *Reduce speed when moving over rough terrain.* Cotton harvesters are top-heavy and can turn over on rough ground.

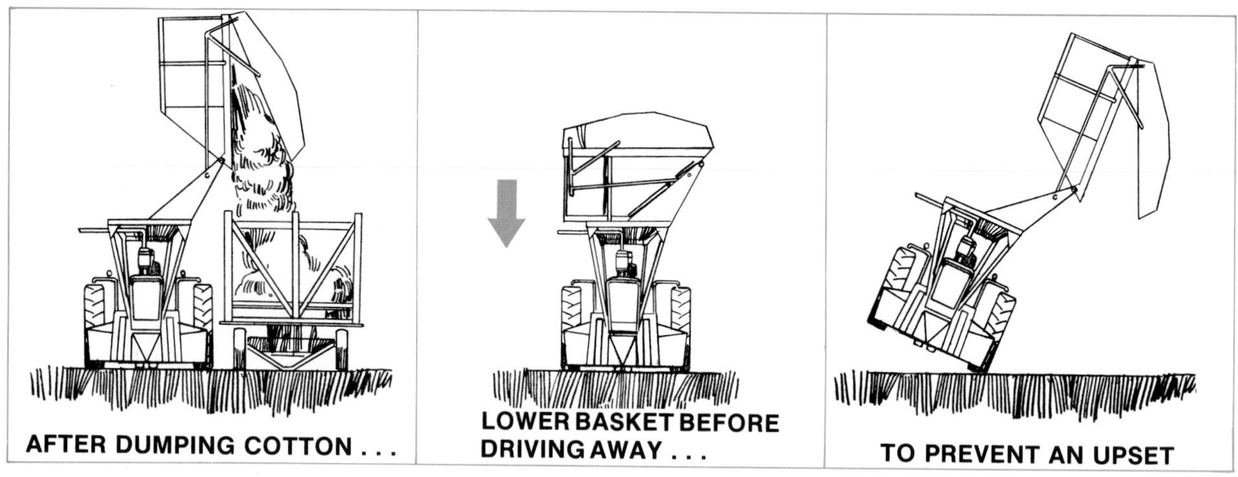

Fig. 5—After Dumping Cotton, Lower Basket Before Driving Away To Prevent An Upset

3. *Keep your harvester safely away from ditches, creeks, and other steep, sloping ground.* Keep end rows smooth and firm. Steep slopes and plowed turnrows make turning difficult, and may cause an upset.

4. *Reduce engine speed before braking or turning.* Quick stops with the high-profile cotton pickers can result in the machine nosing over. Fast turns can result in upsets.

5. *Remain seated when raising or lowering the basket on a tractor-mounted stripper.* A sudden drop of the basket could result in a serious head injury to anyone standing in the wrong place on the operator's platform.

6. *Keep everyone away from trailers in the field.* Be sure there is no one in the trailer before dumping. A load of cotton falling into a trailer could seriously injure or suffocate someone trapped inside.

7. *Be sure you're clear of electrical wires before raising or dumping a basket (Fig. 6).* A raised basket may reach a height of 25 feet (7.6 m), and if it contacts a power line, you could be electrocuted.

8. *Wear personal protection equipment.*

CHOKEDOWNS

Many cotton harvester accidents could be prevented by avoiding chokedowns. You can do a more efficient job and reduce the risk of injury by keeping the main harvesting components—the picker drums, stripping rolls, and the reel on finger-type strippers—free of weeds, rocks, and stumps. When the machine is operating properly, you're safely in the operator station.

To prevent chokedowns:

1. *Practice good weed control and keep fields clear*

Fig. 6—Before Dumping, Be Sure Basket Cannot Touch Electrical Wires When It Is Raised Or Lowered

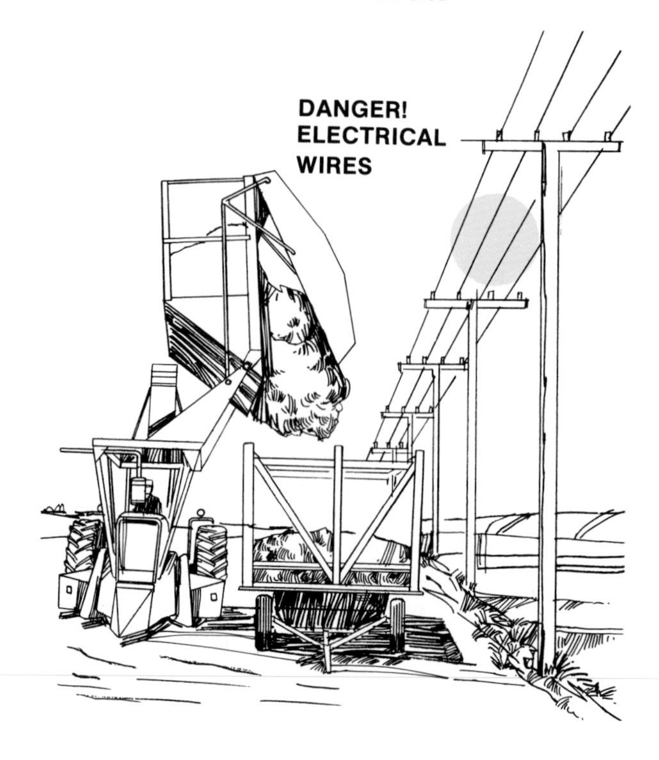

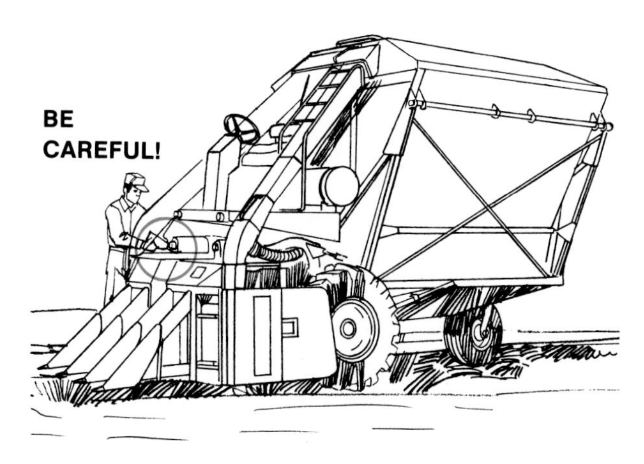

Fig. 7—Use Extreme Care When Lubricating Picker Bar Cams And Sun Gears If Operator's Manual Recommends Running Machine While Lubricating

Fig. 8—Use Handrail and Ladder When Working On Top Of Basket

of rocks and stumps. A cotton harvester operating in a clean field is less likely to choke down.

2. *Adjust all belts and chains on the harvesting unit according to instructions in the operator's manual.* A belt that slips or a chain that jumps can allow the harvesting unit to choke down.

When a picker drum chokes, *disengage all power and shut off the engine.* Wait until all parts have stopped moving, and *then* remove the obstruction. Rotate the doffer by hand until the obstruction can be removed. Never try to start the picker drum turning by engaging the power and pushing the spindles with your foot. If the obstruction suddenly breaks free and the picker unit turns under power, it will endanger any part of the body near the spindles. After removing the obstruction, return to the operator's station, start the engine, and engage the power. If the clutch continues to slip, check for other obstructions or for bent picker bars or doffers.

If the stripping unit chokes down, *disengage all power and stop the engine.* If you attempt to unchoke the stripping rolls, augers, or reel while they're moving, you could be seriously injured.

SERVICE

Cotton harvesters must be serviced regularly. These machines are unique in some ways, and proper maintenance is required for effective and safe operation. Follow these guidelines:

1. *Wear close-fitting clothing and stay alert while lubricating picker bar cams, cam follower bearings, and spindle drive gears (Fig. 7).* This job must be done while the picking unit is running on some machines. Check the operator's manual for the correct procedure.

2. *Disengage power, shut off the engine, take the key, and wait for all parts to stop moving before removing dust from v-belts on cotton harvesters.* These belts often become covered with a fine dust glaze when harvesters are operated under dry conditions. They should be cleaned frequently.

3. *Use the ladder and handrails when working on top of the basket (Fig. 8).* A fall from the top of the basket could injure you seriously. If the basket lid is opened, secure it to keep it from closing accidentally.

4. *Use safety stops and block the picking unit before lubricating the lower doffer bearings (Fig. 9).* Use a solid block that will support the unit. Do not depend on hydraulic pressure to hold up the picking unit.

5. *Install cylinder lock-outs before working under a raised picker basket.*

Fig. 9—Use Safety Stops And Block Picking Unit Before Getting Under It

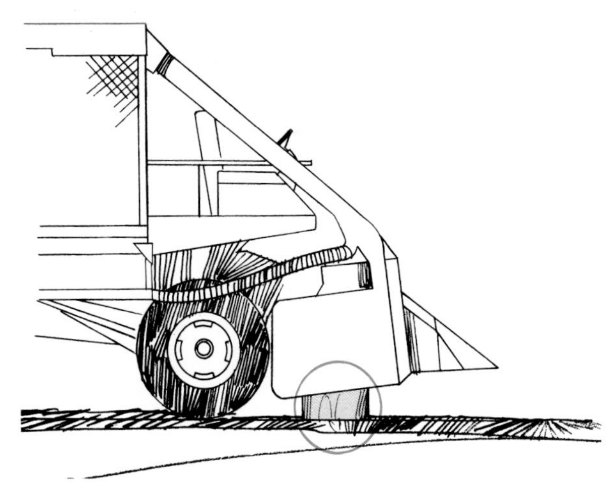

5. *After servicing, replace all shields and guards before starting the machine.* Shields and guards are designed to protect you from moving parts, but they can do their job *only* if they're kept in place.

FIRE HAZARDS

Before cotton is harvested, the plant is killed, either chemically or by frost. This means that highly flammable leaf trash and dead plant parts get mixed in with the lint. Cotton lint is also flammable. When such materials accumulate, fires can start easily.

Take these precautions to prevent fires during cotton harvesting:

1. *Keep the engine clean (Fig. 10).* Lint, leaf trash, and other dry materials on a hot engine could catch fire. Clean the area between hot engine parts and the hood periodically. Check exhaust pipes and muffler often for leaks. Hot exhaust gases or sparks could start a fire.

2. *Use only water or the manufacturer's recommended product as a moistening agent.* Petroleum-based moistening agents increase the fire hazard.

3. *Always dump the basket downwind into the trailer.* Dumping it into the wind could result in cotton blowing back onto the engine and starting a fire.

Fig. 10—To Avoid Fires, Keep Engine Free Of Lint, Dust, And Trash. Shut Off Engine First!

4. *Keep the doffer area free of lint and trash to avoid fire caused by friction between the spindles and trash (Fig. 11).*

5. *Prevent arcing at electrical terminals.* Cotton lint or trash around the battery terminals and other electrical contacts can catch fire. Keep all terminals and contacts clean and tight.

6. *Mount multi-purpose dry chemical fire extinguisher where it will be readily available in case of fire.* Use the pressurized, dry-chemical type approved by Underwriters Laboratory — five to ten pounds or larger. When using an extinguisher, aim it at the source of the fire — not at the flames!

7. Shut off equipment before doing any work.

Fig. 11—Keep Doffer Area Clean To Prevent Plugging In Suction Doors And Air Pipes And To Prevent Picker From Dropping Cotton

POTATO HARVESTERS

Whenever more than one person is involved in the operation of a complex machine, such as a potato harvester, a coordinated effort on the part of each individual is required for safe operation. These key factors are involved:

- **Know the machine.**
- **Train workers.**
- **Make preoperational checks.**
- **Use approved operating procedures.**
- **Transport the harvester correctly.**

See the first section of this chapter for general recommendations.

KNOW THE MACHINE

The first requirement is to *know the harvester.* You have to be well acquainted with each adjustment and component, because no one combination of adjustments and optional parts will work for all soil and weather conditions. Study your operator's manual, especially the safety instructions, before the potato harvest begins (Fig. 12). Even if you are already quite familiar with the machine's operation, you will find that information in the manual becomes more meaningful as you gain experience. The ability to quickly diagnose and correct problems is the best way to eliminate problems that can lead to mistakes and serious injuries.

Several common adjustments affect the performance of your potato harvester:

- **Ground speed**
- **Drive speed**
- **Digger blade depth**
- **Agitation**
- **Elevator speed**
- **Conveyor tilt**
- **Coulter position**

You must understand these adjustments in order to make safe and efficient use of your harvester. Most of these adjustments should be made only after proper shutdown procedures have been performed. Check the operator's manual to find out which components are designed for adjustment during operation and which are not.

Ground Speed

Ground speed is usually adjusted by changing gears on the tractor while maintaining the same throttle setting. This can be accomplished with little interruption to the harvesting operation. Change gears smoothly when altering speed to keep workers who are standing on the harvester from losing their balance.

Fig. 13—Change Speed Smoothly To Avoid Injury To Sorters

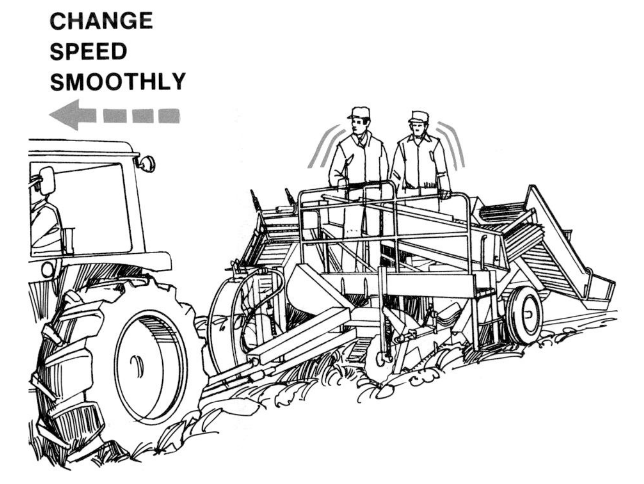

Fig. 12—Study The Operator's Manual To Get To Know The Harvester

Drive Speed

Varying the drive speed of the potato harvester must also be done smoothly. If your harvester is equipped with its own engine, the drive speed can be modified by simply adjusting the harvesting throttle.

If the harvester is operated by a tractor PTO, drive speed must be varied by changing the throttle setting on the tractor. Because this will also change the ground speed, a gear change may be required. Here again, keep the sorters in mind. Jerky or erratic driving could make them lose their balance on the harvester and contact moving parts or fall from the machine (Fig. 13).

Digger Blade Depth

The hazards of making some adjustments may not be obvious—adjusting the digger blade, for example. If it is adjusted while the machine is running:

- **The elevator spud chains could catch your clothing or limbs.**
- **The PTO shaft could catch on clothing.**
- **Tools could fall and be carried into the machine.**
- **The tractor could slip into gear, causing an injury.**

If you disengaged the power to the harvester but still left the engine or tractor running, the possibility of an accident would be reduced, but the potential for all of these hazards would still exist. Disengage

Fig. 14—Disengage The PTO And Shut Off The Tractor Engine Before Adjusting The Digger Blade

Fig. 15—Put Controls In Convenient Places

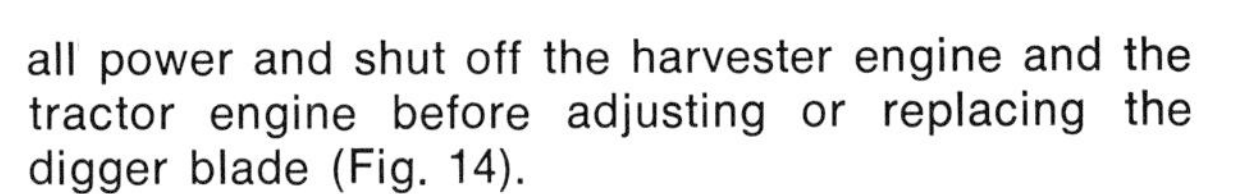

all power and shut off the harvester engine and the tractor engine before adjusting or replacing the digger blade (Fig. 14).

Agitation

The elevator on a potato harvester comes equipped with a plain roller that gives the right amount of elevator shake to separate the dirt from the potatoes under most conditions. But conditions vary greatly with the type of soil and weather, and you may have to replace the plain rollers with special three-point shaker sprockets to get better separation. Make sure all power is disengaged and engines are shut off before servicing these parts.

Fig. 16—Proper Coulter Adjustment Reduces Clogging Problems

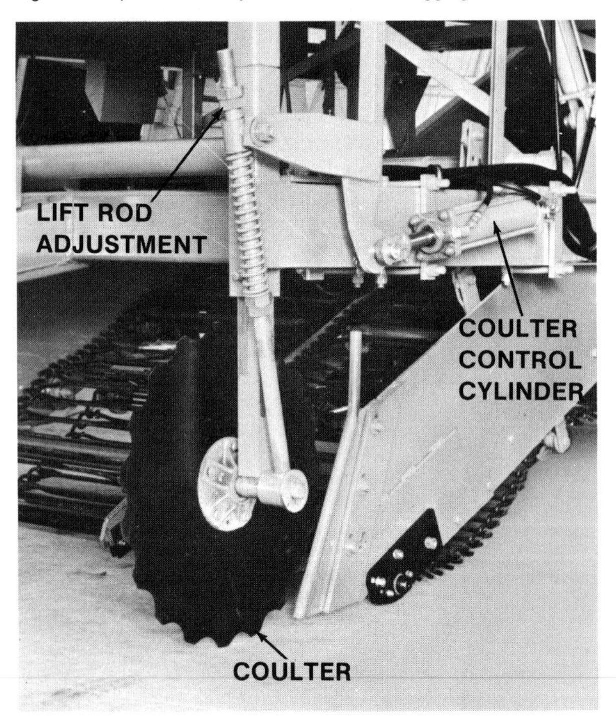

Elevator Speed

Another variable that affects the proper separation of dirt and potatoes is the speed of the elevator. To change elevator speed, disengage the power and turn off all power units, and then change the size of the drive sprocket. Remove any extra links and store them for future use. If the chain fits properly, your harvester will perform more reliably, the chances of a breakdown will be reduced, and there will be less opportunity for you to be injured.

Conveyor Tilt

The tilt of the rear, side, and sorting elevators may be adjusted by hydraulics, ratchet, or hand controls, or may require tools (Fig. 15). Some ratchet and hand controls may also be adjusted without interrupting the harvesting operation. Others, however, should not be operated while the machine is running. In general:

1. *Sorters must not leave the platform while the harvester is in motion.*

2. *Don't reach around or near moving parts.*

3. *Disengage all power and shut off the engine before servicing or adjusting the machine.* Never make adjustments while the harvester is running, unless this is recommended in the operator's manual.

Coulter Position

The main function of the coulters is to cut vines so they will not cause plugging or collect at the sides of the harvester (Fig. 16). If clogging occurs, disengage the power and shut off the engine before attempting to untangle the vines. If clogging occurs inside the machine, make sure the tractor is turned off. *Never* reach into the harvester with hands or other objects unless *all* power is disengaged and the engine is turned off.

TRAIN WORKERS

Everyone connected with the operation of the harvester and tractor should be properly trained (Fig. 17). Before going into the field, review safety precautions with the people who will be operating the harvester and tractor, and tell them exactly what their duties will be. Here are some safety guidelines for operators and workers:

1. *Wear close-fitting clothing that won't get caught in moving parts.*

2. *Wear shoes or boots with slip-resistant soles.* Avoid smooth neoprene, leather, or other materials that tend to become slippery and may lead to falls.

3. *Get on and off the harvester only when it's stopped.*

4. *Wear personal protection to protect yourself from dust and noise.*

Fig. 17—Teach Safe Practices

MAKE PREOPERATIONAL CHECKS

Breakdowns are the greatest enemy of safe and efficient operation. Patience and good judgment are often forgotten or valuable time lost when a harvester is not working properly. Make seasonal and daily inspections as outlined in your operator's manual, and inspect your potato harvester and correct these problem areas before each use:

- **Loose bolts and connections**
- **Poor tire condition, or improper inflation**
- **Leaky hydraulic hoses or connections**
- **Greasy or cluttered sorter's platform**
- **Loose, broken, or missing handrails on the sorter's platform**
- **Loose or missing safety shields**

Remember that a safe operation is also an efficient operation. The extra attention you pay to safety and servicing will be well rewarded by the improved performance of your harvester and the health and well-being of everyone involved.

USE APPROVED OPERATING PROCEDURES

Once you have properly prepared the potato harvester and trained the people who will be operating the machinery, you have taken the two most important steps toward a safe operation.

Chapter 5 outlines the points that a tractor operator should consider. Many of these are even more important when operating a potato harvester. For example, jackrabbit starts, quick turns, and stops that are hazardous for a tractor operator are even more serious for the sorters, who are standing on a moving machine with many moving parts. Unless he performs his job skillfully, the tractor operator could make them lose their balance, possibly causing an accident.

Fig. 18—Safety Features Can Protect Operators Only If Properly Maintained

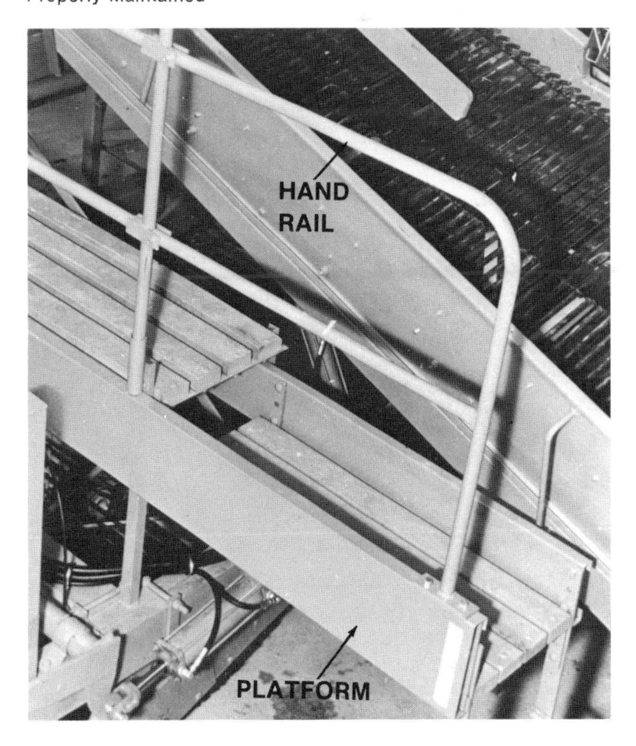

Fig. 19—The Tractor Operator Is The Key Man For Safe Potato Harvesting

The tractor operator:

- **Controls the ground speed**
- **Controls the drive speed**
- **Controls the harvester digging depth**
- **Controls the harvester steering**
- **Raises and lowers the coulters and digger at turns**

In addition, he must be alert to problems encountered by the sorters so he can react quickly for their safety. Because the tractor operator has complete control over the movement of the harvester and the power, he is the key man in both the performance of the harvester and the safety of everyone involved (Fig. 19). To help insure the safety of other workers, the tractor operator should:

1. *Stop the tractor and harvester when a sorter needs to get off the machine. Sorters use prearranged hand signals to communicate.*

1. *Stop the tractor and harvester when a sorter needs to get off the machine.*

2. *Never allow adjustments to be made on the harvester unless proper shutdown procedures have been followed.*

3. *Make turns cautiously.* Remember to raise and lower the coulters and digger on the harvester.

4. *Change gears and adjust the throttle smoothly.*

5. *Be alert to unusual noises that may indicate problems.*

6. *Maintain the machine so it runs smooth and reliably.*

The sorters should be shown how to recognize malfunctions in the harvester such as broken conveyor flights, lodged rocks, or clogging. They should also know how to signal the operator when a problem occurs. Common hand signals are illustrated in Chapter 1.

TRANSPORT HARVESTER CORRECTLY

Potato harvesters, like many other farm machines, are designed for field work, *not* high-speed transport.

Before attempting to transport your harvester:

1. *Install braces on harvester tie rods.*

2. *Use lights, reflectors, and an SMV emblem (Fig. 20).*

3. *Don't transport sorters on the harvester.* Provide another means of transportation for them.

4. *Raise the coulters and digger for ample ground clearance.*

5. *Lower the bulk loader extension.*

6. *Use braces as provided during shipping to support the digger and coulters if the hydraulic system is not connected.*

Follow these six steps. It will reduce hazards during transportation.

SUGAR BEET HARVESTERS

There are two types of beet harvesters: the lifter loader harvester and the tank harvester (Figs. 21 and 22). Both are powered by a tractor's power takeoff system. General safety recommendations for them are given in the first section of this chapter.

Although each of these loaders has a few unique components and operating hazards, they are basically alike. Two potentially hazardous components they have in common are:

- **Rienk screen**
- **Cross conveyor**

Fig. 20—Use Reflectors And An SMV Emblem On Your Potato Harvester

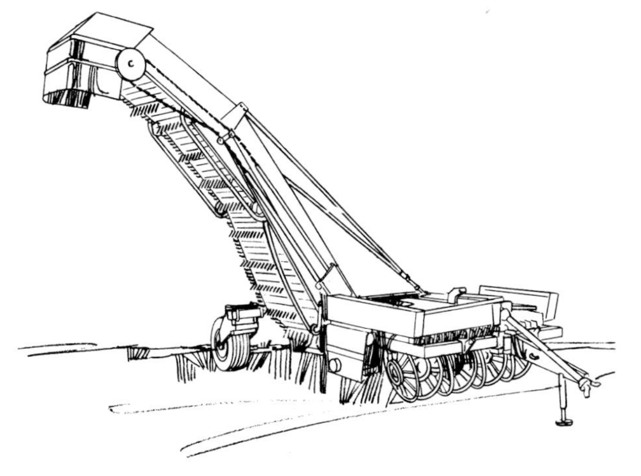

Fig. 21—Lifter Loader Sugar Beet Harvester

RIENK SCREEN

The rienk screen consists of star wheels that move the beets. To avoid getting caught in these star wheels, always disengage the PTO, shut off the tractor engine, and take the key before servicing, adjusting, or cleaning the screen.

Check the rienk screen periodically, and clean up any dirt or mud that might have accumulated (Fig. 23).

CROSS CONVEYOR

The cross conveyor on a tank harvester may be either a spud chain conveyor or a set of grab rolls. Cross conveyors on lifter loaders also consist of grab rolls.

Fig. 22—Tank Sugar Beet Harvester

Always disengage the PTO and shut off the tractor engine before servicing or adjusting these mechanisms. Grab rolls, especially, can catch a hand or a piece of loose clothing, possibly causing a serious injury.

LIFTER LOADER HARVESTERS

The elevator that lifts the beets into the truck is a unique feature of the lifter loader harvester.

Fig. 23—Clean Dirt And Mud Off The Rienk Screen Frequently

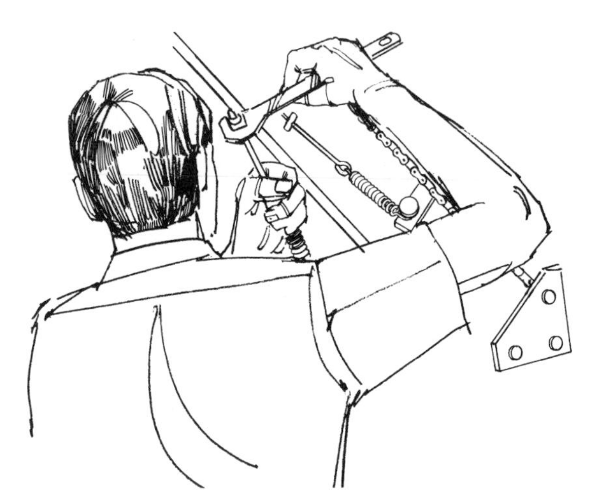

Fig. 24—Disengage The PTO And Shut Off The Engine Before Working On The Elevator

Fig. 25—Don't Get Caught In The Auger—Disengage The PTO, Shut Off The Engine, And Take The Key Before Cleaning Or Adjusting It

The conventional elevator is made up of links of spud chain with flights attached. It is usually operated from the tractor by an electric clutch.

The top section of the elevator chain drops dirt through to the ground. Check this chain for worn links and proper tension. Worn chain can unhook or break, and the chain could fall or jam. Shut off the engine, disengage the PTO, and take the key before servicing the elevator (Fig. 24).

Some lifter loaders have a vertical auger and a horizontal spud-chain conveyor to reach over trucks. Stay alert when servicing the auger and spud chain. Check the bearings on the auger to make sure they're in good condition. Check the horizontal conveyor for worn or bent links.

Fig. 26—Check The Vertical Elevator Often For Bent Or Worn Links

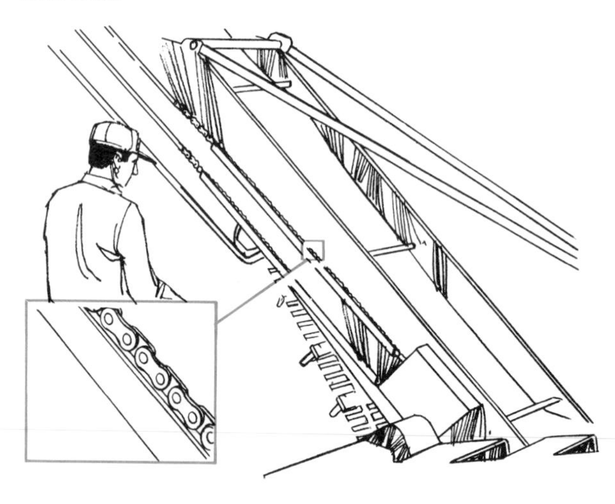

Always disengage the power takeoff, shut off the engine, and take the key when servicing the auger-chain combination to prevent amputation injuries (Fig. 25).

TANK-TYPE HARVESTERS

The tank-type harvester can be used as a lifter-loader-type harvester, or it can store beets in a tank for unloading at the ends of the rows. This type of harvester has two unique parts:

- **Vertical (rear) elevator**
- **Tank**

VERTICAL ELEVATOR

The vertical elevator lifts the beets discharged by the rienk screen to the top of the tank and dumps them onto the upper cross conveyor. Shut off the tractor engine and PTO before inspecting this elevator for worn or bent links (Fig. 26).

TANK

The tank is usually square and funnel-shaped. It may be lined with plywood to reduce the shock of the beets as they are dumped over the edge of the upper cross conveyor.

Stay out of the tank while the engine is running (Fig. 27). A beet could fall off the upper cross conveyor and injure a person in the tank.

Be sure to disengage the power takeoff and shut off the tractor engine before you scrape dirt, mud, or muddy beets from the tank in wet conditions.

PREPARING THE BEET HARVESTER

A beet harvesting season can be much safer and more successful if you make the proper preparations before starting for the field. The harvest period is frequently short, due to weather conditions or other reasons, so make sure your harvesting equipment is in top-notch shape to get maximum capacity from it.

Beet harvester components often wear excessively, because they are exposed to dirt, mud, and stones. You should get your harvester out during a slack period at least four weeks before harvest and go over it carefully, so it will be ready when you need it. This will also help prevent breakdowns that could result in injury when making repairs.

PREPARING THE TRACTOR

The tractor should be prepared before it is hitched to the beet harvester. Proper preparation will result in an easier harvesting operation. Check the operator's manual for your particular machine to get the specific information you need to prepare the tractor.

It may be wise to take the three-point-hitch arms off the tractor (Fig. 28). They can get tangled up in the hydraulic hoses and electric wires when you turn at the row ends. This can lead to hose breakage and other problems that could present hazardous situations. You may have to install a hitch support on the tractor for certain beet harvesters.

ATTACHING BEET HARVESTER TO TRACTOR

The harvester is easily attached when it is done correctly. Refer to the operator's manual for specific instructions. Be alert for pinch points such as those involving a hydraulic cylinder and/or some other moving parts. Shut off the tractor and set the brake before dismounting to hook up the harvester.

Fig. 27—Stay Out Of The Harvester Tank When The Engine Is Running

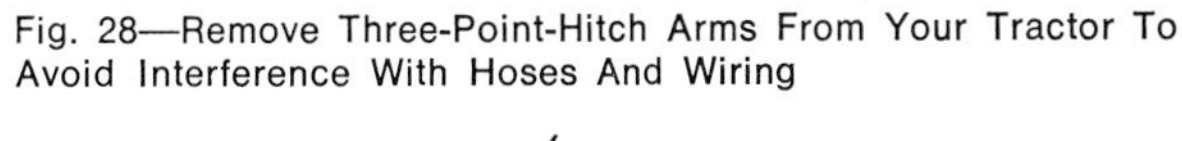

Fig. 28—Remove Three-Point-Hitch Arms From Your Tractor To Avoid Interference With Hoses And Wiring

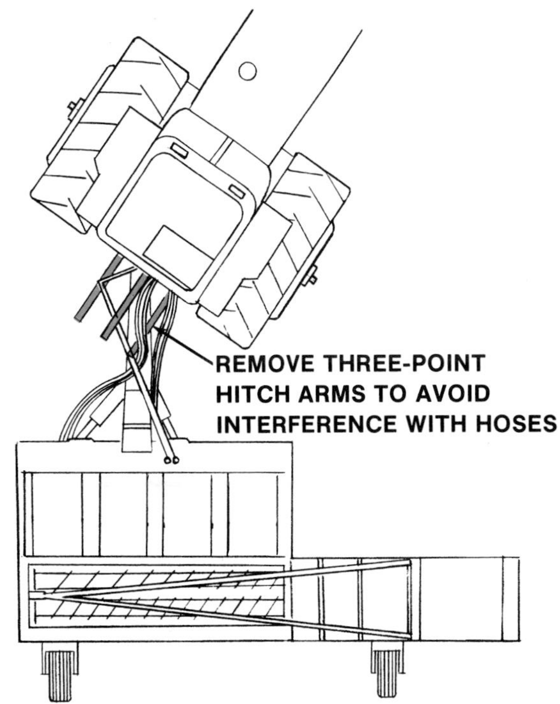

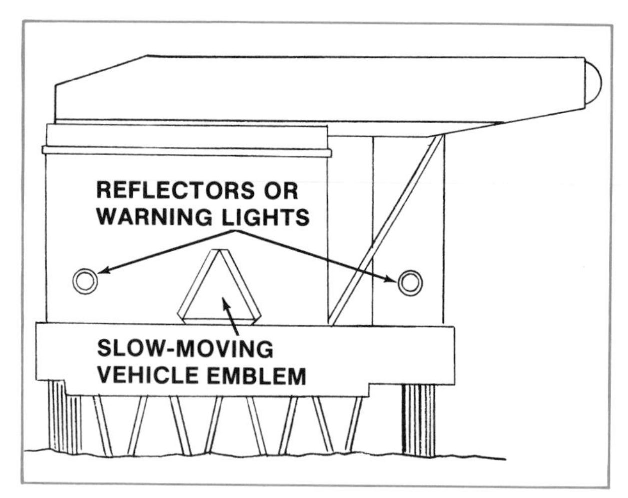

Fig. 29—Equip Your Harvester With Adequate Reflectors And An SMV Emblem

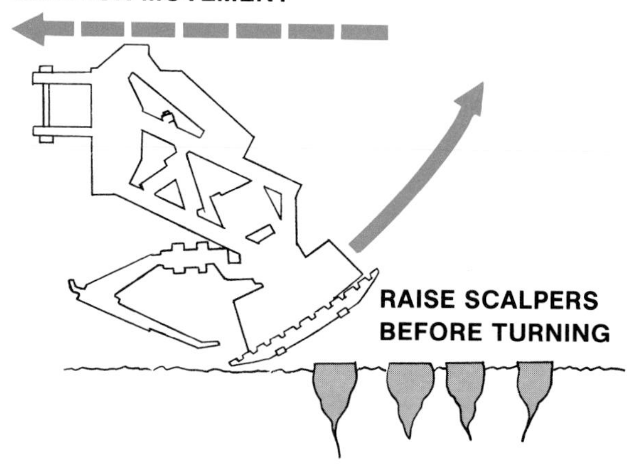

Fig. 30—Raise Scalpers Or Toppers Before Turning

TRANSPORTING THE HARVESTER

Before transporting a beet harvester, equip it with a slow-moving-vehicle emblem and reflectors (Fig. 29).

Remember that the unloading elevator projects out and may hit utility poles, trees, or buildings. Keep an eye out for low power lines and tree limbs. Swing the unloading elevator around into transport position if your harvester is designed to permit it.

Always lift scalpers or toppers when turning or backing up (Fig. 30). Failure to do this can result in serious damage to the equipment. These components can be hazardous if proper practices are not followed. Toppers and defoliators can hit stones or rocks and throw them outward. Shut off the engine when servicing or when someone is approaching.

OPENING A FIELD

When opening a beet field, always harvest the end rows first. A field should have at least 12 end rows to allow for easy turning. Make wide turns while harvesting end rows so that the harvester equipment isn't damaged. Check ditch banks for washouts and loose soil. Remember that the unloading elevator sticks out on the side and may get in the way of poles or trees on field edges (Fig. 31).

HARVESTING SUGAR BEETS

While harvesting, *listen* to your harvester. If a rock or stone gets caught inside, a slip clutch will usually make a clicking sound. If this happens, stop quickly and shut off the PTO. Then try to ease the PTO back into gear. If the clutch still slips, turn off the engine, disengage the PTO, and get off the tractor to locate the problem.

An easy way to get a jammed rock out of the rienk screen is to turn the machine backwards a little bit. This can be done by disengaging the PTO, placing a bar in a universal joint on the harvester power-shaft, then turning it backwards.

A stuck spud chain may have to be disassembled before it can be made operable again. Sometimes a link will slide behind the chain runners and stop the machine. A bent flight could also get hung up on

Fig. 31—The Unloading Elevator Projects Out To The Side—Be Careful!

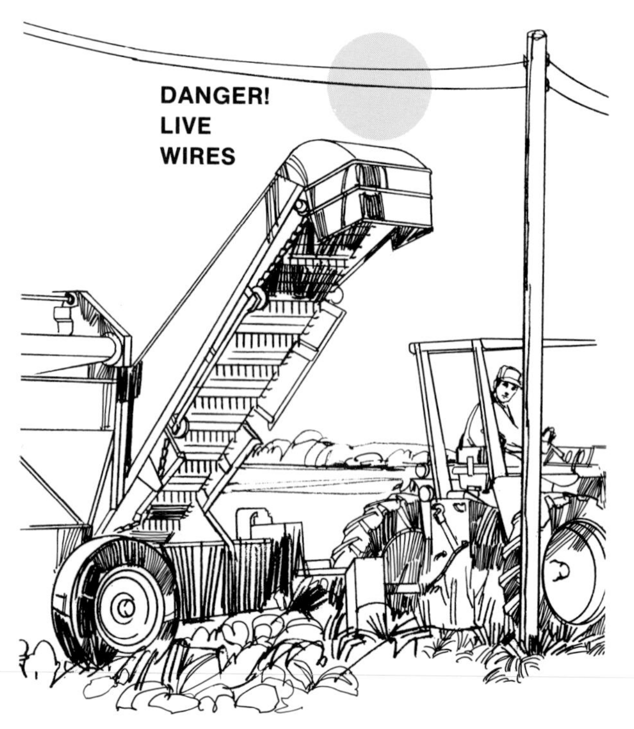

something and cause a clutch to slip.

If a drive chain breaks or comes off, you may have to unplug the harvester manually before it will operate with a new or repaired chain. Watch out for falling beets and pinch points when servicing the harvester.

UNLOADING ON-THE-GO

Unloading on-the-go requires the alertness of both the harvester operator and the truck or wagon operator.

To unload on-the-go, let the truck pull up to the harvester, which should be holding a steady speed. The truck drive can then judge how close to get to the harvester. Hand signals should be used so the two operators can communicate with each other (see Chapter 1).

This technique puts extra responsibility on the harvester operator, because he has to watch the harvester and where it's going, and must watch the truck at the same time. A lot of things are happening at once, and each operation must be observed carefully. Be cautious, stay alert, and take a short break every hour or so to stay that way.

CHAPTER QUIZ

1. True or false? A cotton harvester is most stable and less apt to upset when the basket is raised.

2. List two ways to prevent cotton harvester chokedowns.

3. Which of the following practices will help reduce the fire hazard when working with a cotton harvester?

a) Keep the engine free of lint and trash.

b) Check exhaust pipes and mufflers regularly for leaks.

c) Use a petroleum-based moistening agent.

d) Keep battery terminals clean.

4. True or false? If you are already familiar with a machine, reading the operator's manual is not important.

5. Why is it especially important to change gears and drive speed smoothly when operating a potato harvester?

6. Why should you disengage all power and shut off the engine when servicing a machine?

7. True or false? The tractor operator is the key person responsible for the safety of a potato harvesting operation.

8. Name a potential hazard of standing in the tank of a sugar-beet harvester while the engine is running.

9. What part of a beet harvester projects out and may cause problems during transportation?

a) Rienk screen

b) Tank

c) Unloading elevator

d) Scalpers

10. How does unloading on-the-go place an extra burden on the harvester operator?

11
Target: Materials Handling Equipment

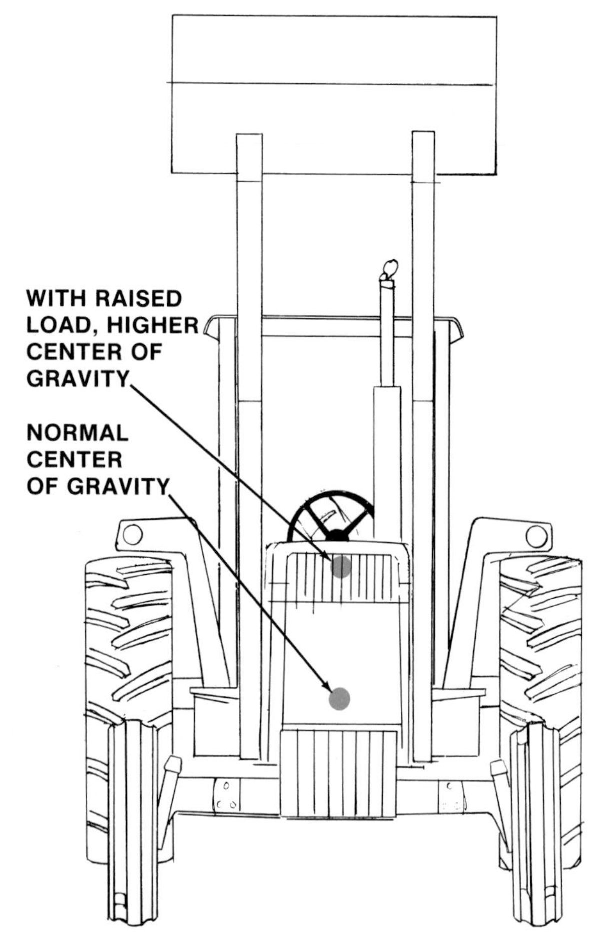

Fig. 1—Loader Height Changes The Center Of Gravity And Stability Of The Tractor-Loader Combination

INTRODUCTION

More materials are being moved on today's farms than ever before. This increased volume demands increased mechanization of materials handling.

Many types of materials are moved, and many devices are used to move them. This chapter discusses safe operating procedures for some of the more common types of materials handling equipment:

- **Front-end loaders**
- **Forklifts**
- **Manure spreaders**
- **Grain bins**
- **Portable elevators and augers**
- **Farm wagons**
- **Silo unloaders**
- **Crop dryers**
- **Grinder-mixers**

FRONT-END LOADERS

Loaders are versatile — they handle feed, manure, bedding, and other materials. Hydraulic fingertip controls permit the operator to load, lift, and control-dump large quantities of material with ease and efficiency.

Loaders are potentially dangerous when not operated properly because they change the center of gravity and stability of the tractor-loader combination. This presents a potential tipping hazard to the operator. For example, large round bales of hay raise the center of gravity significantly when raised by a front-end loader. Use a bale clamp to help secure the bale so there is less danger.

The way the center of gravity moves as the load is raised is illustrated in Fig. 1. Once the center of gravity is moved outside its base of stability, tipping will occur. A few general precautions can reduce the chance of this happening:

1. *Keep the bucket low while carrying loads and while operating on inclines.* Raising the bucket moves the center of gravity up and out. Bumps, holes, rocks, loose fill, or stumps can easily upset a tractor if the bucket is carried high. It also places undue stress on the loader frame. Drive loaded buckets uphill rather than downhill, and stay off steep slopes to prevent bouncing and loss of control (Fig. 2).

2. *Keep travel speed slow.* Speed and height are the two most critical factors when working with loaders. Keep speed down when loading and transporting. Excess momentum causes the tractor to overturn in situations where there would ordinarily be little dan-

Fig. 2—On Steep Inclines, Keep The Bucket Uphill From The Tractor To Prevent Upsets

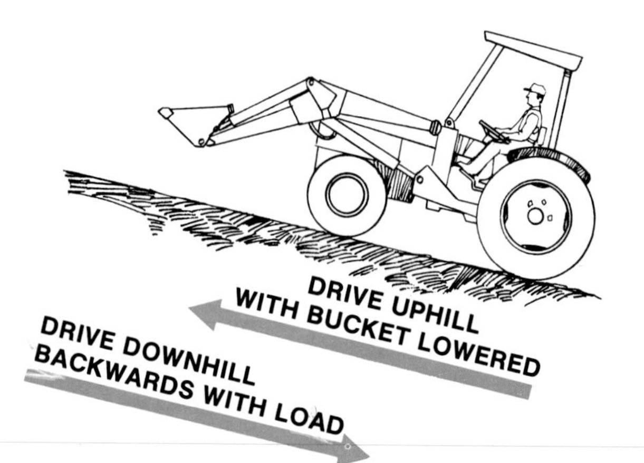

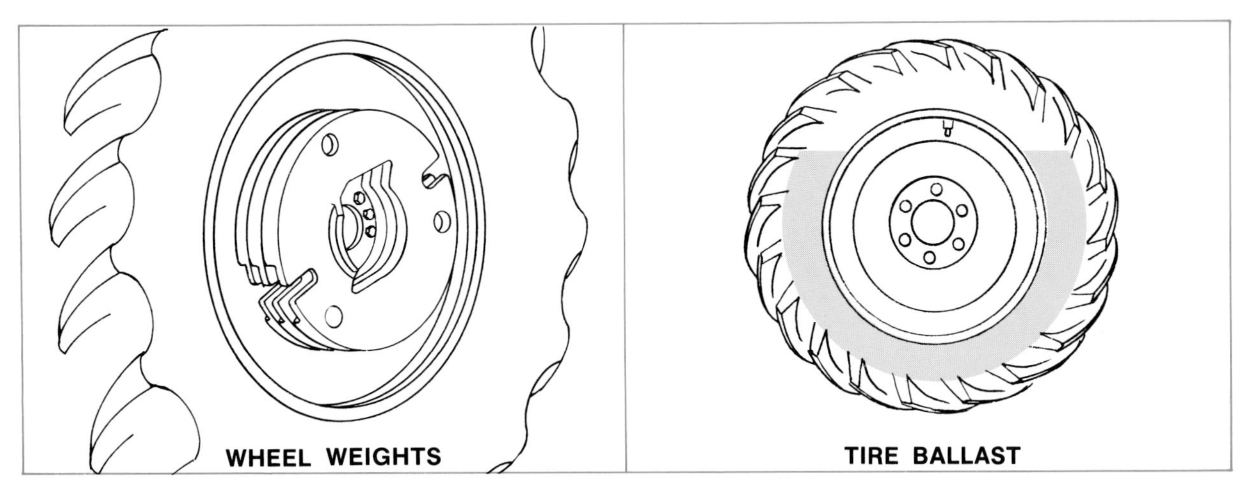

Fig. 3—Use Ballast For Added Tractor Stability

ger. In addition to the danger of overturning, you may damage the equipment, especially on rough ground. Be especially careful when making turns with the loader raised.

3. *Use the recommended amount of ballast to give your tractor extra stability (Fig. 3).* This will counterbalance the extra weight on the front of the tractor and help prevent tractor upsets. Check your operator's manual for specific recommendations for your tractor. More information can be found in Chapter 5.

4. *Load the bucket evenly from side to side and keep within the normal capacity of the tractor and loader.* This will help prevent upsets. The amount of material that can safely be loaded in the bucket will vary tremendously with its density. While it might be safe to fill the bucket with sawdust, the same volume of wet sand could cause a tractor upset. The capacity of your loader is specified in the operator's manual.

5. *Never tow a tractor by attaching a tow chain or cable to the loader.* Towing by the loader can overturn the tractor.

6. *Remove the loader when it's not in use.* Don't use a tractor with a loader when performing tillage operations — it's harder to handle, visibility is bad, and fuel is wasted.

7. *Be especially careful if you use the machine to compact ensilage piles.* Ensilage is slippery and can easily cause an upset.

Operator controls are designed to be used by the operator while sitting in the operator's seat. Operating the controls from the ground or from behind the tractor causes erratic engagement of controls, and could result in an operator getting crushed or entangled in the mechanisms.

Remain at the controls until the hydraulic cycle is complete, and do not allow riders on the tractor or loader. Riders could fall off the loader while the tractor is moving. There usually is not enough time to get the tractor stopped before they are run over — to say nothing of the injuries received in the fall.

Do not use front-end loaders or hydraulic lifts for people. Loaders are too unstable, buckets can tilt, hydraulics can fail, and accidental control movement dump people out.

Lower the bucket to the ground before leaving a tractor. Then the bucket can't be lowered accidentally (Fig. 4).

Fig. 4—Operator Should Have Lowered The Bucket Before Leaving Tractor

CHILDREN PLAYING WITH CONTROLS CAN ACCIDENTALLY LOWER BUCKET!

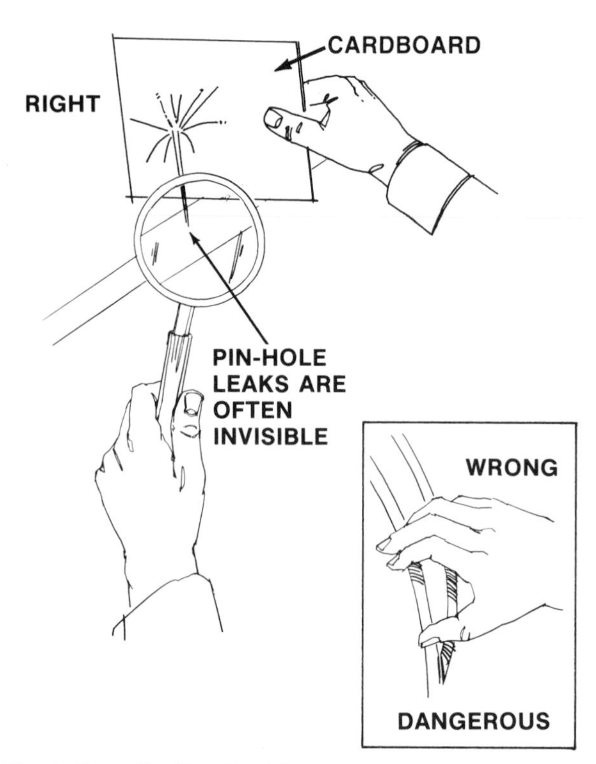

Fig. 5—Never Use Your Hand To Detect Pinhole Hydraulic Leaks

Fig. 6—Check PTO Shield Often To Determine If It Turns Freely

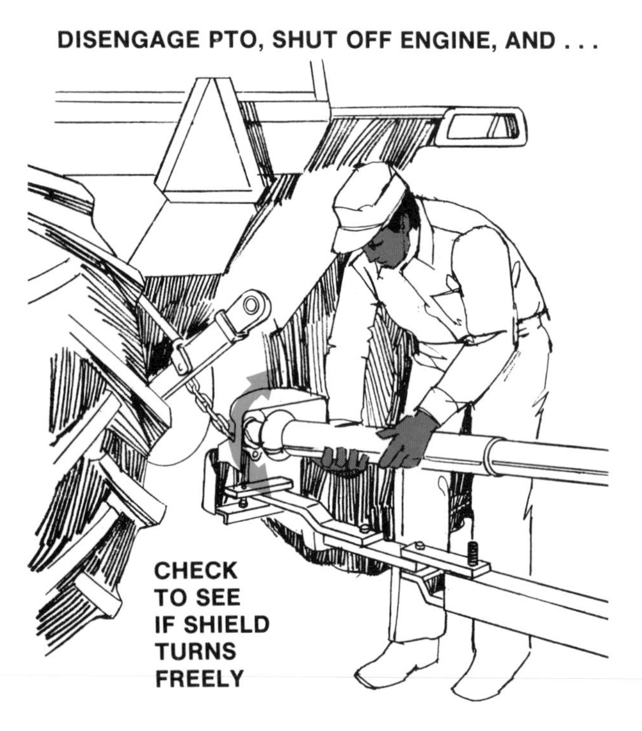

Keep hydraulic connections tight. Repair leaky hoses or connections as soon as you discover them. A high-pressure stream of escaping hydraulic fluid can penetrate skin and cause serious infection or reaction, in addition to physical damage to internal flesh. Use a piece of cardboard or wood to detect pinhole leaks—not your hand (Fig. 5).

Loaders are not designed to be used as battering rams. Using a loader in this way will greatly shorten its life, and may cause an injury if something breaks.

FORKLIFTS

Lifting and moving large quantities of material is done with forklifts on many specialized farms. Fertilizers, seed, fruits, and vegetables are examples of some of the materials moved in this way. Both self-propelled and tractor-attached units are available.

Evidence suggests that many operators are not aware of the potential dangers involved in operating forklifts. The following precautions can help you avoid serious accidents:

1. *Use of some type of guard and load backrest is recommended for forklifts.* This safety equipment will keep the operator from being crushed under heavy loads. It is difficult to balance loads on the forks unless they can be tilted against some kind of backrest. Trying to balance a load while transporting it over rough terrain is hazardous, both in terms of operator safety and danger to the material being transported. Spillage can be costly as well as dangerous. Spreading the forks as wide as possible and using the backrest will help reduce this danger.

2. *Don't overload the forklift or load unstable loads.* Operator's manuals specify the maximum loads that should be carried on the forks. Overloading will eventually result in breakage, reduced operating life, or accident. Avoid very wide or high loads.

3. *Pick up the load in the center so that it is properly balanced on the forks.* Analyze your load in terms of weight and stability, and take appropriate precautions. This may involve reducing the size of the load, restacking, or fastening the load to the lift.

4. *Use extreme caution when handling loads that reduce visibility.* Face the direction of travel, and have a second person use hand signals where necessary (see Chapter 1). Look carefully in all directions before moving to see that the way is clear.

5. *Start moving the load only after it is lifted five or six inches off the ground.* Don't raise the load any higher than necessary, and keep it low while traveling. Carrying loads too high affects the stability of the unit because the center of gravity is changed. Two or three feet should be considered a maximum

Fig. 7—Heavy Objects May Be Thrown Over 100 Feet (30 m)

height for transporting a load. Complete the lifting process when you get to the point where the load is to be deposited.

6. *When turning the forklift, allow for rear-end swing.* Remember that the rear wheels do not follow the track of the front wheels on turns, and may cause damage to material stacked along the aisles. Avoid upsets by never attempting to make turns on ramps, even when the forks are empty.

7. *Avoid sudden stops, starts, and changes in speed when transporting loads.* This calls for alert advance planning of all movements. Use the clutch and brake gently, and don't use reverse as a brake, because the tractor could tip and you could lose control. Move slowly and use the horn and other safety devices, especially near blind corners, to reduce the risk of accidental collisions.

8. *Don't operate a forklift inside a closed building.* Open *all* doors and windows to provide ventilation to prevent carbon monoxide poisoning.

9. *When parking the lift, lower the forks to the ground and shut off the engine.* Use the specified parking procedure—lock the brake and/or put the gear in park position, and remove the key. This will help prevent injury to others while you're away from the forklift.

MANURE SPREADERS

Manure spreaders can be classified into three types:

- **Beater**
- **Flail**
- **Liquid**

All three types are generally drawn behind a tractor and powered by the PTO. The general safe operating rules for PTO equipment given in Chapter 5 apply to manure spreaders, too.

Only one person is needed to operate the machine. When two or more people are working together, one may turn on the machine and accidentally cause an injury to the other. Always check to see if anyone is nearby and warn him of your intentions before starting the engine or engaging the PTO.

Operate only when seated properly on the tractor. Never allow anyone to ride on the drawbar or hitch because of the great danger of falling or getting caught in the PTO. And never use the PTO shield as a step for mounting the tractor.

Be concerned about anyone within 100 feet (30 m) of the sides and rear of the spreader. Ask them to move away before you engage the PTO so that they will not be hit by possible thrown objects.

Check often to see that PTO shields turn freely (Fig. 6). Never make this check when the PTO is in operation. You must make sure that the shield is not somehow bound to the shaft and become a hazard in which you could get caught. Keep hands, feet, and clothing away from all moving parts, and correct the situation immediately.

Like most machines, manure spreaders get plugged from time to time, presenting a situation that could be dangerous.

Excessive plugging can be prevented if you:

1. *Keep stones, boards, and other solid objects out of the spreader.* These not only may plug the spreader, but could cause breakage or accidental injury to someone if hurled from the machine (Fig. 7).

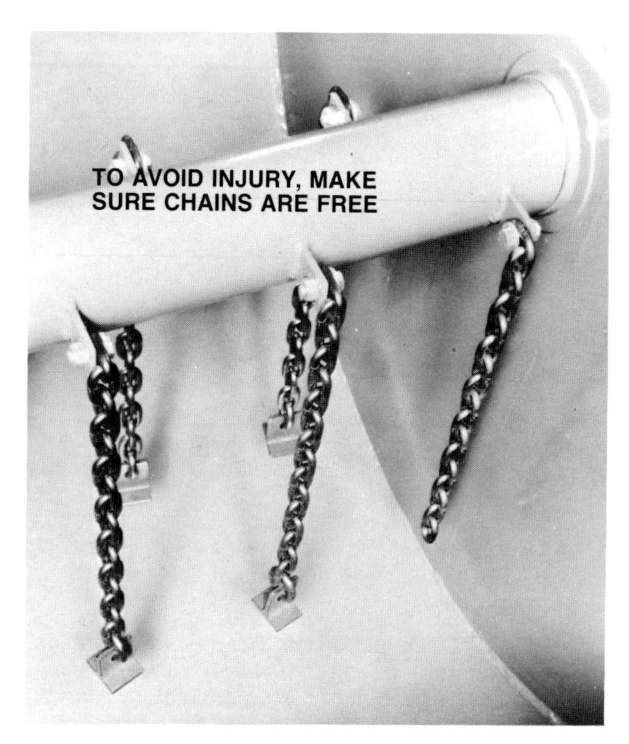

Fig. 8—Loosen Chains Before Loading Flail-Type Spreaders In Cold Weather To Prevent Chances Of Injury Later

2. *On flail-type spreaders, make sure flails and chains are loose before loading in freezing weather (Fig. 8).* If frozen, loosen them with a prybar. Don't try to break them loose by repeatedly engaging the power supply. It is wise to follow this precaution even if the spreader looks okay, because you might find that chains and flails are frozen and attempt to free them without taking safety precautions, such as disengaging the PTO and turning off the engine.

3. *On beater-type spreaders, keep chains and beater mechanisms in good working order.* Replace stretched chains. When dropping loads into the spreader, try not to hit the beater mechanism and chains with the bucket or drop heavy loads of frozen manure from excessive heights. Never get in a spreader to clean it if it's running. The gathering chains can grab you and pull you into the beaters.

Good maintenance helps prevent problems that could lead to accidents. If the spreader *does* become plugged:

1. *Stop the tractor and park properly.* Use the brakes or parking gear to keep the tractor in place. Never rely on gears to keep a heavy load from moving. On an incline, the weight of the tractor and spreader could create a force strong enough to turn the engine over and get the tractor moving.

2. *Disengage the PTO and shut off the engine before doing any work around the spreader.* This will prevent accidental starting of the machine.

3. *Determine and correct the cause of plugging.* Perhaps a rock or a chunk of frozen manure has lodged in the beater mechanism. A drive chain could have broken. Or maybe the spreader was overloaded. Clear the beaters and make sure they turn freely. If you have to remove any shields or guards to make the necessary repairs, replace them before operating the spreader again. Never make any of these repairs without first shutting off the engine.

Maintain rotary flails. The peripheral speed and centrifugal force could cause rotating flails to come loose unless they're properly maintained. Replace flails when they become worn. Keep them bolted tightly, and check their condition regularly to detect danger (Fig. 9). Never approach the machine while it's in operation.

Fig. 9—Check Flail Fasteners Regularly. Keep Flails From Coming Loose And Injuring Someone

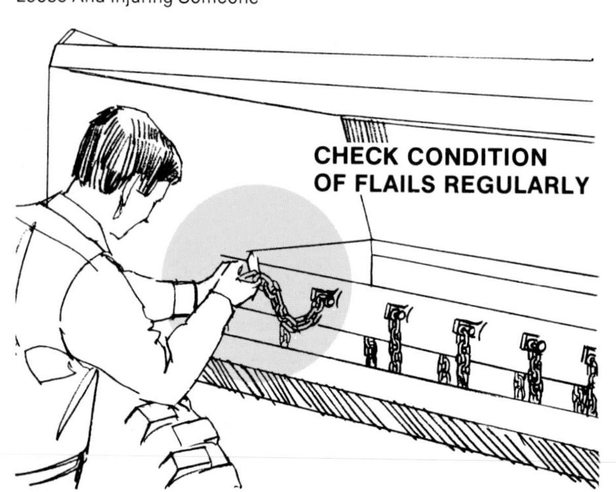

Fig. 10—Always Use A Safety Hitch Pin

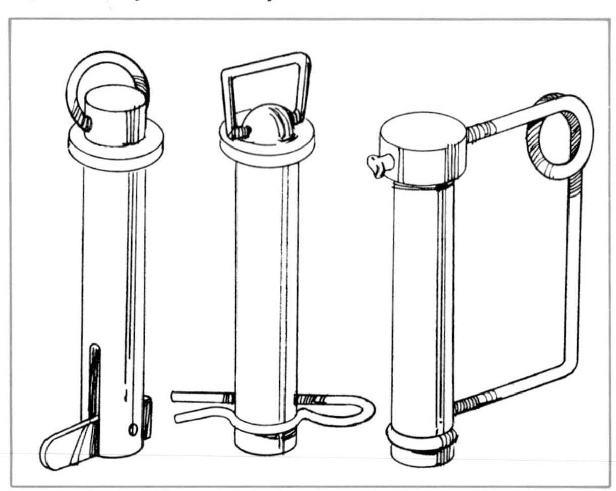

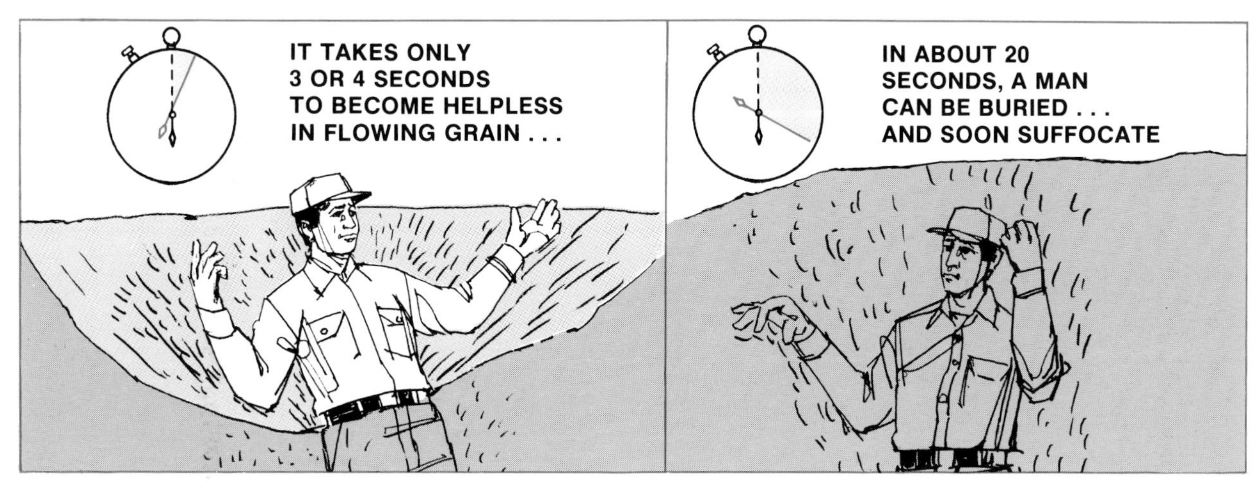

Fig. 11—A Man Occupies Only About 7 Cubic Feet Of Space And Could Be Completely Covered By Grain In About 20 Seconds

Always use a locking safety hitch pin when hitching the spreader to a tractor (Fig. 10). The weight of a heavy load cannot be relied upon to keep the drawpin in place, even on level ground.

Keep the jack stand in good working order. Before unhitching the spreader from the tractor, make sure the jack is locked securely in place. Never rely on an unlocked jack to hold the spreader. Even a slight movement of the tractor could make the spreader fall.

LIQUID MANURE SPREADERS

Liquid manure spreaders generally use a PTO-powered pump to fill large portable tanks and to spread manure. Here are some recommended procedures for their use:

1. *Flush the tank with water.* Never use gasoline. Fumes from gasoline may remain in the tank and can be ignited readily by a spark or discarded smoking materials.

2. *Make sure the relief valve is operative to avoid excessive pressure build-up.* Don't tamper with it.

Additional dangers in handling liquid manure center around the manure storage system. Stored liquid manure produces toxic gases. They are most likely to become a problem when the manure is agitated, so provide plenty of ventilation whenever it is necessary to agitate liquid manure, especially if the agitator has been off for several hours.

Never enter a storage tank unless you are wearing a self-contained breathing apparatus and have a rope tied around you and held by someone outside the danger area.

See Chapter 7 for more information on working with liquid manure.

GRAIN BINS

The large grain storage bin has become a common sight on the American farm. Though grain bins appear to involve little risk, there are a number of potential hazards you should recognize.

SUFFOCATION IN GRAIN

One of these dangers is accidental suffocation in grain bins. This often happens during the unloading of the bin when the grain may bridge and you have to break the bridge to get the grain flowing again.

The unsuspecting farmer who enters a bin with the unloader running may sink in the flowing grain before he realizes what has happened. *It only takes three or four seconds to become helpless,* a few seconds more to be submerged in the grain, and suffocation soon follows (Fig. 11).

The average man takes up only about 7 cubic feet (.20 cm^3) of space. It only takes about 10 seconds with an average auger to fill this space. That's how long it takes to be covered by flowing grain.

Here are some safety measures to follow to prevent these accidents:

1. *If grain bridges, shut off the unloader and use a pipe or some other long object to break the bridge and get the grain flowing again.* Never enter a grain bin while the unloading operation is going on (Fig. 12). Bridged grain may look perfectly safe from the top, but it could hide a cavity that you could fall through and be submerged almost immediately. It only takes a few inches of grain covering you to cause suffocation.

2. *If you must enter a bin, disconnect the power source and make sure no one can turn it on while you're inside.* If the grain should start flowing for some reason, stay near the outer wall and keep walking until the grain flow stops or the bin is empty.

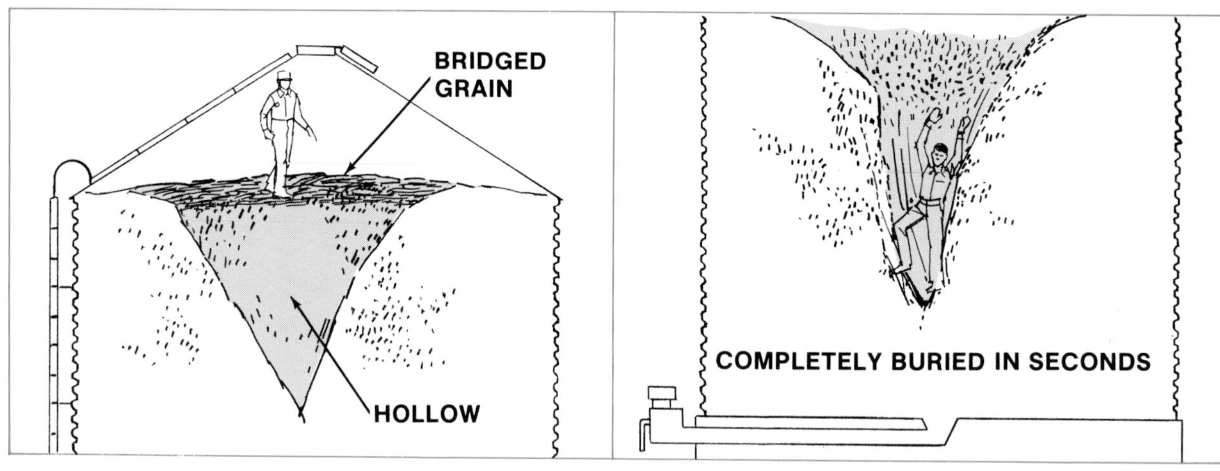

Fig. 12—A Partially Unloaded Bin May Be A Trap, Because Bridged Grain Could Hide A Dangerous Cavity

3. *Install ladders and safety ropes in all bins.* This will provide an exit if you need to get out, as well as a safe way of getting in. But remember that even if there is a ladder in the bin, you must be able to get to it. If you walk out to the center of the bin and get caught in flowing grain, you may not be able to reach a ladder on the side. Hang a safety rope from the center.

4. *If you must enter a bin, tie yourself with a rope and harness so you have a way of getting out.* Have two extra people handy in case something happens—one to hold the rope and one to get extra help if necessary (Fig. 13).

Another problem is being overcome by carbon dioxide. Carbon dioxide is given off by wet grain as it ferments. The CO_2 pushes oxygen out of the bin. Forced ventilation and breathing equipment can help.

5. *If someone is buried without a rope, shut off the auger, and open the emergency dump doors. If there are no emergency dump doors, knock a hole in the tank with a tractor. The victim will need artificial respiration.*

DUST AND FUMES

Spoiled grain or grain in poor condition often gives off a lot of dust that can be dangerous to workers, especially those with certain allergies or asthma. Cleaning out a grain bin can produce enough dust to cause such a reaction. Use a filter respirator that will remove fine dust particles. Keep it handy, and wear it when working in a grain bin, even if it seems uncomfortable.

PORTABLE ELEVATORS AND AUGERS

Elevators and augers are specialized devices used to lift and transport materials like grain, hay, and silage. The simplicity of elevators and augers often leads operators into thinking that they are safe. Unfortunately, these machines have become a major source of farm accidents and fatalities.

OPERATING ELEVATORS AND AUGERS

The primary danger in operating elevators and au-

Fig. 13—Wear A Harness And Rope And Have Someone To Help You When Entering A Bin

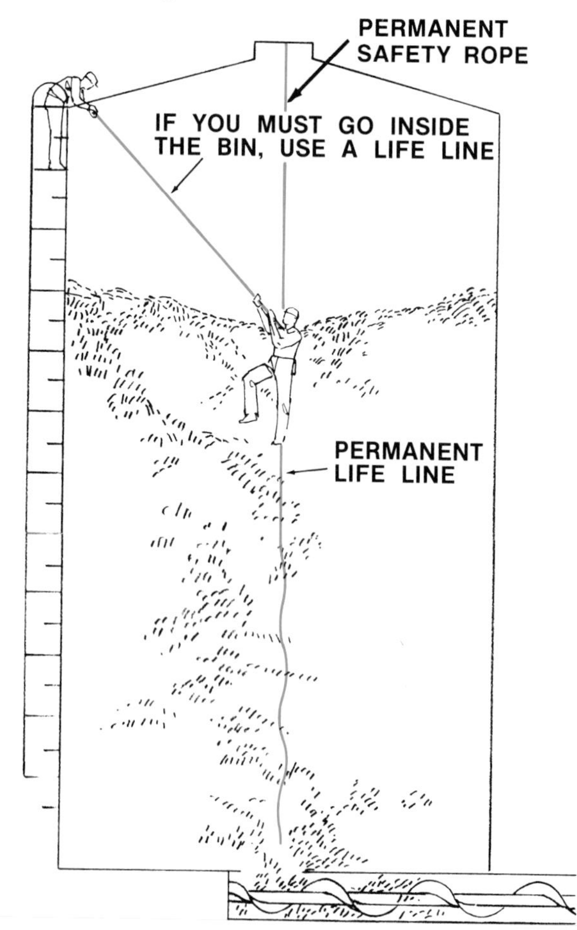

gers is getting caught in moving parts, such as augers, belts, or chains.

Allow workers to operate this equipment only after they are made aware of the potential dangers and safety practices. A few minutes spent in examining the equipment and operator's manuals to learn the potential hazards is well spent time.

Keep children away from the elevator, whether it's stored or in use. Elevators are not intended as slides or seesaws, and children should never be allowed to climb on them.

Don't wear loose, floppy clothing when working around elevators or augers. Such clothing is easily caught and can pull the wearer into the machine before he has time to stop the machine or release himself.

Operate the machine only with guards or covers in place. This minimizes the chance of getting caught in moving parts or allowing objects to fall into the mechanisms. If your machine does not have guards over potentially dangerous moving parts, order them from the manufacturer or build them yourself. The investment could save your life.

Run your elevator or auger no faster than necessary to move the material efficiently from the inlet. For example, on a bale elevator, the bales should be able to be fed into the elevator end-to-end. That is, one bale touching the next. Using the elevator at excessive speeds with only one or two bales on it at a time puts unnecessary wear on the elevator and

Fig. 15—Keep Elevator Low When Passing Under Power Lines—Don't Get Electrocuted

Fig. 14—Allow Plenty Of Space For Turning

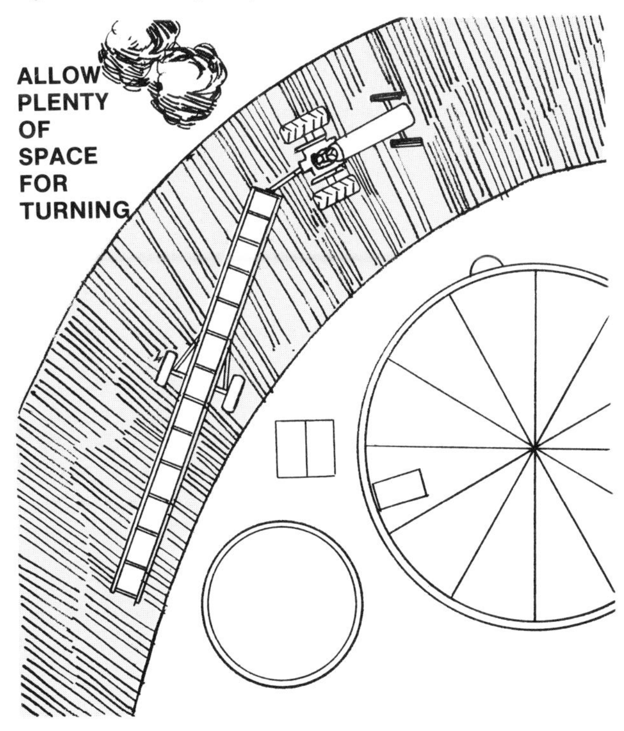

increases the chances of workers getting caught in the mechanism. The slower speed may also allow the elevator to be stopped before causing damage if a bale gets caught. Keeping the bales spaced evenly also helps to balance the load on the undercarriage and keep the elevator from upending.

TRANSPORTING AND POSITIONING ELEVATORS AND AUGERS

Use extreme care when transporting portable elevators and augers. Transport them only in a lowered position with the safety locking device in place. Elevators become top-heavy when they are raised (the same as trying to carry a ladder in an upright position), so the proper way to maneuver them is by keeping them attached to the tractor. Keep in mind that the wheels do not follow in the tracks of the tractor on turns. In addition to possibly hitting a post or building, there is also the danger of the tractor tires hitting the elevator. So avoid sharp turns when towing an elevator or auger. Allow plenty of turning space (Fig. 14).

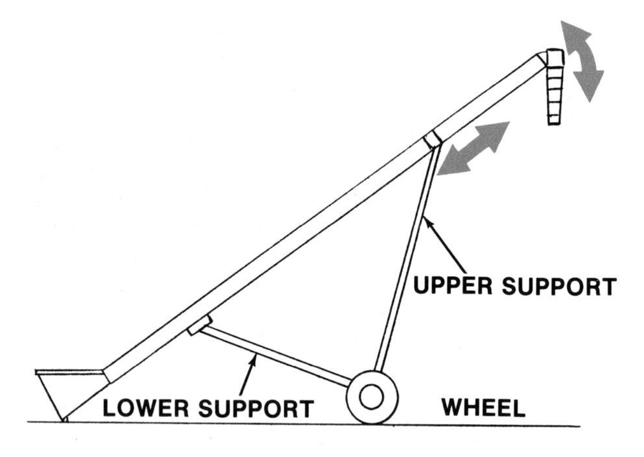

Fig. 16—Basic Construction Of A Portable Elevator

Fig. 17—How Elevator Collapses Can Occur

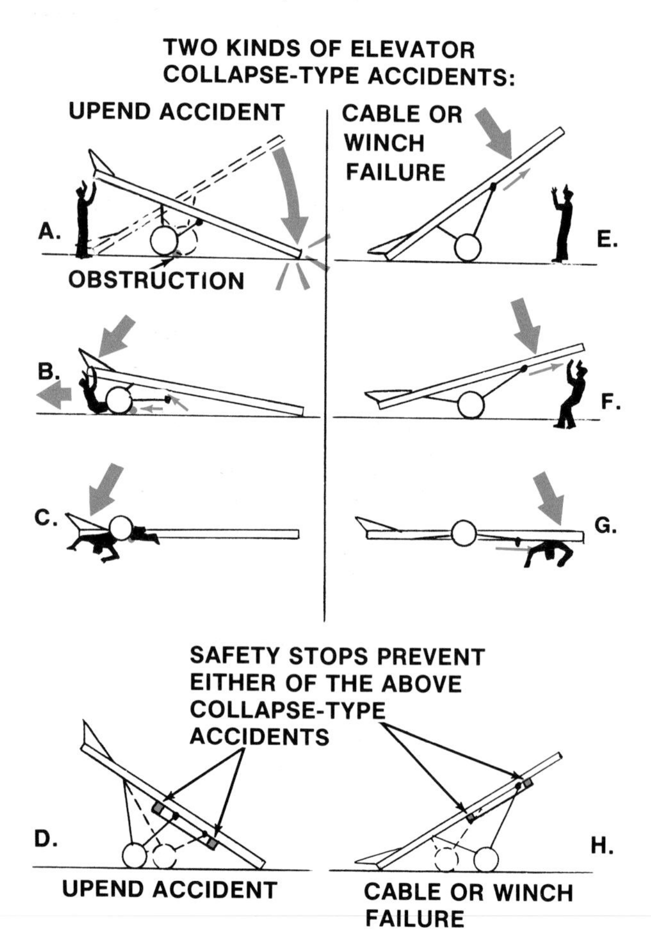

On public roads don't exceed safe travel speeds for the road and equipment condition. Use a flag to mark the end of the elevator, and follow local traffic regulations concerning the use of lighting or reflectors when transporting elevators on public roads. Contact local or state police authorities if you have questions.

Check for power lines when positioning a raised elevator or auger (Fig. 15). Keep the elevator clear of power lines. When rubber tires are the only point contacting the ground, *you* could become the perfect conductor to complete the circuit if the elevator contacts a power line.

Check the condition of the cables regularly. When cables fray or are partially cut, they may suddenly give way and allow the elevator to fall and cause injury to the operator or others. Determine and correct the cause of the faulty cable and replace it. Make sure the cable clamps are tight, and always keep at least two turns of cable on the windlass. This will help reduce tension on the cable clamps and reduce the chance of cable failure.

When raising or lowering an elevator, make sure the cable wraps properly and does not wind off the windlass. And see that the support arms are positioned properly. If one wheel is lower than the other when the elevator is lowered into position, the frame may tilt to one side so that when you raise it, the arms may not align properly. If proper care is not taken, the elevator may fall off the side of the frame and damage the elevator or injure the operator.

Most elevators and augers are equipped with a safety device to keep the elevator from being raised too high. Raising the elevator too high increases chances of upsetting. *Do not alter this safety feature.* If your elevator doesn't have this feature, build a safety stop or warn other workers of this potential hazard. Red paint on the cable or elevator marking the danger point where the elevator may collapse could be used as a warning.

In addition to contact with power lines, collapse of the undercarriage is one of the most common causes of fatal accidents involving portable elevators.

Construction of the undercarriage of a typical elevator is shown in Fig. 16. The lower support is attached to the elevator and is free to rotate about its attaching point. The upper end of the upper support is moved along the bottom of the elevator to raise or lower the elevator to the desired height.

Two types of collapses can occur if the elevator handler is not careful:

- **Upend accident**
- **Cable or winch failure**

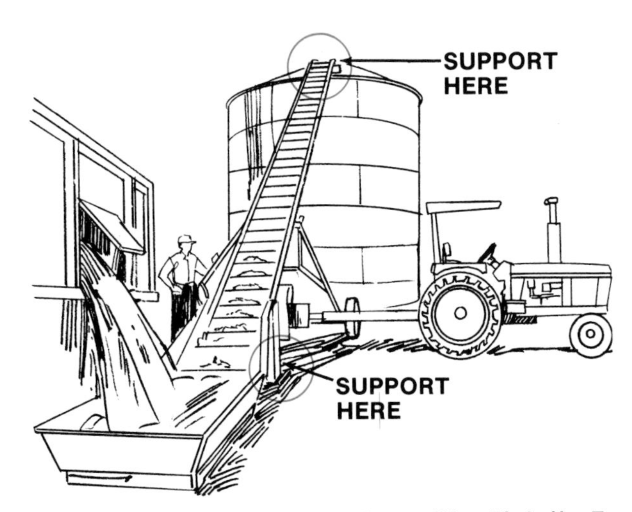

Fig. 18—Support Both Ends Of The Elevator When It's In Use To Prevent Tipping Or Loss Of Control

Here's a look at how these accidents can occur.

The upend collapse-type accident usually occurs when the elevator is being moved by hand. For example, when the elevator is raised too high, as shown in Fig. 17A, the upper end becomes top-heavy. If the wheels hit a hole or obstruction, the handler may lose his grip on the elevator and the upper end will fall to the ground as shown in Fig. 17B. The supporting undercarriage collapses, as illustrated in Fig. 17B and C, when the wheels roll to the rear. Elevator safety stops, as shown in Fig. 17D, can help prevent such a collapse.

The cable or winch failure collapse-type accident occurs when the cable or winch fails to hold the elevator up, as shown in Fig. 17E. This can also happen if the elevator upends. The upper support slides out toward the end and allows the elevator to fall, as shown in Fig. 17F and G. Elevator safety stops, as shown in Fig. 17H, can help prevent such a collapse.

You can also help avoid such accidents by being very careful when handling portable elevators.

If an elevator begins to upend, do not hang on and try to stop it. Get out of the way. Otherwise you may be crushed under a collapsed elevator. Safety tracks that don't allow the trough to be separated from the undercarriage can also prevent collapsing. Keeping the elevator attached to the tractor while maneuvering it will also prevent such an accident.

Remove any objects blocking the wheels when attempting to raise or lower an elevator. When the elevator is being raised or lowered, the wheels must be free to move if the elevator is attached to a tractor or truck. When the elevator is in position for operation, support both ends and block the wheels to prevent slippage or teetering that could cause the elevator to shift position or to nose over when the material reaches the top (Fig. 18).

Another danger involved in raising or lowering some elevators is the accidental release of the cable crank. Be very careful not to allow the crank to be released and spin freely. If this happens, arms and fingers can be broken or serious head injuries can result. It is nearly impossible to get out of the way in time to avoid being hit. Never try to stop a spinning crank (Fig. 19). Many elevators are equipped with safety clutches to keep this from happening. Make sure this clutch is maintained properly. Replace any worn or broken parts and check adjustment periodically as recommended in the operator's manual.

Do not ride or climb on the trough of a bale or grain elevator. The lack of handholds, coupled with metal surfaces that become slippery with wear, provides a hazard that could lead to a slip or fall. And your weight at the top of the elevator could cause it to tip forward. Add these to the danger of getting fingers, feet, or clothing caught in moving parts, and you have a set of circumstances just right for an accident.

Disengage the power source before removing chaff or caught bales from the ends of the elevator. Position the elevator so that the chaff or bales fall freely away from the end to eliminate the danger of getting entangled in the chains and sprockets at the end of the elevator.

Fig. 19—Don't Try To Stop A Spinning Crank—You Could Be Seriously Injured

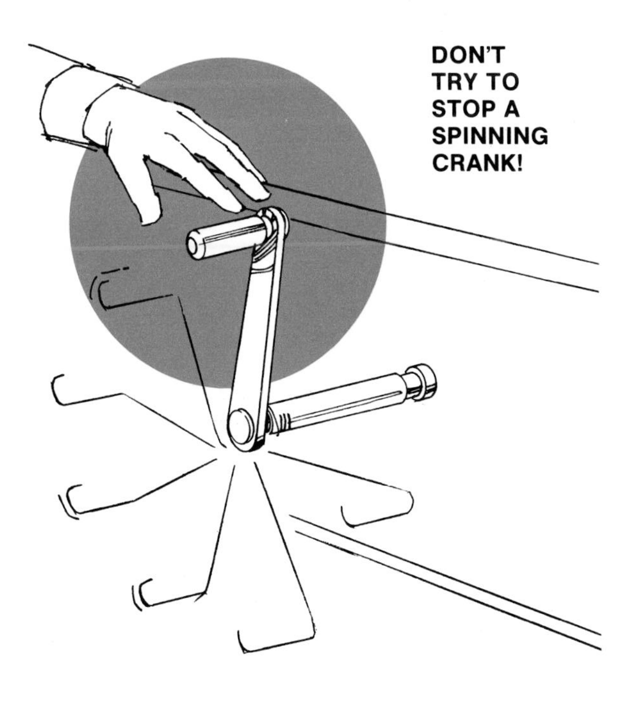

Fig. 20—Know Load Limits And Don't Exceed Them

Augers, in general, are reputed to be among the most potentially hazardous types of farm equipment. Injuries often involve the loss of a hand or foot. Extreme caution must be taken to prevent feet or hands from getting into an auger. The best way to do this is to use shields. If you remove shields for repair or maintenance, replace them immediately.

Take precautions to prevent falling or slipping into the auger. Keep tools picked up and away from the auger, and keep the area clear of ice, mud, grease, or grain that could cause a fall. Never use your hand or foot to push material into a plugged auger. Always use a stick or rod. If a machine gets plugged, shut off the power before attempting to remove the obstruction.

Fig. 21—Stay Away From Banks And Ditches

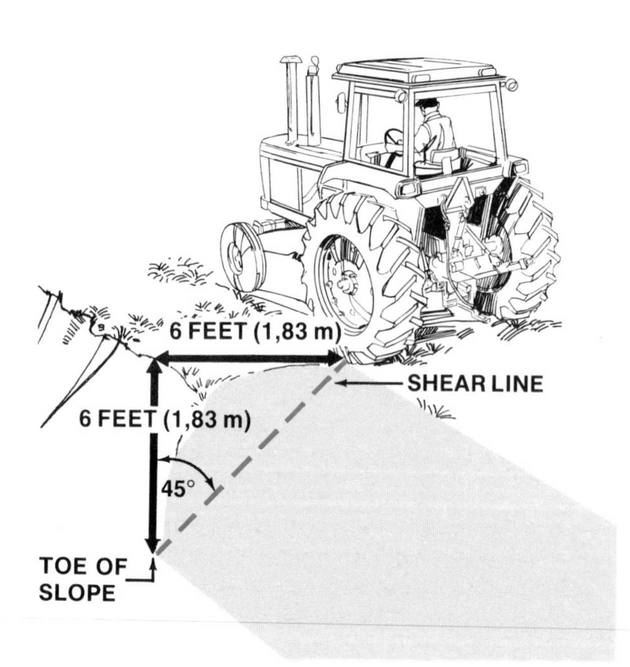

FARM WAGONS

There are several kinds of farm wagons from flatbed wagons used for hand stacking and hauling bales, to PTO-powered, self-unloading forage wagons.

Use a safety locking hitch pin. Many wagon accidents happen when a pin jumps out and releases a wagon. The wagon may turn over or climb a tractor tire and kill the operator. Also use safety chains, especially on public roads.

When loading wagons, keep the load distributed evenly, from side to side and front to rear. An evenly distributed load is easier to control and reduces strain on the frame.

Be especially careful when hauling loads out of ensilage piles. A wagon can easily start sliding and cause a serious upset.

Know the recommended load limits listed in your operator's manual and don't exceed them. These limits allow you to maintain control of the equipment. Excessive loads cause you to lose full control of the wagon.

Keep within the load limits posted on bridges (Fig. 20). A large gravity box filled with shelled corn may weigh more than 19,000 pounds (8,626 kg) and could exceed the load limits for a bridge or the equipment. Keep the wheel bolts tight and use recommended tires and tire pressures.

Slow down when towing heavy loads to maintain steering and breaking control and to avoid fishtailing and upset. Be especially careful on slopes even when the wagon is equipped with brakes. Reduce speed on rough ground and at corners and keep the tractor in gear when going downhill. A general rule is to use a gear no higher than the one you would use to pull the same load up the hill.

Stay a safe distance away from ditches and steep banks (Fig. 21). The bank sides could cave in and let the wagon slide in.

FEED AND FORAGE WAGONS

Shafts, belts, pulleys, beaters, chains, and the other working mechanisms on feed and forage wagons can be hazardous if you fail to follow safe operating practices. Keep people away from wagons while they're in operation. Even when wagons are properly shielded they are not foolproof. Some parts must be exposed to do work. Keep hands and feet away from all moving parts.

Emergency shutoff devices on wagons can't prevent contact with all moving parts. But good shutoffs can help you avoid more serious injury if you should get caught. Stay at the controls when the wagon is operating (Fig. 22). Never enter a wagon with the tractor engine running or the wagon operating.

Before attempting to unclog any part of the machine, disengage all power, shut off the engine, and take the key. Don't just turn off the part you are trying to unclog. You or another person could accidentally turn it back on or get caught in another mechanism while trying to clean out the clog. Make sure the engine is off whenever you lubricate, adjust, or repair a machine.

You will reduce hazards if you maintain your wagon properly. Some things that are potentially hazardous are safety-trip mechanisms that don't work properly and guards that aren't in place. Keep safety mechanisms in perfect operating condition. Always keep guards and shields in place.

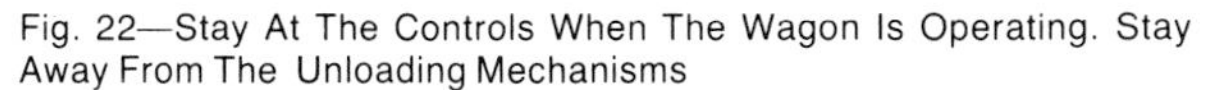

Fig. 22—Stay At The Controls When The Wagon Is Operating. Stay Away From The Unloading Mechanisms

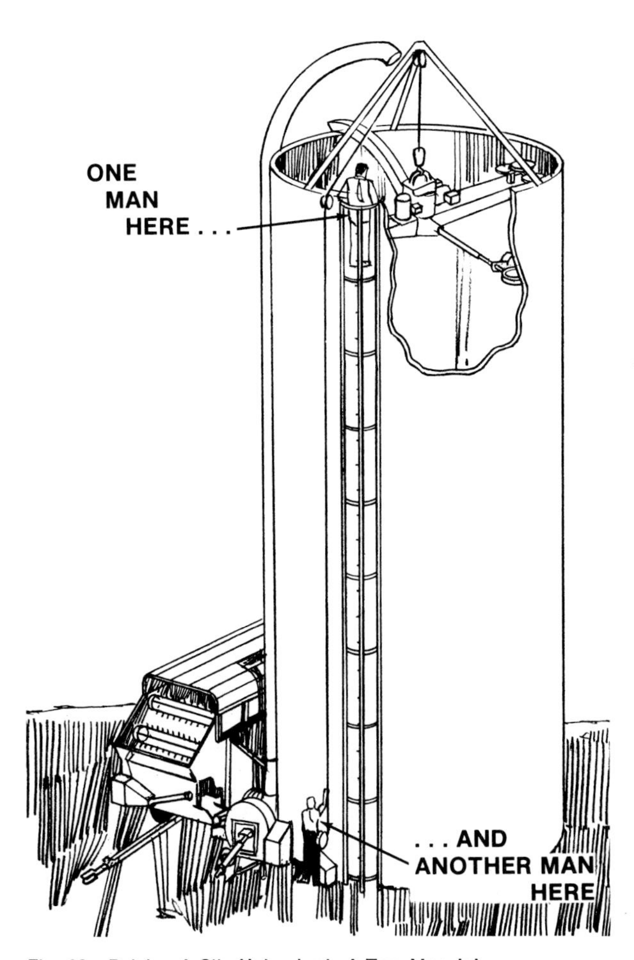

Fig. 23—Raising A Silo Unloader Is A Two-Man Job

SELF-UNLOADING GRAIN WAGONS

Gravity boxes and hydraulically operated lift boxes are used to transport and dump large amounts of grain quickly. Keep children away from grain wagons while they are unloading.

Unload wagons on a level surface. A small amount of additional weight from grain on one side may be enough to upset the wagon if it's not level. Don't drive up on blocks to try to make the grain unload faster.

When hydraulic cylinders are used to dump the load, safely block the box if it is necessary to work on the cylinders while the wagon box is raised.

SILO UNLOADERS

Before filling a silo, you have to raise the unloader to the top of the silo. This should be considered a two-man operation—one man to lift the unloader and the other to watch the rising unloader and signal if something goes wrong (Fig. 23).

The person observing the unloader should position himself at the top of the silo and watch the cable pulleys and electrical cables to make sure they don't tangle or get caught. Do not enter the silo while the unloader is being raised or suspended from the top, and never crawl out onto a suspended loader for any reason. It is generally unstable, and you could easily fall.

The person raising the unloader should not leave the controls for any reason while the machine is in operation.

Do not remove the safety catch device from the winch. Without it, a slip could cause a spinning lift crank that could break an arm if an attempt was made to stop it as the unloader went out of control and fell to the bottom of the silo.

If the silo is not covered, try to raise the unloader on dry and relatively calm days. This helps prevent accidental slips and falls that are more likely to happen on wet or windy days.

Have electrical installations performed by a qualified electrician. Power company officials should be consulted to guarantee an adequate power supply.

Safety controls should be kept in repair and properly used. They include:

- **Overload protection—manual reset type**
- **Emergency switch to shut off equipment—located on equipment or in feed room**
- **Coordinated switches that will stop the rest of the units automatically when one part of the system stops**

To prevent accidental starting by someone else:

1. *Remove fuses and put them in your pocket when working in the silo.*

Fig. 25—Never Enter A Chute To Unplug It—Use A Long Pole To Loosen Silage

Fig. 24—Be Able To Get Assistance If Necessary

2. *Lock the main switch.*

3. *Make sure all circuit protection devices are of the manual reset type.*

Perform seasonal checks of the wiring for cracks, fraying, or bare wires, and replace defective wires.

The unloader may occasionally become plugged or buried, and you must go into the silo to unplug the machine. Before entering the silo, shut off the power supply. Never try to unplug an unloader while it's running—you could be injured. Some farmers find it convenient to have a set of walkie-talkie radios, one of which should be left with someone nearby for use in case assistance is needed (Fig. 24). However, because of the noise while the unloader is running, you may not be able to hear the other person.

If you must start the unloader while in the silo, take precautions to avoid getting caught in the silo or chute. This can happen if you start the unloader while in the silo and the feeder below fails to remove the silage from the bottom of the chute. Have a second person make sure that everything on the outside is working properly.

Sometimes the chute may plug when you're not around. If this happens, use a long pole to loosen the silage. Do not climb up the chute (Fig. 25). The silage could slide down unexpectedly, engulfing you or knocking you off the ladder before you can get down.

It is also important to be aware of and avoid the dangerous gases that can build up in a silo after filling time and for two weeks afterward. Carbon dioxide and nitrogen dioxide can build up to lethal levels, and both are practically undetectable.

If you must enter a silo where gases could have accumulated, run the blower for 15 to 20 minutes to clear the air, and leave it running slowly to provide fresh air. Even silos without roofs can accumulate dangerous gases, because these gases are heavier than air. Respirators will not help, because they don't have an oxygen supply. You will have to use a self-contained breathing apparatus.

If you notice any signs of nausea or headache, get to fresh air, then see a doctor. Even if the exposure doesn't seem serious, problems can develop some time after exposure and cause lung congestion.

CROP DRYERS

Reduce fire hazards with crop dryers by using periodic maintenance checks. Your fuel supplier can guide you in maintaining fire safety.

Keep fuel storage tanks a safe distance from buildings and the dryer unit. A minimum of 25 feet (7.6 m) is generally recommended for tanks with a 500- to 2,000 gallon (1,893 to 7,570 L) capacity, and 50 feet (15 m) for 2,000- to 30,000 gallon (7,570 to 113,550 L) tanks. Be alert for odor suggesting gas leaks in the fuel supply system. In addition to wasting fuel, a small leak can cause gas to accumulate in low areas around buildings which the dryer could ignite, causing a fire or explosion. Ventilate bins and dryer thoroughly before igniting the dryer.

Guards and shields should be secured in place before the dryer is started. Make a preoperational check to see that all fans move freely. Corrosion could seize motors and provide a situation where overheating could cause a fire or explosion.

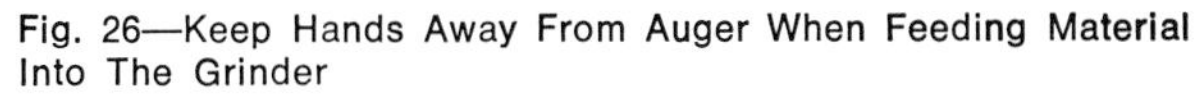
Fig. 26—Keep Hands Away From Auger When Feeding Material Into The Grinder

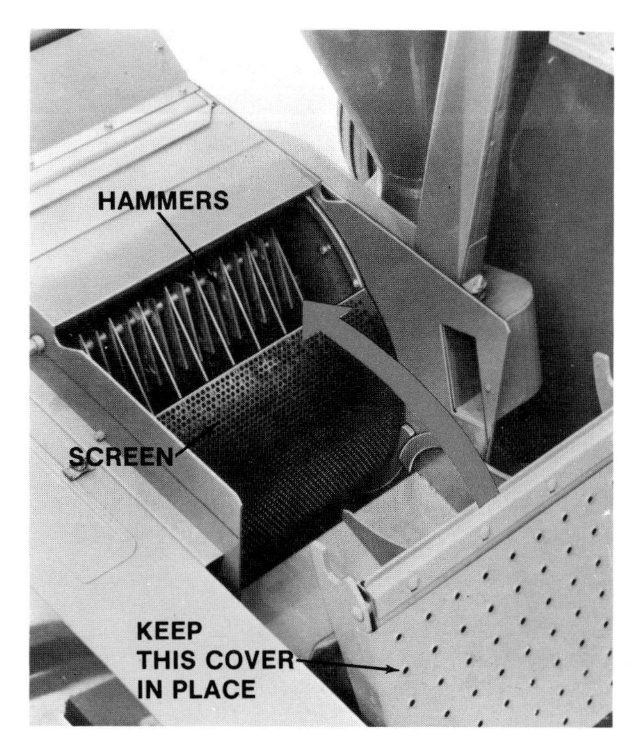

Fig. 27—Keep The Hammermill Cover In Place To Avoid Injury

Preseason checks of controls can save costly down time when the harvest gets underway. Any malfunctioning units should be repaired or replaced by a reputable serviceman.

GRINDER-MIXERS AND TUB GRINDERS

As the mechanization of feed handling increases, a large number of livestock farmers are using grinder-mixers to prepare their own feeds (Fig. 26). These grinder-mixers present many of the same potential hazards as other PTO-driven equipment (see Chapter 5).

Beyond using standard safe PTO operating procedures, there are a few other practices that can help reduce potential hazards of grinder-mixers.

Hammermills are provided for grinding various types of grain and hay into small particles of a desired size. These hammers revolve at extremely high speeds. Keep the hammermill cover in place whenever the hammermill is moving to avoid getting caught in the mechanism or being struck by thrown material (Fig. 27). Generally, the hammermill keeps on running for a short time after the power supply is shut off, so make sure it has stopped completely before opening the cover. This will eliminate the danger of being hit by material or getting caught in the unit.

danger of being hit by material or getting caught in the unit.

Observe additional precautions when feeding material into the grinder. Keep all shields in place, including the shield that keeps material from being thrown back out of the grinder. Failure to do this could result in injuries to the face or eyes. If no guard is provided over the auger in the concentrate-loading hopper, use extreme care to keep your hands away from the auger. Also, keep feed bags from getting caught and plugging the unit. Do not remove these auger guards unless repairs are necessary, and then replace the guards before operating the unit.

If a bag, flake of hay, or any other material gets caught in the machine, don't try to pull it out while the grinder is running. The machine can pull in your arm faster than you can let go. Disengage the power and shut off the engine before removing the obstruction.

If you use your grinder regularly, buy a bale feed attachment that feeds the hay into the grinder slowly to reduce the danger of pulling a hand or arm into the grinder (Fig. 28). Don't try to grind thick slices of hay that could plug the grinder. And never push plugged material into the grinder with your hand. The plugged material could give way suddenly, allowing you to slip into the grinding mechanism. Use a push stick to push the material to the grinder.

To help avoid serious injury use a tub tilt lockout when you work around the tub.

Fig. 28—Use a Bale Feeding Attachment To Feed Hay Into A Grinder Safely

CHAPTER QUIZ

1. The position of a tractor-loader's center of gravity:

a) Is always in the same place

b) Changes, depending on how level the tractor's position is

c) Changes as the position of the loader changes

2. Loaded manure buckets should be driven:

a) Downhill and backed up

b) Uphill and backed down

c) On the side of the hill so front and rear wheels remain level

3. The two most important dangers when using a loader are:

a) Getting caught in moving parts

b) Traveling at excessive speed

c) Overloading

d) Traveling with the loader too high

e) Breakage of hydraulic lines

4. True or false? If the chains in a manure spreader are frozen down, the best procedure for loosening them is to engage the PTO slightly a number of times to break them loose.

5. Many grain bin fatalities result from:

a) Electrocution

b) Auger injuries

c) Suffocation

6. Which precautions can you ignore if you must enter a grain bin?

a) Stay near the edge of the bin.

b) Carry a rope with you.

c) Lay a board on the grain to walk on.

d) Have someone watch you from outside the bin.

e) Wear a dust mask.

7. What are two conditions that could cause an elevator to collapse?

8. What are two electrical safety devices that should be used with silo unloaders?

9. True or false? Silo gases are easily detected when entering silos if they are in sufficient quantity to be dangerous.

10. Filter respirators give adequate protection in which of the following cases:

a) Entering grain bins

b) Opening upright silos

c) Entering liquid manure pits

12 Target: Farm Maintenance Equipment

Fig. 1—Watch For Obstacles, And People, Especially Children

INTRODUCTION

A number of labor-saving devices are available for general utility and farm maintenance operations. Examples include:

- **Rotary mowers**
- **Posthole diggers**
- **Post drivers**
- **Chain saws**
- **Lawn and garden tractors**

Many of these items are used only occasionally, with many months passing between uses. It's important to consider the safety precautions necessary for this equipment, because operators tend to forget procedures that aren't followed repeatedly.

ROTARY MOWERS

Rotary mowers are used for cutting and shredding stalks, clipping pastures, clearing underbrush, and mowing, roadbanks, lawns, and waterways (Fig. 1). Specially designed mowers can condition hay (see Chapter 8). Others can cut brush up to three inches in diameter. Know the job you're going to do and use the right type of rotary mower for it. This helps you do the job safely and will prolong mower life.

SAFETY PRACTICES FOR ROTARY MOWERS

There are many common guidelines for mowing, whether you use a lawn and garden tractor or a row-crop tractor. When handled improperly or carelessly, either type can become involved in a serious accident.

When cutting tall grass, weeds, or brush, watch for objects like tree stumps, tin cans, rocks, and other obstacles that could be hit or thrown by the mower blades (Fig. 2).

Keep people out of the area in which you're working. If anyone comes near the area, shut off the mower and tractor. The peripheral blade speed may be as much as 140 miles per hour, which is approximately one-fourth the speed of a rifle bullet. At that speed, an object hurled by a blade could cause serious or fatal injury.

Make sure the PTO is disengaged before starting the tractor (Fig. 3). Engage the clutch and increase engine speed slowly when starting the mower. Don't exceed recommended PTO speeds. This could result in damage to the mower and danger to you if a part fails and is hurled out.

Any time you hit an obstruction or hear a strange noise, disengage the PTO, shut off the tractor engine, and take the key before investigating the cause. Be sure to set the brakes before dismounting. Many rotary mowers have blades that continue to rotate for some time after the PTO is disengaged. To avoid injury, be sure the blades have stopped before approaching the mower.

Be very careful when turning sharp corners (Fig. 4). The rear tractor wheels could catch the mower frame and throw the mower toward you. Use front-wheel weights when operating three-point-hitch mounted mowers on rough terrain and during transport. If weights aren't used, the front wheels may bounce and cause you to lose steering control.

CAUTION: Do not overspeed 540 rpm mowers with a converter coupling on a 1000 rpm PTO.

You can increase your safety and mowing efficiency by making sure the mower is level and that the rear tires of the tractor are set wider than small mower frames. Wide-set tires will not press down the material about to be cut as much and will provide greater tractor stability (Fig. 5).

Don't try to cut brush with a rotary mower designed only for forage, because you may be exposed to hazards caused by machine failure. When you buy a mower, be sure to find out what materials it can safely cut. You'll need heavy-duty blades for any type of saplings. Small models with heavy blades may not be able to do the job, because the wood may clog the mower as it goes through the chute. Your job may require a mower especially designed for clearing brush.

SAFETY MAINTENANCE

Before operating your mower, become familiar with its operation, adjustments, and maintenance procedures. The best way to learn is to study the operator's manual carefully. Be sure you know all the recommended safety precautions.

Begin your preoperational check by making sure the PTO is disengaged and the engine is shut off. Look for loose nuts and bolts. Inspect the blades of lawn mowers and small garden tractors for nicks and dullness (Fig. 6). You should be able to feel a sharp edge along the cutting portion of the blade. If they feel smooth and rounded, take time to sharpen them. Blade sharpness is a key to efficient mowing. Stalk

Fig. 2—Watch Carefully For Objects That Could Be Thrown By Mower Blades

Fig. 3—Make Sure The PTO Is Disengaged Before Starting The Tractor

Fig. 4—Do Not Turn Sharply With Pull Mowers. The Rear Tractor Wheels Catch The Mower Frame And Throw The Mower

Fig. 5—Increase Tractor Stability By Setting The Tires To Maximum Width

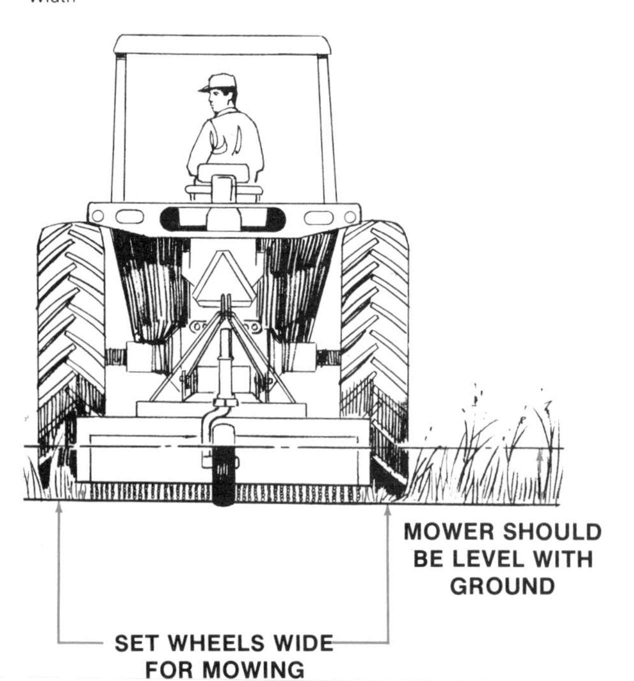

cutting dulls blades quickly because of the abrasive action of soil. When the blades become too dull for additional sharpening, replace them because they are ineffective, and they can be dangerous since mowing will be more difficult and hazardous conditions may develop as a result of such problems.

Rotary mowers are often equipped with runners and safety chain guards (Fig. 7). To avoid excessive wear when cutting to short heights, keep the mower just high enough off the ground so it doesn't ride on the runner shoes. The chain guards reduce the possibility

Fig. 6—Check Rotary Mower Blades Regularly For Loose Nuts And Bolts

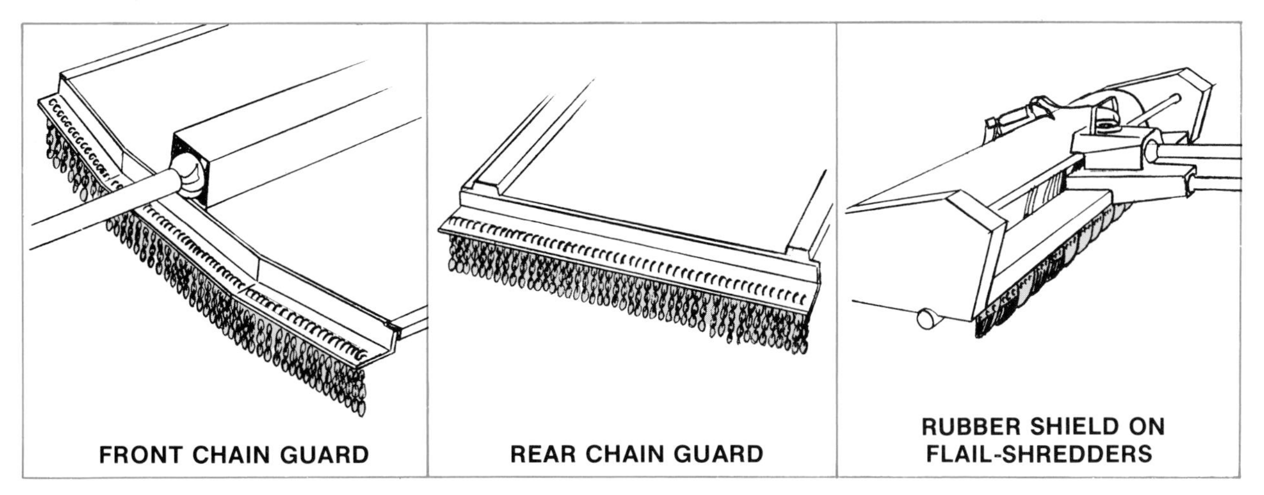

Fig. 7—Keep Shields And Safety Chain Guards In Place

of objects being thrown from under the mower and seriously injuring or killing someone in the area. Make sure chain guards are maintained and kept in place. If they must be removed or raised for certain crop conditions, make a special effort to keep persons out of the area when mowing. Be sure to replace or readjust them as soon as you finish the crop.

Power transmission shafts must be protected by shields or guards (Fig. 8). Keep shields and guards in place on the machine—don't leave them in the shop. Always replace shields or guards after maintenance or repair jobs are completed.

POSTHOLE DIGGERS

Posthole diggers are tractor-powered augers that drill holes 6 to 12 inches (15 to 30 cm) or more in diameter. Special auger sizes are also available from most manufacturers. Let's look at the best ways to dig straight, clean, and uniform holes and to do it safely. Here are some basic practices for using posthole diggers:

1. *Shift the transmission into park or neutral.*

2. *Set tractor brakes before digging.* If the tractor rolls, the auger shaft will be bent.

3. *Run the digger as slowly as possible to keep it under control.*

4. *Dig the hole in small steps (Fig. 9).* Dig down several inches, then bring the auger up to let the soil clear. Repeat this procedure until the desired depth is reached. This allows better control of the auger and can prevent difficulties that could lead to an accident.

LODGED AUGERS

Removing a lodged auger can be dangerous work. Be careful. If the auger gets stuck in wet clay, stones, or roots, disengage the PTO immediately and turn off the engine. Turn the auger backwards with a

Fig. 8—Keep Powershaft Shields In Place

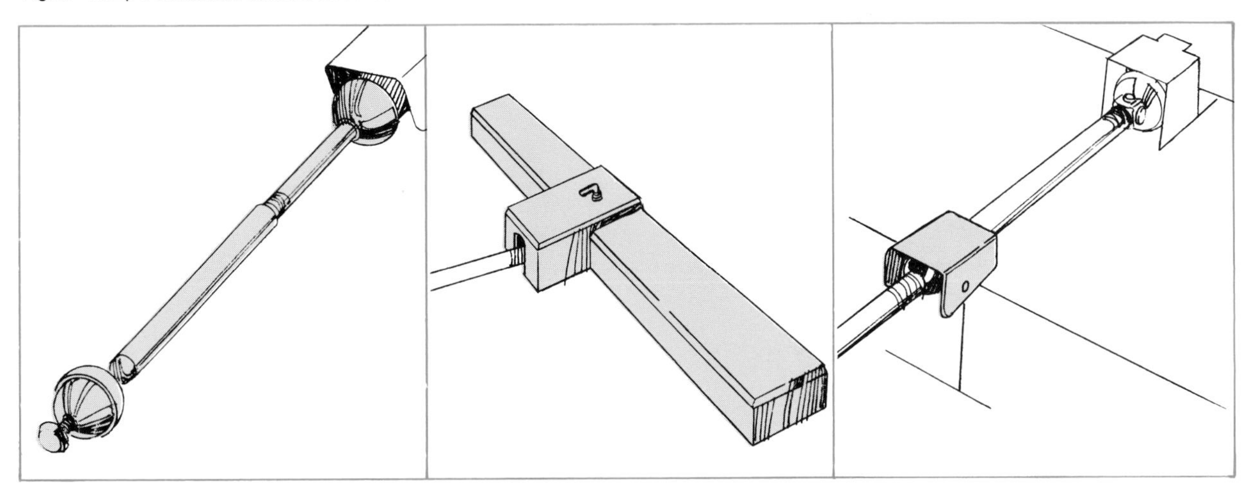

Fig. 9—Dig The Hole In Small Steps, Bringing The Auger Up To Let The Soil Clear

Fig. 10—When Stuck, Turn The Auger Backwards With A Large Wrench

large wrench (Fig. 10). Then attempt to raise the auger with the hydraulic system. Do not attempt to raise the auger while turning it with a wrench. You could be injured if the PTO was accidentally engaged or the hydraulic system suddenly raised the auger!

DIGGER LIFE

The usefulness of your digger can be extended by proper operation and maintenance. There are many types of augers. Choose the one that's right for the job. If you plan to work at different types of jobs, get one that's versatile.

An important safety device on your digger is the shear pin (Fig. 11). It's designed to break when there's too much force on the digger, much as a fuse blows when there's too much electric current on a circuit. It can prevent a sudden failure that could injure you. Be sure to use genuine shear pins—*not* hardened bolts.

POST DRIVERS

The post driver is becoming more popular because it eliminates the time-consuming job of tamping soil around each post. However, it can be hazardous if it's not operated properly (Fig. 12). Follow the manufacturer's instructions carefully to prevent injury.

Here are some general safety precautions that you should follow when operating a post driver:

1. *Always shut off the engine and lower the hammer before attempting to adjust or lubricate the driver.*

2. *Never place your hand between the top of the post and the hammer.* You could be severely injured if the hammer fell or was tripped accidentally.

Fig. 11—The Shear Bolt Is An Important Safety Device—Use Only The Correct One

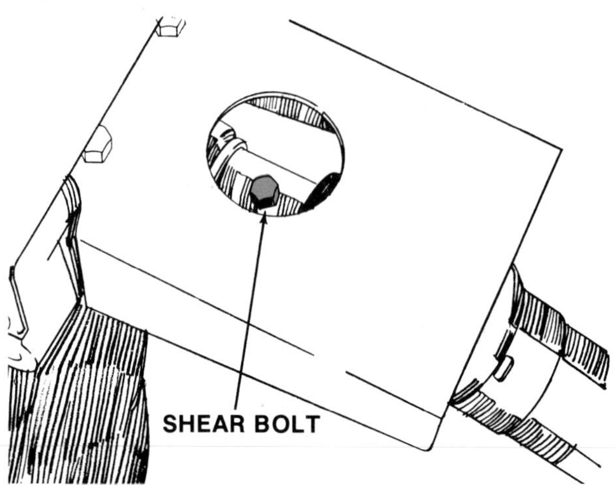

3. *Never exceed recommended hammer strokes per minute (usually no more than 20).* Faster operation may damage the driver and injure the operator.

4. *When driving posts, use a safety fork or guide to steady the post and prevent injury if the post should break or be deflected.*

5. *Keep your hands away from posts that are about to be driven.*

6. *Adjust or lubricate the driver only when the hammer is in the lowest position to avoid being injured by the hammer if it falls or is tripped.*

7. *Put all shields in place before operating the driver.*

8. *Store your driver safely with the hammer in the lowered position, or use safety locks to prevent accidental lowering.*

9. *Wear safety glasses during operation.*

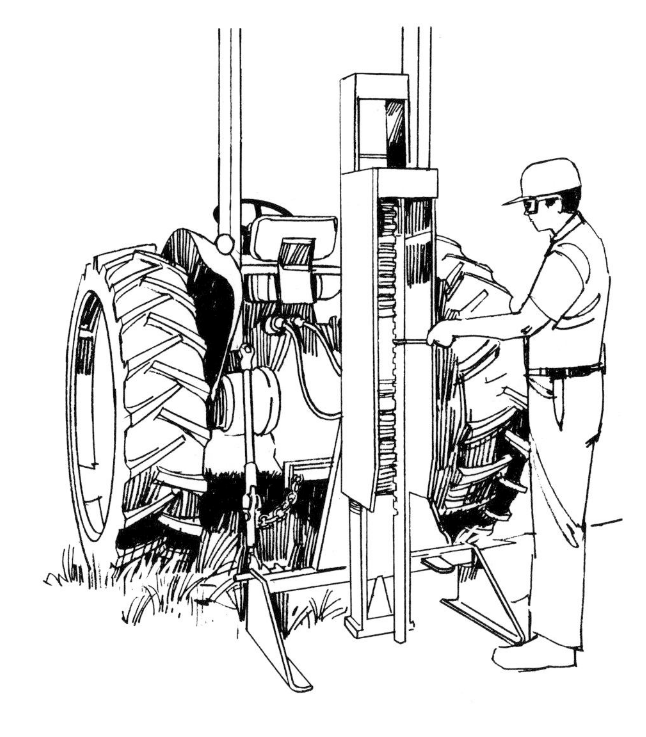

Fig. 12—Post Drivers Can Be Hazardous If Not Operated Properly

CHAIN SAWS

Modern chain saws have become common on farms, as well as in urban areas. Lightweight chain saws, often weighing less than 10 pounds (4.5 kg) are popular for trimming trees, cutting fireplace wood, and clearing storm damage (Fig. 13). There are a number of hazards in the use of chain saws of which users need to be aware.

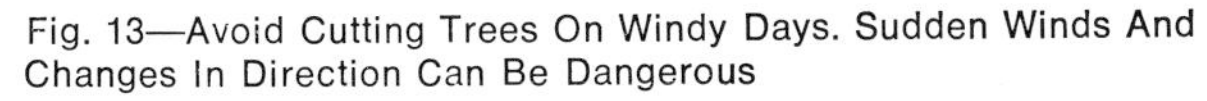

Fig. 13—Avoid Cutting Trees On Windy Days. Sudden Winds And Changes In Direction Can Be Dangerous

Fig. 14—Refuel On Bare Ground To Reduce The Fire Hazard

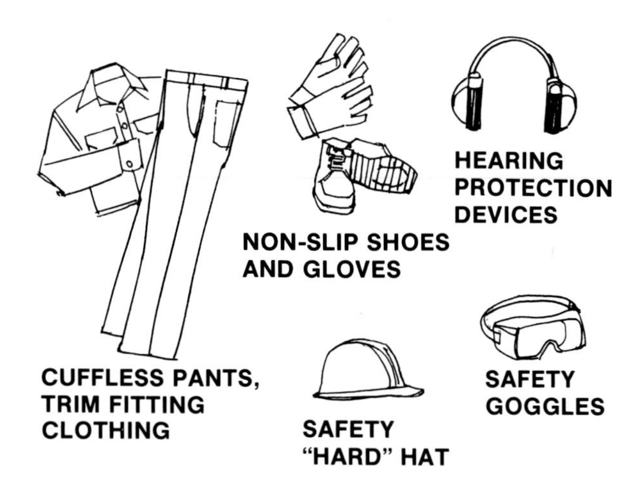

Fig. 15—Use Protective Equipment When Operating A Chain Saw

Consider the following:

- **The weather**
- **Fire**
- **The tree**
- **The saw**
- **The operator**

THE WEATHER

Wind can create very serious hazards when cutting down trees. Even on sunny days, the wind can come up suddenly or change direction unexpectedly, causing a tree to fall in the wrong direction—perhaps onto power lines, buildings, or people. Avoid cutting trees on windy days, or use these days for limbing or trimming only (Fig. 13).

Rain, snow, and ice may lead to slips and falls. Whenever the weather presents a hazard, wear protective clothing and work slowly and carefully. If possible, put the job off until conditions improve.

FIRE

Periods of hot, dry weather make leaves and grass a fire hazard, especially when it's windy out. Spilled fuel adds to the danger. Refuel on bare ground (Fig. 14). Gradually release pressure in fuel tank before removing the cap completely for refueling. Move at least 10 feet (3 m) away from the fuel source before starting the saw. A faulty muffler can provide a spark that could set off a fire. Don't let dry combustible material contact a hot muffler.

Fig. 16—Use Proper Equipment To Help You Do The Job Safely

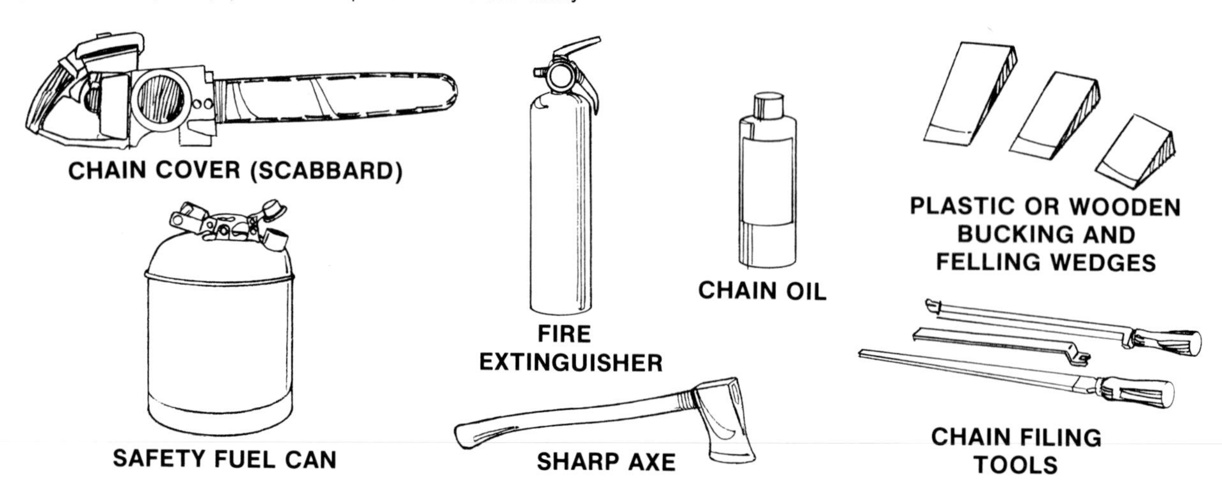

THE TREE

Certain trees can be dangerous to inexperienced operators. Lumberjacks have coined some expressions that describe problem trees:

• *Widowmaker*—This is a tree with broken or dead limbs. A limb doesn't have to be very big or very high on a tree to be capable of causing a serious injury if it suddenly fell on you.

• *Spring pole*—This is a sapling that's bent and held down under tension by another tree. If the spring pole is cut or the other tree is removed from it, the sapling can snap up with tremendous force and seriously injure anyone nearby.

• *Schoolmarm*—This is a tree with a prominent fork in the trunk, or two trees grown together at the base, making it difficult to predict which way it will fall.

Until you've had plenty of experience or instruction, don't attempt to cut trees like these. And don't try to cut any tree with a diameter greater than the length of your chain saw blade. This requires special techniques, and you could be injured if the saw kicked back at you.

THE SAW

With extended use, chain saw noise and vibration can cause hearing loss, fatigue, and swelling of the hands. To reduce these potentially harmful effects:

1. *Wear ear protection.*

2. *Take periodic rest breaks.*

3. *Select a saw that has low noise and vibration characteristics.*

4. *Use a saw with a safety chain brake and tip guard.*

A chain saw must be properly maintained to be safe. This includes sharp teeth, correct chain tension, proper lubrication, and a properly tuned engine. If you notice that the chain tends to walk sideways while cutting or the cut shows a fine powder instead of wood chips, your saw needs sharpening. Keep the engine adjusted so it remains running but the chain stops moving when the throttle is released. Your operator's manual is the best source of maintenance information—use it.

All chains stretch with use. Most of this stretch will occur during the first half hour of operation. Your operator's manual will show you the best way to break in your saw for efficient cutting action and longer chain life. Keep the chain properly tensioned to keep it from jumping off the guide and injuring you.

THE OPERATOR

The most important ingredient in chain saw safety is the operator. Operate the saw safely and use personal protective equipment (Fig. 15). This protective equipment should be provided for the head, ears, eyes, feet, and hands.

Fig. 17 — The Chain Saw Will Move As Soon As The Engine Starts. Make Sure You Hold The Front Handle With The Left Hand And Place Your Right Foot Through The Rear Handle.

When using a chain saw, take along the following equipment, as shown in Fig. 16:

• *A carrying case for the saw or a scabbard for the guidebar*

• *A supply of fuel in a UL-approved can (this is the safest storage container)*

• *Oil for the chain oiler*

• *Some plastic or wooden wedges (not hard metal ones)*

• *A sharp single-blade ax*

• *Tools for chain maintenance*

• *A fire extinguisher or shovel*

Your saw should be equipped with an exhaust muffler, and a spark arrestor muffler, if required by law or when dry conditions are encountered.

The correct position for starting a chain saw is shown in Fig. 17. Never allow another person to help you start a chain saw. If either of you slips or lets go, the other could be cut.

Use extreme caution to be sure the chain does not: contact limbs or logs other than the one you want to cut, strike nails or stones, or touch the ground when it's operating. The saw will jump back if the chain at the nose of the bar touches anything. This is called *kickback*.

CAUSES OF SAWBLADE "KICKBACK"

BLADE NOSE STRIKES ANOTHER OBJECT

IMPROPER STARTING OF BORE

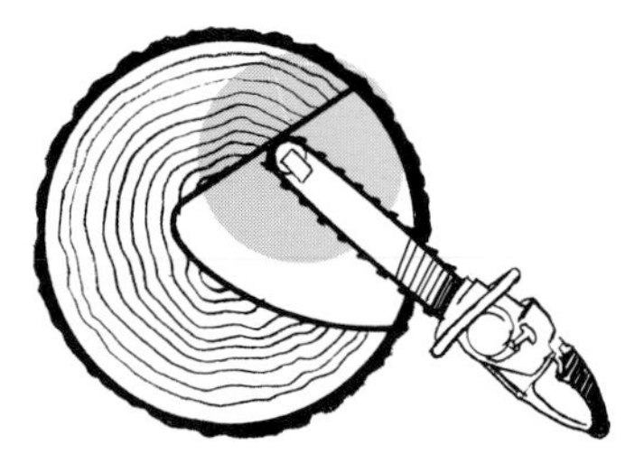

TOP OR BLADE NOSE TOUCHES BOTTOM OR SIDE OF KERF DURING REINSERTION

ANTI-KICKBACK DESIGN AND CHAIN CATCH

SPECIAL CHAIN AND TIP GUARD

CHAIN CATCH IF CHAIN BREAKS

Fig. 18 — Avoid Situations Which Can Cause Kickback. Use Safety Features When Available

Stand to the side to avoid being cut by kickback. Never stand directly in back of the chain saw while cutting wood.

Kickback has many causes. A chain that's misfiled or loose is more likely to kick back. Kickback is also more likely if you start a cut with the saw chain moving too slowly. Other situations in which kickback is possible are shown in Fig. 18.

Use specially designed anti-kickback chains and anti-kickback chain tips to reduce the risk of kickback.

Anti-kickback features for saws include specially designed chains with a ramp at the front of each cutter, and a guard on the top of the saw to protect the end of the blade from pinching and kickback (Fig. 18).

Along with handguards and chain brakes, saws can be equipped with a chain catch that is mounted on the power head of the saw. The chain catch is designed to shorten any backward swing of the chain if the chain breaks (Fig. 18).

Be sure the bumper is against the tree while sawing, or the chain riding across the tree may jerk the saw out of your hands.

When you must control the direction of the fall, notch the tree first. Then use a pulley and rope to control fall (Fig. 19). This is a situation where you should work with another person.

Plan a safe path of retreat before making the felling cut (Fig. 20). When the tree starts to fall, shut off the saw, put it down, and move away in the planned direction. Always carry the saw with the bar to the rear and the muffler away from you.

Never carry a chain saw when it is running. If you should fall, the saw could spin around and cut you severely.

Before attempting the following operations, get the feel of your saw. Make a few trial cuts on small logs supported off the ground so the chain clears the ground. Let the saw do the cutting. You don't need to apply extra pressure.

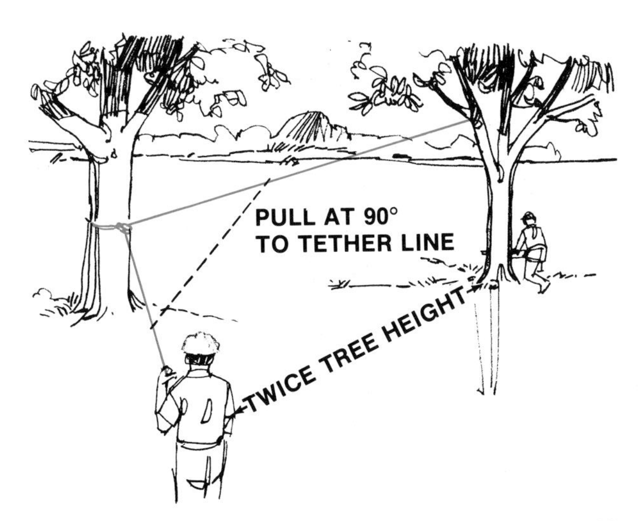

Fig. 19—Use A Pulley And Rope To Control The Direction Of A Fall

Fig. 20—Plan A Safe Retreat Before Felling A Tree

FELLING

Only after you have mastered steady and even cutting should you attempt to fell a tree.

Check the situation carefully before felling a tree. Take note of the larger branches and wind direction to determine how the tree will fall. Be sure you have a clear area around the tree in which to work, and an open pathway from the tree for an escape route.

Remove dirt and stones from the trunk of the tree where the cut will be made.

Examine trees for loose or dead limbs before felling. If such limbs appear to be a hazard, remove them before felling the tree.

When felling a tree:

1. *Cut through trees less than 8 inches (203 mm) thick with one cut.*

2. *On larger trees, make the notch cut on the side of the tree on which it is expected to fall (Fig. 21).* It should have a depth of approximately one-third the diameter of the tree. Make the lower notch cut first. This keeps the chain from binding and being pinched by the wedge of wood while the notch cut is made.

3. *Make the felling or back cut at least 2 inches (50 mm) higher than the horizontal notching cut.* The felling cut should be kept parallel with the horizontal notching cut. Cut it so that wood fibers are left to act as a hinge, keeping the tree from twisting and falling in the wrong direction.

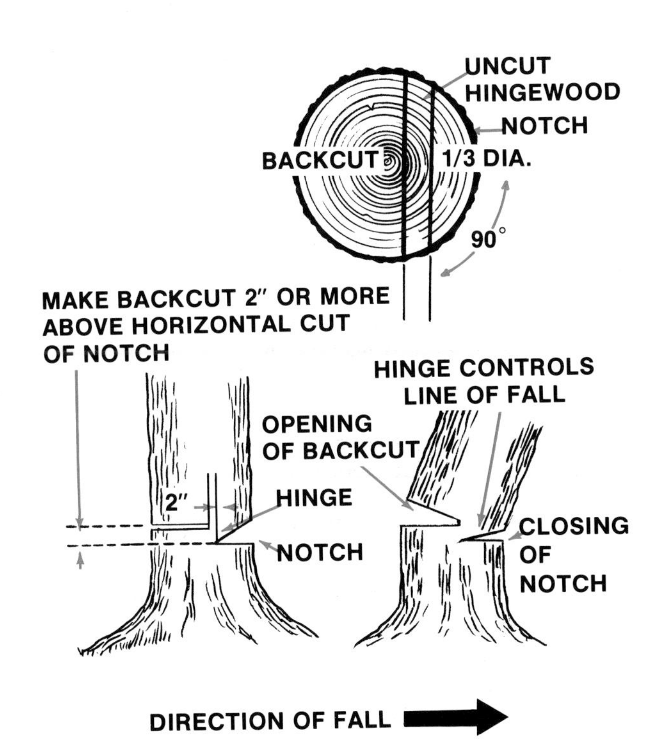

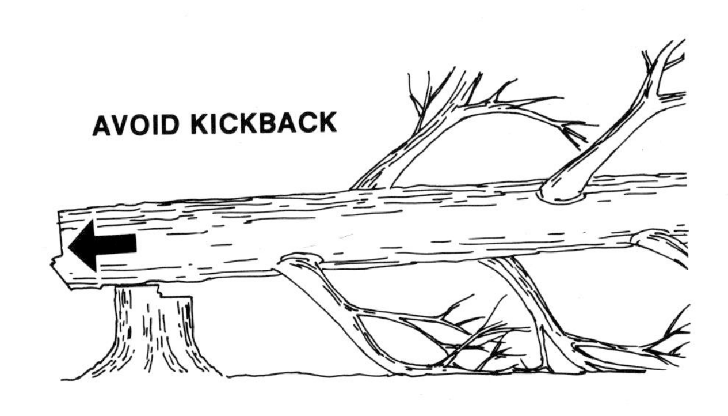

Fig. 21 — Notch And Felling Cuts Must Be Made Correctly To Make The Tree Fall In The Desired Direction. Avoid Base Kickback.

4. *Keep the guidebar in the middle of the cut so the cutters returning in the top groove don't recut the wood.* Don't twist the guidebar in the groove. *Guide* the saw into the tree—don't force it. The rate of feed will depend on the size and type of timber.

5. *Remove the saw from the cut and shut it off before the tree falls.* The tree will begin to fall as the felling cut approaches the hinge fibers. Move to a safe spot.

6. *Do not cut through the hinge fibers.* The tree could fall in any direction—maybe in the direction in which you are retreating.

7. *Avoid Kickback.* Stay clear of the front and rear fell path of the tree.

LIMBING

Most chain saw accidents happen during limbing operations. Leave the larger lower limbs to support the log off the ground to aid bucking cuts (Fig. 22). Prune the smaller limbs in one cut by starting at the bottom end of the tree. Undercuts should be used on limbs supported by branches to keep from binding the chain.

CAUTION: Never cut limbs from the side of the tree where you are standing. Springback may occur to limbs under tension. Springback tension should be reduced before directly cutting limbs.

Fig. 22—Let The Lower Limbs Support The Log For Limbing

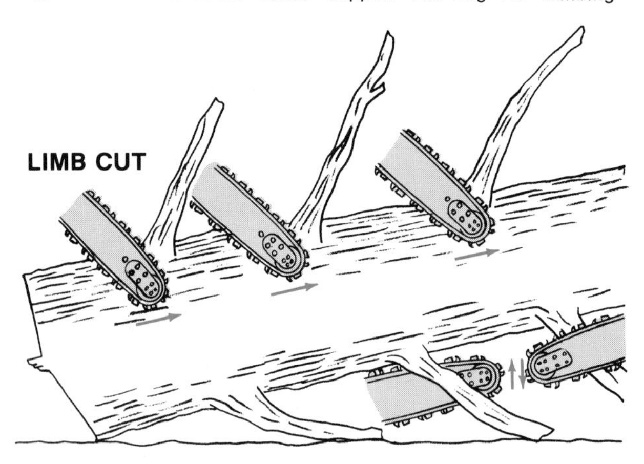

WORK SUPPORTED ALONG ENTIRE LENGTH

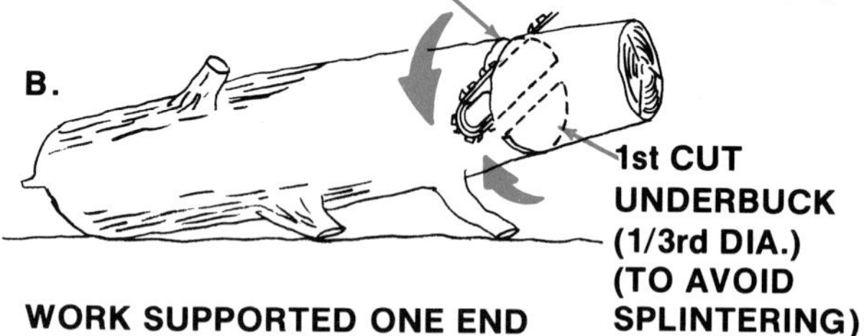

WORK SUPPORTED ONE END

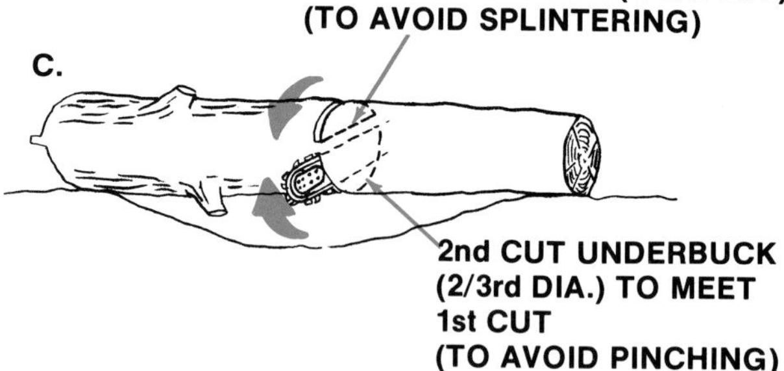

WORK SUPPORTED ON BOTH ENDS

Fig. 23—Use These Bucking Procedures For Safety

BUCKING

Make sure you have good footing and can get out of the way if the log should start to roll. On sloping ground, stand above the log rather than below it. If possible, raise the log clear of the ground by using limbs, logs, or chocks. To avoid pinching the guidebar and saw chain in the cut and splintering the log at the finish of the cut, use the following procedures (Fig. 23):

1. *When the log is supported along its entire length, cut it from the top (overbuck) (Fig. 23A).*

2. *When the log is supported on one end, cut one-third of the diameter from the underside (underbuck).* Then make the final cut by overbucking the upper two thirds to meet the underbucking cut (Fig. 23B).

3. *When the log is supported on both ends, cut one-third of the diameter from the top (overbuck).* Then make the final cut by underbucking the lower two-thirds to meet the overbucking cut (Fig. 23C).

TRIMMING

When removing a limb on a standing tree, hoist the saw with a rope. Don't carry the saw while climbing (Fig. 24). You need to use both hands to climb safely.

It is best to use a safety rope around the tree, fastened securely to your waist to avoid falls.

TOPPING

Topping is a technique for cutting off the top part of a tree while it's still standing (Fig. 25). It's a difficult procedure, and should be attempted only by highly skilled lumbermen.

LAWN AND GARDEN TRACTORS

Lawn and garden tractors were first used primarily for lawn mowing, but have several attachments for doing other farm maintenance chores. One of the most common accidents with these tractors involves upsets.

UPSETS

Most upsets happen when operators:

- *Stop or start suddenly when going up a hill*
- *Turn quickly at high speeds or when on steep slopes*
- *Mow across the face of slopes*

Fig. 24—When Removing A Limb On A Standing Tree, Hoist The Saw With A Rope. Don't Carry The Saw While Climbing

Fig. 25—Topping A Tree Is A Job For Professional Lumbermen—Don't Try It!

To avoid upsets:

1. *Establish a reasonable speed when traveling up a hillside in order to avoid sudden stops and starts.* Slow down before making sharp turns or turning on a slope.

2. *Always mow up and down on slopes or hills—not across the face or along the side (Fig. 26).* Mowing across the face increases the possibility of tipping over. If a hillside is very steep (generally anything steeper than 15 degrees), use a push-type power mower instead (Fig. 27). Wear cleated shoes, if possible, to prevent slips. Golf shoes or baseball spikes are excellent for this task. Mow *across* slopes with a hand mower to maintain better control.

Fig. 26—Mow Up And Down On Steep Slopes With Lawn And Garden Tractors—Not Across

Fig. 27—Use Hand Mowers On Areas Too Steep For The Safe Operation Of A Lawn And Garden Tractor

ATTACHMENTS

In the following sections we will cover operator safety considerations for the following attachments for lawn and garden tractors:

- *Rotary lawn mowers*
- *Hydraulic buckets*
- *Blades*
- *Tiller*

ROTARY LAWN MOWERS

Most accidents and injuries when using rotary mower attachments on lawn and garden tractors result from:

- *Unclogging the mower with the power on*
- *Getting on and off the mower with blades engaged*
- *Working in areas where children are playing*

Clogged Mowers

Rotary mowers can clog when operated in very tall or wet grass. Before trying to unclog a rotary mower:

1. *Disengage the power from the mower blades.* This allows the mower blades to be moved by hand without turning over the engine. This makes it easier to unclog some things.

2. *Stop the engine.* This insures that the blades will not revolve if the power to them is accidentally left engaged, and keeps the power unit from moving when you're trying to unclog it.

3. *Remove the spark plug wire or ignition key (Fig. 28).* This is a good way to prevent accidental starting of the engine and powering of the blades while you're unclogging the mower.

4. *Set parking brake and leave the tractor in gear.* This keeps the lawn and garden tractor from moving when you work on or under the mower. This simple precaution could prevent serious injury.

Mounting And Dismounting

Getting on or off a tractor while the mower is running is dangerous. There are 1½ to 4 inches (38 to 101 mm) of clearance between the ground and the mower blades to permit proper cutting. You could get the tip of your foot under the mower housing and exposed to the rotating blades. Many toes and feet are lost each year because operators fail to stop the blades before getting off the lawn and garden tractor. Play it safe—disengage power to the mower blades and shut off the engine before getting off.

Always wear safety shoes when mowing.

Children In The Area

Don't allow children in the area where you are mowing. They could be hit by an object hurled by the mower blades. If you're mowing in the front yard, send them to the back yard, and vice versa. Never leave the tractor unattended while its running. If you leave the machine, always shut off the engine and remove the key. Never give rides!

HYDRAULIC BUCKETS—FRONT-END LOADERS

Several manufacturers offer hydraulic front-end loaders as attachments for their larger lawn and garden tractors. The low overall height and width allows farm operators to scrape and load in spaces too small for larger farm tractors with front-end loaders and even some skid-steer loaders. Owners of older livestock and dairy confinement buildings can benefit most from this attachment for lawn and garden tractors.

Use the following safety practices when using front-end loaders on lawn and garden tractors:

1. *Avoid overloading.* Check the operator's manual for the load capacity of your particular unit (Fig. 29). Overloading can cause loss of steering control. Check the operator's manual to see if rear tire weights are needed for extra stability.

2. *Back down ramps and steep inclines and keep the bucket low.* Otherwise, a hole or sudden bump could cause the center of gravity to shift, causing a forward rollover or loss of control.

3. *Watch where you're going.* Don't watch the bucket. Watching for spills should not be necessary if you have loaded the bucket properly and are using a safe speed for the ground you're moving over.

4. *Don't attempt to operate the hydraulic loader controls from beside the tractor.* You could be injured by the loader cylinders and lever arms when operating the controls in this manner.

5. *Use the loader only for materials it was designed to handle.* Do not allow people to ride in the bucket. Their weight could overload the unit or they could fall and be injured. Never use the loader for such things as removing fence posts. Do not hook a chain to the bucket to pull something.

Fig. 30—Lower The Bucket Before Leaving The Machine

6. *Lower the bucket to the ground when you leave the machine.* This will keep the bucket from accidentally falling onto unsuspecting children or others (Fig. 30).

7. *Observe the same precautions concerning hydraulic pressure and fluid as outlined in Chapter 5.*

Fig. 28—Remove The Spark Plug Wire Or Ignition Key Before Trying To Unclog A Mower

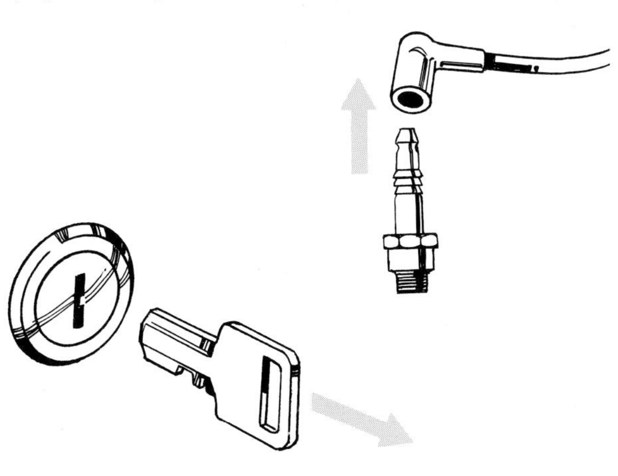

Fig. 29—Follow Recommended Bucket Load Capacity In The Operator's Manual

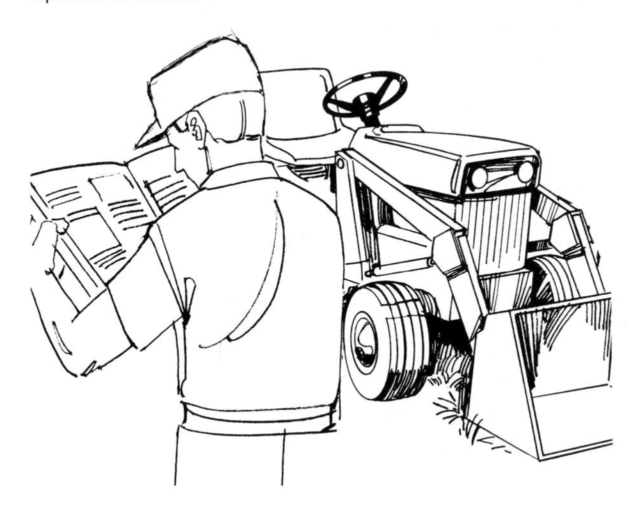

Fig. 31—Block The Blade Before Making Repairs Or Adjustments In The Raised Position

BLADES

In farm maintenance, a blade can be used to push gravel or light fill dirt, snow, or manure across concrete. Blades can be mounted at the front, rear, or center of a lawn and garden tractor. They can be controlled manually or with a hydraulic assist. Most lawn and garden tractor blades are designed for light work, and operators should consider the following practices when using them:

1. *Watch for obstructions.* Avoid pipes, curbs, roots, and other obstructions that could be hit during operation. Hitting heavy objects can cause loss of control and throw the operator from the tractor seat.

2. *Be prepared to stop.* This will help reduce blade damage and prevent throwing the operator if you hit a hidden obstruction.

3. *Keep operating speeds moderate.* This is critical, especially when your blade is locked into a solid position for digging. A stiff blade that hits an object results in a sudden, jolting stop. Most blades can move up and down to follow the terrain in a float position. Others have a spring trip that allows the blade edge to pivot in a small arc, reducing the shock to the operator and preventing damage to the blade. Both of these safety features require an alert operator who is prepared to stop.

4. *Lower the blade when you leave the tractor to prevent accidental lowering that might occur when children or others tamper with the controls.*

5. *Remember that normal operation of a front-mounted blade can affect your steering ability.* On loose surfaces like manure or snow, a blade can reduce steering control, and you could have an accident. Good operators keep this in mind and lower their speed to maintain control of travel direction.

6. *Block the blade if it must be raised while you make adjustments.* Hands and feet can be injured or crushed if the blade falls on them during adjustment or repair (Fig. 31).

7. *Maintain recommended tire pressure to give the best traction.* Equal tire pressure on the rear drive wheels also helps give uniform traction. Different tire pressures on the drive wheels can cause the tractor to pull to one side, and precision and control are reduced. You may also need tire chains and wheel weights for extra traction. Check your operator's manual for specific recommendations.

THE LAWN AND GARDEN TRACTOR AS A POWER UNIT

As power units, lawn and garden tractors are similar to larger tractors. They have many of the same features—PTO shafts, three-point hitches, drawbars, etc. Refer to Chapter 5 for more details on the safe use of some of these features.

Remember that the lawn and garden tractor *does* have some unique safety considerations:

1. *Size is never a very good measure of power.* The small-sized lawn and garden tractor may not appear as potentially hazardous as a larger tractor, but it offers a potential for accidents in the hands of an uninformed or careless operator.

2. *Riders should not be allowed.* A rider can easily fall off, get entangled in moving parts, and be seriously injured. Safe operators insist on "no riders." Also, a lawn and garden tractor usually weighs only 375 to 800 pounds (170 to 363 kg) additional weight, like an extra rider hanging on the back, causes a large weight transfer onto the rear axle. Upsets may result because of this serious load imbalance.

3. *Most lawn and garden tractors are now built to meet established safety standards.* When you purchase these tractors and their attachments, look for evidence that they meet safety standards. The OPEI safety seal shown in Fig. 32 is an example.

Fig. 32—When Buying Lawn And Garden Tractors And Attachments, Make Sure They Meet Safety Standards

CHAPTER QUIZ

1. True or false? Objects thrown by rotary-type mowers are generally moving too slowly to be dangerous.

2. List two advantages of having the rear tires of the tractor set wider than the mower frame.

3. True or false? Holes should be dug in small steps when using a posthole digger.

4. Which of the following are recommended safety practices to use when operating a post driver?

a) Shut off the engine and lower the hammer before adjusting or lubricating the driver.

b) Don't place your hand between the top of the post and the hammer.

c) Always keep the driver operating at over 20 hammer strokes per minute.

d) Adjust or lubricate the driver only when it is locked in the raised position.

e) Use a safety fork or guide when driving posts.

5. What is hazardous about the type of tree called a "widowmaker"?

6. List five safety items that should be kept handy when using a chain saw.

7. True or false? You must control the direction a tree is to be felled.

8. One of the most common accidents with lawn and garden tractors involves:

a) Tire blowouts

b) Upsets

c) Brake failure

9. True or false? It's safest to drive a lawn and garden tractor equipped with a hydraulic bucket forward up steep slopes and inclines.

Glossary and Laws
Suggested Readings

GLOSSARY

A

ABSORPTION RATE—Rate at which a chemical enters the body.

ACCIDENT FREQUENCY RATE—Number of accidents per hour of use of a specified machine. Gives a more complete accident picture than total numbers of accidents only.

ACCUMULATOR—A container which stores fluids under pressure as a source of hydraulic power. It may also be used as a shock absorber.

ACUTE CHEMICAL EXPOSURE—Serious, one-time case of chemical poisoning.

AMMONIA GAS (NH_3)—Colorless gas given off by liquid manure. It can cause coughing, irritate the eyes, lungs, or throat, and may be fatal.

ANEMOMETER—Instrument used to determine wind speed. Helpful when using pesticide sprays.

ANHYDROUS AMMONIA—Pressurized ammonia without water, used as a nitrogen fertilizer. It can cause skin irritation, burns, and blindness. Prolonged inhalation can lead to suffocation.

ANTIDOTE—Remedy that counteracts the harmful effects of a poison.

AQUEOUS AMMONIA (NH_4OH)—Ammonia dissolved in water, a popular liquid fertilizer. Can cause skin burns and blindness.

ARTIFICIAL RESPIRATION—Method of trying to restore normal breathing to someone who has stopped breathing.

AVICIDE—Agent that destroys birds.

B

BACTERICIDE—Substance that destroys bacteria.

BODY PROTECTION—Special clothing designed to protect a worker. Includes aprons, rubber or plastic garments, and certain types of padding.

C

CARBON DIOXIDE (CO_2)—Colorless, odorless gas given off by liquid manure. Small concentrations (3-6%) can cause drowsiness and headache. Stronger concentrations (30%) can cause death by suffocation.

CARBON MONOXIDE (CO)—Colorless, odorless, deadly gas present in engine exhaust.

CAUTION—Signal word indicating a hazard of minor importance. Also used on chemical labels to indicate low to moderate toxicity.

CENTER OF GRAVITY—Point about which all parts exactly balance one another. On conventional two-wheel-drive tractors, it is located above the axle and behind the midpoint of the tractor's length.

CENTRIFUGAL FORCE—Force that resists any change in direction from traveling in a straight line. Centrifugal force is a major factor in tractor upsets.

CHOLINESTERASE TEST—Blood test to determine activity of cholinesterase, an enzyme that affects nerve impulse transmission. Regular tests are recommended for workers who handle organophosphates.

CHRONIC CHEMICAL EXPOSURE—Accumulation of a chemical in the body that causes a mild, slow poisoning.

CLASS A FIRE—Fire involving the burning of ordinary combustibles like paper or wood.

CLASS B FIRE—Fire in which flammable liquids like gasoline or grease are burned.

CLASS C FIRE—Electrical fire.

CONTAMINATION—Pollution of a person or object by a chemical or some other substance.

CRUSH POINTS — Are created when two objects move toward each other or one object moves toward a stationary object.

CUTTING POINTS—Are created by a single object moving rapidly or forcefully enough that it cuts a relatively soft object.

D

DANGER—Signal word that indicates a very serious hazard. Also, a word used on the labels of highly toxic chemicals.

DECIBEL (dB) — A unit for measuring the relative intensity of sound.

DERMAL CHEMICAL EXPOSURE—A chemical entering the body by absorption through the skin.

DISCOMFORT INDEX — Formula that relates temperature to humidity, used to determine acceptable working conditions.

DRIFT—Spray particles carried on the wind, often creating a hazard.

E

EAR PROTECTION — Any device used to protect the ears from loud noises — generally earplugs or earmuffs.

ELECTRIC SHOCK—Passage of an electric current through a person's body.

EMETIC—A substance that induces vomiting.

EYE PROTECTION — Devices used to protect the eyes from dust, fumes, chemicals, etc. Some of these devices are safety glasses, goggles, and face shields.

F

FACE PROTECTION—Shield to protect the face from splashing, dust, chaff, sparks, etc.

FAIR LABOR STANDARDS ACT—Law that prohibits youth under 16 from working at certain farm jobs. See APPENDIX.

FERTILIZER—Material (manure or chemical) used to make soil more fertile. See ANHYDROUS AMMONIA, AQUEOUS AMMONIA, MANURE.

FERTILIZER BURN — Skin irritation caused by the hygroscopic action of dry fertilizer.

FIELD OF VISION—Area a person is able to see from one position, normally about 170 degrees.

FISHTAIL—Back-and-forth weaving action of trailed equipment.

FOOT PROTECTION—Special safety shoes designed to protect feet from sharp or falling objects.

FREE-WHEELING PARTS: INERTIA—The parts continue to move, after the power is shut off, because of their own inertia or the inertia of other moving parts connected to them.

FUNGICIDE—Agent used to kill fungi.

H

HAND PROTECTION—Various types of gloves designed to protect the hands.

HAND SIGNALS—Standardized system of gestures and hand movements for use in machinery operation, published by the American Society of Agricultural Engineers.

HAZARD—Dangerous object or situation that has the potential to cause injury.

HEAD PROTECTION—Headgear designed to protect the head against bumps and falling objects. Two common types are available—hard hats and bump caps.

HERBICIDE—Substance used to kill weeds.

HERTZ (Hz) — Cycles per second.

HYDROGEN SULFIDE (H_2S)—Flammable, poisonous, bad-smelling gas given off by liquid manure, especially during agitation. Very small amounts can cause nausea, dizziness, and a dulled sense of smell, and 1-percent concentrations can cause unconsciousness or death.

HYGROSCOPIC — Readily taking up and retaining moisture. A characteristic of dry fertilizer that can cause skin irritation.

I

INHALATION CHEMICAL EXPOSURE—Chemical entering the body through the respiratory system.

INSECTICIDE—Agent used to kill insects.

J

JACKKNIFE—A tractor and a hitched implement doubling up to form an agle of 90 degrees or less.

K

KICKBACK — Sudden, violent kicking action of a chain saw.

L

LD50 VALUE—Number of milligrams of pesticide per kilogram of body weight that kills 50 percent of a group of test animals. The lower the LD50 value, the more toxic the chemical.

M

MAN-MACHINE COMMUNICATION—By use of controls, gauges, observation, etc., man can understand how his machine is performing.

MAN-MACHINE SYSTEM—Interaction between man and his machine in a work situation.

MANURE—Livestock excreta, often stored as a liquid. Liquid manure may give off poisonous gases, especially when agitated. See AMMONIA GAS, CARBON DIOXIDE, HYDROGEN SULFIDE, METHANE.

METHANE (CH_4)—Colorless, odorless gas given off by liquid manure. Methane is nontoxic, but highly flammable.

MITICIDE—Agent that destroys mites.

O

ORAL CHEMICAL EXPOSURE — Chemical entering the body through the mouth.

OVERBUCK—Cut a limb or log from above.

P

PERSONAL PROTECTIVE EQUIPMENT — Clothing and devices designed to protect workers from bodily injury. See BODY PROTECTION, EAR PROTECTION, EYE PROTECTION, FACE PROTECTION, FOOT PROTECTION, HAND PROTECTION, HEAD PROTECTION, RESPIRATORY PROTECTION.

PESTICIDE—Agent used to destroy pests—insects, rodents, weeds, etc. Most pesticides are poisonous to humans, and some can be fatal. See AVICIDE, BACTERICIDE, FUNGICIDE, HERBICIDE, INSECTICIDE, MITICIDE, RODENTICIDE.

PESTICIDE SAFETY TEAMS—Group organized to help deal with serious chemical emergencies. It maintains a 24-hour answering service. The number is 800-424-9300.

PINCH POINT—Place where two parts of a mechanism move toward one another, creating a potentially hazardous situation.

PINHOLE LEAK—Liquid under high pressure escaping through an extremely small opening.

PULL-IN POINTS—Are created when attempting to remove material while the machine is running—people may be pulled into the moving parts of the machine and seriously injured.

R

REACTION TIME—Interval of time between a stimulus or signal and a person's response to it.

RESPIRATORY PROTECTION—A device worn over the nose or mouth to protect the respiratory tract. Some devices also supply the wearer with oxygen. Respiratory protective devices include various types of respirators and self-contained breathing devices.

RODENTICIDE—Substance that kills rodents.

ROLLOVER PROTECTIVE STRUCTURE (ROPS) — Protective frame or cab designed to limit most tractor upsets to 90 degrees, and to protect the operator in upsets beyond 90 degrees.

S

SAFETY-ALERT SYMBOL — Symbol used on farm machines to draw attention to potential hazards, and in operator's manuals to emphasize safety tips. Consists of an exclamation point within a triangle.

SCHOOLMARM—Tree with a prominent fork in the trunk, or two trees grown together at the base.

SHEAR POINTS—Are created when the edges of two objects are moved toward or next to one another closely enough to cut a relatively soft material.

SHEAR LINE—Point at which the ground near a ditch or river begins to cave in when subjected to a heavy weight, such as a tractor or self-propelled machine.

SHIELD — Protective device or covering to protect people from moving machine parts or other hazardous areas.

SILO GAS—Dangerous gases (primarily carbon dioxide and nitrogen dioxide) that can build up to lethal levels inside silos and are almost impossible to detect.

SLOW-MOVING-VEHICLE EMBLEM (SMV)—Triangular sign attached to the rear of tractors and other farm machines used on public roads. Identifies these machines as traveling slower than 25 mph (40 kw/h).

SPONTANEOUS COMBUSTION — Fire caused by chemical action within the substance that catches fire.

SPRING POLE—Sapling bent over and held down under tension by another tree.

STABILITY—Balance and resistance to upsets of a tractor or machine.

STORED ENERGY—Energy confined just waiting to be released.

SUFFOCATION—Death caused by lack of oxygen.

SYNERGISM (CHEMICAL)—Combined action of two compounds that produces an effect greater than the individual effects of the compounds added together. May lead to a serious hazard if two pesticides are mixed by an inexperienced person.

T

TORQUE—A twisting or turning force that is a factor in tractor upsets.

TOXICITY—Poisonous effect of a chemical or some other substance.

TRANSPORT POSITION—Narrowed machine position for highway travel.

U

UNDERBUCK—Cut a limb or log from underneath.

UNIVERSAL SYMBOLS—Picture symbols used on machine controls throughout the world.

V

VISIBLE COLOR SPECTRUM — A series of colors from red to violet that merge into one another.

W

WARNING—Signal word indicating a moderately serious hazard. Also used on labels of chemicals of medium toxicity.

WIDOWMAKER—Tree with dead or broken limbs that are apt to fall on a worker.

WRAP POINT — Exposed machine component that rotates or rotating shaft where anything may be pulled in and begin wrapping.

ABBREVIATIONS

ASAE—American Society of Agricultural Engineers (sets standards for many components for agricultural use)

cps—cycles per second

CSA—Canadian Standards Association

GFI—Ground-fault interrupter

Kg—Kilogram

Mg—milligram

mph—miles per hour

NIFS—National Institute for Farm Safety

NSC—National Safety Council

NFPA—National Fire Protection Association

OPEI—Outdoor Power Equipment Institute

OSHA—Occupational Safety and Health Act of 1972 (see page 322)

psi—pounds per square inch (of pressure)

PTO—power takeoff

ROPS—rollover protective structure

SAE—Society of Automotive Engineers (sets standards for many components for both automotive and agricultural use) (also see ASAE)

SMV—Slow Moving Vehicle

UL—Underwriters Laboratory

USDA—United States Department of Agriculture

TWO FEDERAL LAWS* YOU NEED TO KNOW ABOUT:

The Occupational Safety And Health Act (OSHA) And The Hazardous-Occupations Order For Youth Under 16

INTRODUCTION

Farm accidents are avoided by safe practices and safe working conditions.

Be alert for potential hazards and hazardous habits. Then make efforts to correct safety problems. By the time you read this section of the book, you will be able to recognize common and specific farm machinery hazards and you will be trying to reduce them. Also, you will know and use safe work practices, like those found in the previous chapters.

Laws are supplemental to sound safety efforts, because much cannot be written into safety standards. But laws can be significant too, so let's look at two important federal laws that concern operators and their employers.

U. S. PUBLIC LAW 91-596 KNOWN AS OSHA

The Williams-Steiger Occupational Safety and Health Act of 1970 seeks:

". . . to assure so far as possible every working man and woman in the Nation safe and healthful working conditions and to preserve our human resources . . ."

This is carried out through the establishment of safety standards, applying to agriculture, inspection of farms with employees to determine if the employer is complying with the safety standards, and issuing citations for safety violation which include fines up to $1000.00 for very serious violations. This law covers nearly all employers including farmers and ranchers with one or more employees at any time during the year.

Farm employers should be concerned with general health and safety duties which may include such things as:

- Be aware that you have a general duty to provide a place of employment free from recognized hazards and to comply with occupational safety and health standards promulgated under the Act.
- Familiarize yourself with occupational safety and health standards.
- Make sure your employees know about OSHA.
- Check conditions in your workplace to make sure they conform to applicable safety and health standards.

These and other general duties, although not specifically identified in the law, are part of the employers responsibility in providing safe and healthful working conditions.

Some specific responsibilities are recognized in the law, including the following items:

1. *Comply with established occupational safety and health standards.*

2. *Inform workers about safety.*

Post, in the workplace, the OSHA poster informing employees of their rights and responsibilities.

3. *Maintain Records.*

Keep required OSHA records of work-related injuries and illnesses (if you have eight or more employees), and post the annual summary during the entire month of February each year. Report to the nearest OSHA area office each injury or health hazard that results in a fatality or hospitalization of five or more employees.

4. *Permit Inspection.*

Permit representatives to participate in the inspection walk-around. If there are not such representatives, allow a reasonable number of employees to confer with the compliance officer during the inspection.

Do not discriminate against employees who properly exercise their rights under the Act.

Post OSHA citations of violations of standards at the worksite involved.

Operators of farm machinery and their employees also are required to comply with established safety and health standards and to obey all rules, regulations and orders issued under the Act that apply to their conduct while at work.

Here's a check list for *farm machinery operators* and all other agricultural employees. As an *employee* you should:

- Comply with any applicable OSHA standards.
- Follow all of your employer's safety and health standards and rules.
- Wear or use prescribed protective equipment.
- Report hazardous conditions to your supervisor.
- Report any job-related injuries or illnesses to your employer and seek treatment promptly.
- Cooperate with the OSHA compliance officer conducting an inspection if he inquires about conditions at your jobsite.

*This section is intended to explain in broad terms the concept and effect of two important federal laws (OSHA and Hazardous Occupations Order in Agriculture). It is not intended as a legal interpretation of the laws and should not be considered as such.

• Use your rights under the Act responsibly.

Using safe work practices and observing OSHA standards benefits you, your family and your future.

The following standards have been established specifically for agricultural operations:

The first four standards established specifically for agricultural occupations were on:

• *Anhydrous Ammonia.* Covers the container construction, location and installation, valves and fittings and safety relief devices. Standards most applicable to farmers are those on nurse tanks on farm vehicles and on the applicators.

• *Temporary Labor Camps.* Covers site selection, building construction, space, ventilation, and heating. It also prescribes sanitation requirements for cooking and eating space, water supply, laundry and bathing facilities, toilets, refuse disposal, and insect and rodent control.

• *Pulpwood Logging.* Covers environmental conditions, clothing and personal protective devices, first aid, hand tools, explosives, stationary and mobile equipment, machinery guards, mufflers, and guylines.

• *Slow-Moving Vehicles.* Prescribes use of the triangular SMV emblem on the rear of all farm vehicles or towed equipment while traveling at speeds of 25 miles per hour or less on public roads.

• *Roll-Over Protective Structures.* Establishes the general requirements for roll-over protective structures and seat belts on agricultural tractors. Specifies ROPS test procedures, performance requirements, and employee operating instructions.

• *Guarding.* Covers the application of guards and shields for protection from the hazards associated with moving machinery parts on farm field equipment, farmstead equipment, and cotton gins. Includes a list of safe operating practices for periodic employee instruction.

Copies of current standards can be obtained from the nearest regional OSHA office. (See list at end of this section.)

OSHA provides for each state to develop and operate its own job safety and health programs. If your state has such a program, Federal OSHA personnel may be in a monitoring position at this time and the state is the enforcement agency. Contact your state's Department of Labor and/or Department of Health for details.

THE HAZARDOUS-OCCUPATIONS ORDER FOR YOUTH UNDER 16

The Hazardous Occupations Order makes it unlawful to hire or even permit without pay any youth under 16 years of age to do any of the jobs listed as hazardous, unless:

1. The youth is working on a farm owned or operated by the youth's parent or legal guardian, or

2. The youth has a training certificate which provides exemption from specific hazardous jobs, or

3. The youth is employed under a cooperative student-learner program in which there is a written agreement between the employer and the school.

AGRICULTURAL WORK CLASSIFIED AS HAZARDOUS FOR YOUTH UNDER 16 IS:

*1. **Tractor**—Operating a tractor of over 20 PTO horsepower, or connecting or disconnecting an implement or any of its parts to or from such a tractor.

*2. **General Machinery**—Operating or assisting to operate (including starting, stopping, adjusting, feeding, or any other activity involving physical contact associated with the operation of) any of the following machines: Corn picker, cotton picker, grain combine, hay mower, forage harvester, hay baler, potato digger, mobile pea viner, feed grinder, crop dryer, forge blower, auger conveyor, the unloading mechanism of a nongravity-type self unloading wagon or trailer, power post-hole digger, power post driver, or nonwalking type rotary tiller.

3. **Specialized Machinery**—Operating or assisting to operate (including starting, stopping, adjusting, feeding, or any other activity involving physical contact associated with the operation) any of the following machines: Trencher or earthmoving equipment, forklift, potato combine, power-driven circular, band, or chain saw.

4. **Livestock**—Working on a farm in a yard (pen, or stall occupied by a bull, boar or stud horse maintained for breeding purposes, or sow with suckling pigs, or cow with newborn calf (with umbilical cord present).

5. **Woodlot**—Felling, bucking, skidding, loading, or unloading timber with butt diameter of more than 6 inches.

6. **Ladder and Scaffold**—Working from a ladder or scaffold (painting, repairing, or building structures, pruning trees, picking fruit, etc.) at a height of over 20 feet.

7. **Transport** — Driving a bus, truck, or automobile when transporting passengers, or riding on a tractor as a passenger or helper.

8. **Toxic Atmosphere**—Working inside: a fruit, forage, or grain storage designed to retain an oxygen deficient or toxic atmosphere; an upright silo within 2 weeks after silage has been added or when a top

*Youth 14-16 years of age can receive a certificate of training from 4-H, FFA or Vo-Ag classes or clinics which allow them to seek employment in catagories 1 and 2.

☆ U.S. GOVERNMENT PRINTING OFFICE: 1970 – 378-063

CERTIFICATE OF TRAINING

This is to certify that: Certificate No. CES__________

Date__________

(Name) (Date of Birth)

(Address) (Zip Code)

is 14 years of age or more and has successfully completed__________ the training program and examination in: (4-H or Vo-Ag)

(Cross out one) 1. Tractor Operation

2. Tractor and Machinery Operation

as specified by the U. S. Department of Labor in the Agricultural Hazardous-Occupations Order (Subpart E-1 of 29 CFR, Part 1500) pertaining to the employment of youth under 16 years of age.

Certifying Authority (Extension Agent or Vocational Agriculture teacher only) Person who conducted the training program

TO THE TRAINEE

1. You may apply for and accept employment as (a) a tractor operator as described in item 1 on the reverse side of this sheet; or (b) as both a tractor and machinery operator as described in items 1 and 2 on the reverse side of this sheet, if certified for both the tractor and machinery operation. This certificate is acceptable by the U. S. Department of Labor as proof of training.
2. You may not be employed in any of the occupations listed in items 3 through 11 described on the reverse side of this sheet.
3. This certificate does not certify training in adjustment of equipment or your proficiency in its operation, nor does it certify that you know how to safely operate any particular make or model of tractor or machine.
4. It is the responsibility of your employer to instruct you on the safe and proper operation of the equipment you are to use.
5. The U.S. Department of Labor requires that your employer "maintain close supervision where feasible, or where not feasible, in work such as cultivating, the employer or his representative check on your progress at least midmorning, noon, and midafternoon. . ."
6. Any accident resulting in an injury to you should be reported by your employer to the proper authority including the certifying authority who signed this certificate.
7. The copy of this certificate marked "Employer's Copy" at the lower right corner, must be kept on file by your employer.

Trainee's Copy

Reverse Side of Sheet

OCCUPATIONS IN AGRICULTURE PARTICULARLY HAZARDOUS FOR THE EMPLOYMENT OF CHILDREN BELOW THE AGE OF 16

Item (1). Operating a tractor of over 20 PTO horsepower, or connecting or disconnecting an implement or any of its parts to or from such a tractor.

Item (2). Operating or assisting to operate (including starting, stopping, adjusting, feeding, or any other activity involving physical contact associated with the operation) any of the following machines:

(i) Corn picker, cotton picker, grain combine, hay mower, forage harvester, hay baler, potato digger or mobile pea viner;

(ii) Feed grinder, crop dryer, forage blower, auger conveyor, or the unloading mechanism of a nongravity-type self-unloading wagon or trailer; or

(iii) Power post-hole digger, power post driver, or non-walking type rotary tiller.

Item (3). Operating or assisting to operate (including starting, stopping, adjusting, feeding, or any other activity involving physical contact associated with the operation) any of the following machines:

(i) Trencher or earthmoving equipment;

(ii) Fork lift;

(iii) Potato combine; or

(iv) Power-driven circular, band, or chain saw.

Item (4). Working on a farm in a yard, pen, or stall occupied by a

(i) Bull, boar or stud horse maintained for breeding purposes; or

(ii) Sow with suckling pigs, or cow with newborn calf (with umbilical cord present).

Item (5). Felling, bucking, skidding, loading, or unloading timber with butt diameter of more than 6 inches.

Item (6). Working from a ladder or scaffold (painting, repairing, or building structures, pruning trees, picking fruit, etc.) at a height of over 20 feet.

Item (7). Driving a bus, truck, or automobile when transporting passengers, or riding on a tractor as a passenger or helper.

Item (8). Working inside:

(i) A fruit, forage, or grain storage designed to retain an oxygen deficient or toxic atmosphere;

(ii) An upright silo within 2 weeks after silage has been added or when a top unloading device is in operating position;

(iii) A manure pit; or

(iv) A horizontal silo while operating a tractor for packing purposes.

Item (9). Handling or applying (including cleaning or decontaminating equipment, disposal or return of empty containers, or serving as a flagman for aircraft applying) agricultural chemicals classified under the Federal Insecticide, Fungicide, and Rodenticide Act (7 U.S.C. 135 et seq) as Category I of toxicity identified by the word "poison" and the "skull and crossbones" on the label; or Category II of toxicity, identified by the word "warning" on the label;

Item (10). Handling or using a blasting agent, including but not limited to, dynamite, black powder, sensitized ammonium nitrate, blasting caps, and primer cord; or

Item (11). Transporting, transferring, or applying anhydrous ammonia.

Fig. 1—Cooperative Extension Service 4-H Program Certificate of Training

unloading device is in operating position; a manure pit; or a horizontal silo while operating a tractor for packing purposes.

9. **Chemicals**—Handling or applying (including cleaning or decontaminating equipment, disposal or return of empty containers, or serving as a flagman for aircraft applying) agricultural chemicals classified under the Federal Insecticide, Fungicide, and Rodenticide Act (7 U.S.C. 135 et seq.) as Category I of toxicity, identified by the word "poison" and the "skull and crossbones" on the label; or Category II of toxicity, identified by the word "warning" on the label.

10. **Blasting**—Handling or using a blasting agent, including but not limited to dynamite, black powder, sensitized ammonium nitrate, blasting caps, and primer cord.

11. **Fertilizers**—Transporting, transferring, or applying anhydrous ammonia.

Exception: A. The Hazardous-Occupations Order in Agriculture does not apply to youth under 16 employed by his parent or by a person standing in place of his parent (legal guardian) on a farm owned and operated by such parent or person.

B. If the youth is employed under a cooperative student-learner program in which there is a written agreement between the employer and the local school.

Fourteen and fifteen year old youth who want to work with or operate a tractor over 20 PTO horsepower or the general machinery listed above as being hazardous must enroll and successfully complete the approved training programs offered by the Cooperative Extension Service 4-H Program in your state or by the Vocational Agriculture Department in your local school system.

When you successfully complete either of these training programs you will receive a certificate of training (Fig. 1). Then this valid exemption certificate allows a fourteen or fifteen year old to be legally employed in only Items 1 and/or 2 listed on the back of the certificate.

OSHA OFFICES

Region I: Connecticut, Maine, Massachusetts, New Hampshire, Rhode Island, Vermont

18 Oliver Street
Boston, Massachusetts 02110

Area Offices:

Custom House Building, State Street
Boston, Massachusetts 02109

450 Main Street—Rm. 617
Hartford, Connecticut 06103

55 Pleasant Street—Rm. 425
Concord, New Hampshire 03301

District Office:

U.S. Courthouse—Rm. 503A
Providence, Rhode Island 02903

Region II: New York, New Jersey, Puerto Rico, Virgin Islands, Canal Zone

1515 Broadway (1 Astor Plaza)
New York, New York 10036

Area Offices:

90 Church Street—Rm. 1405
New York, New York 10007

700 East Water Street—Rm. 203
Syracuse, New York 13210

370 Old Country Road
Garden City, L.I., New York 11530

970 Broad Street—Rm. 635
Newark, New Jersey 07102

605 Condado Avenue—Rm. 328
Santurce, Puerto Rico 00907

Region III: Delaware, District of Columbia, Maryland, Pennsylvania, Virginia, West Virginia

15220 Gateway Center

3535 Market Street
Philadelphia, Pennsylvania 19104

Area Offices:

1317 Filbert Street—Suite 1010
Philadelphia, Pennsylvania 19107

3661 Virginia Beach Blvd.—Rm. 111
Norfolk, Virginia 23502

400 N. 8th Street—Rm. 8018
Richmond, Virginia 23240

31 Hopkins Plaza—Rm. 1110A
Baltimore, Maryland 21201

Room 802, Jonnet Building
4099 William Penn Highway
Monroeville, Pennsylvania 15146

Region IV: Alabama, Florida, Georgia, Kentucky, Mississippi, North Carolina, South Carolina, Tennessee

1375 Peachtree Street, N.E.—
Suite 587 Atlanta, Georgia 30309

Area Offices:

1371 Peachtree Street, N.E.—
Rm. 723 Atlanta, Georgia 30309

3200 E. Oakland Park Blvd.—
Rm. 204
Fort Lauderdale, Florida 33308

2809 Art Museum Drive
Suite 4
Jacksonville, Florida 32207

600 Federal Place—Rm. 561
Louisville, Kentucky 40202

118 North Royal Street—Rm. 801
Mobile, Alabama 35502

1361 East Morehead Street
Charlotte, North Carolina 28204

1600 Hayes Street—Suite 302
Nashville, Tennessee 37203

2047 Canyon Road, Todd Mall
Birmingham, Alabama 35216

6605 Abercorn Street—Suite 201
Savannah, Georgia 31405

Region V: Illinois, Indiana, Michigan, Minnesota, Ohio, Wisconsin

300 South Wacker Drive—
Rm. 1201 Chicago, Illinois 60606

Area Offices:

300 South Wacker Drive
Chicago, Illinois 60606

700 Bryden Road—Rm. 224
Columbus, Ohio 43215

633 W. Wisconsin Ave.—Rm. 400
Milwaukee, Wisconsin 53203

46 East Ohio Street—Rm. 423
Indianapolis, Indiana 46204

1240 East Ninth Street—Rm. 847
Cleveland, Ohio 44199

220 Bagley Avenue—Rm. 626
Detroit, Michigan 48226

110 South Fourth Street—Rm. 437
Minneapolis, Minnesota 55401

550 Main Street—Rm. 5522
Cincinnati, Ohio 45202

234 N. Summit Street—Rm. 734
Toledo, Ohio 43604

Region VI: Arkansas, Louisiana, New Mexico, Oklahoma, Texas

1512 Commerce Street, 7th Floor
Dallas, Texas 75201

Area Offices:

1100 Commerce Street—Rm. 6B1
Dallas, Texas 75202

1205 Texas Avenue—Rm. 421
Lubbock, Texas 79401

420 South Boulder—Rm. 512
Tulsa, Oklahoma 74103

307 Central National Bank Bldg.
Houston, Texas 77002

546 Carondelet Street—4th Floor
New Orleans, Louisiana 70130

District Office:

U.S. Custom House Bldg.—Rm. 325
Galveston, Texas 77550

Region VII: Iowa, Kansas, Missouri, Nebraska

823 Walnut Street—Rm. 300
Kansas City, Missouri 64106

Area Offices:

1627 Main Street—Rm. 1100
Kansas City, Missouri 64108

210 North 12th Boulevard—Rm. 554
St. Louis, Missouri 63101

City National Bank Building—Rm. 803
Harney and 16th Streets
Omaha, Nebraska 68102

Region VIII: Colorado, Montana, North Dakota, South Dakota, Utah, Wyoming

1961 Stout Street—Rm. 15010
Denver, Colorado 80202

Area Offices:

8527 W. Colfax Avenue
Lakewood, Colorado 80202

455 East 4th South—Suite 309
Salt Lake City, Utah 84111

2812 1st Avenue North—Suite 525
Billings, Montana 59101

Region IX: Arizona, California, Hawaii, Nevada, Guam, American Samoa, Trust Territory of the Pacific Islands

450 Golden Gate Avenue—
Rm. 9470
San Francisco, California 94102

Area Offices:

100 McAllister Street—Rm. 1706
San Francisco, California 94102

2721 North Central Avenue—Suite 910
Phoenix, Arizona 85004

19 Pine Avenue—Rm. 514
Long Beach, California 90802

333 Queen Street—Suite 505
Honolulu, Hawaii 96813

Region X: Alaska, Idaho, Oregon, Washington

506 Second Avenue—Rm. 1808
Seattle, Washington 98104

Area Offices:

506 Second Avenue—Rm. 1906
Seattle, Washington 98104

610 C Street—Rm. 214
Anchorage, Alaska 99501

921 S.W. Washington Street—Rm. 526
Portland, Oregon 97205

Chemical Safety

Safety Considerations for Ethanol Production and Use — David E. Baker, University of Missouri, 1980
Safety Guidelines for Alcohol Fuel Plants — William D. Hanford, National Safety Council (NSC), 1980
Production and Use of Ethanol — University of Arkansas
Personal Protective Equipment for Agriculture — NSC Bulletin 699.41-4

Electrical Safety

National Fire Protection Assn. (NFPA) 70 — National Electrical Code
NFPA 78 — Lightning Protection

Machine Guarding

ASAE S318.6 — Safety for Agricultural Equipment
ASAE S354.1 — Safety for Farmstead Equipment
NSC Bulletin 699.41-16 — Guarding and Shielding Farm Equipment

Toxic Gases

Carbon Dioxide — NSC, Data Sheet 682
Entry Into Grain Bins and Food Tanks — NSC, Data Sheet 663

Fire Prevention and Control

NFPA 30 — Flammable and Combustible Liquids Code
NFPA 321 — Classification of Flammable Liquids
NFPA 497 — Classification of Class 1 Hazardous Locations
NFPA 385 — Tank Vehicles for Flammable and Combustible Liquids
NFPA 69 — Explosion Prevention System
NFPA 518 — Fire Prevention in Use of Cutting and Welding Processes
NFPA 327 — Cleaning or Safeguarding Small Tanks and Containers
Cleaning Small Containers That Have Held Combustibles — NSC Data Sheet 432.

SUGGESTED READINGS

TEXTS AND PERIODICALS

University of Minnesota, Ag Extension Office. Extensive list of safety brochures and training materials. For a complete list write to: (Texts, cassettes, films). Robert A. Aherin, Extension Safety Program Specialist, 230 Ag Engineering Building, University of Minnesota, St. Paul, Minnesota 55108.

Power Tool Safety and Operation; Hoerner, Thomas A. and Bettis, Mervin D., HoBar Publications, 1305 Tiller Lane, St. Paul, Minnesota 55112.

4-H Tractor Program; National 4-H Service Committee, Inc. Connecticut Avenue, Washington, D.C. 20015.

Farm Tractor Maintenance; Brown, rlen D. and Morrison, Ivan G.; The Interstate Printers and Publishers, Danville, Illinois, 1958.

Operating Farm Tractors And Machinery; Publications Distribution, Iowa State University, Ames, Iowa 50010, 1974.

Tractor Operation and Daily Care; American Association for Vocational Instruction Materials, Engineering Center, Athens, Georgia 30601.

Vocational Agriculture Training Program; Bobbitt and Doss; Michigan State University, East Lansing, Michigan 48823, 1973.

Fundamentals of Machine Operation: series of agricultural machinery texts which includes the Safety text which makes up this book; John Deere Service Training, Dept. 333, John Deere Road, Moline, Illinois 61265.

*Fundamentals of Service:*series on agricultural power mechanics; John Deere Service Training, Dept. 333, John Deere Road, Moline, Illinois 61265.

Farm and Ranch Safety Guide; National Safety Council, 444 North Michigan Avenue, Chicago, Illinois 60611. (There is a cost for this item).

Practical safety information pieces on a variety of subjects to assist farmers and ranchers in preventing accidents in their operations are available from: Farm Department, National Safety Council 444 Michigan Avenue, Chicago, Illinois 60611. (Send a self-addressed stamped envelope for more information.)

Personal Protective Equipment for Agriculture; NSC Bulletin 699.41-4

MANUFACTURER'S LITERATURE

Safety rules on specific agricultural machines are covered in the operator's manuals for these machines. Refer to these manuals, which are sent with each new machine. If these manuals are not available, send your request to the manufacturer's technical literature or publications department. Be sure to specify the make, model, year of manufacture, and serial number of the machine for which you wish an operator's manual.

FILMS, VIDEOS

For The Rest Of Your Life. Shows hazards to eyes in handling anhydrous ammonia. By Society For Prevention Of Blindness. Excellent film depicting the hazards associated with anhydrous ammonia, National Society for Prevention of Blindness, 79 Madison Avenue, New York, New York 10016; or Media Library Audiovisual Center, University of Iowa, Iowa City, Iowa 52242.

Chance of a Lifetime. Tells importance of safe operation of modern complex farm machinery. 16 mm color. Ontario Dept. of Agriculture and Food, School of Engineering, University of Guelph, Ontario, Canada.

Cotton Picker Safety. See your John Deere dealer.

Second Chance. Shows effect on family and importance of using guards on farm equipment. 16 mm color. Ontario Dept. of Agriculture and Food, Information Branch, Toronto, Ontario, Canada.

Seeds of Safety. Story of an accident. 16 mm color. Indiana Farm Bureau, 130 East Washington, Indianapolis, Indiana 46204.

Within The Frame of Safety. Shows protective frames for farm tractors. 16 mm color. International Harvester Company, 401 North Michigan Avenue, Chicago, Illinois 60611.

Agricultural Accidents and Rescue. Pennsylvania State University, Audio Visual Services, University Park, Pennsylvania 76002. Program rent charge. Preview is available.

A Mowing Safety Lesson from John Deere. An 11 minute review of safety lessons. Free use. Modern Talking Picture Service, 5000 Park Street North, St. Petersburg, Florida 33709.

SLIDE SETS

Agricultural Machinery Safety Slide Set (FMO-182S). 35 mm. color. Matching set of 240 slides for illustrations in FMO "Agricultural Machinery Safety" text. John Deere Service Publications, Dept. F., John Deere Road, Moline, Illinois 61265.

INSTRUCTOR'S KITS

Power Tool Safety and Operation; Instructor's Packet. Includes one copy of manual plus teaching activities, transparency masters and safety exams for 24 common power tools. Ho Bar Publications, 1305 Tiller Lane, St. Paul, Minnesota 55112.

IH Farm Equipment Safety Teaching Kit; Includes subject-matter references, problems to solve, list of questions, suggested activities, and six transparency masters. International Harvester, 401 North Michigan Avenue, Chicago, Illinois 60611.

WORKBOOK

Agricultural Safety Workbook (FMO-18603W). Contains chapter units with chapter quizzes, study and discussion activities and laboratory exercises. Corresponds to *FMO Agricultural Safety* text.

INSTRUCTOR'S GUIDE

Agricultural Safety Instructor's Guide. Contains chapter units with teaching tips, answers to chapter quizzes, transparency masters and suggested script for transparencies. Corresponds to *FMO Agricultural Safety* Text.

A

B

C

C (continued)

H (continued)

I

J

K

L

M

M (continued)

N

O

P

P (continued)

R

R (continued)

S

S (continued)

T

T (continued)

U

V

W